MW00845334

Process Safety for Engineers

This book is one in a series of process safety guidelines and concept books published by the Center for Chemical Process Safety (CCPS). Please go to www.wiley.com/go/ccps or www.aiche.org/ccps/publications for a full list of titles in this series. A few are listed below.

- Guidelines for Hazard Evaluation Procedures
- Guidelines for Revalidating Process Hazard Analysis
- Layer of Protection Analysis - Simplified Process Risk Assessment
- Guidelines for Consequence Analysis of Chemical Releases
- Bow Ties in Risk Management
- Guidelines for Safe Process Operations and Maintenance
- Conduct of Operations and Operational Discipline
- Management of Change for Process Safety
- Guidelines for Asset Integrity Management
- Guidelines for Chemical Reactivity Evaluation and Application to Process Design
- Guidelines for Inherently Safer Chemical Processes: A Life Cycle Approach
- Guidelines for Integrating Process Safety into Engineering Projects
- Performing Effective Pre-Startup Safety Reviews
- Guidelines for Investigating Process Safety Incidents
- Guidelines for Risk Based Process Safety
- Guidelines for Defining Process Safety Competency Requirements
- Incidents that Define Process Safety
- More Incidents that Define Process Safety
- Recognizing and Responding to Normalization of Deviance
- Essential Practices for Creating, Strengthening, and Sustaining Process Safety Culture
- Process Safety Leadership from the Boardroom to the Frontlines

Process Safety for Engineers:
An Introduction

Second Edition

CENTER FOR CHEMICAL PROCESS SAFETY

of the

AMERICAN INSTITUTE OF CHEMICAL ENGINEERS

120 Wall Street, 23rd Floor • New York, NY 10005

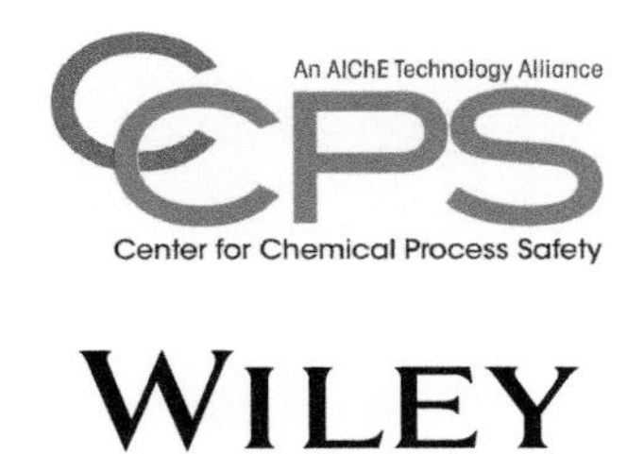

WILEY

A Joint Publication of the American Institute of Chemical Engineers and John Wiley & Sons, Inc.

Registered Office
John Wiley & Sons, Inc., 111 River Street, Hoboken, NJ 07030, USA

Editorial Office
111 River Street, Hoboken, NJ 07030, USA

For details of our global editorial offices, customer services, and more information about Wiley products visit us at www.wiley.com.

Wiley also publishes its books in a variety of electronic formats and by print-on-demand. Some content that appears in standard print versions of this book may not be available in other formats.

Library of Congress Cataloging-in-Publication Data is Applied for:
ISBN: 9781119830986

Cover Design: Wiley
Cover Image: © US Chemical Safety and Hazard Investigation Board (CSB) video, public domain

SKY10033782_031522

Process Safety for Engineers: An Introduction

Is dedicated to

Pete Lodal

Pete has supported CCPS for thirty-four years, and counting, through his forty-two year career at Eastman Chemical Company and currently as a CCPS Staff Consultant. He has been recognized as a Fellow of Eastman Chemical Company, a CCPS Fellow, and an AIChE Fellow in addition to being a registered PE in the state of Tennessee and a CCPSC. Pete is the author of many papers and we all look forward to his honest, enlightening, and humorous presentations at the Global Congress on Process Safety. He has supported many CCPS and AIChE committees. His leadership of the CCPS Planning Committee has been instrumental in creating impactful products to further process safety understanding. Through his current membership on the AIChE Board of Directors, Pete is an influential advocate for process safety.

Has the realm of process safety benefited from Pete's insightful support? In the words of Pete's hero, Curly Howard, "Why soitenly!"

LIST OF FIGURES

LIST OF TABLES

ACRONYMS AND ABBREVIATIONS

ACC	American Chemistry Council
AEGL	Acute Exposure Guideline Level
AIChE	American Institute of Chemical Engineers
ALARP	As Low as Reasonably Practicable
API	American Petroleum Institute
ASME	American Society of Mechanical Engineers
BLEVE	Boiling Liquid Expanding Vapor Explosion
BMS	Burner Management System
CCPS	Center for Chemical Process Safety (of AIChE)
CFR	Code of Federal Regulations
CMA	Chemical Manufacturers Association
COMAH	Control of Major Accident Hazards (U.K. Regulation incorporating the EU Seveso Directive requirements)
COO	Conduct of Operations
CPQRA	Chemical Process Quantitative Risk Assessment
CSB	Chemical Safety Board (US)
DDT	Deflagration to Detonation Transition
DIERS	Design Institute for Emergency Relief Systems
EHS	Environmental, Health, and Safety (sometimes written as SHE or HSE)
ERS	Emergency Relief System
EPA	U.S. Environmental Protection Agency
ERPG	Emergency Response Planning Guideline
ESD	Emergency Shutdown System
EU	European Union
FCCU	Fluidized Catalytic Cracking Unit
FEL	Front End Loading
FFS	Fitness For Service
FMEA	Failure Modes and Effects Analysis
FMECA	Failure Modes, Effects, and Criticality Analysis
FTA	Fault Tree Analysis

GHS	Globally Harmonized System
HAZID	Hazard Identification Study
HAZMAT	Hazardous Materials
HAZOP	Hazard and Operability Study
HEART	Human Error Assessment and Reduction Technique
HIRA	Hazard Identification and Risk Analysis
HRA	Human Reliability Analysis
HSE	Health and Safety Executive (U.K.)
HTHA	High Temperature Hydrogen Attack
HRO	High Reliability Organization
I&E	Instrument and Electrical
IDLH	Immediately Dangerous to Life and Health
IEC	International Electrotechnical Commission
IOGP	International Association of Oil & Gas Producers
IOW	Integrity Operating Window
IPL	Independent Protection Layer
ISD	Inherently Safer Design
ISO	International Organization for Standardization
Isom	Isomerization Unit
ITPM	Inspection Testing, and Preventive Maintenance
JSA	Job Safety Analysis
KPI	Key Performance Indicator
LFL	Lower Flammable Limit
LNG	Liquefied Natural Gas
LOPA	Layer of Protection Analysis
LOPC	Loss of Primary Containment
LOTO	Lock Out Tag Out
LPG	Liquefied Petroleum Gas
LSIR	Location Specific Individual Risk
MAWP	Maximum Allowable Working Pressure

MCC	Motor Control Center
MIE	Minimum Ignition Energy
MOC	Management of Change
MOC	Minimum Oxygen Concentration
MOOC	Management of Organizational Change
NASA	National Aeronautics and Space Administration
NDT	Nondestructive Testing
NFPA	National Fire Protection Association
OD	Operational Discipline
OIMS	Operational Integrity Management System (ExxonMobil)
OSHA	U.S. Occupational Safety and Health Administration
PAC	Protective Action Criteria
PFD	Process Flow Diagram
PFD	Probability of Failure on Demand
PHA	Process Hazard Analysis
P&ID	Piping and Instrumentation Diagram
PLC	Programmable Logic Controller
PRA	Probabilistic Risk Assessment
PRD	Pressure Relief Device
PRV	Pressure Relief Valve
PSE	Process Safety Event
PSI	Process Safety Information
PSI	Process Safety Incident
PSM	Process Safety Management
PSO	Process Safety Officer
PSSR	Pre-Startup Safety Review
QRA	Quantitative Risk Analysis
RAGAGEP	Recognized and Generally Accepted Good Engineering Practice
RBPS	Risk Based Process Safety
RMP	Risk Management Plan

RP	Recommended Practice
SACHE	Safety and Chemical Engineering Education
SCAI	Safety Controls Alarms and Interlocks
SDS	Safety Data Sheet
SHIB	Safety Hazard Information Bulletin
SIF	Safety Instrumented Function
SIL	Safety Integrity Level (as per IEC 61508 / 61511 standards)
SIS	Safety Instrumented System
SOL	Safe Operating Limits
TEEL	Temporary Emergency Exposure Limit
THERP	Technique for Human Error Rate Prediction
TQ	Threshold Quantity
UFL	Upper Flammable Limit
U.K.	United Kingdom
U.S.	United States
UST	Underground Storage Tank
VCE	Vapor Cloud Explosion

GLOSSARY

ALARP	As Low As Reasonably Practicable – As low as reasonably practicable; the concept that efforts to reduce risk should be continued until the incremental sacrifice (in terms of cost, time, effort, or other expenditure of resources) is grossly disproportionate to the incremental risk reduction achieved. The term as low as reasonably achievable (ALARA) is often used synonymously.
Asset integrity	The condition of an asset that is properly designed and installed in accordance with specifications and remains fit for purpose.
Atmospheric storage tank	A storage tank designed to operate at any pressure between ambient pressure and 3.45kPa gage (0.5 psig).
Audit	A systematic, independent review to verify conformance with prescribed standards of care using a well-defined review process to ensure consistency and to allow the auditor to reach defensible conclusions.
Autoignition temperature	The lowest temperature at which a fuel/oxidant mixture will spontaneously ignite under specified test conditions.
Barrier	A control measure or grouping of control elements that on its own can prevent a threat developing into a top event (prevention barrier) or can mitigate the consequences of a top event once it has occurred (mitigation barrier). A barrier must be effective, independent, and auditable. See also Degradation Control. (Other possible names: Control, Independent Protection Layer, Risk Reduction Measure).
Block flow diagram	A simplified drawing representing a process. It typically shows major equipment and piping and can include major valves.
Boiling-Liquid-Expanding-Vapor Explosion (BLEVE)	A type of rapid phase transition in which a liquid contained above its atmospheric boiling point is rapidly depressurized, causing a nearly instantaneous transition from liquid to vapor with a corresponding energy release. A BLEVE of flammable material is often accompanied by a large aerosol fireball, since an external fire impinging on the vapor space of a pressure vessel is a common cause. However, it is not necessary for the liquid to be flammable to have a BLEVE occur.
Bow tie model	A risk diagram showing how various threats can lead to a loss of control of a hazard and allow this unsafe condition to develop into a number of undesired consequences. The diagram can show all the barriers and degradation controls deployed.
Change	Any addition, process modification, or substitute item (e.g. person or thing) that is not a replacement-in-kind.
Checklist analysis	A hazard evaluation procedure using one or more pre-prepared lists of process safety considerations to prompt team discussions of whether the existing safeguards are adequate.

Chemical process industry	The phrase is used loosely to include facilities which manufacture, handle, and use chemicals.
Chemical reactivity	The tendency of substances to undergo chemical change.
Combustible dust	A finely divided combustible particulate solid that presents a flash fire hazard or explosion hazard when suspended in air or the process specific oxidizing medium over a range of concentrations.
Conduct of Operations (COO)	The embodiment of an organization's values and principles in management systems that are developed, implemented, and maintained to (1) structure operational tasks in a manner consistent with the organization's risk tolerance, (2) ensure that every task is performed deliberately and correctly, and (3) minimize variations in performance.
Consequence	The undesirable result of a loss event, usually measured in health and safety effects, environmental impacts, loss of property, and business interruption costs.
Deflagration	A combustion that propagates by heat and mass transfer through the un-reacted medium at a velocity less than the speed of sound.
Degradation factor	A situation, condition, defect, or error that compromises the function of a main pathway barrier, through either defeating it or reducing its effectiveness. If a barrier degrades then the risks from the pathway on which it lies increase or escalate, hence the alternative name of escalation factor. (Other possible names: **Barrier Decay Mechanism, Escalation Factor, Defeating Factor**).
Degradation control	Measures which help prevent the degradation factor impairing the barrier. They lie on the pathway connecting the degradation threat to the main pathway barrier. Degradation controls may not meet the full requirements for barrier validity. (Other possible names: **Degradation Safeguard**, **Defeating Factor Control, Escalation Factor Control, Escalation Factor Barrier**).
Detonation	A release of energy caused by the propagation of a chemical reaction in which the reaction front advances into the unreacted substance at greater than sonic velocity in the unreacted material.
Endothermic chemical reaction	A reaction involving one or more chemicals resulting in one or more new chemical species and the absorption of heat.
Exothermic chemical reaction	A reaction involving one or more chemicals resulting in one or more new chemical species and the evolution of heat.
Explosion	A release of energy that causes a pressure discontinuity or blast wave.

Failure Modes and Effects Analysis (FMEA) A hazard identification technique in which all known failure modes of components or features of a system are considered in turn, and undesired outcomes are noted.

Fireball The atmospheric burning of a fuel-air cloud in which the energy is in the form of radiant and convective heat. The inner core of the fuel release consists of almost pure fuel whereas the outer layer in which ignition first occurs is a flammable fuel-air mixture. As buoyancy forces of the hot gases begin to dominate, the burning cloud rises and becomes more spherical in shape.

Fitness for Service (FFS) A systematic approach for evaluating the current condition of a piece of equipment in order to determine if the equipment item is capable of operating at defined operating conditions (e.g., temperature, pressure).

Flammability limits The range of gas or vapor amounts in air that will burn or explode if a flame or other ignition source is present. Importance: The range represents a gas or vapor mixture with air that may ignite or explode. Generally, the wider the range the greater the fire potential. See also Lower Explosive Limit / Lower Flammable Limit and Upper Explosive Limit / Upper Flammable Limit.

Flammable liquids "An ignitable liquid that is classified as a Class I liquid. A Class I liquid is a liquid that has a closed-cup flash point below 100 °F (37.8 °C), as determined by the test procedures described in NFPA 30 and a Reid vapor pressure not exceeding 40 psia (2068.6 mm Hg) at 100°F (37.8 °C), as determined by ASTM D323, Standard Method of Test for Vapor Pressure of Petroleum Products (Reid Method). Class IA liquids include those liquids that have flash points below 73 °F (22.8 °C) and boiling points below 100 °F (37.8 °C). Class IB liquids include those liquids that have flash points below 73°F (22.8 °C) and boiling points at or above 100 °F (37.8 °C). Class IC liquids shall include those liquids that have flash points at or above 73 °F (22.8 °C), but below 100 °F (37.8 °C)." (adapted from NFPA 30).

Flash fire A fire that spreads by means of a flame front rapidly through a diffuse fuel, such as a dust, gas, or the vapors of an ignitable liquid, without the production of damaging pressure.

Flash point temperature The minimum temperature at which a liquid gives off sufficient vapor to form an ignitable mixture with air within the test vessel used (Methods: ASTM 502). The flash point is less than the fire point at which the liquid evolves vapor at a sufficient rate for indefinite burning.

Frequency Number of occurrences of an event per unit time (e.g., 1 event in 1000 yr = 1×10^{-3} events/yr).

Front End Loading (FEL) The process for conceptual development of projects. This involves developing sufficient strategic information with which owners can address risk and make decisions to commit resources in order to maximize the potential for success. (Other possible names: **pre-project planning, front-end engineering design, feasibility analysis, early project planning**).

Hazard An operation, activity or material with the potential to cause harm to people, property, the environment or business; or simply, a potential source of harm.

Hazard analysis The identification of undesired events that lead to the materialization of a hazard, the analysis of the mechanisms by which these undesired events could occur and usually the estimation of the consequences.

Hazard and Operability Study (HAZOP) A systematic qualitative technique to identify process hazards and potential operating problems using a series of guide words to study process deviations. A HAZOP is used to question every part of a process to discover what deviations from the intention of the design can occur and what their causes and consequences may be. This is done systematically by applying suitable guide words. This is a systematic detailed review technique, for both batch and continuous plants, which can be applied to new or existing processes to identify hazards.

Hazard Identification Part of the Hazard Identification and Risk Analysis (HIRA) method in which the material and energy hazards of the process, along with the siting and layout of the facility, are identified so that a risk analysis can be performed on potential incident scenarios.

Hazard Identification and Risk Analysis (HIRA) A collective term that encompasses all activities involved in identifying hazards and evaluating risk at facilities, throughout their life cycle, to make certain that risks to employees, the public, or the environment are consistently controlled within the organization's risk tolerance.

Hot work Any operation that uses flames or can produce sparks (e.g., welding).

Human factors A discipline concerned with designing machines, operations, and work environments so that they match human capabilities, limitations, and needs. Includes any technical work (engineering, procedure writing, worker training, worker selection, etc.) related to the human factor in operator-machine systems.

Human reliability analysis A method used to evaluate whether system-required human-actions, tasks, or jobs will be completed successfully within a required time period. Also used to determine the probability that no extraneous human actions detrimental to the system will be performed.

Impulse The area under the overpressure-time curve for explosions. The area can be calculated for the positive phase or negative phase of the blast.

Incident	An event, or series of events, resulting in one or more undesirable consequences, such as harm to people, damage to the environment, or asset/business losses. Such events include fires, explosions, releases of toxic or otherwise harmful substances, and so forth.
Incident investigation	A systematic approach for determining the causes of an incident and developing recommendations that address the causes to help prevent or mitigate future incidents.
Independent Protection Layer (IPL)	A device, system, or action that is capable of preventing a scenario from proceeding to the undesired consequence without being adversely affected by the initiating event or the action of any other protection layer associated with the scenario. Note: Protection layers that are designated as "independent" have specific functional criteria. A protection layer meets the requirements of being an IPL when it is designed and managed to achieve the following seven core attributes: Independent; Functional; Integrity; Reliable; Validated, Maintained and Audited; Access Security; and Management of Change.
Individual risk	The risk to a person in the vicinity of a hazard. This includes the nature of the injury to the individual, the likelihood of the injury occurring, and the time period over which the injury might occur.
Inherently Safer Design (ISD)	A way of thinking about the design of chemical processes and plants that focuses on the elimination or reduction of hazards, rather than on their management and control.
Inspection, Testing, and Preventive Maintenance (ITPM)	Scheduled proactive maintenance activities intended to (1) assess the current condition and/or rate of degradation of equipment, (2) test the operation/functionality of equipment, and/or (3) prevent equipment failure by restoring equipment condition.
Jet fire	A fire type resulting from the discharge of liquid, vapor, or gas into free space from an orifice, the momentum of which induces the surrounding atmosphere to mix with the discharged material.
KSt value	The deflagration index of a dust cloud. It is a dust-specific measure of the explosibility, in units of bar-m/s. Not that it is not a physical property of a substance, but dependent on particle size, test conditions, etc. The equation is the so-called cubic /cube root law.
Interlock	A feature that makes the state of two mechanisms or functions mutually dependent. It may be used to prevent undesired states in a finite-state machine, and may consist of any electrical, electronic, or mechanical devices or systems. (Wikipedia)
Intrinsically safe	Equipment in which any spark of any thermal effect produced. Including normal operation and specified fault conditions, are not capable of causing ignition of a given explosive gas atmosphere.

Lagging metric	A retrospective set of metrics based on incidents that meet an established threshold of severity.
Layer of protection	A concept whereby several independent devices, systems, or actions are provided to reduce the likelihood and severity of an undesired event.
Layer of Protection Analysis (LOPA)	An approach that analyzes one incident scenario (cause-consequence pair) at a time, using predefined values for the initiating event frequency, independent protection layer failure probabilities, and consequence severity, in order to compare a scenario risk estimate to risk criteria for determining where additional risk reduction or more detailed analysis is needed. Scenarios are identified elsewhere, typically using a scenario-based hazard evaluation procedure such as a HAZOP Study.
Leading metric	A forward-looking set of metrics that indicate the performance of the key work processes, operating discipline, or layers of protection that prevent incidents.
Likelihood	A measure of the expected probability or frequency of occurrence of an event. This may be expressed as an event frequency (e.g., events per year), a probability of occurrence during a time interval (e.g., annual probability) or a conditional probability (e.g., probability of occurrence, given that a precursor event has occurred).
Lockout/tagout	A safe work practice in which energy sources are positively blocked away from a segment of a process with a locking mechanism and visibly tagged as such to help ensure worker safety during maintenance and some operations tasks.
Loss of Primary Containment (LOPC)	An unplanned or uncontrolled release of material from primary containment, including non-toxic and non-flammable materials (e.g., steam, hot condensate, nitrogen, compressed CO_2 or compressed air).
Management of Change (MOC)	A management system to identify, review, and approve all modifications to equipment, procedures, raw materials, and processing conditions, other than replacement-in-kind, prior to implementation to help ensure that changes to processes are properly analyzed (for example, for potential adverse impacts), documented, and communicated to employees affected. This includes management of organizational change.
Management of Organizational Change (MOOC)	Management of change as it applies to a change in position or responsibility within an organization or any change to an organizational structure, policy, or procedure that affects process safety.
Management review	A PSM program element that provides for the routine evaluation of other PSM program management systems/elements with the objective of determining if the element under review is performing as intended and producing the desired results as efficiently as possible. It is an ongoing due diligence review by management that fills the gap between day-to-day work activities and periodic formal audits.

Management system	A formally established set of activities designed to produce specific results in a consistent manner on a sustainable basis.
Mechanical integrity	A management system focused on ensuring that equipment is designed, installed, and maintained to perform the desired function.
Minimum Ignition Energy (MIE)	The minimum amount of energy released at a point in a combustible mixture that caused flame propagation away from the point, under specified test conditions. The lowest value of the minimum ignition energy is found at a certain optimum mixture. The lowest value is usually quoted as the minimum ignition energy.
Minimum Oxygen Concentration (MOC)	The concentration of oxidant, in a fuel-oxidant-diluent mixture below which a deflagration cannot occur under specified conditions. Limiting Oxidant Concentration (LOC) is synonymous with the term Minimum Oxygen Concentration (MOC).
Mitigation barrier	A barrier located on the right-hand side of a bow tie diagram lying between the top event and a consequence. It might only reduce a consequence, not necessarily terminate the sequence before the consequence occurs. (Other possible names: **Reactive Barrier, Recovery Measure**).
Near miss	An event in which an accident (that is, property damage, environmental impact, or human loss) or an operational interruption could have plausibly resulted if circumstances had been slightly different. (Other possible names: **Near Hit).**
Normalization of deviance	A gradual erosion of standards of performance as a result of increased tolerance of nonconformance.
Operating procedures	Written, step-by-step instructions and information necessary to operate equipment, compiled in one document including operating instructions, process descriptions, operating limits, chemical hazards, and safety equipment requirements.
Operational Discipline (OD)	The performance of all tasks correctly every time.
Operational readiness	A management system element associated with efforts to ensure that a process is ready for start-up/restart. This element applies to a variety of restart situations, ranging from restart after a brief maintenance outage to restart of a process that has been mothballed for several years.
OSHA Process Safety Management (OSHA PSM)	A U.S. regulatory standard that requires use of a 14-element management system to help prevent or mitigate the effects of catastrophic releases of chemicals or energy from processes covered by the regulations 29 CFR 1910.119.
Overpressure	Any pressure above atmospheric caused by a blast.

Performance standard Measurable statement, expressed in qualitative or quantitative terms, of the performance required of a system, equipment item, person or procedure (that may be part or all of a barrier), and that is relied upon as a basis for managing a hazard. The term includes aspects of functionality, reliability, availability, and survivability.

Physical explosion The catastrophic rupture of a pressurized gas/vapor-filled vessel by means other than reaction, or the sudden phase-change from liquid to vapor of a superheated liquid.

Piping and Instrumentation Diagram (P&ID) A diagram that shows the details about the piping, vessels, and instrumentation.

Pmax The maximum pressure occurring in a closed vessel during the explosion of an explosible dust atmosphere determined under specific test conditions.

Pressure Relief Valve (PRV) A pressure relief device which is designed to reclose and prevent the further flow of fluid after normal conditions have been restored. (Other possible names: **Pressure Safety Valve**).

Pre-Startup Safety Review (PSSR) A systematic and thorough check of a process prior to the introduction of a highly hazardous chemical to a process. The PSSR must confirm the following: Construction and equipment are in accordance with design specifications; Safety, operating, maintenance, and emergency procedures are in place and are adequate; A process hazard analysis has been performed for new facilities and recommendations and have been resolved or implemented before startup, and modified facilities meet the management of change requirements; and training of each employee involved in operating a process has been completed.

Prevention barrier A barrier located on the left-hand side of bow tie diagram and lies between a threat and the top event. It must have the capability on its own to completely terminate a threat sequence. (Other possible name: **Proactive Barrier**).

Preventive maintenance Maintenance that seeks to reduce the frequency and severity of unplanned shutdowns by establishing a fixed schedule of routine inspection and repairs.

Primary containment A tank, vessel, pipe, transport vessel or equipment intended to serve as the primary container for, or used for the transfer of, a material. Primary containers may be designed with secondary containment systems to contain or control a release from the primary containment. Secondary containment systems include, but are not limited to tank dikes, curbing around process equipment, drainage collection systems into segregated oily drain systems, the outer wall of double-walled tanks, etc.

Probabilistic Risk Assessment (PRA) A commonly used term in the nuclear industry to describe the quantitative evaluation of risk using probability theory.

Probability The expression for the likelihood of occurrence of an event or an event sequence during an interval of time, or the likelihood of success or failure of an event on test or on demand. Probability is expressed as a dimensionless number ranging from 0 to 1.

Process Flow Diagram (PFD) A diagram that shows the material flow from one piece of equipment to the other in a process. It usually provides information about the pressure, temperature, composition, and flow rate of the various streams, heat duties of exchangers, and other such information pertaining to understanding and conceptualizing the process.

Process Hazard Analysis (PHA) An organized effort to identify and evaluate hazards associated with processes and operations to enable their control. This review normally involves the use of qualitative techniques to identify and assess the significance of hazards. Conclusions and appropriate recommendations are developed. Occasionally, quantitative methods are used to help prioritize risk reduction.

Process knowledge management A management system element that includes work activities to gather, organize, maintain, and provide information to other management system elements. Process safety knowledge primarily consists of written documents such as hazard information, process technology information, and equipment-specific information. Process safety knowledge is the product of this management system.

Process safety A disciplined framework for managing the integrity of operating systems and processes handling hazardous substances by applying good design principles, engineering, and operating practices.

Note: Process safety focuses on efforts to reduce process safety risks associated with processes handling hazardous materials and energies. Process safety efforts help reduce the frequency and consequences of potential incidents. These incidents include toxic or flammable material releases (loss events), resulting in toxic effects, fires, or explosions. The incident impact includes harm to people (injuries, fatalities), harm to the environment, property damage, production losses, and adverse business publicity.

Process safety culture The common set of values, behaviors, and norms at all levels in a facility or in the wider organization that affect process safety.

Process Safety Incident/Event An event that is potentially catastrophic, i.e., an event involving the release/loss of containment of hazardous materials that can result in large-scale health and environmental consequences.

Process Safety Information (PSI) Physical, chemical, and toxicological information related to the chemicals, process, and equipment. It is used to document the configuration of a process, its characteristics, its limitations, and as data for process hazard analyses.

Process Safety Management (PSM) A management system that is focused on prevention of, preparedness for, mitigation of, response to, and restoration from catastrophic releases of chemicals or energy from a process associated with a facility.

Process Safety Management Systems Comprehensive sets of policies, procedures, and practices designed to ensure that barriers to episodic incidents are in place, in use, and effective.

Protective Action Criteria (PAC) Protective Action Criteria are essential components for planning and response to uncontrolled releases of hazardous chemicals. These criteria, combined with estimates of exposure, provide the information necessary to evaluate chemical release events for the purpose of taking appropriate protective actions.

Protective Action Criteria includes AEGL, ERPG, and TEEL and is available in 3 levels for over 3100 chemicals.

Pool fire The combustion of material evaporating from a layer of liquid at the base of the fire.

Reactive chemical A substance that can pose a chemical reactivity hazard by readily oxidizing in air without an ignition source (spontaneously combustible or peroxide forming), initiating or promoting combustion in other materials (oxidizer), reacting with water, or self-reacting (polymerizing, decomposing or rearranging). Initiation of the reaction can be spontaneous, by energy input such as thermal or mechanical energy, or by catalytic action increasing the reaction rate.

Recognized and Generally Accepted Good Engineering Practice (RAGAGEP) A term originally used by OSHA, stems from the selection and application of appropriate engineering, operating, and maintenance knowledge when designing, operating and maintaining chemical facilities with the purpose of ensuring safety and preventing process safety incidents.

It involves the application of engineering, operating or maintenance activities derived from engineering knowledge and industry experience based upon the evaluation and analyses of appropriate internal and external standards, applicable codes, technical reports, guidance, or recommended practices or documents of a similar nature. RAGAGEP can be derived from singular or multiple sources and will vary based upon individual facility processes, materials, service, and other engineering considerations.

Responsible Care©	An initiative implemented by the Chemical Manufacturers Association (CMA) in 1988 to assist in leading chemical processing industry companies in ethical ways that increasingly benefit society, the economy, and the environment while adhering to ten key principles.
Reliability-Centered Maintenance (RCM)	A systematic analysis approach for evaluating equipment failure impacts on system performance and determining specific strategies for managing the identified equipment failures. The failure management strategies may include preventive maintenance, predictive maintenance, inspections, testing, and/or one-time changes (e.g., design improvements, operational changes).
Replacement-in-kind	An item (equipment, chemicals, procedures, organizational structures, people, etc.) that meets the design specifications, if one exists, of the item it is replacing.
Risk	A measure of human injury, environmental damage, or economic loss in terms of both the incident likelihood and the magnitude of the injury or loss. A simplified version of this relationship expresses risk as the product of the Frequency and the Consequence of an incident (i.e., Risk = Frequency x Consequence).
Risk analysis	The estimation of scenario, process, facility and/or organizational risk by identifying potential incident scenarios, then evaluating and combining the expected frequency and impact of each scenario having a consequence of concern, then summing the scenario risks if necessary to obtain the total risk estimate for the level at which the risk analysis is being performed.
Risk assessment	The process by which the results of a risk analysis (i.e., risk estimates) are used to make decisions, either through relative ranking of risk reduction strategies or through comparison with risk targets.
Risk management	The systematic application of management policies, procedures, and practices to the tasks of analyzing, assessing, and controlling risk in order to protect employees, the general public, the environment, and company assets, while avoiding business interruptions. Includes decisions to use suitable engineering and administrative controls for reducing risk.
Risk Based Process Safety (RBPS)	The Center for Chemical Process Safety's (CCPS) PSM system approach that uses risk-based strategies and implementation tactics that are commensurate with the risk-based need for process safety activities, availability of resources, and existing process safety culture to design, correct, and improve process safety management activities.
Risk Management Plan (RMP) Rule	U.S. EPA's accidental release prevention Rule, which requires covered facilities to prepare, submit, and implement a risk management plan.

Risk tolerance criteria	A predetermined measure of risk used to aid decisions about whether further efforts to reduce the risk are warranted.
Root cause	A fundamental, underlying, system-related reason why an incident occurred that identifies a correctable failure(s) in management systems. There is typically more than one root cause for every process safety incident.
Safe haven	A building or enclosure that is designed to provide protection to its occupants from exposure to outside hazards.
Safe work practices	An integrated set of policies, procedures, permits, and other systems that are designed to manage risks associated with non-routine activities such as performing hot work, opening process vessels or lines, or entering a confined space.
Safe Operating Limits (SOL)	Limits established for critical process parameters, such as temperature, pressure, level, flow, or concentration, based on a combination of equipment design limits and the dynamics of the process.
Safeguard	Design features, equipment, procedures, etc. in place to decrease the probability or mitigate the severity of a cause-consequence scenario.
Safety critical element	Any part of an installation, plant or computer program whose failure will either cause or contribute to a major accident, or the purpose of which is to prevent or limit the effect of a major accident. Safety Critical Elements are typically barriers or parts of barriers. In the context of this book, safety includes harm to people, property, and the environment. (Other possible names: **Safety and Environmental Critical Element, Safety Critical Equipment, Safety Critical Control**).
Safety Instrumented Function (SIF)	A system composed of sensors, logic servers, and final control elements for the purpose of taking the process to a safe state when predetermined conditions are violated.
Safety Integrity Level (SIL)	Discrete level (one out of four) allocated to the SIF for specifying the safety integrity requirements to be achieved by the SIS.
Safety Instrumented System (SIS)	A separate and independent combination of sensors, logic solvers, final elements, and support systems that are designed and managed to achieve a specified safety integrity level. A SIS may implement one or more Safety Instrumented Functions (SIFs).
Scenario	An integrated set of policies, procedures, permits, and other systems that are designed to manage risks associated with non-routine activities such as performing hot work, opening process vessels or lines, or entering a confined space.
Shelter-in-place	A process for taking immediate shelter in a location readily accessible to the affected individual by sealing a single area (an example being a room) from outside contaminants and shutting off all HVAC systems.

Societal risk A measure of risk to a group of people. It is most often expressed in terms of the frequency distribution of multiple casualty events.

Top event The main system failure of interest in a fault tree analysis.

Toxicity The quality, state, or degree to which a substance is poisonous and/or may chemically produce an injurious or deadly effect upon introduction into a living organism. (Merriam-Webster Dictionary)

Toxicology A science that deals with poisons and their effect and with the problems involved (such as clinical, industrial, or legal problems). (American Industrial Hygiene Association)

Vapor Cloud Explosion (VCE) The explosion resulting from the ignition of a cloud of flammable vapor, gas, or mist in which flame speeds accelerate to sufficiently high velocities to produce significant overpressure.

ACKNOWLEDGMENTS

The American Institute of Chemical Engineers (AIChE) and the Center for Chemical Process Safety (CCPS) express their appreciation and gratitude to all members of the *Process Safety for Engineers: An Introduction, Second Edition* and their CCPS member companies for their generous support and technical contributions in the preparation of this book.

The collective industrial experience and know-how of the subcommittee members makes this book especially valuable to all who strive to learn from incidents, take action to prevent their recurrence and improve process safety performance.

Project Writer:

This manuscript was written by Cheryl Grounds who thanks the Subcommittee Members and Peer Reviewers for their content contribution to this book and their dedication to teaching process safety. Final technical editing was completed by CCPS staff – Jennifer Bitz leading with support from Bruce Vaughen and Anil Gokhale.

Subcommittee Members:

Jerry Forest	Celanese, CCPS Project Chair
Cheryl Grounds	CCPS Staff Consultant and Writer
Dan Crowl	Michigan Technological University
Kobus Diedericks	Nova Chemicals
Ken First	CCPS Staff Consultant
Warren Greenfield	WG Associates LLC
Barry Guillory	Louisiana State University
Jack McCavit	JL McCavit Consulting, LLC
Robin Pitblado	DNV

Before publication, all CCPS books are subjected to a thorough peer review process. CCPS gratefully acknowledges the thoughtful comments and suggestions of the peer reviewers. Their work enhanced the accuracy and clarity of these guidelines. Although the peer reviewers have provided many constructive comments and suggestions, they were not asked to endorse this book and were not shown the final manuscript before its release.

Peer Reviewers:

Brian Farrell	CCPS Staff Consultant
Jeff Fox	CCPS Emeritus
Jerry Fung	Canadian Natural Resources Limited
Jim Klein	ABS Consulting
Ray Mentzer	Purdue University
Hocine Ait Mohamed	Rio Tinto
Greg Nesmith	Dow Chemical Company
Bala Raman	Ecolab
Mark Setterfield	Tronox
Jonathan Slater	3M
Rajagopalan Srinivasa	Indian Institute of Technology Madras
Ron Unnerstall	University of Virginia
Bruce Vaughen	CCPS Staff Consultant
Ronald J. Willey	Northeastern University

ONLINE MATERIALS ACCOMPANYING THIS BOOK

Some figures in this book are reduced in size to enhance readability. They are available online in full size.

To access this online material, go to:

www.aiche.org/ccps/publications/PSIntro

Enter password: 2ndEdition

PREFACE

The Center for Chemical Process Safety (CCPS) was established in 1985 to protect people, property and the environment from major chemical incidents by bringing best practices and knowledge to industry, academia, the government and the public around the world. As part of this vision, CCPS has focused on developing and disseminating technical information through collective wisdom, tools, training and expertise from experts within the Chemical Engineering Industry. The primary source of this information is a series of guideline and concept books to assist industry in implementing various elements of process safety and risk management. This book is part of this series.

As a not-for-profit organization, CCPS has published over 100 books, written by member company representatives who have donated their time, talents, and knowledge. Industry experts, and contractors that prepare the books, typically provide their services at a discount in exchange for the recognition received for their contributions in preparing these books for publication.

The integration of process safety into the engineering curricula is an ongoing goal of the CCPS. To this end, CCPS created the Safety and Chemical Engineering Education (SACHE) committee which develops training modules for process safety. The CCPS Technical Steering Committee initiated the creation of this book to assist colleges and universities in meeting this challenge and to aid chemical engineering programs in meeting recent accreditation requirements for including the hazards associated with chemical processing into the undergraduate chemical engineering curricula. This second edition updates the content and realigns the flow to facilitate the understanding of process safety by university students and early career engineers.

1
Introduction and Regulatory Overview

1.1 Purpose of this Book

This book provides an introduction to process safety. It is intended to be used either to accompany a stand-alone process safety course or as a resource of supplemental material for existing curricula. The book provides both text and a toolkit in the form of references, tools, lecture materials, and links to learning materials. (Refer to Online Materials Accompanying this Book.) This book is intended to familiarize the undergraduate student or early career engineer with important process safety concepts. As an overview, is not intended to provide the technical details on the topics which are available through various Center for Chemical Process Safety (CCPS) publications and other sources.

The overall learning objectives of this book are:

- State the importance of process safety as illustrated by process safety incidents,
- Use common process safety definitions and terminology,
- Summarize Risk Based Process Safety (RBPS) as defined by the CCPS,
- Participate in and contribute to basic hazard identification and risk analysis,
- Locate basic process safety information, resources, and tools,
- Apply process safety concepts to design and operation of engineered systems, and
- Contribute, in a supporting role, to the management of process safety hazards and risks.

1.2 Target Audience

The primary audience for this publication is third year to graduate level chemical engineering students and those engineers new to process safety in the workforce. This book can also be used by other engineering disciplines since process safety is important to all fields of engineering.

1.3 Process Safety – What Is It?

Process safety can be defined as follows.

Process Safety - A disciplined framework for managing the integrity of operating systems and processes handling hazardous substances by applying good design principles, engineering, and operating practices.

Note: Process safety focuses on efforts to reduce process safety risks associated with processes handling hazardous materials and energies. Process safety efforts help reduce the frequency and consequences of potential incidents. These incidents include toxic or flammable material releases (loss events), resulting in toxic effects, fires, or explosions. The incident impact includes harm to people (injuries, fatalities), harm to the environment, property damage, production losses, and adverse business publicity. (CCPS Glossary)

Such events happen at chemical facilities, refineries, onshore and offshore oil and gas facilities, and in other industries that handle or process flammable, combustible, toxic, or reactive materials and hazardous energies. For the rest of this book, the term "process facility" or "facility" will be used to mean the previously mentioned facilities and any other operation that handles or processes flammable, combustible, toxic, or reactive materials.

After an explosion in a BP Texas City refinery in 2005 that led to 15 fatalities and injured over 170 others, an independent commission was created to examine the process safety mindset, or culture, of BPs refinery operations, this commission came to be known as the Baker Panel. (Baker 2007) The Chemical Safety Board quoted the Baker Panel as follows.

> "Process safety hazards can give rise to major accidents involving the release of potentially dangerous materials, the release of energy (such as fires and explosions), or both. Process safety incidents can have catastrophic effects and can result in multiple injuries and fatalities, as well as substantial economic, property, and environmental damage. Process safety refinery incidents can affect workers inside the refinery and members of the public who reside nearby. Process safety in a refinery involves the prevention of leaks, spills, equipment malfunctions, over-pressures, excessive temperatures, corrosion, metal fatigue, and other similar conditions. Process safety programs focus on the design and engineering of facilities, hazard assessments, management of change, inspection, testing, and maintenance of equipment, effective alarms, effective process control, procedures, training of personnel, and human factors." (CSB 2007)

Process safety applies to all phases of a facility life cycle. Process safety programs cover the operation of pilot facilities during the research phase. This also includes the selection of the chemistry and unit operations chosen to achieve the design intent of the process. Process safety is included in the design and engineering phase in choices about the type of unit operations and equipment items to use, the facility layout, and the equipment design. Process safety during operation includes, as was mentioned in the CSB quote, "hazard assessments, management of change, inspection, testing, and maintenance of equipment, effective alarms, effective process control, procedures, and training of personnel". The engineering decisions made during research and development, pilot work, engineering design, and facility operations can impact process safety performance.

1.4 Process Safety, Occupational Safety, and Environmental Impact

The terms "process safety", "occupational safety", and "environmental impact" are different, even though they are often genericized to simply "safety". All three deserve attention. Assuming that addressing one will manage the other is a mistake. The Baker Panel report noted that while BP had systems to support occupational safety; the company misunderstood that good occupational safety performance does not indicate good process safety performance. (Baker 2007) This concept is illustrated in this example: when choosing an airline, should you select the one that has a good record on employee slips, trips, and falls or the one that has fewer crashes?

In the chemical, petrochemical, and most other industries, companies are required to have an occupational safety program. This program may be required by federal or local regulations. It can apply to workers in a manufacturing plant, a research laboratory, pilot plant, and even in office locations. The focus of these programs is to prevent harm to workers from workplace accidents such as falls, cuts, sprains and strains, being struck by objects, repetitive motion injuries, and so on. These are good and very necessary programs. They typically focus on preventing harm to an individual from occupational illnesses and injuries.

They are not, however, what process safety is about. Process safety focuses on acute, though unlikely events whereas personal safety focuses on typically less acute, but more likely events.

Process safety focuses on preventing harm from acute events that may harm multiple people (or the environment) in a single incident.

Process safety environmental impact is equally as important. Interestingly, sometimes environmental drivers can come into conflict with process safety drivers. For example, carbon canisters used to control air emissions from hydrocarbon storage tanks can catch fire if not designed, installed, and maintained properly. (Zerbonia, et al, 2001) Consider both topics and balance the risks appropriately.

Engineers will likely be faced with decisions impacting process safety and thus preventing harm to people and the environment. Although engineering is based on scientific principles, there are typically uncertainties involved in calculations and limits to any engineer's expertise. In the workplace, engineering is just one of the factors influencing business decisions. Other factors include financial resources, business goals, political constraints, organizational culture and decision-making processes. Engineering ethics is important in guiding an engineer in making decisions in the workplace.

1.5 History of Process Safety

Organizations in the process industries have a long-standing concern for process safety. (See the inset about the manufacture of nitroglycerine as an example.) Organizations originally had safety reviews for processes that relied on the experience and expertise of the people in the review. In the middle of the 20th century, more formal review techniques began to appear in the process industries. These included the Hazard and Operability Study (HAZOP), developed by ICI in the 1960s (CIA 1977), Failure Modes and Effects Analysis (FMEA), Checklist and What-If reviews. These are qualitative techniques for assessing the hazards of a process or operation.

Quantitative analysis techniques, such as Fault Tree Analysis (FTA), which had been in use by the nuclear industry, Quantitative Risk Assessment (QRA), and Layer of Protection Analysis (LOPA) also began to be used in the process industries in the 1970s, 1980s and 1990s. Modeling techniques were developed for analyzing the consequences of spills and releases, explosions, and toxic exposures. The Design Institute of Emergency Relief Systems (DIERS) was established within the American Institute of Chemical Engineers (AIChE) in 1976 to develop methods for the design of emergency relief systems to handle runaway reactions. By the mid to late 1970s, process safety was a recognized technical specialty. The American Institute of Chemical Engineers (AIChE) formed the Safety and Health Division in 1979.

In 1976, a runaway reaction occurred near Seveso, Italy that resulted in the release of 2,3,7,8-tetrachlorodibenzo-p-dioxin, commonly known as dioxin, into surrounding areas. Dioxin is a toxic chemical. Many people developed chloracne, a skin disease, and a 17 km^2 (6.6 sq. mi.) area was made uninhabitable. This incident eventually led to stricter regulations for the process industries in the European Economic Community in 1982, under what is known as the Seveso Directive (EC).

Nitroglycerine. The manufacture of nitroglycerine is an example of how process safety has evolved. Alfred Nobel began manufacturing nitroglycerine in 1864. The process was to add fuming nitric acid and sulfuric acid to glycerin while keeping the temperature at about 20-25 °C (68-77 °F). Keeping the temperature under control was critical to the safety of the process. Figure 1.1 shows a picture of an early nitroglycerine process, in which an operator is charged with observing the temperature and stopping feeds to the reactor if the temperature got too high. Note that he is sitting on a one-legged stool. The reasoning was that, if he fell asleep (watching a temperature indicator for 8 to 10 hours a day is not the most interesting thing to do), he would fall and wake up. Also, note the size of the reactor. This represents a huge amount of nitroglycerine in one place, so if it did explode, the damage would be considerable. Explosions were not uncommon. Alfred Nobel's own brother did not survive such an explosion.

Enough explosions occurred that some locations banned the use of nitroglycerine. Alfred Nobel's major breakthrough was discovering that when mixed with an inert carrier, it became safer to handle. This form of nitroglycerine is called dynamite.

The process for making nitroglycerine evolved becoming inherently safer. The large, original batch reactors were replaced with small, continuous reactors (Figure 1.2) that are a fraction of the size of the original reactors. This makes the reaction easier to control, because the heat removal capability is better (notice the cooling coils for the reactor in Figure 1.2) and the mixing intensity of the smaller reactor is higher. It also reduces the extent of the damage if an explosion did occur. Automated controls mean that no one has to stand in front of the reactor anymore.

A defining moment in the chemical industry occurred in 1984. A release of methylisocyanate, a toxic and flammable material, occurred at a chemical plant in Bhopal, India. More than 3,000 fatalities resulted from this release. (This incident will be described in more detail in Chapter 6.) The original chemical plant design had many safeguards against this event. These safeguards were not maintained, and at the time of the incident many were not functioning. The Bhopal event drove home the point that technical expertise alone was not enough, and that hazards or risk management was as important as the technical aspect of process safety. The Bhopal incident led to the formation of the Center for Chemical Process Safety (CCPS) in 1985. The CCPS is a not-for-profit organization that is part of the U.S. based American Institute for Chemical Engineers (AIChE) with a vision to protect people, property, and the environment [by] bringing the best process safety knowledge and practices to industry, academia, the government, and the public around the world through collective wisdom, tools, training, and expertise.

Figure 1.1. Nitroglycerine reactor in the 19th century
(Nobelprize.org)

Figure 1.2. Continuous nitroglycerine reactor
(Biazzi SA)

The concept of process safety management evolved as the chemical process industries matured their focus on process safety. A management system is a formally established and documented set of activities and procedures designed to produce specific results in a consistent manner on a sustainable basis.

Process Safety Management (PSM) is a management system that is focused on prevention of, preparedness for, mitigation of, response to, and restoration from releases of chemicals or energy from a process associated with a facility. (CCPS Glossary)

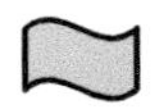

In this book the term process safety management normally refers to the CCPS Risk Based Process Safety twenty element model. Process Safety Management, (PSM), refers to the fourteen element U.S. OSHA regulatory model. Both are described in this section.

In 1985 the Chemical Manufacturers Association (CMA), which later became the American Chemical Council (ACC) issued process safety management guidelines (CMA 1985). By 1989, the CCPS introduced a set of 12 process safety management elements (CCPS 1989). The American Petroleum Institute (API) also issued process safety management guidelines in 1990 (API RP 750). The CCPS studied the various approaches at the time and gleaned the 12 characteristics from interactions with its member companies and traditional business process consulting firms that had significant experience in evaluating management systems. Those guidelines were the first generic set of principles to be compiled for use in designing and evaluating process safety management systems.

In 1992 the U.S. Occupational Safety and Health Administration (OSHA) issued the "Process Safety Management of Highly Hazardous Chemicals" (OSHA PSM 1910.119 regulation, which had its own, although similar, set of process safety management elements. (OSHA) The U.S. Environmental Protection Agency (EPA) issued its own version in 1995 under the authority of the Clean Air Act. This regulation is commonly referred to as RMP, or Risk Management Plan, since the regulation requires the development and submittal of a risk plan based on the regulatory definitions and requirements. (EPA) Both of these regulations are described in further detail in Chapter 3.

OSHA PSM regulations cover the impacts of hazards to workers on-site; EPA RMP covers the impacts of hazards off-site. Applicability for both regulations is triggered by having more than a specified amount, commonly referred to as a Threshold Quantity (TQ), of specified chemicals, usually called Highly Hazardous Chemicals, or by more than 4,500 kilograms (10,000 pounds) of a flammable material in one location. PSM as defined by the CCPS, which does not depend on threshold quantities, is much broader than OSHA PSM and EPA RMP.

Process safety practices and formal safety management systems have been in place in some companies for many years. Process safety management is widely credited for reductions in major accident risk and in improved process safety performance in the process industry. Nevertheless, many organizations continue to be challenged by inadequate management system performance, resource pressures, and stagnant process safety results. To promote process safety excellence and continuous improvement throughout the process industries, the CCPS created risk-based process safety (RBPS) as the framework for the next generation of process safety management (CCPS 2007). RBPS has the format of a good practice and not a regulatory requirement. A comparison of CCPS RBPS elements with OSHA PSM and EPA RMP elements is provided in Figure 3.3.

1.6 Basic Process Safety Definitions

The terms "process safety" and "process safety management" have been defined previously in this section. Some other process safety terms are important to understand and use correctly. These include the following.

> A **hazard** is an operation, activity, or material with the potential to cause harm to people, property, the environment, or business; or simply, a potential source of harm. (CCPS Glossary)

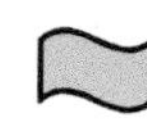

Reinforcing this definition, a hazard is something with the potential to cause harm. Process safety is all about reducing the chance, the risk, that the harm will occur. A typical dictionary definition of risk, for example the Merriam-Webster online dictionary, is "the possibility of loss or injury" or "someone or something that creates or suggests a hazard". The CCPS definition considers the hazard (what can go wrong), the magnitude (how bad can it be), and the likelihood (how often can it happen).

> **Risk** is a measure of human injury, environmental damage, or economic loss in terms of both the incident likelihood and the magnitude of the injury or loss. A simplified version of this relationship expresses risk as the product of the Frequency and the Consequence of an incident (i.e., Risk = Frequency x Consequence). (CCPS Glossary)

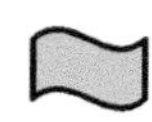

Although the formula for risk is simple, the concept of risk is complex. The risk level that a person is willing to tolerate is influenced by their perspective. Consider the risk of dying during a recreational activity. That is a voluntary risk: a risk that is willingly taken because the risk is outweighed by the benefit of the enjoyment. Transportation risks are also voluntary; however, the activity is seen as a necessity. Working is a voluntary risk where the risk is outweighed by the compensation received. Some non-voluntary risks include living near a chemical plant (and not being a worker at that plant). Each type of risk has different risk levels. Base jumping risk is higher risk than scuba diving. Driving is higher risk than flying. Logging is higher risk than farming. A variety of risks are illustrated in Figure 1.3 and Figure 1.4. which show fatality numbers and rate, respectively.

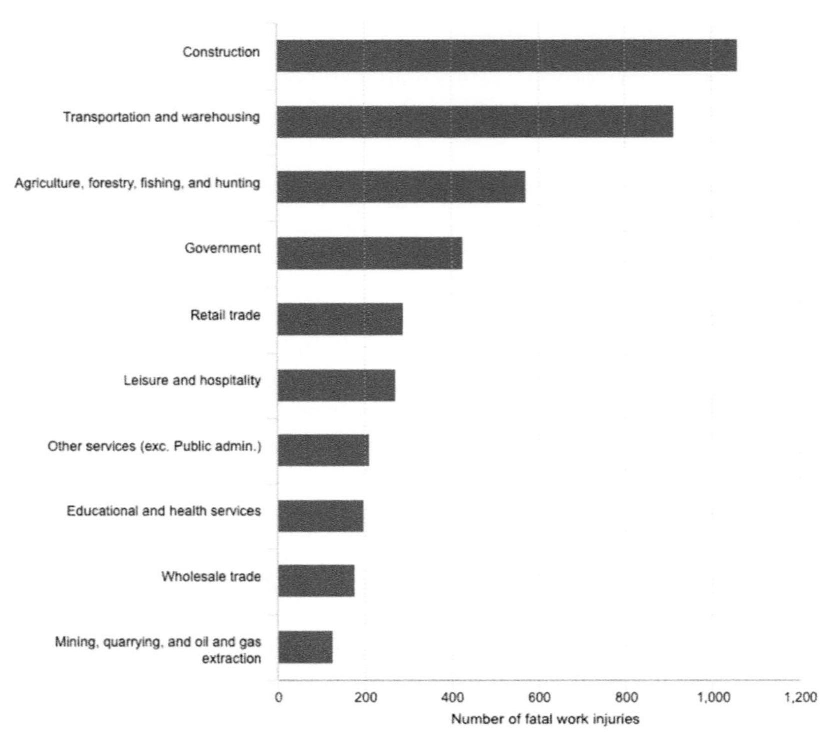

Figure 1.3. Number of fatal work injuries, by industry sector, 2019 (BLS 2019)

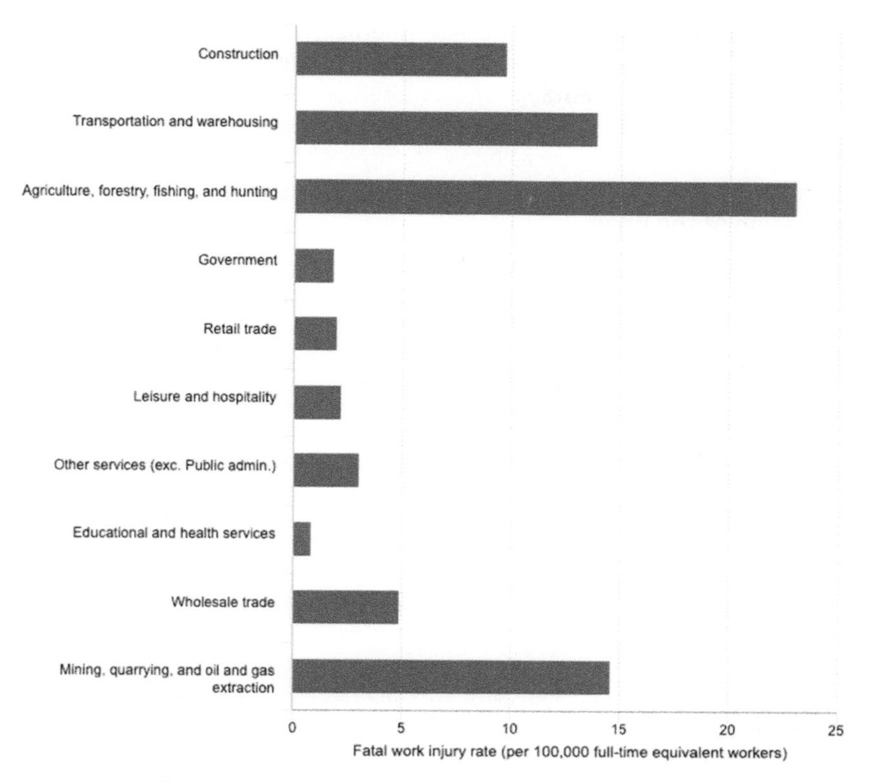

Figure 1.4. Fatal work injury rate by industry sector, 2019
(BLS 2019)

It is important in industry to manage risk at a level that the company decides it can tolerate. Individuals all have a different perception of tolerable risk; companies need to define a risk tolerance level understood by all. As employees make decisions on how to conduct work, the implicit risk level should be consistent such that a single employee's decisions do not increase risks level beyond tolerability and thus, potentially impact all employees by degrading the company's performance.

Process safety is about managing risks.

Fundamentally, process safety is about answering these questions.

1. What can go wrong? (human injury, environmental damage, or business economic loss).
2. How bad could it be? (magnitude of the loss or injury).
3. How often might it happen? (likelihood of the loss or injury).
4. How can the risk be prevented, or its impact minimized?

Not all processes and operations have the same amount of risk. Understanding risk helps a company decide how to prioritize its efforts to manage process safety. Resources, that is, money and people, are finite. When designing, operating, and maintaining a facility, the engineer has a wide range of options of which process safety management activities to implement at the facility (with the minimum requirement being complying with local and federal regulations). Process safety should be risk-based; a process with low risk does not warrant the same amount of resources as one with high risk.

1.7 Organization of the Book

This book moves from understanding and identifying hazards to analyzing and managing risk and concludes with a brief discussion on systems used to manage process safety.

Chapters 1, 2, and 3 answer the question ***what is process safety and why should I care about it?*** This includes defining basic terms, describing CCPS Risk Based Process Safety, and identifying prevalent process safety regulations, codes, and standards.

Chapters 4 through 9 answer the question ***what are the hazards?*** Basic concepts of fire, explosion, toxics, reactive chemicals, and other hazards are described along with where to find data on hazardous properties of materials and metrics used to classify incidents involving those materials.

Chapters 10 through 15 answer the question ***how do I address the hazards in my design?*** This is aligned with the three questions noted in the previous section: *what can go wrong, how bad can it be, and how often might it happen?* These chapters go through the basics of incorporating process safety in design and identifying what can go wrong in terms of equipment failure. Consequence analysis and risk assessment concepts are introduced along with methods to mitigate risk.

Chapters 16 through 20 answer the question ***how do I manage risk in operations?*** The topics of human factors, operational readiness, management of change, operating procedures, conduct of operations, safe work practices, and emergency management are addressed.

Chapters 21 through 23 answer the question ***how do I sustain the focus on process safety?*** These chapters address the RBPS elements that focus on the management of process safety and how all the aspects come together to support process safety culture.

The chapters begin with a seminal process safety incident. Many of the incident investigations revealed that deficiencies in multiple process safety topics contributed to the incidents. This book highlights the key points in each incident that are relevant to that chapter. Most of the process safety incidents cited in this book come from these sources.

Incidents that Define Process Safety (CCPS 2008)
More Incidents that Define Process Safety (CCPS 2019)
U.S. Chemical Safety Board Investigations and videos (CSB)

A list of CSB investigations and videos, organized by chapters in this book, is provided in Appendix G. Additionally, refer to Appendix H for a table listing root causes for selected major process safety incidents.

Most chapters conclude with:

- Problem sets,
- Other relevant incidents,
- What a new engineer might do, and
- References

The following symbols are used to bring attention to important points in the text.

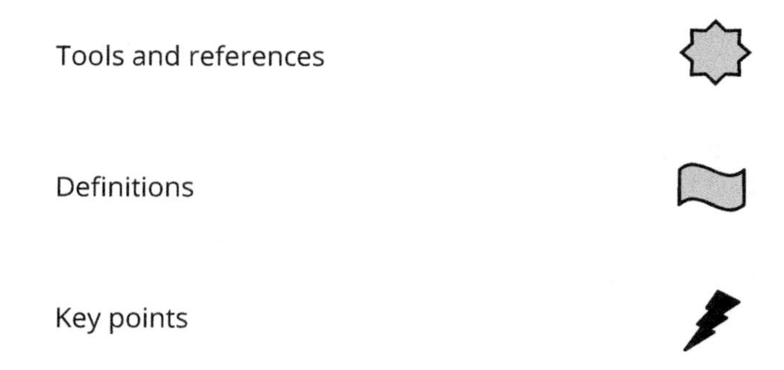

1.8 Use of this Book in University Courses

This book is intended to support both the teaching of an introductory level process safety course and also as a reference for the inclusion of process safety topics in typical engineering courses. To support the latter, a matrix relating the chapters in this book with typical engineering courses is included in Appendix B.

This book focuses on general process safety concepts while the textbook by Crowl and Louvar, *Chemical Process Safety, Fundamentals with Applications, 4th Edition* (Crowl 2019), provides a more fundamental engineering approach to process safety. These books can be used in combination.

1.9 Exercises

1. Describe the differences in focus between process safety and occupational safety.
2. Name two major incidents and the process safety regulations that were prompted by the incidents.
3. Explain the relationship between "hazard" and "risk".
4. Figure 1.4. lists risks associated with various industries. Select 5 of these industries and identify what hazards may be inherent in that industry.
5. Identify 3 risks you take routinely. For each,
   a) Identify if it is voluntary or non-voluntary.
   b) Identify what can go wrong.
   c) Identify how bad it could be.
   d) Identify how often the bad outcome might happen.

1.10 References

ACC, https://www.americanchemistry.com.

AIChE, https://www.aiche.org.

API RP 750, "Management of Process Hazards", American Petroleum Institute, Washington, D.C., 1990.

Baker 2007, Baker James et al., "The Report of the BP U.S. Refineries Independent Safety Review Panel", https://www.csb.gov/assets/1/20/baker_panel_report1.pdf?13842.

Biazzi SA, www.Biazzi.com.

BLS 2019, "Graphics for Economic News Releases", U.S. Bureau of Labor Statistics, https://www.bls.gov/charts/census-of-fatal-occupational-injuries/number-and-rate-of-fatal-work-injuries-by-industry.htm.

CCPS 1989, *Guidelines for Technical Management of Chemical Process Safety*, Center for Chemical Process Safety, John Wiley & Sons, Hoboken, N.J.

CCPS 2007, *Guidelines for Risk Based Process Safety*, Center for Chemical Process Safety, John Wiley & Sons, Hoboken, N.J.

CCPS 2008, *Incidents that Define Process Safety*, Center for Chemical Process Safety, John Wiley & Sons, Hoboken, N.J.

CCPS 2019, *More Incidents that Define Process Safety*, Center for Chemical Process Safety, John Wiley & Sons, Hoboken, N.J.

CCPS Glossary, "CCPS Process Safety Glossary", Center for Chemical Process Safety, https://www.aiche.org/ccps/resources/glossary.

CIA 1977, *A Guide to Hazard and Operability Studies*, Chemical Industries Association, U.K.

CMA 1985, "Process Safety Management (Control of Acute Hazards)", Chemical Manufacturers Association, Washington, D.C.

CSB, U.S. Chemical Safety and Hazard Investigation Board, Washington, D.C., https://www.csb.gov/.

CSB 2007, "Refinery Explosion and Fire", Investigation Report, Report No. 2005-04-I-TX, U.S. Chemical Safety and Hazard Investigation Board, Washington, D.C., March.

Crowl 2019, Daniel A. Crowl and Joseph F. Louvar, *Chemical Process Safety, Fundamentals with Applications*, 4th edition, Pearson, NY.

EC, "The Seveso Directive – Technological Disaster Risk Reduction", viewed September 24, 2020, https://ec.europa.eu/environment/seveso.

EPA, U.S. Environmental Protection Agency, https://www.epa.gov/rmp.

Nobelprize.org 2015, "Alfred Nobel in Scotland". Nobel Media AB 2014, http://www.nobelprize.org/alfred_nobel/biographical/articles/dolan/.

OSHA, U.S. Occupational Safety and Health Administration, https://www.osha.gov/laws-regs/regulations/standardnumber/1910/1910.119.

Zerbonia 2001, Robert A. Zerbonia , Cybele M. Brockmann , Paul R. Peterson & Denise Housley, "Carbon Bed Fires and the Use of Carbon Canisters for Air Emissions Control on Fixed-Roof Tanks", *Journal of the Air & Waste Management Association*, 51:12, 1617-1627, DOI: 10.1080/10473289.2001.10464393.

2
Risk Based Process Safety

2.1 Learning Objectives

The learning objectives of this chapter are:

- Summarize the four pillars and the twenty elements of Risk Based Process Safety,
- Explain the "risk based" aspect of RBPS, and
- State who created RBPS and why.

2.2 Incident: BP Refinery Explosion, Texas City, Texas, 2005

2.2.1 Incident Summary

An explosion occurred within the isomerization unit (Isom) of BP's Texas City Refinery during a startup after a turnaround in March 2005. Fifteen contractors were fatally injured and over 170 people harmed. The Isom unit and adjacent plants and equipment suffered major damage.

The contractors that were fatally injured were located in portable plant buildings used to support an adjacent plant turnaround. They were in an area operated as an uncontrolled area, i.e., a safe area without any hot work permit/electrical controls imposed (CCPS 2008).

Key Points:

Hazard Identification and Risk Analysis – Locate people out of harm's way. The portable buildings were located in close proximity to the Isom unit and were occupied during the Isom startup, a more hazardous phase of operation.

Conduct of Operations – Human factors should be part of the operational system analysis. This includes fatigue. The operators had worked 12-hour shifts for more than 29 consecutive days at the time of the incident.

2.2.2 Detailed Description

See Figure 2.1 for a diagram of the Isom system. The following excerpt from the CSB report describes the incident.

> "During the startup, operations personnel pumped flammable liquid hydrocarbons into a distillation tower for over three hours without any liquid being removed, which was contrary to startup procedure instructions. Critical alarms and control instrumentation provided false indications that failed to alert the operators of the high level in the tower. Consequently, unknown to the operations crew, the 52 m (170 ft) tall tower was overfilled and liquid overflowed into the overhead pipe at the top of the tower.
>
> The overhead pipe ran down the side of the tower to pressure relief valves located 45 m (148 ft) below. As the pipe filled with liquid, the pressure at the bottom rose

rapidly from about 145 kPa gauge (21 psig) to about 441 kPa gauge (64 psig). The three pressure relief valves opened for six minutes, discharging a large quantity of flammable liquid to a blowdown drum with a vent stack open to the atmosphere. The blowdown drum and stack overfilled with flammable liquid, which led to a geyser-like release out the 34 m (113 ft) tall stack. This blowdown system was an antiquated and unsafe design; it was originally installed in the 1950s and had never been connected to a flare system to safely contain liquids and combustible flammable vapors released from the process.

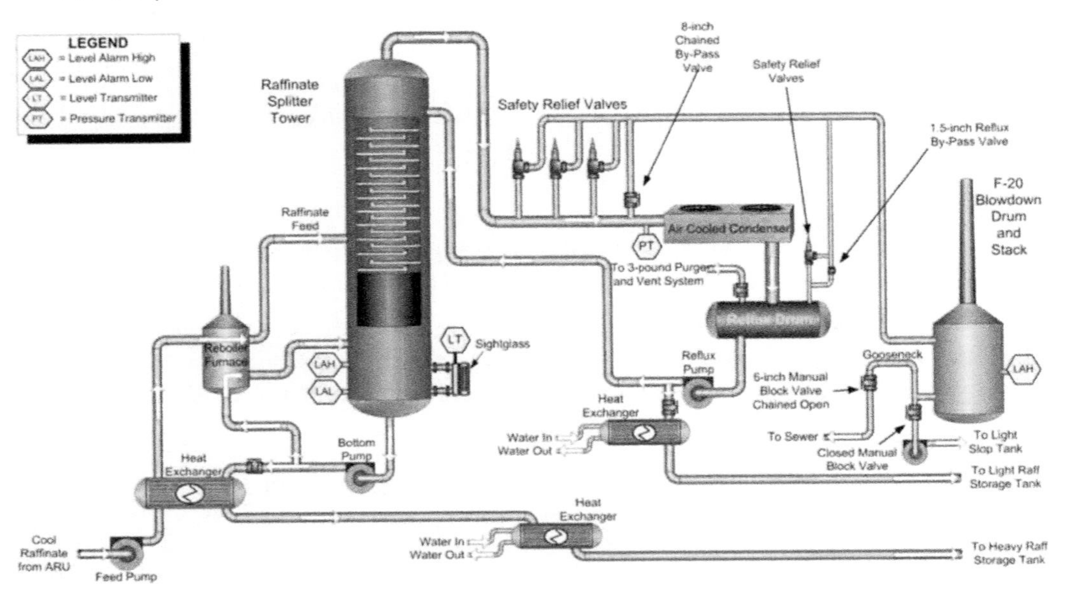

Figure 2.1. Process flow diagram of the raffinate column and blowdown drum (CSB 2007)

The released volatile liquid evaporated as it fell to the ground and formed a flammable vapor cloud. The most likely source of ignition for the vapor cloud was backfire from an idling diesel pickup truck located about 7.6 m (25 ft) from the blowdown drum. The 15 employees who did not survive the explosion were contractors working in and around temporary trailers that had been previously sited by BP as close as 37 m (121 ft) from the blowdown drum." (CSB, 2008)

Figures 2.2 and 2.3 show the damage to the unit and the portable buildings, respectively. In Figure 2.2, the red arrow points to the top of the blowdown stack. In Figure 2.3, the red arrow in upper left points to the blowdown drum.

The BP investigation concluded "that while many departures to the startup procedure occurred, the key step that was instrumental in leading to the incident was the failure to establish Heavy Raffinate rundown to tankage, while continuing to feed and heat the tower. By the time the Heavy Raffinate flow was eventually started, the Splitter bottoms temperature was so high, and the liquid level in the tower so high, that this intervention made matters worse by introducing significant additional heat to the feed." (CCPS 2008).

Figure 2.2. Texas City isomerization unit aftermath (CSB 2007)

Figure 2.3. Destroyed trailers west of the blowdown drum (CSB 2007)

The investigation team concluded that the splitter was overfilled and overheated because "the Shift Board Operator did not adequately understand the process or the potential consequences of his actions or inactions on March 23."

2.2.3 Lessons

Many Risk Based Process Safety Elements were involved in the BP Texas City explosion. Five are listed here. The bulleted findings are reproduced from the CSB and the Baker Panel reports. (CSB 2007)

Process Safety Culture. Process safety culture is the first of the 20 risk based process safety elements. Most striking of the CSB findings are those with respect to the process safety culture at BP and the Texas City plant. Selected CSB findings regarding BP's process safety culture are included in the following list. Some of these findings could easily apply to other companies. The CSB recommended that BP create an "independent panel of experts to examine BP's corporate safety management systems, safety culture, and oversight of the North American refineries." This became known as the Baker Panel. The Baker Panel report focused on safety management systems at BP and resulted in ten recommendations to the BP Board of Directors.

Selected CSB findings:

- "Cost-cutting, failure to invest and production pressures from BP Group executive managers impaired process safety performance at Texas City.
- The BP Board of Directors did not provide effective oversight of BP's safety culture and major incident prevention programs. The Board did not have a member responsible for assessing and verifying the performance of BP's major incident hazard prevention programs.
- Reliance on the low occupational injury rate at Texas City as a safety indicator failed to provide a true picture of process safety performance and the health of the safety culture.
- A "check the box" mentality was prevalent at Texas City, where personnel completed paperwork and checked off on safety policy and procedural requirements even when those requirements had not been met."

Selected Baker Panel finding:

- "BP has not instilled a common, unifying process safety culture among its U.S. refineries. Each refinery has its own separate and distinct process safety culture. While some refineries are far more effective than others in promoting process safety, significant process safety culture issues exist at all five U.S. refineries, not just Texas City. Although the five refineries do not share a unified process safety culture, each exhibits some similar weaknesses. The Panel found instances of a lack of operating discipline, toleration of serious deviations from safe operating practices, and apparent complacency toward serious process safety risks at each refinery."

BP acquired the Texas City refinery as part of its merger with Amoco in 1999. Neither Amoco (the previous facility operator) nor BP replaced blowdown drums and atmospheric stacks, even though a series of incidents warned that this equipment was unsafe. In 1992,

OSHA cited a similar blowdown drum and stack as unsafe, but the citation was withdrawn as part of a settlement agreement and therefore the drum was not connected to a flare as recommended. Amoco, and later BP, had safety standards requiring that blowdown stacks be replaced with equipment such as a flare when major modifications were made. In 1997, a major modification replaced the Isom blowdown drum and stack with similar equipment, but Amoco did not connect it to a flare. In 2002, BP engineers proposed connecting the Isom blowdown system to a flare, but a less expensive option was chosen.

Hazard Identification and Risk Analysis. The people fatally injured were not involved in the process and did not need to be located so close to a hazardous unit operation. This incident led to a major revision of API code regarding facility siting: API Recommended Practice 752: *Management of Hazards Associated with Location of Process Plant Permanent Buildings*. It also led to the creation of API Recommended Practice 753: *Management of Hazards Associated with Location of Process Plant Portable Buildings*.

Training and Performance Assurance. A lack of supervisory oversight and technically trained personnel during the startup, an especially hazardous period, was an omission contrary to BP safety guidelines. An extra board operator was not assigned to assist, despite a staffing assessment that recommended an additional board operator for all Isom startups. Supervisors and operators poorly communicated critical information regarding the startup during the shift turnover; BP did not have a shift turnover communication requirement for its operations staff. Isom operators were likely fatigued from working 12-hour shifts for 29 or more consecutive days.

The operator training program was inadequate. The central training department staff had been reduced from 28 to eight, and simulators were unavailable for operators to practice handling abnormal situations, including infrequent and high hazard operations such as startups and unit upsets.

Management of Change. BP Texas City did not effectively assess changes involving people, policies, or the organization that could impact process safety. For example, the control room staff was reduced from 2 people to one, who was overseeing three units

Local site Management of Change rules required that where a portable building was to be placed within 100 meters (350 ft) of a process unit a Facility Siting Analysis had to be carried out. However, this location had already been used many times for these trailers. Not doing an effective MOC put all the people in the portable buildings at unnecessary risk (CCPS 2008).

Asset Integrity and Reliability. The process unit was started despite previously reported malfunctions of the tower level indicator, level sight glass, and a pressure control valve. Deficiencies in BP's mechanical integrity program resulted in equipment with failed instrumentation being started up.

2.3 Risk Based Process Safety

Chapter 1 defined process safety and process safety management. This chapter describes the elements of Risk Based Process Safety (RBPS) which is a management system created by the Center for Chemical Process Safety (CCPS) in 2007 and described in the *Guidelines for Risk Based Process Safety* (CCPS 2007). It is based upon the CCPS and OSHA process safety management systems created in the early 1990s improving upon them by incorporating lessons learned over the years.

RBPS is a performance-based approach. It is not a prescriptive management system nor is it a regulatory obligation. It sets out a structure that addresses key topics supporting process safety. Companies should create their own structure conforming to regulations, meeting their own objectives, and learning from RBPS.

> The RBPS approach is based on the premise that all hazards and risks are not equal and that higher risks warrant greater resources. Through this approach, "limited company resources can be optimally apportioned to improve both facility safety performance and overall business performance". (CCPS 2007)

The RBPS structure is composed of twenty management system elements grouped under four pillars.

1. Commit to Process Safety
2. Understand Hazards and Risk
3. Manage Risk
4. Learn from Experience

All twenty elements, in combination, support good process safety management. It is challenging to be an expert in all of the elements. Many process safety experts choose to focus on a few elements at various points in their career. It can take many years to develop breadth and depth of knowledge across all the elements.

The pillar structure including the management system elements is provided in Figure 2.4. Following this, each pillar and its elements are described briefly.

Pillar: Commit to Process Safety
- Process safety culture
- Comply with standards
- Process safety competency
- Workforce involvement
- Stakeholder outreach

Pillar: Understand Hazards and Risk
- Process knowledge management
- Hazard identification and risk analysis

Pillar: Manage Risk
- Operating procedures
- Safe work practices
- Asset integrity and reliability
- Contractor management
- Training and performance assessment
- Management of change
- Operational readiness
- Conduct of operations
- Emergency management

Pillar: Learn from Experience
- Incident investigation
- Measurement and metrics
- Auditing
- Management review and continuous improvement

Figure 2.4. RBPS structure
(CCPS 2021)

Commitment to process safety is the cornerstone of process safety excellence. Organizations generally do not improve without strong leadership and solid management commitment. All involved should commit to process safety in order to create a strong process safety culture.

For process safety, management needs to recognize that process safety is not the same as occupational safety and move beyond occupational safety programs. A company's commitment to process safety is demonstrated by its actions.

Organizations that **understand hazards and risk** are better able to allocate limited resources in the most effective manner. Industrial experience has demonstrated that businesses using hazard and risk information to plan, develop, and deploy stable, lower-risk operations are much more likely to enjoy long-term success.

Managing risk focuses on four issues: (1) prudently operating and maintaining processes that pose the risk, (2) managing changes to those processes to ensure that the risk remains tolerable, (3) maintaining the integrity of equipment and assuring quality of materials, fabrications, and repairs, and (4) preparing for, responding to, and managing incidents that do occur. Managing risk helps a company or a facility deploy management systems that help sustain long-term, incident-free, and profitable operations.

Learning from experience involves monitoring, and acting on, internal and external sources of information. Despite a company's best efforts, operations do not always proceed as planned, accidents and near misses occur. A *near miss* is an event in which an accident (that is, property damage, environmental impact, or human loss) or an operational interruption could have plausibly resulted if circumstances had been slightly different. Organizations must be ready to turn their mistakes, and those of others, into opportunities to improve process safety efforts.

The twenty elements are described in more detail in the following sections. As a new engineer or someone new to process safety, some of these elements will have a more direct impact on you than others, but all have some impact. For example, learning about the Codes and Standards that affect your process and location will be an important part of your first few years in industry, whereas you may not be involved in Stakeholder Outreach. Nevertheless, the effort expended by the organization on stakeholder outreach may have a direct impact on how you will have to approach process safety at your locale.

2.4 Pillar: Commit to Process Safety

The aim of the first pillar is to ensure that the foundation for process safety is in place and embedded throughout the organization.

RBPS Element 1: Process Safety Culture

Process safety culture is a commonly held set of values, norms, and beliefs. It can be stated as "How we do things around here," "What do we expect here," and "How we behave when no one is watching."

This element describes a positive environment where employees at all levels are committed to process safety. It starts at the highest levels of the organization and is shared by all. Process safety leaders nurture this process. Process safety culture is differentiated from occupational safety culture as it addresses less frequent major incident prevention cultures as well as occupational safety.

Process safety culture highlights the necessary role of leadership engagement to drive the process. Safety culture should not be thought of as a passive outcome of a specific work environment, rather it is something that can be managed and improved.

RBPS Element 2: Compliance with Standards

Regulations, codes, and standards document learnings and good practices. Compliance requires identifying, developing, and implementing them as appropriate. Standards should be developed for both new construction and existing operations. These can be internal and external standards, national and international codes and standards, and local, state and federal regulations and laws.

Requirements issued by regulators and consensus standards organizations may need interpretation and implementation guidance. The element also includes proactive development activities for corporate, consensus, and governmental standards.

RBPS Element 3: Process Safety Competency

Process safety competency requires creating, developing, and maintaining technical knowledge, continuously improving that knowledge and competency, and ensuring that appropriate process safety information is available to those who should use it.

This element addresses skills and resources that a company should have in the right places to manage its process safety hazards. It includes verification that the company collectively has these skills and resources and that this information is applied in succession planning and management of organizational change.

As several investigations have shown, excellent performance in occupational safety does not guarantee similar performance in process safety. Personnel may have a good understanding of the precursors to occupational incidents and the barriers and behaviors that prevent them, but not necessarily the same level of understanding/knowledge for more complex and rarer process safety events.

Companies need to assure themselves that personnel at all levels understand how to apply process safety principles. This competency requirement applies across the life cycle (project design and construction, operation, and dismantling), including routine, nonroutine, and emergency tasks. A system for verification of this competency is necessary.

RBPS Element 4: Workforce Involvement

Workforce involvement means active participation of company and contractor workers in the design, development, implementation, and continuous improvement of process safety in the workplace.

This element addresses the need for broad involvement of operating and maintenance personnel in process safety activities to make sure that lessons learned by the people closest to the process are considered and addressed.

Workforce involvement in process safety reviews (e.g. risk assessments, management of change, etc.) ensures their knowledge of potential problems and operation of key safety systems is included and considered in potential risk reduction enhancements. Their involvement also helps build an understanding in personnel of major hazards and how barriers are deployed to make these safe. Most regulators require workforce involvement for process safety.

RBPS Element 5: Stakeholder Outreach

Stakeholder outreach. strives to make relevant process safety information available to a variety of organizations including the neighboring community, local emergency responders, and other companies in the industry.

This element covers activities with the community, contractors, and nearby facilities to help neighbors, outside responders, regulators, and the public understand the asset's hazards and potential emergency scenarios.

This activity may be coordinated by local regulators but usually is driven by the company for larger integrated facilities.

2.5 Pillar: Understand Hazards and Risk

This pillar addresses process knowledge management and the identification of hazards and management of risks. Process knowledge must be readily available and kept up to date. Risk management is an extension of some regulatory requirements, but in practice most companies conduct hazard identification and add risk ranking as part of the analysis. Inherently safer design considerations are applied here. RBPS also notes that companies may choose to go beyond qualitative risk ranking to some form of quantitative risk assessment (LOPA or QRA).

RBPS Element 6: Process Knowledge Management

Process knowledge management involves activities associated with compiling, cataloging, and making available process safety information (PSI). It also includes understanding the information, not simply compiling data.

Process Safety Information - Physical, chemical, and toxicological information related to the chemicals, process, and equipment. It is used to document the configuration of a process, its characteristics, its limitations, and as data for process hazard analyses. (CCPS Glossary)

It includes verification of the accuracy of this information and confirmation that this information is kept up to date. This information must be readily available to those who need it to safely perform their jobs.

This knowledge and documentation should be accessible to those designing, operating, and maintaining a facility. Important knowledge includes incident lessons and datasets, updated engineering standards, equipment drawings and specifications, operational experience and upsets, and new or updated process safety tools.

During engineering projects, teams can change at each stage, and relevant process safety knowledge should be transmitted along with the design.

RBPS Element 7: Hazard Identification and Risk Analysis

Hazard identification and risk analysis (HIRA) encompasses all activities involved in process hazard analysis (PHA) and evaluating risks at facilities, throughout their life cycle, to make certain that risks to employees, the public, and the environment are managed within the organization's risk tolerance.

Hazard identification and risk assessment are complementary activities that initially focus on identifying process safety hazards, then determining their potential consequences, and finally estimating the scenario risks. HIRA includes recommendations to reduce or eliminate hazards, reduce potential consequences, or reduce frequency of occurrence. Analysis may be qualitative or quantitative depending on the level of risk. HIRA is a core process safety activity and the main focus of this book.

HIRA analyses vary from simple to complex. In addition to basic topics such as identifying responses to upsets, potential leak scenarios, and important barriers and integrity, it must also take into account environmental conditions, material composition changes, and both routine and non-routine operations. The potential impacts considered include harm to people (workers and people located next to the facility), to the environment, and to the asset.

Many different tools are used for HIRA. Hazard identification tools include simple checklists, What-If analysis and HAZOP analysis. Risk assessment tools include fire hazard analysis and explosion studies, LOPA, and QRA. Inherent safety methods and functional safety assessments fall within HIRA.

That which has not been identified cannot be prevented or mitigated. HIRA results should be tracked using a risk register or other tracking system. This is to ensure that no identified issue is inadvertently neglected.

2.6 Pillar: Manage Risk

This pillar addresses many important topics for operational safety and management of risks. These include operating procedures, safe work practices, contractor management, training, operational readiness and conduct of operations. This pillar also addresses asset integrity, management of change, and emergency management.

RBPS Element 8: Operating Procedures

Standard operating procedures (SOP) requires written instructions for all phases of operation including routine, non-routine, startup, shutdown, and emergency. Good procedures also describe the process, hazards, tools, protective equipment, and controls in sufficient detail that operators understand the hazards, can verify that controls are in place, and can confirm that the process responds in an expected manner.

These procedures describe how the operation is to be carried out safely, define safe operating limits, explain the consequences of deviation from safe operating limits, identify key safeguards, and address special situations and emergencies.

Operating procedures have improved substantially from the past approach of simply taking start-up procedures from the design contractor. Presently, procedures are designed with operating personnel engagement, are periodically updated based on feedback and any modifications, and use modern layouts with graphics and photographs to convey key safety messages. Risks from deviations are highlighted – e.g. if equipment purging is required before start-up, the procedure highlights safety risks with shorter duration purging. Barrier management is an important aspect of process safety and the procedures highlight relevant barriers potentially affected by the procedure.

RBPS Element 9: Safe Work Practices

Safe work practices are requirements established to control hazards and are used to safely operate, maintain, and repair equipment and conduct specific types of work. They include control of work (job safety analysis (JSA), permits and oversight), opening pipework or vessels, energy isolation, and other activities. These practices are used when developing detailed work plans, ensuring that requirements are met, and the appropriate safeguards have been or will be implemented for the work. They cover non-routine work and are often supplemented with permits. These fill the gap between operating and maintenance procedures and the hazards and risks specific to the work being conducted at the time.

Typically, several parties are involved in safe work practices including the owner and its contractors. Interface documents dictate what safe work practices are used and specify who approves the work.

RBPS Element 10: Asset Integrity and Reliability

Asset integrity and reliability is the systematic implementation of inspections, tests, and maintenance to ensure that equipment, and safety critical devices will be functional for their intended application throughout their life.

This includes proper selection of materials; inspection, testing, and preventive maintenance; and design for maintainability. During the design stage, potential asset integrity problems can be anticipated and significantly mitigated.

Equipment and control systems can be affected by harsh environments. Some equipment can be hard to inspect, particularly in remote or offshore installations.

Many existing facilities are operating beyond their intended design life and are managing aging issues which can impact asset integrity.

RBPS Element 11: Contractor Management

Contractor management is a system of controls to ensure that contracted services support both safe facility operations and the company's process safety and occupational safety performance goals. This element includes the selection, acquisition, use, and monitoring of such contracted services.

These controls ensure that contract workers perform their jobs safely, and that contracted products and services do not add to or increase safety risks.

Contractors are prominent in both operations and maintenance activities. They have specialized knowledge and equipment to enable challenging tasks to be performed safely and efficiently. It is necessary to align the process safety program of the company with its contractors to ensure that all aspects are addressed and that everyone knows their responsibilities.

RBPS Element 12: Training and Performance Assurance

Training and performance. assurance involves practical instruction in job and task requirements and methods. Performance assurance provides a means by which workers demonstrate that they have understood the training and can apply it in practical situations.

Training and performance assurance applies to operators, maintenance workers, supervisors, engineers, leaders, and process safety professionals. Performance assurance verifies that the trained skills are being practiced proficiently.

Work is challenging, and a high degree of skill is needed to perform tasks correctly. Many incidents identify weaknesses in training and job execution as underlying causes. Defining the training that is required to perform a task successfully helps underpin a training program. This should include process safety hazards and how to participate in or interpret risk analysis studies, as appropriate.

Formal testing of knowledge and skills is an important part of this element to assure that participants have understood the material. It includes on-the-job task verification.

RBPS Element 13: Management of Change (MOC)

MOC strives to ensure that changes to a process do not inadvertently introduce new hazards or unknowingly increase risks. This includes identification of a change, review of potential hazards, and authorization of the change. Changes can include those to the process, equipment, operations, and key personnel or organization. The MOC occurs prior to implementation and ensures that potentially affected personnel are notified of the change, and that documentation is kept up to date. Documentation includes drawings, operating and maintenance procedures, and training material. The objective is to prevent or mitigate incidents prompted by unmanaged change.

Many past incidents have been due to changes that were not properly assessed, and which defeated existing safeguards or introduced new hazards.

The MOC is a formal process involving similar tools to the initial hazard identification and risk assessment. A newer aspect of MOC is the recognition that organizational change can also create process safety issues and a specific Management of Organizational Change (MOOC) procedure has been developed (CCPS 2013).

RBPS Element 14: Operational Readiness

Operational readiness evaluates the process before start-up to ensure the process can be safely started. It applies to restart of facilities after being shut down or idled as well as after process changes and maintenance, and to start-up of new facilities. A pre-startup safety review (PSSR) is a common tool in verifying operational readiness.

An important aspect is to verify that all barriers identified in design reviews and captured in action tracking systems have been implemented and/or any outstanding actions are approved for later close-out.

RBPS Element 15: Conduct of Operations

Conduct of operations (CCPS, 2011) is the means by which management and operational tasks are carried out in a deliberate, consistent, and structured manner. Managers and co-workers ensure tasks are carried out correctly and prevent deviations from expected performance. Conduct of operations is closely tied to an organization's culture. Workers at every level are expected to perform their duties with alertness, due thought, full knowledge, sound judgment, and a proper sense of pride and accountability.

This element aims to ensure the organization has operational discipline, in other words, that operating practices are applied correctly and fully every time. In doing this, normalization of deviance can be avoided (i.e. hazardous shortcuts that succeed once and become the standard way of doing the activity). (CCPS 2018) It also identifies an operational envelope beyond which procedures no longer apply. Personnel and contractors must recognize these conditions and either stop the activity or call in more experienced personnel to address the deviation.

RBPS Element 16: Emergency Management

Emergency management includes planning for possible emergencies, providing resources to execute the plan, practicing and improving the plan, training or informing employees, contractors, neighbors, and local authorities, and effectively communicating with stakeholders in the event an incident does occur.

Firefighting philosophies can vary from provision of site emergency responders and firefighting equipment to the use of mutual aid support, or a combination of the two.

Personnel evacuation typically can be provided through local emergency response capabilities; however, in remote locations, specific arrangements should be made.

2.7 Pillar: Learn from Experience

This pillar addresses how to learn from experience - from incidents and near misses, from leading and lagging metrics, and from audit reports. The aim of all of these is to identify deeper systemic causes and to implement corrective actions that solve the wider issue, not just the specific instance. The final element of management review leads to continual improvement. This goes beyond prescriptive requirements. Good process safety should seek improvements by learning from experience.

RBPS Element 17: Incident Investigation

Incident investigation is the process of reporting, tracking, and investigating incidents and near misses to identify root causes so that corrective actions are taken, trends are identified, and learnings are communicated to appropriate stakeholders.

It is essential that learnings are shared with management and workers at the facilities through regular safety talks or in formal training. These personnel have the most to gain from the information and can ensure that the learnings drive the necessary changes. The management review (see Element 20), normally conducted annually, reviews incident investigations and ensure the recommendations are implemented, including actions to address broader systemic concerns.

An important aspect is that evidence from incidents needs to be preserved until the investigation can occur.

RBPS Element 18: Measurement and Metrics

The measurement and metrics element establishes performance and efficiency indicators to monitor the effectiveness of the RBPS management system and its constituent elements and work activities. It addresses which leading and lagging indicators to consider, how often to collect data, and what to do with the information to help ensure responsive, effective RBPS management system operation.

Guidance is available on suitable metrics from API RP 754 "Process Safety Performance Indicators for the Refining and Petrochemical Industries" for downstream (API RP 754) and IOGP 456 "Process safety - recommended practice on key performance indicators for upstream" (IOGP 456). Contractors often perform much of the work at a site, so their process safety performance should be included in the company collected metrics.

API 754 and IOGP 456 are closely aligned and have 4 tiers of process safety events. Tiers 1 and 2 are actual losses of containment, fatality, or multiple injury events. Tiers 3 and 4 address demands on safety systems, near miss events, delayed inspections and maintenance, and deficient safety management systems.

RBPS Element 19: Auditing

Auditing provides for a periodic review of process safety management system performance. The aim is to identify gaps in performance and identify improvement opportunities, and subsequently to track closure of these gaps to completion. Auditors may be second party (company personnel but independent of operational roles at the site) or third party (independent company) depending on company policies or local regulations. Auditors must have the requisite skills for facility or activities under audit.

RBPS Element 20: Management Review and Continuous Improvement

Management review and continuous improvement is the routine evaluation of whether management systems are performing as intended and producing the desired results as efficiently as possible. It is an ongoing "due diligence" review by management that fills the gap between day-to-day work activities and periodic formal audits.

The review formalizes the link between goals set and results achieved. Where goals are not achieved then additional actions or investment can be included in the annual plan to correct the deficiency. Where goals are achieved, then new actions or investments can be agreed upon, so that process safety performance is not static but continually improves over time.

2.8 What a New Engineer Might Do

A new engineer may be expected to have a basic understanding of process safety and may be involved in entry-level work in many of the RBPS elements. Having a familiarity with the RBPS elements will enable a new engineer to understand the context of how the work they are doing relates to other ongoing work. Further details on what a new engineer might do is presented later in this book as the specific RBPS elements are discussed. Table 2.1 lists RBPS activities that a new engineer, chemical or other discipline, might be involved in.

Table 2.1. Process safety activities for new engineers

Element	Activity	Engineering Discipline
Commit to Process Safety		
Process Safety Culture	Learn responsibilities for process safety roles	All
Compliance with Standards	Learn and apply standards and regulations that apply to processes and equipment	All
Process Safety Competency	Take advantage of training opportunities	All
Workforce Involvement	Contribute Listen to the input from operators, technicians, etc.	All All
Stakeholder Outreach	Become aware of the organization's outreach efforts	All

Table 2.1 continued

Element	Activity	Engineering Discipline
Understand Hazards and Risk		
Process Knowledge Management	Ensure accuracy of process safety information (PSI) Be familiar with Safety Data Sheets Develop reactivity matrix Develop and update design basis calculations Develop and update Piping and Instrumentation Diagrams (P&IDs) Verify electrical area classification Develop automation logic diagrams and cause and effect charts. Maintain equipment files (design and fabrication information)	All All All Chemical Chemical I&E Engineers I&E & Control Mechanical
Hazard Identification and Risk Assessment	Participate in HIRAs Assemble required PSI Review applicable RAGAGEPs Compile industry incidents Review of seismic threats and structural response to impulse loading from overpressure events, etc.	All Chemical All All Civil
Manage Risk		
Operating Procedures	Write new and update operating procedures	Chemical
Safe Work Practices	Write Permits Approve permits	Safety Safety & Chemical
Asset Integrity and Reliability	Identify equipment covered by Inspection, Testing and Preventive Maintenance (ITPM) program Write ITPM procedures Analyze inspection results Approve and monitor repairs Maintain ITPM records Testing of SCAI and Safety Instrumented Systems Review structural integrity	Chemical & Mechanical Mechanical Mechanical Mechanical Mechanical & I&E & Control Civil Structural
Contractor Management	Approve contractors with respect to safety Train contractors	Safety All
Training and Performance Assurance	Generally not applicable for new engineers	
Management of Change	Identify changes Participate in MOCs Assemble required PSI Review applicable RAGAGEPs	All All All All
Operational Readiness	Participate in readiness reviews Assemble required information	All All
Conduct of Operations	Perform all tasks reliably in accordance with policies and procedures	All
Emergency Management	Learn emergency response plans for your area	All

Table 2.1 continued

Element	Activity	Engineering Discipline
Learn from Experience		
Incidents	Participate in investigations Recognize and identify near misses	All All
Metrics	Record and maintain data Analyze data	All All
Auditing	Assemble requested information for audits	All
Management Review	Generally not applicable for new engineers	

2.9 Summary

The RBPS elements work together like the interlocking pieces of a jigsaw puzzle. Let's use a hypothetical situation to illustrate this.

An incident or near miss occurs in a chemical facility. An *incident investigation* is conducted. Near misses are much more likely to be noticed and reported in a company with a good *process safety culture*. A good investigation will use appropriate resources in the facility and other expertise within or outside of the company as needed (*workforce involvement*). Recommendations will come from the investigation and the results will be shared with similar facilities in the company so everyone can learn from the event.

Recommendations from the investigation will involve changes in what the facility is doing. These can range from procedural changes to the addition of new process safety controls or interlocks to a redesign of the process, including new process equipment. In any case, something will change (*management of change*). A MOC review is held, again with people with the appropriate expertise. If it is an extensive change, the HIRA (*Hazard Identification and Risk Analysis*) is updated to identify any new hazards. As changes are made, they comply with standards (*compliance with standards*). New or updated procedures are written (*operating procedures*). If new equipment, even as simple as a new safety interlock, is added it is put on a maintenance and inspection schedule (*asset integrity*). Operators and maybe maintenance personnel are trained on the new procedure, process, etc. (*training and performance assurance*). Changes to piping and equipment are made in a safe manner (*safe work practices*). In addition to information about the event itself, new information, such as piping and instrumentation diagrams, equipment or technology descriptions are added to the existing process safety information documentation (*process knowledge management*). Before you restart the process with the changes implemented a pre-startup safety review is held (*operational readiness*). When the entire project is complete, the company may choose to do a review of the project to be sure all of its procedures and processes were complied with, and if not, why (*measurement and metrics*).

This hypothetical example did not cover every element, but it did cover more than half of them, illustrating how all the pieces come together.

The RBPS provides a useful basis for a company-specific process safety management system. . Many other management systems, both in regulatory systems and in company practice, have varying numbers of elements. Some combine topics into fewer elements, others expand these out giving more elements. The number and arrangement are not important so long as the topics are covered. RBPS highlights some key topics not included in other systems.

As is noted in the pillar Learn from Experience, those interested in process safety should strive to learn from their own experience as well as the experience of others. All the aspects of RBPS should be considered as an opportunity to learn and to improve process safety performance, whether or not they are regulated at your specific operating location.

2.10 Exercises

1. List 3 RBPS elements evident in the BP Texas City explosion summarized at the beginning of this chapter. Describe their shortcomings as related to this accident.
2. Considering the BP Texas City explosion, what actions could have been taken to reduce the risk to the people in the facility?
3. Of the 20 Risk Based Process Safety elements, which do you think is the starting point? Why?
4. Describe the relationship between OSHA Process Safety Management regulations and CCPS Risk Based Process Safety.
5. Describe how the term "risk based" is applied in Risk Based Process Safety.
6. Briefly describe the focus of the four pillars of RBPS.
7. Identify 5 stakeholders impacted by the BP Texas City explosion.
8. Why is hazard identification so fundamental to the management of process safety?
9. Why is management of change so important to the management of process safety?
10. An intrastate natural gas pipeline failed catastrophically in a residential area resulting in fatalities and property damage. The pipeline was constructed over 60 years before the incident. The length of the pipeline that failed had not been adequately inspected for many years. Identify 2 RBPS elements and explain how they are evident in this incident.
11. Refer to Table 2.1. Identify 5 activities that you might be able to perform on the first days of your new job in a facility.

2.11 References

API RP 754, "Process Safety Performance Indicators for the Refining and Petrochemical Industries", American Petroleum Institute, Washington, D.C., 2016.

Baker 2007, Baker James et al., "The Report of the BP U.S. Refineries Independent Safety Review Panel", www.csb.gov.

CCPS Glossary, "CCPS Process Safety Glossary", Center for Chemical Process Safety, https://www.aiche.org/ccps/resources/glossary.

CCPS 2007, *Guidelines for Risk Based Process Safety*, Center for Chemical Process Safety, John Wiley & Sons, Hoboken, N.J.

CCPS 2008, *Incidents that Define Process Safety*, Center for Chemical Process Safety, John Wiley & Sons, Hoboken, N.J.

CCPS 2011, *Conduct of Operations and Operational Discipline*, Center for Chemical Process Safety, John Wiley & Sons, Hoboken, N.J.

CCPS 2013, *Managing Process Safety Risks During Organizational Change*, Center for Chemical Process Safety, John Wiley & Sons, Hoboken, N.J.

CCSP 2018, *Recognizing and Responding to Normalization of Deviance*, Center for Chemical Process Safety, John Wiley & Sons, Hoboken, N.J.

CCPS 2021, *Process Safety in Upstream Oil and Gas*, Center for Chemical Process Safety, John Wiley & Sons, Hoboken, N.J.

CSB 2007, "Refinery Explosion and Fire", Investigation Report, Report No. 2005-04-I-TX, U.S. Chemical Safety and Hazard Investigation Board, Washington, D.C., March.

CSB 2008, Video – "Anatomy of a Disaster", U.S. Chemical Safety and Hazard Investigation Board http://www.csb.gov/videos.

IOGP 456, "Process safety – recommended practice on key performance indicators for upstream", London, U.K., 2018.

3
Process Safety Regulations, Codes, and Standards

3.1 Learning Objectives

The learning objectives of this chapter are:

- Summarize the major process safety regulations used worldwide, and
- Explain the difference between regulations, codes, and standards.

3.2 Incident: Montreal, Maine & Atlantic Railway Derailment and Fire, Quebec, Canada, 2013

3.2.1 Incident Summary

In the early hours of July 6, 2013, an unattended Montreal, Maine & Atlantic (MMA) Railway train rolled from its overnight parking location on a slope and proceeded over 7 miles into the town of Lac-Megantic where it derailed. The train was carrying crude oil and the resulting fires and explosions fatally injured 47 people and destroyed 40 buildings as well as 53 vehicles. Figure 3.1 shows the scene. Forty-seven counts of criminal negligence were filed against three MMA employees and the company declared bankruptcy as a result of this incident. In the figure, the color indicates relative size of breach: orange = large, yellow = medium, blue = small.

This incident prompted discussions and recommendations by the Transportation Safety Board of Canada on the safe rail transportation of crude oil and the DOT final rule in May 2015 to strengthen safe rail transportation of large volumes of flammable liquids. (NCSL 2015 and TSB 2013)

Key Points:

Compliance with Standards –Take all the advice you can. Standards include the experience, hard learnings, and even expert calculations of many others. Take their advice and follow the standards.

Management of Change – Beware of creeping change. When small changes happen slowly over time, it is easy to overlook them. The problem is eventually the small changes add up to a big change that has not been recognized or the risk managed.

Emergency Management – Is it really 'all clear'? It's human nature to want an emergency to be over - to declare it under control. However, when that emergency involved operating equipment, an expert in the control of that equipment should be consulted to verify that the equipment status is truly safe.

Figure 3.1. Lac-Megantic tank cars with breaches to their shells (TSB 2013)

3.2.2 Detailed Description

The MMA-002 train was traveling from North Dakota to a refinery in New Brunswick, Canada. The train was made up of 72 cars carrying 7.7 million liters (2 million gallons) of crude oil (UN1267). Just before midnight on July 5, 2013, the train was parked in Nantes, Quebec, Canada.

The 1,433 m (4,700 ft) long train that contained 72 tank cars loaded with crude oil from the Bakken fields in North Dakota. (NTSB 2015) The cars were DOT-111 design. With the fracked crude from primarily Texas and North Dakota, the U.S. was producing more crude oil than it had in 30 years. Transportation of crude oil by rail had increased significantly to move the crude to refineries for processing. Carloads carrying oil in 2014 rose by more than 5000 percent when compared with 2008 numbers (NCSL 2015).

The fracked crude oils from formations such as the Bakken tend to be lighter than other crudes. They are of a lower density, flow freely at room temperature, and have a higher proportion of light hydrocarbon fractions resulting in higher API gravities (between 37° and 42°). A Sandia report stated that "No single parameter defines the degree of flammability of a fuel; rather, multiple parameters are relevant." (Sandia 2015) The attention following this incident is continuing to prompt discussion on the safe transport of various classifications of crude oils.

The locomotive engineer stopped the train on a downhill grade on the main track. He used the pneumatic brakes, applied the brakes on the locomotive and the buffer car, and began to apply the hand brakes to some cars (but fewer than recommended in company procedures), and shut down the trailing locomotives. He tested the hand brake by releasing the locomotive automatic brakes but did not release the locomotive independent brakes.

He communicated with the rail traffic controller noting mechanical difficulties he had experienced including excess smoke and a loss of power in the lead engine. They decided to address these issues in the morning. The locomotive engineer went off-duty to stay in a Lac-

Megantic hotel. The taxi driver noted the smoke from the smokestack along with oil droplets. The locomotive engineer stated that he had informed the company of this.

Just before midnight, a fire was reported on a train at Nantes. A track foreman met with the fire department and was told that the emergency fuel cut-off switch had been used to shut down the lead locomotive. This stopped the fuel to the fire. The firefighters also put the locomotive electrical breakers in the off position. The track foreman and the fire department were in conversation with the rail traffic controller. The engineer asked the rail traffic controller if he needed to return to the train to start another engine. He was told that the track manager had dispatched a track foreman to the site. The train was left for the night with no engines running and thus no power to the pneumatic compressor.

Over the course of the next hour, air pressure bled from the brake system and the train then started to roll downhill. It reached a speed of over 105 kph (65 mph) and traveled the 11.6 km (7.2 mi) to the town of Lac-Megantic where 63 railcars derailed, releasing approximately 6 million liters (1 million gallons) of crude oil. The spill flowed to the lake, ignited, and resulted in the death of 47 people.

The MMA procedure for parking of unattended trains required 9 hand brakes to be set for trains of this length and additional hand brakes to be used if the train was parked on a slope of the grade in Nantes. Canadian rail industry best practice would have been to set the brakes on 40% of the railcars (which means brakes should have been set on 28 railcars). Only 7 hand brakes were set on this train and the engineer improperly performed a brake test without releasing the locomotive's air brakes. When the firefighters responded to the train fire in Nantes, they shut down the locomotive per the firefighting procedure; however, they did not follow the procedure addressing parking the train on the grade. Additionally, they did not contact the locomotive engineer. With none of the other locomotives running, the air in the brake system started to deplete and an hour later the train began to roll downhill. The train reached 105 kph (65 mph). The track in the Lac-Megantic switch area was rated for only 24 kph (15 mph).

At the time of the incident, the DOT-111 train car was the standard car for flammable liquids. Many changes happened during the increased production of fracked crudes including the number of cars in a single train, the overall volume of crude transported by train, and the properties of the fracked crude itself. The DOT-111 car was not capable of withstanding the impacts experienced in the Lac-Megantic derailment. A 2015 DOT final rule addressed "High-hazard flammable trains" (HHFT) which means "a continuous block of 20 or more tank cars loaded with a flammable liquid or 35 or more tank cars loaded with a flammable liquid dispersed through a train.". This rule included provisions on enhanced braking, enhanced standards for new and existing tank cars, reduced operating speeds, more accurate classification of unrefined petroleum-based products, and rail routing risk assessment. (DOT 2015) The DOT-117 is the new generation of rail car now used for transportation of HHFTs as shown in Figure 3.2.

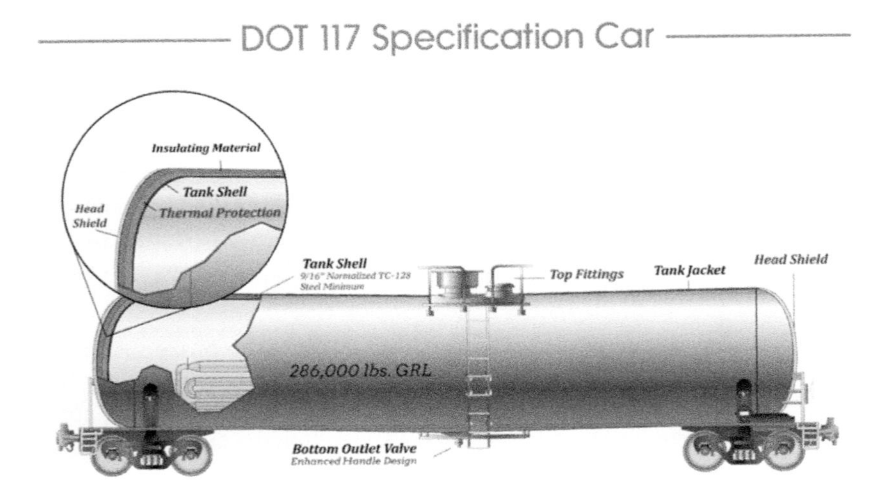

Figure 3.2. DOT-117 train car
(DOT 2015)

3.2.3 Lessons

Compliance with Standards. The MMA SOP required a prescribed number of hand brakes to be set depending on the number of railcars and the grade of the parking location. The MMA-002 train was not in compliance with this requirement. Additionally, the brake effectiveness check was not performed correctly in that the check was conducted with the air brakes set. Standards, whether governmental or company, should be followed. When standards are not followed and work is completed based solely on one's experience or judgment, then the benefit of other person's experiences, hard learnings, and even expert calculations are a resource and opportunity wasted.

Asset Integrity and Reliability. The locomotive that failed had engine problems in October 2012 and a repair was made. Two days before the Lac-Megantic incident the locomotive engineer reported problems with the same engine surging. When the locomotive was parked at Nantes, the smoke and oil spray was noticed by the taxi driver, but the locomotive engineer and the rail traffic controller felt it could wait until morning to be addressed. Nonetheless, this same engine was the only one left running and the sole source of air pressure for the parked train. After the incident, tests showed that the cam bearing had fractured when the mounting bolt was over-tightened after the non-standard repair in October. Repairs should be made following expert direction. Making do with materials on hand and over-tightening bolts are frequently noted in accident reports. Additionally, operational issues with equipment that has just been repaired should be reported and investigated to ensure that it is fit for continued service.

Management of Change. This incident is an example of creeping change in an industry over multiple years. The industry was, for the most part, satisfied with the performance of the DOT-111 cars. However, the number of cars in a single train, the volume of crude oil being transported, and the properties of the crude oil were changing significantly. The impact of this on the risk profile was not effectively addressed until this incident prompted the industry to do so. Likewise, the MMA railroad did not perform an adequate risk assessment when they began transporting large trains of flammables, and in particular, MMA did not assess the risk of changing to single person train operations or the risk of leaving trains unattended on a grade. In addition to procedural changes, a thorough risk assessment could have recommended several engineered safeguards.

Emergency Management. The emergency response to the train on fire at Nantes was also an opportunity to stop the incident before it progressed, but this opportunity was missed since MMA management assigned a person that wasn't trained and qualified as a locomotive engineer to assist the fire department. An emergency scene should not be declared under control until personnel qualified to make that determination are on scene and able to do so. For example, with a house fire, the fire department may work in conjunction with a utility company. With operating equipment, experts in the use and control of that equipment should be consulted before the scene is declared safe.

3.3 Regulations, Codes and Standards

Compliance with Standards is a RBPS element. It is about having a system to identify, develop, acquire, evaluate, disseminate, and provide access to applicable regulations, codes, and standards that affect process safety. The standards system addresses both internal and external standards; national and international codes and standards, industry association guidance and practices; and local, state, and federal regulations and laws. The system makes this information easily and quickly accessible to potential users.

Compliance with Standards system interacts in some fashion with many other RBPS elements. Knowledge of and conformance to *standards* helps a company operate and maintain a safe facility, consistently implement process safety practices, and minimize legal liability. The standards system also forms the basis for the standards of care used in an audit program to determine management system conformance. The standards system provides a communication mechanism for informing management and personnel about the company's obligations and compliance status.

Although the RBPS element is titled Compliance with Standards, it is actually about regulations, codes, and standards. Many of the lessons learned through process safety incidents and experience are codified in regulations, codes, and standards. The three may have different weight in terms of legal enforcement; however, the learnings contained in all of them are worth considering when developing an engineering design or operating a facility. In total, be many, regulations, codes and standards may be applicable to any facility or project.

A **regulation** is an authoritative rule dealing with details or procedure; a rule or order issued by an executive authority or regulatory agency of a government and having the force of law. (Merriam-Webster)

Simply put, a regulation is the law. When designing or operating a facility, at a minimum, compliance with the law is required. The law can be enforced with the threat of penalties and loss of license to operate. Regulations differ between countries, regions, states, and cities.

A **code** is a systematic statement of a body of law especially one given statutory force. (Merriam-Webster)

Codes are written by organizations, such as the National Fire Protection Association, often using a committee comprised of experts in the topic, those who use the code, and those who manufacture equipment addressed in the code. Codes are a consensus set of rules that the authors recommend others to follow.

A **standard** is something established by authority, custom, or general consent as a model or example. (Merriam-Webster)

A standard can be written by a joint industry association, e.g. the Compressed Gas Association (CGA), or by a company. Many companies take a code and amend it with details specific to their business and learnings from their incidents and positive experiences. Standards are more detailed than codes frequently providing the specifics on how to meet a code intent.

As written, codes and standards are voluntary: they are not legal requirements. However, frequently they are adopted by a local authority which makes them legally enforceable.

Codes and standards contain good practices from across an industry. They can be educational and share industry learnings effectively.

Companies can have different ways of ensuring compliance with standards. Some may develop their own internal standards that are based on applicable consensus codes and standards, such as listed in Table 2.2. Others may rely on Subject Matter Experts (SME) who review projects for code compliance, or some combination of both.

3.3.1 Regulations

In the US, the OSHA 1910.119 – "Process safety management of highly hazardous chemicals" is the key regulation. Its stated purpose is for preventing or minimizing the consequences of catastrophic releases of toxic, reactive, flammable, or explosive chemicals. These releases may result in toxic, fire, or explosion hazards. Consistent with OSHA's mission "to provide safe and healthful working conditions for working men and women", OSHA PSM focuses on the worker inside a facility.

The OSHA PSM regulation was promulgated in 1992 following multiple process safety incidents including the Phillips Pasadena Incident in 1989. (OSHA a)

The OSHA PSM regulation includes 14 elements:

- Process Safety Information
- Process Hazard Analysis
- Operating Procedures
- Training
- Contractors
- Mechanical Integrity
- Hot Work
- Management of Change

- Incident Investigation
- Compliance Audits
- Pre startup safety review
- Emergency planning & Response
- Trade secrets
- Employee Participation

CCPS built upon these fourteen elements in creating the Risk Based Process Safety system.

Also, in the US, the EPA "Risk Management Plan (RMP) Rule" is equally a key regulation. It implements Section 112(r) of the 1990 Clean Air Act amendments requiring facilities that use certain hazardous substances to develop a Risk Management Plan. EPA RMP focuses on people and the environment outside of a facility. (EPA a)

In the EU, the SEVESO Directive was introduced in 1982 following the Seveso incident in Italy in 1976 (see section 1.5 for details) and the Flixborough explosion in 1974 (see section 18.2 for details). The Seveso Directive has been updated and currently Seveso-III is in effect. The Seveso-III-Directive (2012/18/EU) aims at the prevention of major accidents involving hazardous substances and at limiting their consequences both to humans and the environment. (EC a) The "Control of Major Accident Hazards (COMAH) Regulation" is the regulation that enforces Seveso Directive in the U.K. (HSE)

Offshore in the U.K., the "Offshore Installations (Safety Case) Regulations" were created in 1992 following the Piper Alpha incident in 1988. (HSE 2015) In the U.S. offshore, following the Deepwater Horizon incident in 2010, the "Safety and Environmental Management Systems Rule", 30 CFR Part 250 Subpart S, was finalized. (BSEE)

Similar process safety management regulations cover operations onshore and offshore in Norway, Australia, Canada, and other countries. In countries where no such regulations exist, many companies choose to follow the regulations cited previously in this section. Examples of process safety regulations around the world are listed in Table 3.1.

Table 3.1. Examples and sources of process safety related regulations

U.S. Federal, State, and Local Laws and Regulations
U.S. OSHA – "Process Safety Management Standard" (29 CFR 1910.119) (OSHA a)
U.S. OSHA - "Occupational Safety and Health Act" – General Duty Clause, Section 5(a)(1) (OSHA b)
U.S. EPA – "Risk Management Program Plan Rule" (20 CFR 68) (EPA a)
U.S. EPA - "Clean Air Act" – General Duty Requirements, Section 112(r)(1) (EPA b)
California "Risk Management and Prevention Program" (OES)
Contra Costa County "Industrial Safety Ordinance" (Contra Costa)
Delaware "Extremely Hazardous Substances Risk Management Act" (DNREC)
Nevada "Chemical Accident Prevention Program" (NDEP)
New Jersey "Toxic Catastrophe Prevention Act" (NJ)
International Laws and Regulations
Australian "National Standard for the Control of Major Hazard Facilities" (NOHSC)
Canadian Environmental Protection Agency – "Environmental Emergency Planning", CEPA, 1999 (Section 200) (EC)
Canadian Transportation Safety Board Regulations (TSB)
Chinese "Guidelines for Process Safety Management of Petrochemical Corporations" (SAWS)
European Commission "Seveso III Directive" (EC a), implemented separately in all member states
Korean "Occupational Safety and Health Act" (MOEL)
Malaysia – Department of Occupational Safety and Health (DOSH) Ministry of Human Resources Malaysia, Section 16 of Act 512 (DOSH)
Mexican Norma Oficial Mexicana (NOM)
Singapore "Workplace Safety and Health (Major Hazard Installations) Regulations" (SG)
United Kingdom, Health and Safety Executive "COMAH Regulations" (HSE)
United Kingdom, Health and Safety Executive, "Offshore Installations (Safety Case) Regulations" (HSE 2015)

3.3.2 Codes and Standards

Codes and standards are typically written by joint industry organizations or committees based on their research, knowledge, and experience. In the refining, petrochemical, and oil and gas sectors, some of the dominant codes and standards include those listed in Table 3.2. Codes and standards can focus on a specific topic, e.g. NFPA 30 "Combustible and Flammable Liquids Code", or can put forward a management system, e.g. ISO 9001 "Quality Management Systems".

Table 3.2. Sources of process safety related codes and standards and selected examples

Source	Examples
American Chemistry Council	"Responsible Care® Management System" and RC 12001 (ACC)
American National Standards Institute (ANSI)	ANSI/ISA-84.00.01-2004 Parts 1-3 (IEC 61511 Mod) "Functional Safety: Safety Instrumented Systems for the Process Industry Sector".
	ISA-84.91.xx - a series of normative standards and guidelines regarding safety controls, alarms, and interlocks.
American Petroleum Institute Recommended Practices (API)	API 520, "Sizing, Selection, and Installation of Pressure Relief Devices"
	API 521, "Pressure Relieving and Depressuring Systems"
	API RP 752, "Management of Hazards Associated with Location of Process Plant Buildings"
	API RP 941, "Steels for Hydrogen Service at Elevated Temperatures and Pressures in Petroleum Refineries and Petrochemical Plants"
American Society of Mechanical Engineers (ASME)	"Boiler and Pressure Vessel Code"
	"Process Piping" B31.3
ASTM	International Test methods for chemical and combustible dust properties
Canadian Standards Group	"Process Safety Management Standard" CSZ Z767 (CSA)
Compressed Gas Association safety standards (CGA)	
Factory Mutual Data Sheets (FMG)	FM Data Sheet 7-82N, "Storage of Liquid and Solid Oxidizing Materials"
International Electrotechnical Commission (IEC)	IEC 61508, "Functional Safety of Electrical/Electronic/Programmable Electronic Safety-related Systems" (E/E/PE, or E/E/PES), 2010.
	IEC 61511, "Functional safety - Safety instrumented systems for the process industry sector, 2003".
International Organization for Standardization (ISO)	ISO 9001 – "Quality Management Systems"
	ISO 14001 – "Environmental Management Systems"
	ISO 31000 – Risk Management
	ISO 45001 – "Occupational Health and Safety Management Systems"

Table 3.2 continued

Source	Examples
NACE	International standards on corrosion management
National Fire Protection Association (NFPA)	NFPA 30, "Flammable and Combustible Liquids Code"
	NFPA 70®, "National Electrical Code®"
	NFPA 652, "Standard on the Fundamentals of Combustible Dust"
Organization for Economic Cooperation and Development –	"Guiding Principles on Chemical Accident Prevention, Preparedness, and Response", 2003 (OECD)
The Chlorine Institute (CI)	Chlorine Customers Generic Safety and Security Checklist
The Instrumentation, Systems, and Automation Society (ISA)	
The Fertilizer Institute	Recommended Practices for Loading/Unloading Anhydrous Ammonia (TFI)

Standards are also written by companies and, again, can be focused on a specific technical topic or can present a management system. Some companies have their own engineering design standards that may take an industry code and amend it with details relevant to their business. They may create a standard for a topic in their business that is not addressed by an industry code or standard.

The ExxonMobil Operations Integrity Management System (OIMS) is a management system. The framework as shown in Figure 3.3 includes 11 elements and is the cornerstone of their Safety, Security, Health and Environmental performance. (EM 2009)

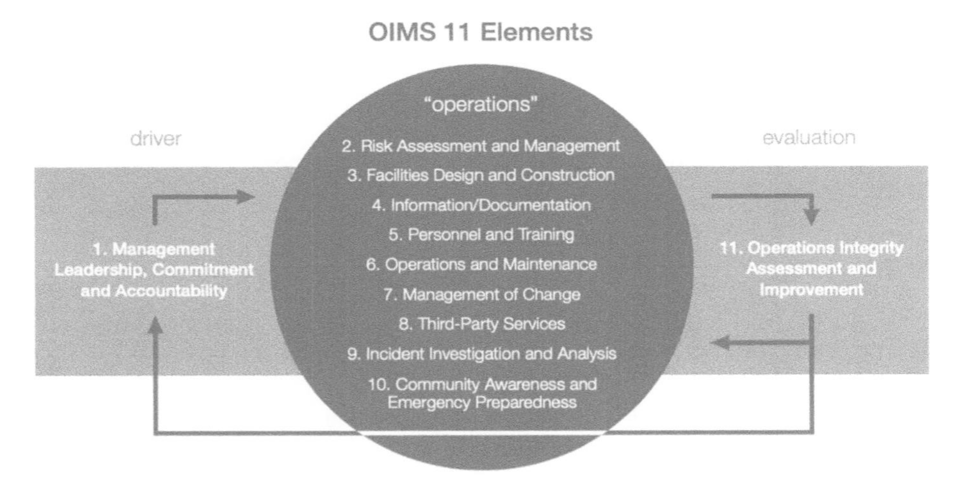

Figure 3.3. ExxonMobil Operations Integrity Management System (EM 2009)

The DuPont "wheel", as it has become known, includes 14 elements grouped by technology, personnel, and facilities. Operational discipline connects all the elements. This is a topic that will be discussed in Chapter 19. (AIChE 2009)

Figure 3.4. DuPont Process Safety and Risk Management model (AIChE 2009)

Shell uses their Hazards and Effects Management Process (HEMP) which provides a systematic method of identifying hazards and managing risks. BP calls their system the Operating Management System. Many other companies have similar systems that integrate management of occupational safety, process safety, and environmental impact. They are frequently seen to be supporting business sustainability. (Shell)

3.3.3 Additional topics

Existing Equipment. An interesting attribute of most regulations, codes, and standards is that they are primarily applicable to new, not existing, designs and equipment. Only when a significant portion of an existing facility is modified (exact level is dependent on the code) do the regulations come into effect. Companies should decide how to address existing equipment. Many companies specify in their company standards which provisions apply to existing equipment and design modifications.

RAGAGEP. Recognized and Generally Accepted Good Engineering Practiceis a term created by OSHA and used in OSHA PSM. Specifically, OSHA PSM references or implies the use of RAGAGEP in statements on equipment design and mechanical integrity . While OSHA has not published a formal list of what they consider to be RAGAGEP, they do cite examples in a

2016 letter of interpretation. (OSHA c) RAGAGEP documents may be codes and standards, either from industry organizations or company-specific documents. An example RAGAGEP list is provided in Appendix C. Once a company creates its RAGAGEP list, these documents become requirements in its operations.

Guidelines. In addition to regulations, codes, and standards, guidelines are available to help manage risk. CCPS has authored many guidelines on process safety topics. The *Guidelines for Risk Based Process Safety* is an example. Guidelines are not legal requirements. They are advice or suggestions on how to approach a topic.

A comparison of CCPS RBPS guidelines with U.S. OSHA PSM and U.S. EPA RMP regulations is provided in Table 3.3. CCPS RBPS goes beyond the prescriptive requirements of the regulations. It is more comprehensive and is performance-based - stating what should be achieved and leaving how to do it is up to the practitioner.

Table 3.3. Comparison of RBPS elements with U.S. OSHA PSM and U.S. EPA RMP elements

CCPS RBPS Element	U.S. OSHA PSM/U.S. EPA RMP Elements
Commit to Process Safety	
1. Process Safety Culture	- -
2. Compliance with Standards	Process Safety Information
3. Process Safety Competency	- -
4. Workforce Involvement	Employee Participation
5. Stakeholder Outreach	Stakeholder Outreach (EPA RMP)
Understand Hazards and Risk	
6. Process Knowledge Management	Process Safety Information
7. Hazard Identification and Risk Analysis	Process Hazard Analysis
Manage Risk	
8. Operating Procedures	Operating Procedures
9. Safe Work Practices	Operating Procedures, Hot Work Permits
10. Asset Integrity and Reliability	Mechanical Integrity
11. Contractor Management	Contractors
12. Training and Performance Assurance	Training
13. Management of Change	Management of Change
14. Operational Readiness	Pre-startup Safety Review
15. Conduct of Operations	- -
16. Emergency Management	Emergency Planning and Response
Learn from Experience	
17. Incident Investigation	Incident Investigation
18. Measurement and Metrics	- -
19. Auditing	Compliance Audits
20. Management Review and Continuous Improvement	- -

3.4 What a New Engineer Might Do

As a new engineer you may be asked to identify applicable regulations, codes, and standards that apply to a design or piece of equipment. It is good to become familiar with the regulations, codes, and standards, including company standards, that apply in your area of expertise. You may be asked to compare various approaches between the codes and standards and to identify key requirements. A challenge can be that codes and standards use language such as "shall" and "should". When you read a statement with the word "should", you may be in a place to make a judgement on whether you should, or should not, apply that statement in your design. Before dismissing that statement, it may be appropriate to dig a little deeper and try to understand why that "should" statement was included. It could be based on a lesson learned from a process safety incident.

3.5 Tools

The tools in the topic area of regulations, codes, and standards are the documents themselves. Sources and examples are listed in Tables 3.1 and 3.2.

Some codes and standards are freely available while others are for purchase individually or through as subscription.

3.6 Summary

Regulations, codes, and standards codify the learnings of many companies over many years. It may be tempting to be satisfied with complying with the legal requirements and not conforming with the voluntary codes and standards. Consider the consequences of adopting this approach. Those learnings are valuable and could help manage risks, even if they are not a legal requirement. As is seen in the Lac Megantic incident, existing regulations may not be sufficient to address current risks.

3.7 Other Incidents

This chapter began with a detailed description of the Lac-Megantic railway disaster that highlighted the importance of regulations. Other incidents that highlight the importance of regulations, codes, and standards include the following.

- PEMEX LPG Explosion, Mexico, 1984
- Hickson Welsh Jet Fire, U.K., 1992
- BSLR Deflagration and Fire, Texas, U.S., 2003
- CTA Acoustics Dust Explosion, Kentucky, U.S., 2003
- Buncefield Storage Tank Overflow and Explosion, U.K., 2005
- T2 Laboratories Runaway Reaction and Explosion, Florida, U.S., 2007
- Imperial Sugar Dust Explosion, Georgia, U.S., 2008
- Varanus Island Pipeline Explosion, Australia, 2008
- ConAgra Foods Explosion, North Carolina, U.S., 2009
- NDK Crystal Vessel Rupture, Illinois, U.S., 2009

- DuPont Belle Phosgene, West Virginia, U.S., 2010
- Kleen Energy Explosion, Connecticut, U.S., 2010
- Hoeganaes Metal Dust Fires, Tennessee, U.S., 2011
- Chevron Richmond Refinery Fire, California, U.S., 2012
- West Fertilizer Company AN Explosion, Texas, U.S., 2013
- Freedom Industries Chemical Spill, West Virginia, U.S., 2014
- Rui Hai International Logistics AN Explosion, China, 2015

3.8 Exercises

1. List 3 RBPS elements evident in the Lac-Megantic railway derailment incident summarized at the beginning of this chapter. Describe the shortcomings as related to this accident.
2. Considering the Lac-Megantic railway derailment incident, what actions could have been taken to reduce the risk of this incident?
3. Describe the differences between a regulation, a code, and a standard.
4. Identify a major process safety regulation in your country. What is its stated purpose or intent? To whom does it apply? When was it last revised?
5. Identify a code or standard relevant to pressure relief valves.
6. Identify a code or standard relevant to LNG tanks.
7. Describe the difference between RBPS and PSM.
8. Describe how existing equipment is addressed in codes and standards.
9. Referring to the BP Texas City explosion described in Chapter 2 and the list of example RAGAGEP practices, which practices may have been applicable in this situation.
10. Who can author codes and standards?
11. A person states "We have complied with the relevant regulations. There is nothing more to be done." It this statement true or false? Explain your answer.

3.9 References

ACC, American Chemistry Council, Arlington, VA., www.americanchemistry.com.

AIChE 2009, "Two centuries of process safety in DuPont", *Process Safety Progress*, American Institute of Chemical Engineers, New York, N.Y., March.

ANSI, American National Standards Institute, New York, N.Y., www.ansi.org.

API, American Petroleum Institute, Washington, D.C., www.api.org.

ASME, American Society of Mechanical Engineers, New York, NY, www.asme.org.

ASTM, ASTM international, Pennsylvania, www.astm.org.

BSEE, https://www.bsee.gov/resources-and-tools/compliance/safety-and-environmental-management-systems-sems.

CGA, Compressed Gas Association, ww.cganet.com.

CI, The Chlorine Institute, Arlington, VA, www.chlorineinstitute.org, http://www.chlorineinstitute.org/pub/ed19b46c-c6a7-acca-ce7f-43f496d0dcae.

Contra Costa, "Contra Costa County Industrial Safety Ordinance", www.co.contra-costa.ca.us/.

CSA, "Process Safety Management Standard", CSA Z767-17, CSA Group, Toronto, Ontario, Canada.

DNREC, "Extremely Hazardous Substances Risk Management Act", Regulation 1201, Accidental Release Prevention Regulation, Delaware Department of Natural Resources and Environmental Control, March 11, 2006.

DOSH, Malaysia – Department of Occupational Safety and Health (DOSH) Ministry of Human Resources Malaysia, Section 16 of Act 512.

DOT 2015, "Rule Summary: Enhanced Tank Car Standards and Operational Controls for High-Hazard Flammable Trains", U.S. Department of Transportation, https://www.transportation.gov/mission/safety/rail-rule-summary.

EC, "Environmental Emergency Regulations" (SOR/2003-307), Environment Canada.

EC, "The Seveso Directive – Technological Disaster Risk Reduction", https://ec.europa.eu/environment/seveso.

EM 2009, "Operations Integrity Management System", ExxonMobil, Irving, Texas, July, https://corporate.exxonmobil.com/-/media/global/files/risk-management-and-safety/oims-framework-brochure.pdf.

EPA, "Accidental Release Prevention Requirements: Risk Management Programs" Under Clean Air Act Section 112(r)(7), 20 CFR 68, U.S. Environmental Protection Agency, June 20, 1996, https://www.epa.gov/rmp.

EPA b, Clean Air Act Section 112(r)(1) – "Prevention of Accidental Releases – Purpose and general duty", Public Law No. 101-529, U.S. Environmental Protection Agency, November 1990.

FMG, FM Global Datasheets, https://www.fmglobal.com/fmglobalregistration.

HSE, "Control of Major Accident Hazards Regulations (COMAH)", U.K. Health and Safety Executive, https://www.hse.gov.uk/comah/.

HSE 2015, "The Offshore Installations (Offshore Safety Directive) (Safety Case etc.) Regulations 2015. Guidance on Regulations", U.K. Health and Safety Executive https://www.hse.gov.uk/offshore/safetycases.htm.

IEC, International Electrotechnical Commission, Geneva, Switzerland, IEC Central Office, https://www.iec.ch/.

ISA, The Instrumentation, Systems, and Automation Society, Research Triangle Park, N.C., www.isa.org.

ISO, International Organization for Standardization, https://www.iso.org/home.html.

Merriam-Webster, https://www.merriam-webster.com/dictionary.

MOEL, "Occupational Safety and Health Act", Korea Ministry of Employment and Labor, 2020

NACE, NACE International, https://www.nace.org.

NCSL 2015, "Transporting Crude Oil by Rail: State and Federal Action", National Conference of State Legislatures, October 30, 2015, http://www.ncsl.org/research/energy/transporting-crude-oil-by-rail-state-and-federal-action.aspx.

NDEP, "Chemical Accident Prevention Program" (CAPP), Nevada Division of Environmental Protection, NRS 259.380, February 15, 2005.

NFPA, National Fire Protection Association, Quincy, MA, www.nfpa.org.

NJ, "Toxic Catastrophe Prevention Act" (TCPA), New Jersey Department of Environmental Protection Bureau of Chemical Release Information and Prevention, N.J.A.C. 7:31 Consolidated Rule Document, April 17, 2006.

NOHSC, "Australian National Standard for the Control of Major Hazard Facilities", Australia National Occupational Health and Safety Commission, 2002.

NOM, Mexican Norma Oficial Mexicana, www.economia-noms.gob.mx.

NTSB 2015, "Improve Rail Tank Car Safety", National Transportation Safety Board, viewed April 11, 2018, https://www.ntsb.gov/safety/mwl/Pages/mwl5_2015.aspx.

OES, "California Accidental Release Program (CalARP) Regulation", CCR Title 19, Division 2 – Office of Emergency Services, Chapter 2.5, June 28, 2002.

OSHA, "Process Safety Management of Highly Hazardous Chemicals" (29 CFR 1910.119), U.S. Occupational Safety and Health Administration, May 1992, https://www.osha.gov/laws-regs/regulations/standardnumber/1910/1910.119.

OECD, "Guiding Principles on Chemical Accident Prevention, Preparedness, and Response, 2nd edition", Organization for Economic Co-Operation and Development, Paris, 2003 www2.oecd.org/ guidingprinciples/index.asp.

OSHA b, Section 5(a)(1) – "General Duty Clause", Occupational Safety and Health Act of 1970, Public Law 91-596, 29 USC 652, Occupational Safety and Health Administration, December 29, 1970.

OSHA c, Occupational Safety and Health Administration, https://www.osha.gov/laws-regs/standardinterpretations/2016-05-11-0.

Sandia 2015. "Literature Survey of Crude Oil Properties Relevant to Handling and Fire Safety in Transport", Sandia National Laboratories, SAND2015-1823, March.

SAWS, Chinese State Administration of Worker Safety, https://www.chinesestandard.net/PDF.aspx/AQT3034-2010.

Shell, https://www.shell.com/business-customers/marine/commitment-and-policy-on-hsse-social-performance.html.

SG, "Singapore Workplace Safety and Health (Major Hazard Installations) Regulations", https://sso.agc.gov.sg/SL/WSHA2006-S202-2017.

TFI, The Fertilizer Institute, https://www.tfi.org/safety-and-security-tools/recommended-practices-loadingunloading-anhydrous-ammonia.

TSB, "Canadian Transportation Safety Board Regulations", https://laws-lois.justice.gc.ca/eng/Regulations/SOR-2014-37/index.html.

TSB 2013, "Transportation Safety Board of Canada Railway Investigation Report R13D0054 Runaway and Main-Track Derailment Montreal, Maine & Atlantic Railway Freight Train MMA-002 Mile 0.23, Sherbrooke Subdivision, Lac-Megantic", Transportation Safety Board of Canada, Quebec, 06, https://www.tsb.gc.ca/eng/enquetes-investigations/rail/2013/r13d0054/r13d0054.html.

4
Fire and Explosion Hazards

4.1 Learning Objectives

The learning objectives of this chapter are:

- Understand fire and explosion hazards,
- Identify chemical properties related to flammability and explosivity,
- Identify and differentiate types of fire, and
- Identify and differentiate types of explosions.

4.2 Incident: Imperial Sugar Dust Explosion, Port Wentworth, Georgia, 2008

4.2.1 Incident Summary

A dust explosion, followed by a series of secondary dust explosions, occurred at the Imperial Sugar refinery in Port Wentworth, Georgia in February 2008 that led to 14 fatalities and injured 36, some permanently. The explosions destroyed the facility (Figure 4.1). Imperial Sugar was fined $8.7 million. This incident provides an important lesson in the importance of understanding the hazard created when combustible dust is released outside of the process equipment and into a building or structure.

Figure 4.1. Imperial Sugar refinery after the explosion
(CSB 2009)

The explosion created national attention to combustible dust hazards in the chemical and agricultural industries and triggered an OSHA National Emphasis Program (NEP) for solids handling facilities. A NEP is a program by OSHA to protect workers in industries that have been determined to present higher risks to people and the environment. In addition to the combustible dust NEP, OSHA has NEPs on Process Safety Management and Isocyanates. A complete report and a video describing the event are available from the Chemical Safety Board (CSB 2009).

Key Points:

Process Safety Competency – Knowledge alone is not enough. You must apply what you know about safe handling of material hazards.

Conduct of Operations – Common saying, "The three most important operations in a plant handling combustible dusts are housekeeping, housekeeping, and housekeeping."

Incident Investigation – Near misses are indications from the process that something in the system is broken. Pay attention to these.

4.2.2 Detailed Description

Imperial Sugar Company purchased the Port Wentworth facility in 1997. The facility refined raw sugar into granulated sugar and sugar products.

The facility was housed inside a four-story building, with the silos extending from the ground to above the top floor (Figure 4.2). Raw cane sugar was converted into granulated and powdered sugar. The refinery had dozens of belt conveyors, screw conveyors, bucket elevators, mills, as well as packaging equipment. Granulated sugar was stored in three large 374 m^3 (13,200 ft^3) silos. The granulated sugar was conveyed from the silos to either the powdered sugar mills, packaging equipment, specialty sugar production, or a bulk sugar building. At the powdered sugar process, belt conveyors and bucket elevators conveyed the granulated sugar to the powdered sugar mills. In 2007, the horizontal belt conveyors were enclosed by steel panels to protect the sugar from contamination.

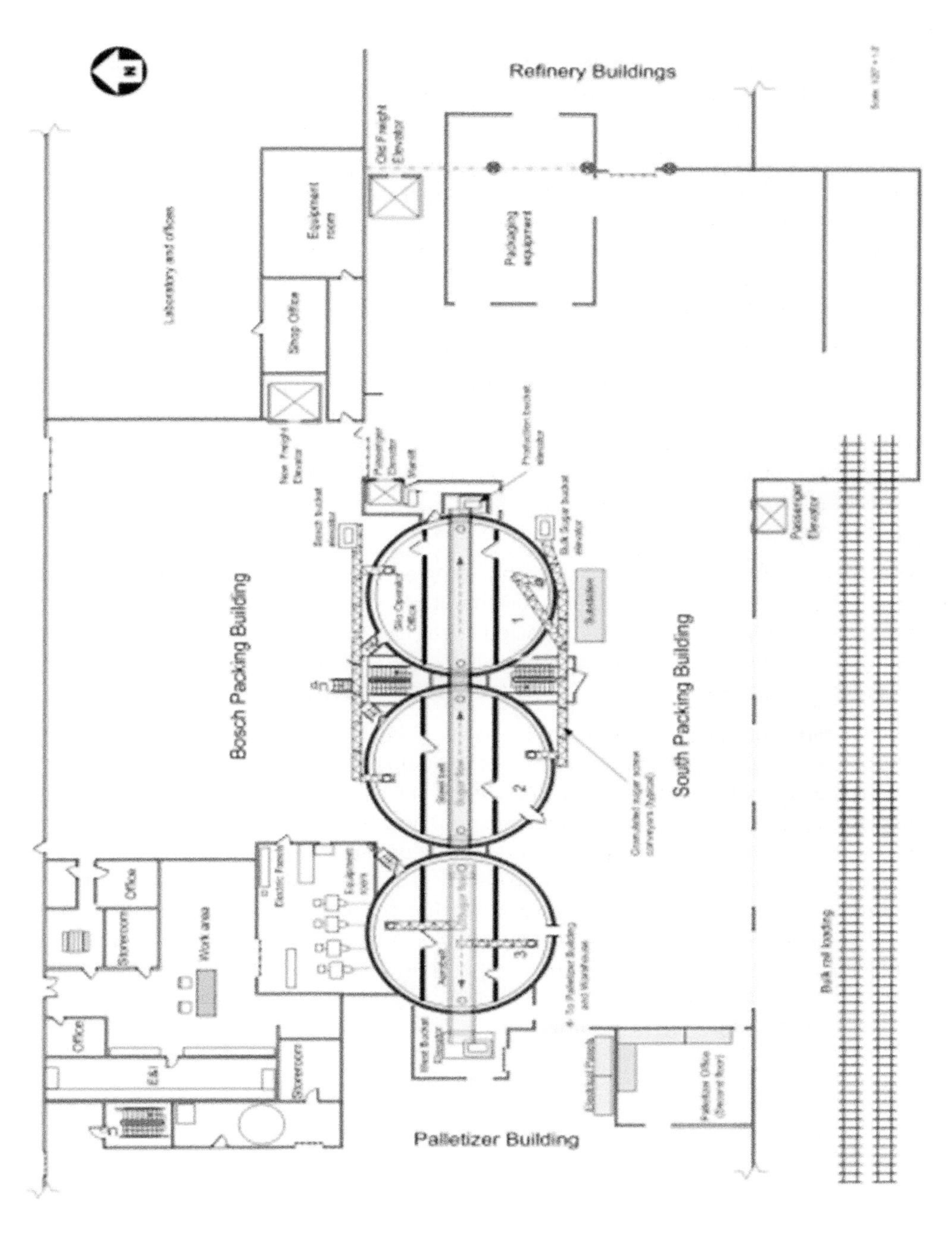

Figure 4.2. Imperial Sugar packing buildings first floor plan

(CSB 2009)

The first explosion most likely started in an enclosed belt conveyor located underneath the silos. The ignition source might have been an overheated bearing or belt support. The belt enclosure allowed the formation of dust clouds above the Minimum Explosion Concentration (MEC) of the sugar dust in the interior of the silo tunnel, providing fuel for the explosion.

The pressure wave from the initial explosion spread throughout the building, dislodging sugar dust that had accumulated in the building due to leaks from the sugar processing equipment. The dislodged dust ignited and created fireballs resulting in several secondary explosions throughout the building. These explosions were powerful enough to buckle the concrete floors and create flying debris. These fireballs continued to occur for 15 minutes after the initial explosion. The CSB report notes that secondary explosions occurred on all four floors. (See Figures 4.1 and 4.3)

When the belt conveyors were enclosed, fugitive dust that had before settled out and accumulated on the floor, were instead contained inside the tunnel enclosure. A lack of housekeeping had allowed a thick layer of dust to accumulate, greater than that suggested by the NFPA dust standard. This allowed the formation of flammable dust clouds which could be ignited (overheated bearings and belt supports are a common source of ignition in solids handling equipment).

The CSB report also states that the sugar handling equipment was not adequately sealed, resulting in large quantities of sugar being spilled onto the floors or escaping into the rooms. An internal inspection noted that "tons of spilled sugar had to be routinely removed from the floors and returned to the refinery for reprocessing". This gives one an idea of the amount of sugar dust regularly spilled. See Figure 4.4 for an example of conditions within the plant.

When handling dusts, leaks can accumulate on surfaces in a process rack or building, such as the floors, beams, and light fixtures. Frequently the dust that leaks out from equipment is the finest (smallest particle size) of the dust being released. With combustible dusts, the explosion severity is usually inversely proportional to the particle size, i.e., smaller particle size has higher explosion severity. Also, smaller particle size dust is usually easier to ignite than the same dust with a larger particle size. An initial event, such as an explosion in an equipment item, creates both a fireball and a pressure wave that can easily disperse and ignite these deposits. This creates a secondary explosion or explosions, see Figure 4.5. These secondary explosions can cause damage and injuries comparable to large vapor cloud explosions. For facilities handling combustible dust, a good housekeeping program is as important, if not more important, as a hot work permit program.

Figure 4.3. Imperial Sugar Refinery after the explosion
(CSB 2009)

Figure 4.4. Motor cooling fins and fan guard with sugar dust, piles of sugar on floor
(CSB 2009)

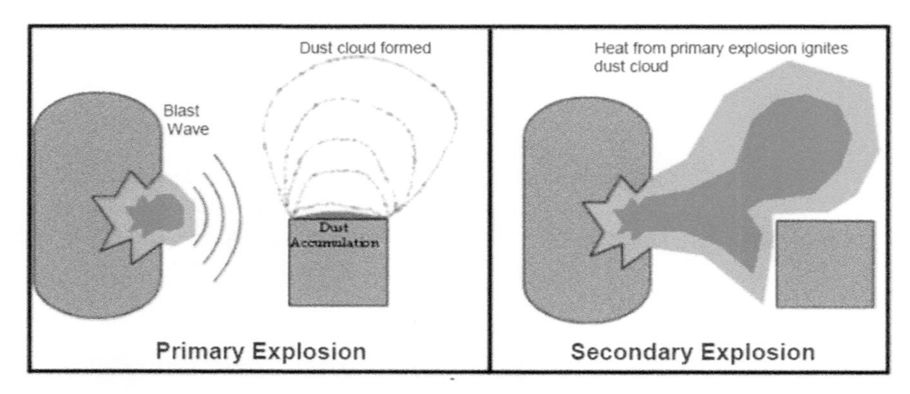

Figure 4.5. Secondary dust explosion
(OSHA a)

4.2.3 Lessons

Compliance with Standards. The facility did not fully comply with NFPA 499 "Recommended Practice for the Classification of Combustible Dusts and of Hazardous (Classified) Locations for Electrical Installations in Chemical Process Areas". Given the amount of dust that was allowed to accumulate on a regular basis many parts of the plant should have been classified Class II, Division 1 or Zone 20/Zone 21. Imperial Sugar did not classify hazardous areas. The CSB notes that although the site had some properly rated electrical devices in hazardous areas, they also had non-rated electrical devices in the same area. Also, the site did not comply with NFPA 499, 654 and the NEC in the handling of combustible dusts, did not classify hazardous areas correctly and used non-rated devices in what should have been classified areas.

Process Safety Competency. Competency is closely linked to the RBPS element *Process Knowledge Management*, but this incident illustrates the difference between knowledge and competency. Evidence indicates the personnel had the *knowledge* of the hazards of combustible dust at the facility. Quality assurance and safety personnel were aware of OSHA's National Emphasis Program on dusts, and an explosion was safely vented from a dust collector 10 days before this incident, and site had fugitive dust collection systems for collecting emissions. Competency implies applying what is known. Neither management nor employees seem to have fully appreciated the hazards of combustible dusts. The housekeeping program was not adequate for the situation, what housekeeping was done was frequently done improperly, e.g. using compressed air to clean dust deposits (a hazardous practice in itself as it creates a flammable dust cloud), and the fugitive duct collection equipment was not properly maintained.

Hazard Identification and Risk Analysis. Hazard reviews were conducted by Imperial Sugar's insurance carrier, but they failed to identify the hazard of dust accumulation.

Asset Integrity and Reliability. The fugitive dust collection system was inadequate and poorly maintained. The dust accumulations resulted in the secondary explosions that destroyed the entire building and led to the fatalities.

Training and Performance Assurance. Initial and annual safety training was done, but it seems to have focused on occupational safety. Safety training had not covered the hazard of dust accumulations since 2005.

Management of Change. The belt conveyor was enclosed without conducting a Management of Change (MOC) review. The lack of hazard awareness, ignoring of near misses, and lack of an MOC review led to the creation of an unprotected enclosure containing combustible dust clouds. An MOC review, performed by competent people knowledgeable about dust explosion hazards, would have evaluated the need for explosion protection such as venting, suppression, or inerting within such an enclosure.

Conduct of Operations. Written housekeeping programs were not effectively implemented. What cleaning was done did not always include elevated surfaces. Dust collection system design and maintenance may also have been contributing factors to the fugitive emissions, but no action was taken to reduce leaks or fix the fugitive dust collection system. Also, there had been many small fires in this and other Imperial Sugar locations, which did not lead to larger fires or explosions. These may have caused the staff to become complacent regarding the hazards of combustible dust. This phenomenon is known as Normalization of Deviance, in this case, thinking that having many small fires was normal and tolerable.

Incident Investigation. It has already been mentioned that this facility, and other Imperial Sugar refineries had many small fires and near misses. For example, in this facility operators noted that buckets in the bucket elevators sometimes broke loose and fell to the bottom of the elevator. In one case this started a fire. An explosion in a dust collector occurred 10 days before this incident. These near misses and the explosion were warning signs that were not heeded.

4.3 Introduction to Fires

Fire is a chemical reaction; it is an oxidation reaction. Fire, however, is a rapid, exothermic oxidation reaction. It generates heat and light (an exception is a hydrogen fire as well as low light emittance in methanol and carbon disulfide fires) and produces smoke as a product of incomplete combustion. Fire requires three things to occur:

- Fuel,
- Oxygen, and
- Ignition source

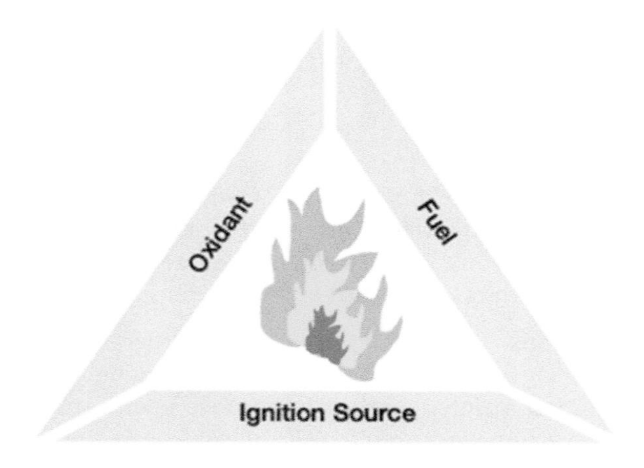

Figure 4.6. Fire Triangle
(Crowl 2012)

This is illustrated in the fire triangle shown in Figure 4.6. All three have to be present for combustion to occur. This simple model facilitates considering how to prevent a fire and how to extinguish a fire. A fire can be prevented by excluding any one of the three legs. For example, blanketing a tank of flammable liquid with nitrogen removes the oxidant leg. Removing sparks from static electricity, electrical equipment, and vehicles excludes the ignition sources leg. Firefighting extinguishes a fire in the same way, e.g. blanketing a pool fire with firefighting foam to exclude oxygen or removing heat by cooling with water.

A fourth dimension has been added to the fire triangle to create a fire tetrahedron. This fourth dimension is the uninhibited chemical chain reaction that produces the heat sufficient to maintain a fire. Some fire suppression agents function by inhibiting the chain reaction which extinguishes the fire.

It is important to understand these definitions and terminology related to fire.

Flash point temperature - The minimum temperature at which a liquid gives off sufficient vapor to form an ignitable mixture with air within the test vessel used (Methods: ASTM 502). The flash point is less than the fire point at which the liquid evolves vapor at a sufficient rate for indefinite burning. (CCPS Glossary)

Flammability limits - The range of gas or vapor amounts in air that will burn or explode if a flame or other ignition source is present. Importance: The range represents a gas or vapor mixture with air that may ignite or explode. Generally, the wider the range the greater the fire potential. See also Lower Explosive Limit / Lower Flammable Limit and Upper Explosive Limit / Upper Flammable Limit. (CCPS Glossary)

Autoignition temperature - The lowest temperature at which a fuel/oxidant mixture will spontaneously ignite under specified test conditions. (CCPS Glossary)

Minimum oxygen concentration - The concentration of oxidant, in a fuel-oxidant-diluent mixture below which a deflagration cannot occur under specified conditions. Limiting Oxidant Concentration (LOC) is synonymous with the term Minimum Oxygen Concentration (MOC). (CCPS Glossary)

Minimum ignition energy (MIE) - The minimum amount of energy released at a point in a combustible mixture that caused flame propagation away from the point, under specified test conditions. The lowest value of the minimum ignition energy is found at a certain optimum mixture. The lowest value is usually quoted as the minimum ignition energy. (CCPS Glossary)

Flash point is the primary characteristic to classify the relative flammability of liquids; however, organizations define flammability differently. Refer to Chapter 7 for further information on chemical hazards data sources. (NFPA, OSHA b, UN)

- NFPA 30, "Flammable and Combustible Liquids Code"
  - A Class I Liquid is a combustible liquid with a closed-cup flash point not exceeding 38°C (100°F.)
  - A Class II Liquid is a combustible liquid with a closed-cup flash point at or above 100°F (37.8°C) but below 140°F (60°C).
- A Class III Liquid is with a closed-cup flash point at or above 140°F (60°C) OSHA and the UN "Globally Harmonized System of Classification and Labeling of Chemicals"
  - Flammable liquid: not more than 93°C (199.4°F)

The relationship between these terms is illustrated in Figure 4.7. If a flammable liquid is above its flash point, it will evolve flammable vapors. The upper and lower flammability limits define levels at which the resulting fuel/oxygen vapor concentration is either too rich or too lean (respectively) to burn. If the fuel/oxygen vapor concentration is in the flammable range and it is exposed to an ignition source meeting the minimum ignition energy for that fuel, then combustion will occur. If the fuel/oxygen vapor concentration is in the flammable range and it is above its autoignition temperature, it will ignite without the presence of an ignition source.

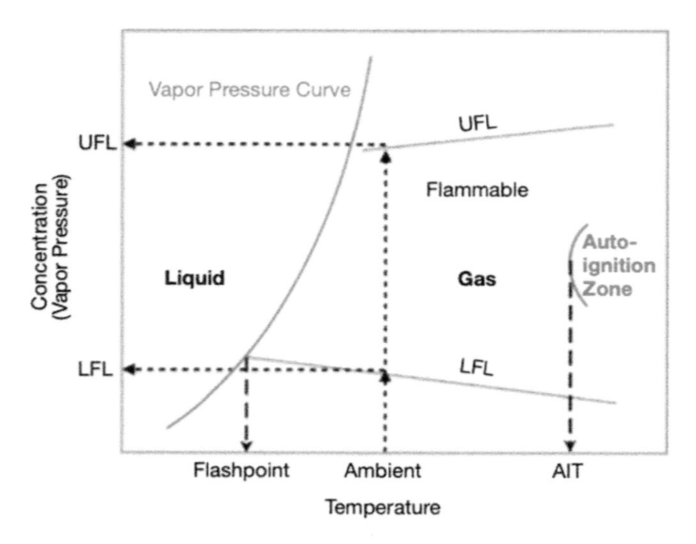

Figure 4.7. Relationship between flammability properties (Crowl 2012)

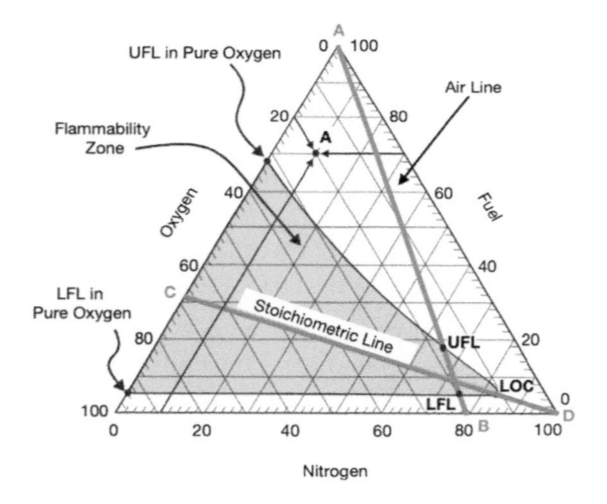

Figure 4.8. Flammability diagram (Crowl 2012)

A flammability diagram, Figure 4.8, also shows the flammable range. In addition, it illustrates that the introduction of an inert gas to the fuel/oxygen mixture can be used to bring the mixture outside of the flammable range. A description of the flammability diagram and how to use it is provided in *Minimize the Risks of Flammable Materials*. (Crowl 2012)

Flammability properties are available in Safety Data Sheets (SDS), National Fire Protection (NFPA) standards, and the CCPS Chemical Reactivity Worksheet (refer to Chapter 5) as well as other sources (refer to Chapter 7). These are important pieces of process safety information (Refer to Section 2.5) Table 4.1 presents flammability properties of selected vapors and liquids.

Many of these properties are typically based on ambient conditions. Also, it is common to handle mixtures of materials, as opposed to pure components, in which case methods such as LeChatelier's rule can be used to estimate flammability limits. LeChatelier's rule represents a weighted average of UEL and LEL by taking the molar concentration on a flammable only basis, of each component of the mixture and the corresponding flammability limits into account. Estimating the properties for mixtures can also be done through computer simulation and laboratory testing.

Table 4.1. Flammability properties (Crowl 2012)

	Species	LFL, vol% fuel	UFL, vol% fuel	Flashpoint, °C	LOC, vol% oxygen	MIE, mJ	AIT, °C
Vapors	Methane	5.0	15.0	−188	12	0.28	600
	Ethane	3.0	12.5	−135	11	0.24	515
	Propane	2.1	9.5	−104	11.5	0.25	450
	Hydrogen	4.0	75.0		5	0.018	400
	Ammonia	16.0	25.0				651
	Carbon Monoxide	12.5	74.0		5.5		609
	Hydrogen Sulfide	4.3	45.0		7.5		260
	Acetylene	2.5	80.0	−18		0.020	305
Liquids	Hexane	1.2	7.5	−23	12	0.248	234
	Ethylene	2.7	36.0	−136	10		450
	Benzene	1.4	7.1	−11	11.4	0.225	562
	Ethanol	4.3	19.0	13	10.5		422
	Methanol	7.5	36.0	11	10	0.140	463
	Formaldehyde	7.0	73.0	−53			430
	Acetone	2.6	12.8	−18	11.5		538
	Styrene	1.1	6.1	32	9.0		490
	Gasoline	1.4	7.6	−43	12		

Minimum ignition energy varies between materials. Table 4.2 provides example values. The following list provides energy values, for reference. Comparing the list and the Table, it is clear that while it is good practice to minimize ignition sources, it isn't possible to completely eliminate them.

- 2 mJ is barely perceptible by a person,
- 10 mJ is "distinctly" perceptible,
- Spark discharge from walking across the room ~22 mJ,
- 250 mJ results in severe shock, and
- 10,000 mJ is potentially lethal.

Table 4.2. Minimum ignition energies for selected materials (Crowl 2019)

Chemical	MIE (mJ)
Acetylene	0.020
Ethylene oxide	0.062
Ethylene	0.080
Propane	0.250
Methane	0.280

4.4 Types of Fires

Each liquid and vapor fire type has its own characteristics. These fire types are listed in order of increasing consequence severity. The type of fire is dependent upon the flammability characteristics determined by process safety information of the fuel.

Pool fire - The combustion of material evaporating from a layer of liquid at the base of the fire. (CCPS Glossary)

Jet fire - A fire type resulting from the discharge of liquid, vapor, or gas into free space from an orifice, the momentum of which induces the surrounding atmosphere to mix with the discharged material. (CCPS Glossary)

Flash fire - A fire that spreads by means of a flame front rapidly through a diffuse fuel, such as a dust, gas, or the vapors of an ignitable liquid, without the production of damaging pressure. (CCPS Glossary)

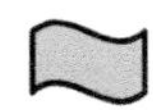

Fireball - The atmospheric burning of a fuel-air cloud in which the energy is in the form of radiant and convective heat. The inner core of the fuel release consists of almost pure fuel whereas the outer layer in which ignition first occurs is a flammable fuel-air mixture. As buoyancy forces of the hot gases begin to dominate, the burning cloud rises and becomes more spherical in shape. (CCPS Glossary)

Pool fires involve a pool of liquid. This may be from a spill of a flammable liquid on the ground or water, from the top of a hydrocarbon tank, or where a spilled liquid has followed grading to a drain. The location of a pool fire may be influenced by grading, berms or dikes, and containment booms. Firefighting foams may be used to blanket a pool fire to control, and potentially, extinguish it. Note direct application of firewater to burning pools is dangerous as this can spread the fire and water will not extinguish the flames as it sinks below the surface. The radiation from a pool fire can cause harm to people and equipment depending on the duration of exposure. Additionally, the products of incomplete combustion can be toxic.

Figure 4.9. Pool fire
(insightnumerics 2018)

Jet fires involve release, and ignition, of flammable materials under pressure. The combination of the heat of the fire and the momentum of the jet stream greatly increase the heat transfer and can do much more damage than a pool fire. Jet fires may result from damaged equipment and piping, flanges, vents, and other sources. With so many potential sources, it is difficult to determine which direction the jet fire will be directed and thus, what equipment to protect.

Figure 4.10. Jet fire
(BakerRisk)

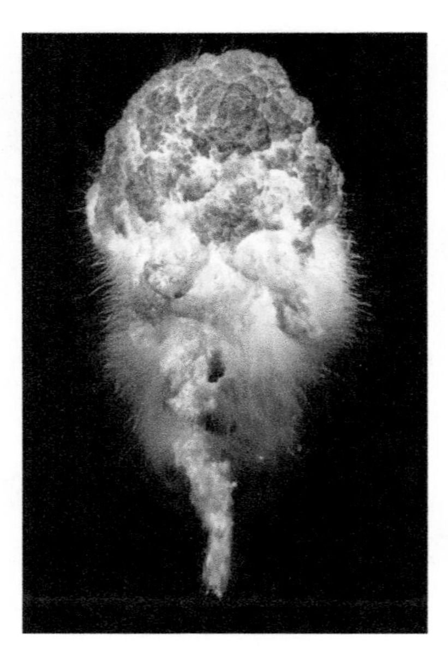

Figure 4.11. Fireball
(Sandia 2020)

When a flammable material is released and a fuel/air mixture disperses downwind finding an ignition source, a flash fire will consume the fuel/air mixture back to the ignition source. The direction of the dispersion is influenced by atmospheric conditions (refer to Chapter 13). A flash fire is short-lived and is primarily of risk to people and light equipment (e.g. electrical and control cables) that are within it. Often a flash fire precedes a pool or jet fire.

Fireballs are lifted into the air as the combustion occurs and consumes the fuel. They are short-lived but sufficient to cause injury to exposed people.

4.5 Types of Explosions

The previous section stated fire is a rapid oxidation reaction that produces heat and light. An explosion increases the speed of that reaction significantly, enough so to create a shock wave. Two major types of explosions are possible: chemical and physical.

Explosion - A release of energy that causes a pressure discontinuity or blast wave. (CCPS Glossary)

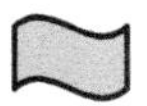

4.5.1 Chemical Explosions

Chemical explosions are caused by chemical reactions and can be uniform or propagating reactions. Uniform reactions occur throughout the space of the reaction mass, such as a runaway reaction in a reactor. A propagating reaction, e.g. combustion, moves through the mass of the reactant, such as in a vapor cloud explosion. When the chemical reaction occurs in the solid or liquid phase, it is a condensed phase explosion.

the solid or liquid phase that is a condensed phase explosion.

Vapor Cloud Explosion - The explosion resulting from the ignition of a cloud of flammable vapor, gas, or mist in which flame speeds accelerate to sufficiently high velocities to produce significant overpressure. (CCPS Glossary)

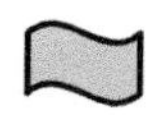

If a fuel/air mixture forms a vapor cloud and finds an ignition source, it could create a flash fire back to the release point. Alternatively, where the conditions are suitable, it could create a vapor cloud explosion. An explosion is described by a pressure-time curve as shown in Figure 4.12. The damage caused by an explosion is due to the pressure and the impulse, described by the area under the pressure-time curve. The curve is at a stationary point at some point away from the explosion center.

An explosion in which the reaction front is less than the speed of sound is a deflagration. Where it is greater than the speed of sound, it is a detonation. The overpressures, and potential resultant damage, are significantly greater for a detonation than for a deflagration.

The speed of the reaction front is influenced by three main factors: fuel reactivity, congestion or obstruction, and confinement. The fuel reactivity is related to the laminar burning velocity. Congestion or obstruction describes the size and number of blockages in the path of the reaction front (Figure 4.13). Confinement describes the limits on how an explosion can propagate. For example, an explosion in an open field can expand in more directions than one confined between two plates or ultimately, one that is confined in a pipe shape, like the barrel of a gun.

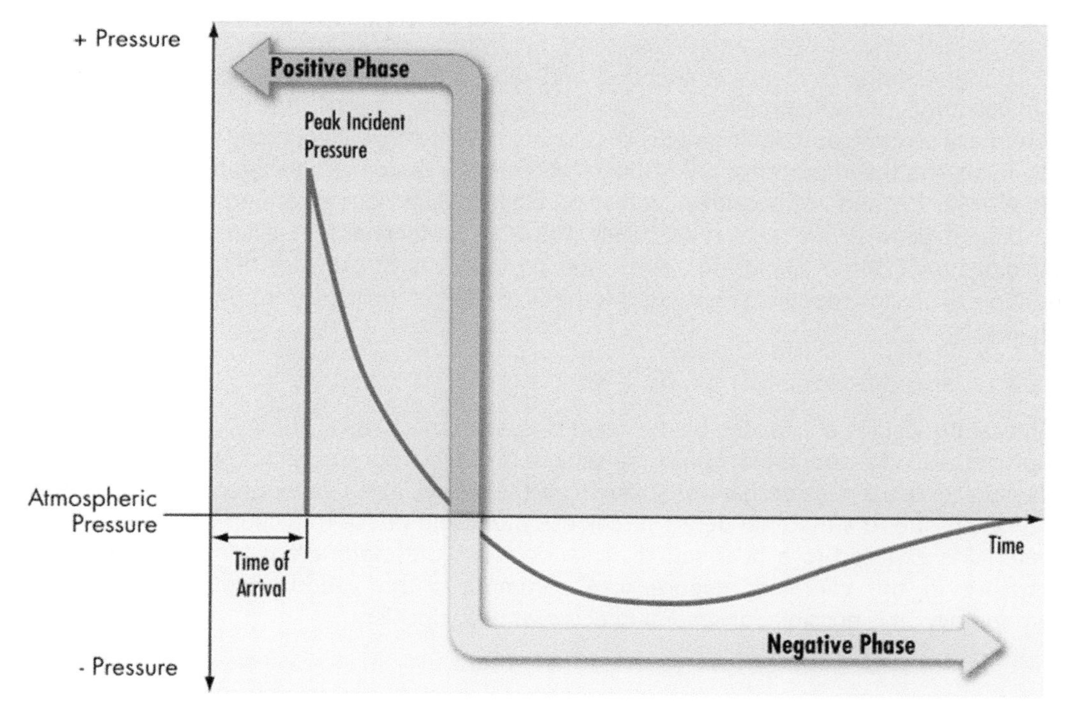

Figure 4.12. Explosion pressure-time curve
(FEMA)

Figure 4.13. Degrees of congestion from low to high (left to right) (BakerRisk 2001)

Overpressure – Any pressure above atmospheric caused by a blast. (CCPS Glossary)

Impulse - The area under the overpressure-time curve for explosions. The area can be calculated for the positive phase or negative phase of the blast. (CCPS Glossary)

Deflagration - A combustion that propagates by heat and mass transfer through the un-reacted medium at a velocity less than the speed of sound. (CCPS Glossary)

Detonation - A release of energy caused by the propagation of a chemical reaction in which the reaction front advances into the unreacted substance at greater than sonic velocity in the unreacted material. (CCPS Glossary)

Understanding, modeling, preventing, and mitigating vapor cloud explosions is an area of significant focus in industrial facilities handling flammable materials due to the harm they can potentially cause. Many significant process safety . events have highlighted the importance of this focus and the importance of locating and designing buildings so as to protect occupants. API RP 752, 753 and 756 provide guidance on blast-resistant design of permanent structures, portable structures, and tents, respectively. The CCPS "Guidelines for Evaluating Process Plants Buildings for External Explosions, Fires, and Toxic Releases" provides detailed guidance on multiple explosion modeling methods. Methods used in modeling explosions are discussed in Chapter 13.

4.5.2 Physical Explosions

Physical explosions are caused by the rapid release of mechanical energy, and include vessel ruptures, BLEVEs and rapid phase transition. A vessel rupture occurs when the internal pressure exceeds the mechanical strength of the vessel. The vessel strength is sometimes weakened due to a material defect or corrosion. A rapid phase transition can occur when a material is exposed to a heat source. This increases the material's volume, increasing the pressure in the container. Figure 4.14 summarizes the various explosion types and terminology. It is possible for several to occur with any incident.

Physical Explosion - The catastrophic rupture of a pressurized gas/vapor-filled vessel by means other than reaction, or the sudden phase-change from liquid to vapor of a superheated liquid

(CCPS Glossary)

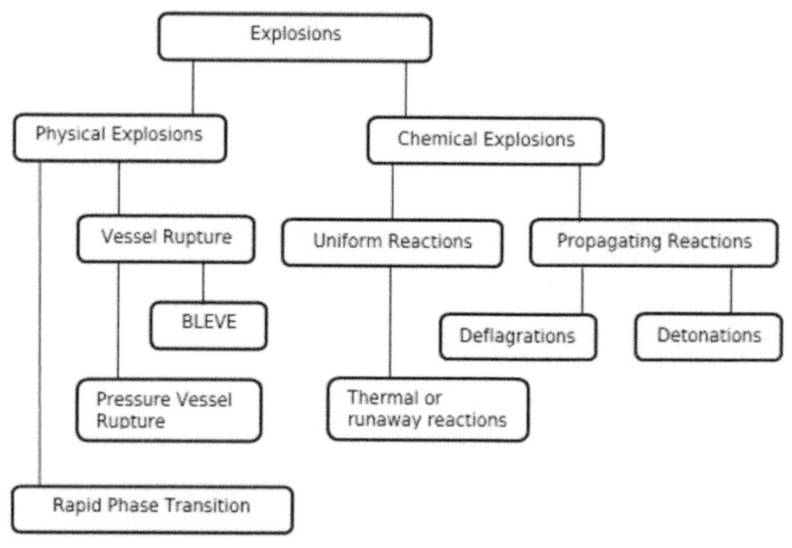

Figure 4.14. Relationships between the different types of explosions (Crowl 2003)

Boiling Liquid Expanding Vapor Explosion (BLEVE) A type of rapid phase transition in which a liquid contained above its atmospheric boiling point is rapidly depressurized, causing a nearly instantaneous transition from liquid to vapor with a corresponding energy release. A BLEVE of flammable material is often accompanied by a large aerosol fireball, since an external fire impinging on the vapor space of a pressure vessel is a common cause. However, it is not necessary for the liquid to be flammable to have a BLEVE occur. (CCPS Glossary)

Table 4.3 provides examples of the types of explosions. You can observe that some incidents can involve multiple types of explosions, for example a vessel rupture leading to a BLEVE.

Table 4.3. Examples of various types of explosions (adapted from Crowl 2003)

Type of Explosion	Examples	Example Incidents
Rapid phase transition	Water being pumped into a vessel containing hot water	
BLEVE	Rupture of a flammable gas railcar exposed to fire	Williams Geismar Heat Exchanger Rupture/Explosion
Vessel rupture	Mechanical failure of a vessel at high-pressure Failure of a relief device during overpressure	NDK Vessel Rupture Williams Geismar Heat Exchanger Rupture/Explosion Spain, 1978: an overfilled road tanker carrying propylene ruptured, and the released gas ignited. The explosion led to 217 fatalities and severely burned 200 others (CCPS-2017 a)
Uniform reaction (aka condensed phase explosions)	Runaway reactions Decomposition reactions	T2 Labs Reaction/Explosion Port Neal Ammonium Nitrate West Fertilizer Tianjin Explosion Port of Beirut Ammonium Nitrate Explosion
Propagating reactions	Combustion of flammable vapors or dust	Imperial Sugar Dust Explosion Hayes Lemmerz, Dust Explosion Varanus Island Pipeline Rupture/Explosion Multiple Natural Gas Explosions Oil Storage Tank Explosion Buncefield Jaipur

4.5.3 Dust Explosions

This chapter began with the description of a dust explosion. Many people are not aware that combustible particulate materials, as a dust, can explode. Just as with other types of explosions, significant overpressure. and damage can result. Dust explosions have occurred in a variety of industries as shown in Figure 4.15. Combustible dusts, except agricultural dusts, are not covered by regulations such as OSHA PSM. Specific attention should be paid to combustible dusts to ensure that these risks are addressed, regardless of regulatory requirements.

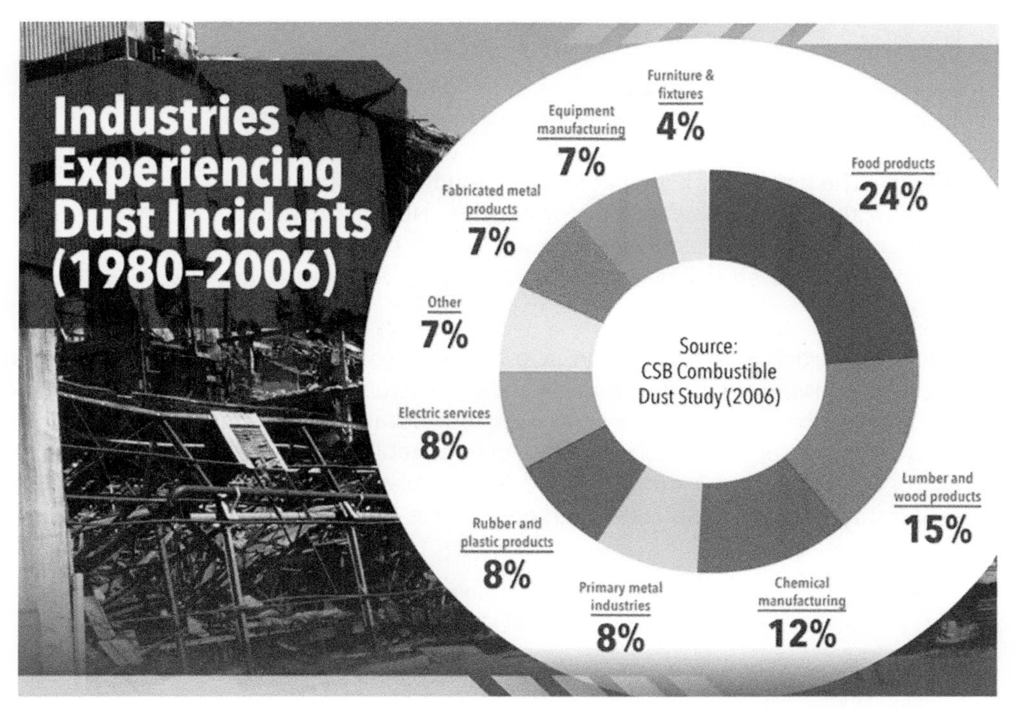

Figure 4.15. Dust explosions in industry
(CSB a)

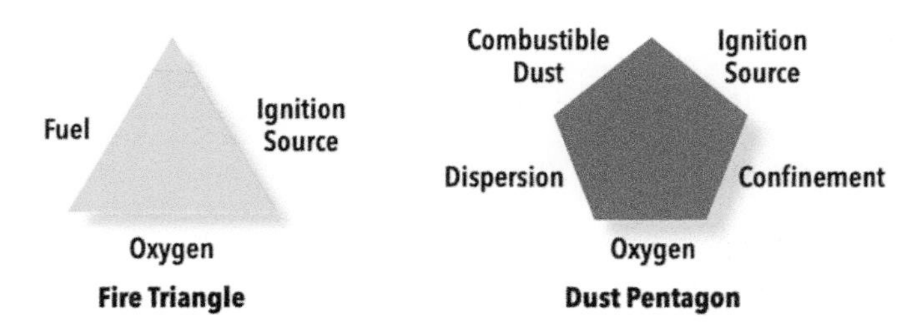

Figure 4.16. Fire triangle and dust pentagon
(CSB b)

Building on the fire triangle, dust explosions also require dispersion and confinement of the dust, thus creating a dust pentagon (Figure 4.16). Dust often settles on the rafters, suspended ceilings, and equipment in manufacturing facilities. Typically, an initial fire or small explosion occurs which dislodges the dust and disperses it in a confined space (above the suspended ceiling, in a confined area of the facility) supporting a subsequent larger explosion. As little as 0.8 mm (1/32 in) accumulated dust can support a dust explosion. Housekeeping, removing any dust buildup, is a key prevention method.

Additional terms relevant to dust explosions include:

Pmax - The maximum pressure occurring in a closed vessel during the explosion of an explosible dust atmosphere determined under specific test conditions. (CCPS Glossary)
KSt Value - The deflagration index of a dust cloud. It is a dust-specific measure of the explosibility, in units of bar-m/s. Not that it is not a physical property of a substance, but dependent on particle size, test conditions, etc. The equation is the so-called cubic /cube root law. (CCPS Glossary)

Particle size, particle size distribution, and moisture content are key parameters when influencing dust explosions. As a dust particle size or moisture content decreases, the Kst and Pmax increase and the minimum explosive dust concentration decreases. The MIE is much higher for dusts than for flammable vapors (refer to Table 4.2). For example, the MIE for powdered sugar is between 10 and 30 mJ and for granulated sugar (sieved to less than 500 µm) is greater than 1000 mJ. (CSB 2009) Methods used in modeling explosions are addressed in Chapter 13.

Table 4.4. Selected combustible dust properties
(OSHA c)

Dust Explosion Class	Kst	Dust	Particle Size µm	Kst (bar m/s)
St 0		Wood dust	33	-
St 1	0 < Kst < 200	Activated Carbon	18	44
St 2	200 < Kst < 300	Cellulose	33	229
St 3	Kst > 300	Aluminum powder	<10	515

Note: The actual class is sample specific and will depend on varying characteristics of the material such as particle size or moisture.

4.6 Fire and Explosion Prevention

The best way to manage a fire or explosion is to prevent it from occurring. Remembering the fire triangle, removing one of the legs will prevent a fire from occurring. Where industries handle flammable and combustible materials, fuel is normally excluded by containing it in pipes and vessels. "Keep it in the pipe" is a mantra that many companies use. However, when a loss of containment occurs, then the fuel is present.

The second leg of the triangle is oxygen which is present in the atmosphere. Oxygen may be excluded through gas blanketing by using either an inert gas or a hydrocarbon rich gas. Nitrogen is typically used in inert gas blanketing systems to maintain the vapor space below the lower flammable limit. Fuel gas or flue gas blanketing systems are based on maintaining the vapor space above the upper flammable limit. The sizing of these blanketing systems, and any vents that may interact with them, is key to maintaining the vapor space outside of the flammable range.

This leaves the heat leg of the triangle which is addressed through ignition source control. Unfortunately, as Trevor Kletz, a founder of process safety, said, "Ignition sources are free" meaning that ignition sources are prevalent. Potential ignition sources and means to control them are provided in Table 4.5. These ignition sources and control methods apply to flammable materials and combustible dusts. Additionally, for dusts, control of the confinement and dispersion legs of the dust pentagon through diligent housekeeping is a prevention measure.

Table 4.5. Ignition sources and control methods

Ignition sources	Control Method
Electrical	Electrical Area Classification (refer to following text) Bonding and grounding
Smoking	Prohibition of use or allowance only in specified areas
Overheated materials	Maintain integrity of equipment including that handling/conveying solids
Hot surfaces	Elimination of surfaces above autoignition temperatures of materials being handled
Burner flames	Facility siting such that heaters are located apart from equipment handling flammable materials
Sparks	Control of hot work through a work permit system
Spontaneous ignition	Control of pyrophoric materials such as ferrous sulfide scale
Cutting and welding	Control of hot work through a work permit system
Static electricity	Bonding and grounding
Chemical Reaction	Understanding and control of chemical processes
Lightning	Provision of lightning protection
Vehicles	Control of vehicular access in areas handling flammable materials

Bonding and grounding serves to maintain separate pieces of equipment at the same potential energy such that a spark does not occur between them. Grounding serves to direct electrical current to the earth where it is dissipated.

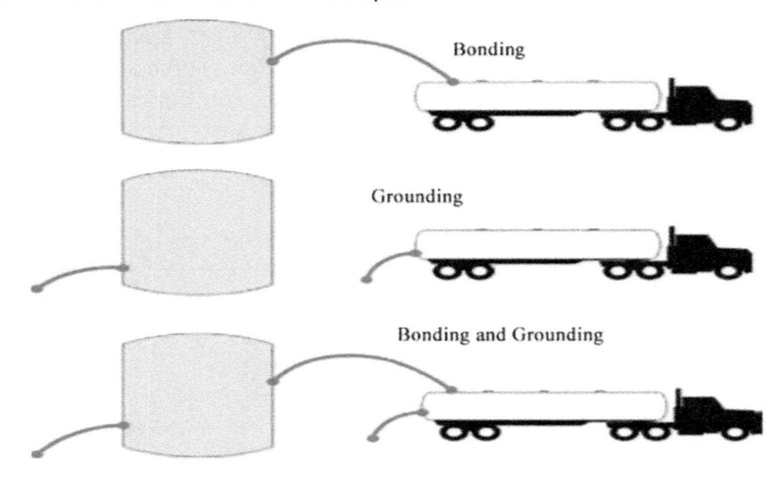

Figure 4.17. Bonding and grounding
(OSHA d)

Electrical area classification serves to group equipment areas based on the potential for fires and explosions and specify the types of electrical installations in the area, typically intrinsically safe equipment, to protect against ignition.

Intrinsically safe - Equipment and wiring which is incapable of releasing sufficient electrical or thermal energy under normal or abnormal conditions to cause ignition of a specific hazardous atmospheric mixture or hazardous layer. A protection technique based upon the restriction of electrical energy within apparatus and of interconnecting wiring, exposed to a potentially explosive atmosphere, to a level below that which can cause ignition by either sparking or heating effects. Because of the method by which intrinsic safety is achieved, it is necessary to ensure that not only the electrical apparatus exposed to the potentially explosive atmosphere, but also other electrical apparatus with which it is interconnected, is suitably constructed. (CCPS Glossary)

In the US, the "National Electrical Code" (NEC) (NFPA 70) provides the guidance on electrical area classification. It uses a Class/Division/Group system where

- Classes - defines the general nature of the hazardous material in the surrounding atmosphere (e.g. flammable gases or vapors; combustible dusts; and ignitable fibers),
- Divisions - defines the probability of hazardous material being present in the surrounding atmosphere, and

- Groups - defines the type of the hazardous material in the surrounding atmosphere (e.g. flammable gas, vapor, or liquid; dust).

In the U.K., BS EN 60079 "Explosive atmospheres classification of areas" is used. (BS 2015) In Europe, the ATEX directives are used: ATEX 214 "equipment" Directive 2014/34/EU and ATEX 137 "workplace" Directive 1999/92/EC. ATEX is an acronym for "Appareils destinés à être utilisés en ATmosphères EXplosives" (French for "Equipment intended for use in EXplosive Atmospheres"). (ATEX)

Both the BS EN and ATEX are based on a zone approach to classification. ATEX uses additional zones to classify combustible dust and fiber.

- Zone 0 – an explosive atmosphere is present continuously or for long periods or frequently.
- Zone 1 – an explosive atmosphere is likely to occur in normal operation occasionally.
- Zone 2 – an explosive atmosphere is not likely to occur in normal operation but, if it does occur, will persist for a short period only.

4.7 What a New Engineer Might Do

A new engineer might be asked to identify fire and explosion hazards of the materials handled in a facility. The new engineer would be expected to understand the terminology and use it appropriately. Not all combustion is an "explosion", as the media frequently mistakes. One way to become familiar with fires and explosions is to view the CSB videos at www.csb.gov/videos that provide not only explanations but also an image of what the fire or explosion might look like and the resulting damage. New engineers might collect and update process safety information related to flammability hazards including flash points, boiling points, UEL/LEL, electrical area classification and others. They might also verify effectiveness of controls such as grounding and bonding continuity surveys, inerting effectiveness, electrical area classification requirements are met, and proper separation of flammable storage.

A new engineer can benefit from reviewing the CSB investigations and videos relevant to this chapter as listed in Appendix G.

4.8 Tools

The following resources are available for helping to understand and protect against fire and explosions.

Chemical properties data. Greater detail on these tools is provided in Chapter 7.

- **Safety Data Sheets (SDS).** The SDS includes chemical properties, hazards, and protective measures.
- **NIH WebWISER.** WISER is a system provided by the National Institutes of Health designed to assist emergency responders in hazardous material incidents. (NIH)

American Petroleum Institute (API) recommended practices. The API is an industry trade association. API committees have generated recommended practices that address many segments of the oil and natural gas industry. Some of these recommended practices address process safety and fire protection. Of note are the following. (API)

- API 520 "Sizing, Selection, and Installation of Pressure-Relieving Devices"
- API 521 "Pressure-Relieving and Depressuring Systems"
- API RP 752 "Management of Hazards Associated with Location of Process Plant Permanent Buildings"
- API RP 753 "Management of Hazards Associated with Location of Process Plant Portable Buildings"
- API RP 756 "Management of Hazards Associated with Location of Process Plant Tents"
- API RP 2001 "Fire Protection in Refineries"
- API RP 2003 "Protections Against Ignitions Arising Out of Static, Lightning and Stray Currents"

CCPS *Guidelines for Combustible Dust Hazard Analysis.* This book describes how to conduct a Combustible Dust Hazard Analysis (CDHA) for processes handling combustible solids. The book explains how to do a dust hazard analysis by using either an approach based on compliance with existing consensus standards, or by using a risk-based approach. Worked examples in the book help the user understand how to do a combustible dust hazards analysis. (CCPS 2017)

CCPS *Guidelines for Consequence Analysis of Chemical Releases.* This Guidelines book provides technical information on how to conduct a consequence analysis to satisfy your company's needs and the U.S. EPA rules. It covers quantifying the size of a release, dispersion of vapor clouds to an endpoint concentration, outcomes for various types of explosions and fires, and the effect of the release on people and structures. (CCPS 1995)

CCPS *Guidelines for Evaluating Process Plant Buildings for External Explosions, Fires, and Toxic Releases, 2nd Edition*. Siting of permanent and temporary buildings in process areas requires careful consideration of potential effects of explosions and fires arising from accidental release of flammable materials. This book, which updates the 1996 edition, provides a single-source reference that explains the American Petroleum Institute (API) permanent (752) and temporary (753) building recommended practices and details how to implement them. New coverage on toxicity and updated standards are also highlighted. Practical and easy-to-use, this reliable guide is a must-have for implementing safe building practices. (CCPS 2012)

CCPS *Guidelines for Siting and Layout of Facilities, 2nd Edition*. This book has been written to address many of the developments since the 1st Edition which have improved how companies survey and select new sites, evaluate acquisitions, or expand their existing facilities. This book updates the appendices containing both the recommended separation distances and the checklists to help the teams obtain the information they need when locating the facility within a community, when arranging the processes within the facility, and when arranging the equipment within the process units. (CCPS 2018)

CCPS ***Guidelines for Fire Protection in Chemical, Petrochemical, and Hydrocarbon Processing Facilities***. This book consolidates fire prevention and protection tools and resources in a single document. (CCPS 2003)

CCPS ***Guidelines for Pressure Relief and Effluent Handling Systems, 2nd Edition.*** Providing in-depth guidance on how to design and rate emergency pressure relief systems, Guidelines for Pressure Relief and Effluent Handling Systems incorporates the current best designs from the Design Institute for Emergency Relief Systems (DIERS) as well as American Petroleum Institute (API) standards. Presenting a methodology that helps properly size all the components in a pressure relief system, the book includes software; the CCFlow suite of design tools and the new SuperChems™ for DIERS Lite software, making this an essential resource for engineers designing chemical plants, refineries, and similar facilities. (CCPS 1998)

CCPS ***Guidelines for Safe Handling of Powders and Bulk Solids.*** Powders and bulk solids are handled widely in the chemical, pharmaceutical, agriculture, smelting, and other industries. They present unique fire, explosion, and toxicity hazards. Substances which are practically inert in consolidated form may become quite hazardous when converted to powders and granules. This book discusses the types of hazards that can occur in a wide range of process equipment and with a wide range of substances and presents measures to address these hazards. (CCPS 2004)

CCPS ***Guidelines for Vapor Cloud Explosion, Pressure Vessel Burst, BLEVE and Flash Fire Hazards, 2nd Edition.*** This guide provides an overview of methods for estimating the characteristics of vapor cloud explosions, flash fires, and boiling-liquid-expanding-vapor explosions (BLEVEs) for practicing engineers. It has been updated to include advanced modeling technology, especially with respect to vapor cloud modeling and the use of computational fluid dynamics. The text also reviews past experimental and theoretical research and methods to estimate consequences. Heavily illustrated with photos, charts, tables, and diagrams, this manual is an essential tool for safety, insurance, regulatory, and engineering students and professionals. (CCPS 2010)

Crowl and Louvar, *Chemical Process Safety Fundamentals with Applications 4th Edition,* This book provides a compilation of many process safety topics. It links quantitative academic concepts to industrial process safety. (Crowl 2019)

Crowl, *Understanding Explosions*. This book describes the fundamentals and types of explosions as well as methods to prevent their occurrence. (Crowl 2003)

FM Global property loss prevention data sheets. FM Global is an insurance company that has used its loss experience to generate data sheets, which are essentially standards, on variety of topics. These data sheets are intended to reduce the chance of property damage. Topics of interest include industrial boilers, gas turbines, and extinguishing systems. (FMG)

National Fire Protection Association (NFPA) codes. The NFPA is a trade association that generates a large number of codes addressing fire and electrical hazards. The codes are often adopted by local authorities having jurisdiction thus making the code legally enforceable in that jurisdiction. These codes are a good source of knowledge addressing fire protection and suppression. Of note are the following. (NFPA)

- NFPA 30 "Flammable and Combustible Liquids Code"
- NFPA 70 "National Electrical Code"
- NFPA 56 "Standard for Fire and Explosion Prevention During Cleaning and Purging of Flammable Gas Piping Systems"
- NFPA 61 "Standard for the Prevention of Fires and Dust Explosions in Agricultural and Food Processing Facilities"
- NFPA 68 "Standard on Explosion Protection by Deflagration Venting"
- NFPA 430 "Code for the Storage of Liquid and Solid Oxidizers"
- NFPA 654 "Standard for the Prevention of Fire and Dust Explosions from the Manufacturing, Processing, and Handling of Combustible Particulate Solids"
- NFPA 704 "Standard System for the Identification of the Hazards of Materials for Emergency Response"

Many other codes address water and foam suppression sprinkler systems, storage systems, and fire pumps.

4.9 Summary

This chapter addresses the basics of fires and explosions from the fire triangle to the combustible dust pentagon. It explains terminology relevant to flammability, ignition, and explosive dust characteristics. The types of fires and explosions are described as well as methods to prevent ignition.

4.10 Other Incidents

This chapter began with a detailed description of the Imperial Sugar dust explosion. Other fire and explosion incidents include the following.

- Pemex LPG Terminal, Mexico City, Mexico, 1984
- BP Grangemouth Hydrocracker Explosion, U.K., 1987
- Celanese Pampa Explosion, Texas, U.S., 1987
- Piper Alpha Platform, North Sea, U.K., 1988
- Phillips Pasadena, Texas, U.S., 1989
- Total FCCU Explosion, France, 1992
- Texaco Oil Refinery Explosion and Fire, U.K., 1994
- Elf Refinery BLEVE, Feyzin, France, 1996
- Esso Longford Gas Plant Explosion, Australia, 1998
- Motiva Enterprises LLC, Delaware, U.S., 2001
- Hayes Lammerz Dust Explosion, Indiana, U.S., 2003

- BP Isomerization Unit Explosion, Texas City, Texas, U.S., 2005
- Buncefield Storage Tank Overflow and Explosion, U.K., 2005
- Olive Oil Storage Tank Explosion, Italy, 2006
- BLSR Deflagration and Fire, Texas, U.S., 2007
- Valero-McKee LPG Refinery Fire, Texas, U.S., 2007
- Imperial Sugar Dust Explosion, Georgia, U.S., 2008
- University of California at Los Angeles Laboratory Explosion, California, U.S., 2008
- Varanus Island Pipeline, Australia, 2008
- CITGO Refinery Fire, Texas, U.S., 2009
- ConAgra Foods, North Carolina, U.S., 2009
- NDK Crystal Vessel Rupture, Illinois, U.S., 2009
- Petroleum Oil Lubricants Explosion, Jaipur India, 2009
- Big Branch Mine Explosion, West Virginia, U.S., 2010
- Kleen Energy Explosion, Connecticut, U.S., 2010
- Pike River Coal Mine Explosion, South Island, New Zealand, 2010
- Texas Tech University Laboratory Explosion, Texas, U.S., 2010
- Hoeganaes Dust Fires, Tennessee, U.S., 2011
- Shell Refinery Fire, Singapore, 2011
- Chevron Richmond Refinery Fire, California, U.S., 2012
- West Fertilizer Company Explosion, Texas, U.S., 2013
- Williams Olefins Heat Exchanger Rupture, Louisiana, U.S., 2013
- University of Hawaii Laboratory Explosion, Hawaii, U.S., 2016
- Port of Beirut Ammonium Nitrate Explosion, Lebanon, 2020

4.11 Exercises

1. List 3 RBPS elements evident in the Imperial Sugar Dust explosion incident summarized at the beginning of this chapter. Describe their shortcomings as related to this accident.
2. Considering the Imperial Sugar Dust explosion incident, what actions could have been taken to reduce the risk of this incident?
3. Search for a Safety Data Sheet from a supplier of t-Butyl Amine. What is the Flash Point (FP), Upper and Lower Flammable Limits (UFL and LFL), and the Autoignition Temperature (AIT)? What is your reference?
4. Find the same flammability properties from a different source. What is the Flash Point, Upper and Lower Flammable Limits, and the Autoignition Temperature? Does this agree with the SDS you referenced?
5. What are the upper and lower flammability limits for diesel fuel, gasoline, and propane? What are their flash points?
6. What type of fire is possible from an atmospheric pressure storage tank storing a flammable liquid? And from an industrial process unit handing flammable materials under high temperature and pressure?
7. What is the difference between a deflagration and a detonation?

8. What must be present for a dust explosion to occur?
9. Find the values for Upper and Lower Flammable Limits in air, Upper and Lower Flammable Limits in oxygen and the Limiting Oxygen Concentration in Figure 4.18.
10. Estimate the Lower Flammable Limit for a vapor mixture of 0.4 mole fraction Methane and 0.6 mole fraction Ammonia. What is your reference for the pure chemical Flammable Limits? Show your work.
11. What was the source of ignition in the BP Texas City explosion and in the Imperial Sugar Dust explosion? How might these sources of ignition have been controlled?
12. How are electrical sources prevented from becoming a source of ignition?

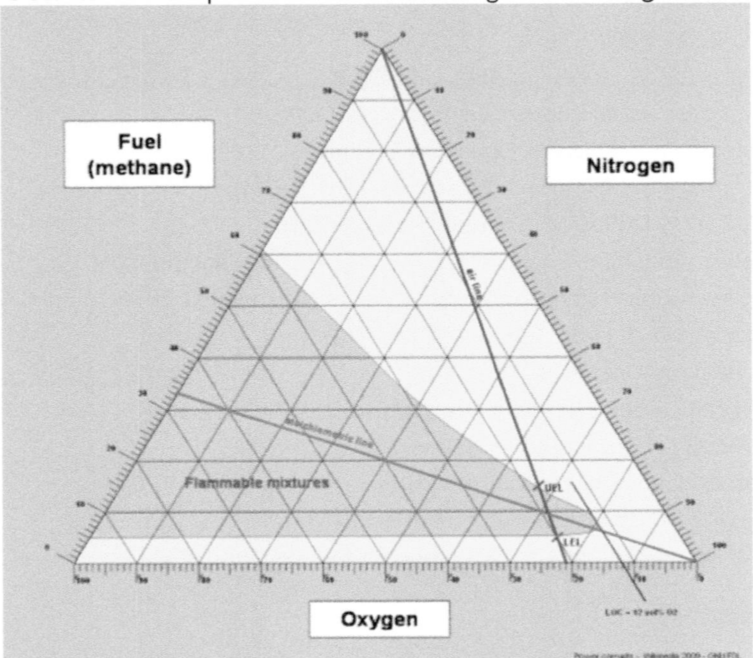

Figure 4.18. Methane flammability diagram (For Problem 4.9) (Wiki)

4.12 References

API, American Petroleum Institute, Washington, D.C., www.api.org.

- API 520 "Sizing, Selection, and Installation of Pressure-Relieving Devices"
- API 521 "Pressure-Relieving and Depressuring Systems"
- API RP 752 "Management of Hazards Associated with Location of Process Plant Permanent Buildings"
- API RP 753 "Management of Hazards Associated with Location of Process Plant Portable Buildings"
- API RP 756 "Management of Hazards Associated with Location of Process Plant Tents"

- API RP 2001 "Fire Protection in Refineries"
- API RP 2003 "Protections Against Ignitions Arising Out of Static, Lightening and Stray Currents"

ATEX, https://eur-lex.europa.eu/legal-content/EN/TXT/

BakerRisk, https://www.bakerrisk.com/services/testing/dispersion-jet-fire/.

BakerRisk 2001, "Vapor Cloud Explosion Prediction Methodology Guide", work performed by BakerRisk for the 2000 Explosion Research Cooperative (Project A150-201).

BS 2015, BS EN 60079 *Explosive atmospheres Classification of areas. Explosive gas atmospheres* https://www.en-standard.eu/bs-en-60079-10-1-2015-explosive-atmospheres-classification-of-areas-explosive-gas-atmospheres/.

CCPS Glossary, "CCPS Process Safety Glossary", Center for Chemical Process Safety, https://www.aiche.org/ccps/resources/glossary.

CCPS 1995 *Guidelines for Consequence Analysis of Chemical Releases,* Center for Chemical Process Safety, John Wiley & Sons, Hoboken, N.J.

CCPS 1998, *Guidelines for Pressure Relief and Effluent Handling Systems, 2nd Edition*, Center for Chemical Process Safety, John Wiley & Sons, Hoboken, N.J.

CCPS 2003, *Guidelines for Fire Protection in Chemical, Petrochemical, and Hydrocarbon Processing Facilities,* Center for Chemical Process , John Wiley & Sons, Hoboken, N.J.

CCPS 2004, *Guidelines for Safe Handling of Powders and Bulk Solids*, John Wiley & Sons, Hoboken, N.J.

CCPS 2010, *Guidelines for Vapor Cloud Explosion, Pressure Vessel Burst, BLEVE and Flash Fire Hazards, 2nd Edition*, Center for Chemical Process, John Wiley & Sons, Hoboken, N.J.

CCPS 2012, *Guidelines for Evaluating Process Plant Buildings for External Explosions, Fires, and Toxic Releases*, 2nd Edition, Center for Chemical Process, John Wiley & Sons, Hoboken, N.J.

CCPS 2017, *Guidelines for Combustible Dust Hazard Analysis*, Center for Chemical Process, John Wiley & Sons, Hoboken, N.J.

CCPS 2018, *Guidelines for Siting and Layout of Facilities*, 2nd Edition, Center for Chemical Process Safety, John Wiley & Sons, Hoboken, N.J.

Crowl, 2003, Daniel A. Crowl, *Understanding explosions*, Center for Chemical Process Safety, John Wiley & Sons, Hoboken, N.J.

Crowl 2011, Daniel A. Crowl and Joseph F. Louvar, *Chemical Process Safety, Fundamentals with Applications*, 3rd ed., Pearson, NY.

Crowl 2012, *Minimize the Risks of Flammable Materials, Chemical Engineering Progress*, American Institute of Chemical Engineers, New York, NY, April.

Crowl 2019, Daniel A. Crowl and Joseph F. Louvar, *Chemical Process Safety, Fundamentals with Applications 4th Edition*., Pearson, NY.

CSB 2009, "Sugar dust explosion and fire", Investigation Report, Report No. 2008-3-I-FL, U.S. Chemical Safety and Hazard Investigation Board, September. http://www.csb.gov/investigations.

CSB a, U.S. Chemical Safety and Hazard Investigation Board, https://www.csb.gov/recommendations/mostwanted/combustibledust/.

CSB b, U.S. Chemical Safety and Hazard Investigation Board, https://www.csb.gov/assets/1/6/csb_cdl_fact_sheet_-_combustible_dust.pdf.

FEMA, https://www.fema.gov/pdf/plan/prevent/rms/426/fema426_ch4.pdf.

FMG, FM Global Datasheets, .

insightnumerics 2018, http://insightnumerics.com/dir/wp-content/uploads/2018/02/Pool-Fire_preview.png.

NFPA, National Fire Protection Association, Quincy, MA, www.nfpa.org.

- NFPA 30 "Flammable and Combustible Liquids Code"
- NFPA 70 "National Electrical Code"
- NFPA 56 "Standard for Fire and Explosion Prevention During Cleaning and Purging of Flammable Gas Piping Systems"
- NFPA 61 "Standard for the Prevention of Fires and Dust Explosions in Agricultural and Food Processing Facilities"
- NFPA 68 "Standard on Explosion Protection by Deflagration Venting"
- NFPA 499 "Recommended Practice for the Classification of Combustible Dusts and of Hazardous (Classified) Locations for Electrical Installations in Chemical Process Areas"
- NFPA 430 "Code for the Storage of Liquid and Solid Oxidizers"
- NFPA 654 "Standard for the Prevention of Fire and Dust Explosions from the Manufacturing, Processing, and Handling of Combustible Particulate Solids"
- NFPA 704 "Standard System for the Identification of the Hazards of Materials for Emergency Response"

NIH, https://webwiser.nlm.nih.gov/getHomeData

OSHA a, "Combustible Dust in Industry: Preventing and Mitigating the Effects of Fire and Explosions", SHIB 07-31-2005, https://www.osha.gov/dts/shib/shib073105.html.

OSHA b, U.S. Occupational Safety and Health Administration, https://www.osha.gov/laws-regs/regulations/standardnumber/1910/1910.119.

OSHA c, "Hazard Communication Guidance for Combustible Dusts", https://www.osha.gov/sites/default/files/publications/3371combustible-dust.pdf.

OSHA d, viewed September 25, 2020 https://www.osha.gov/dts/osta/otm/images/otm_iv_5/otm_iv_5_IV_2.jpg.

Sandia 2020, viewed September 25, 2020, https://www.sandia.gov/news/publications/lab_accomplishments/articles/2020/energy_homeland_security.html.

UN, "Globally Harmonized System Of Classification And Labelling Of Chemicals (GHS) Sixth Edition", United Nations, New York and Geneva, 2011, https://www.un-ilibrary.org/transportation-and-public-safety/globally-harmonized-system-of-classification-and-labelling-of-chemicals-ghs-sixth-revised-edition_591dabf9-en.

5
Reactive Chemical Hazards

5.1 Learning Objectives

The learning objectives of this chapter are:

- Understand the hazards of reactive chemistry,
- Identify the hazardous reactive chemical properties of a material, and
- Identify where to find resources and tools that identify reactive chemistry hazards.

5.2 Incident: T-2 Laboratories Reactive Chemicals Explosion, Jacksonville, Florida, 2007

5.2.1 Incident Summary

A runaway reaction during the production of methylcyclopentadienyl manganese tricarbonyl (MCMT) at T2 Laboratories, Inc. in Florida and resulted in the rupture of the reactor on December 19, 2007. The resulting explosion led to four fatalities T2 employees and injured 32 people including members of the public. Pieces of the reactor were found 1.6 km (1 mi) away. Thirty-two structures were damaged. Figure 5.1 shows a section of the reactor, weighing approximately 907 kg (2,000 lb) that damaged a building 121 m (400 ft) away from the reactor. Figure 5.2 shows the control room. The explosion was heard, and the overpressure felt 24 km (15 mi) away in downtown Jacksonville, FL. (CSB 2009)

Figure 5.1. Portion of 7.6 cm (3 in)-thick reactor
(CSB 2009)

Figure 5.2. T2 Laboratories control room
(CSB 2009)

After the event, the CSB estimated the explosion was equivalent to 635 kg (1,400 lb) of TNT (CSB 2009).

A key outcome of this event was a recommendation by the Chemical Safety Board to Accreditation Board for Engineering and Technology, Inc. (ABET). The ABET now requires programs of Chemical, Biochemical, Biomolecular and Similarly Named Engineering Programs to include "Engineering application of these sciences to the design, analysis, and control of processes, including the hazards associated with these processes..." (ABET 2015) The hazards include chemical reactivity hazards.

Key Points:

Process Safety Competency – Does someone on the job understand the importance of process safety and its proper application? We work with many very intelligent people; but that does not mean that they understand how to relate the technical aspects of the facility to process safety concepts. Without someone on the site asking the right questions, process safety may be lacking.

Hazard Identification and Risk Analysis – What if? It is a very simple and powerful question. It can help to identify a range of hazards and once those hazards are identified, then mitigation and protection measures can be put in place.

Incident Investigation – If an operation yields an unexpected result, ask why did it do that? Through understanding why, you may recognize that you were heading down the path to an incident. More importantly, now you know how to avoid it.

5.2.2 Detailed Description

T2 Laboratories Inc. started in 1996 as a solvent blending business founded by a chemical engineer and a chemist. One of their products was a blend of MCMT, a gasoline additive. In 2004 T2 began producing MCMT, which became their primary product by 2007.

The runaway reaction occurred during the first step of the MCMT process. This was a reaction between methylcyclopentadiene (MCPD) dimer and sodium in diethylene glycol dimethyl ether (diglyme).

MCPD and diglyme were charged to a 9.3 m^3 (2,450 gal) reactor and sodium metal was added manually through a valve at the top of the reactor, (see Figure 5.3). Heat was applied to the reactor using hot oil set at 182 °C (360 °F) to melt the sodium and start the reaction to make methylcyclopentene. Hydrogen was a byproduct, vented through a pressure control valve. At 99 °C (210 °F) the agitator was started (by this time the sodium should have melted). At 149 °C (300 °F) the heat was turned off. The reaction was known to be exothermic and at 182 °C (360 °F) cooling was applied.

After eliminating other possible causes, the CSB concluded that loss of cooling was the immediate cause of the runaway reaction. The reactor was cooled by adding water to the jacket and allowing it to boil off (Figure 5.3).

The cooling system, necessary to control the exothermic reaction, could be totally incapacitated or severely impaired by several single failures: loss of cooling water from supply, a drain valve left open or partially open, failure of the valve actuators, blockage in the supply line, temperature sensor failure, or mineral build up in the jacket. (CSB 2009)

Without cooling, the temperature could continue to rise. Subsequent testing showed that a second exothermic reaction occurred at 199 °C (390 °F). This reaction was more energetic than the first, desired reaction. The owner/operators of T2 Laboratories did not know about this second reaction. This reaction generated enough pressure, very rapidly, to burst the reactor, rated for 41.4 bar (600 psig).

5.2.3 Lessons

Process Safety Culture. In hindsight, it seems the owners of T2 did not have the necessary process safety competency or know how to build a strong process safety culture.

Compliance with Standards. T2 was not in compliance with the OSHA Hazard Communication Standard. No written evidence was found that T2 had a confined space entry, lock-out/tag-out, personal protective equipment program, or employee training program.

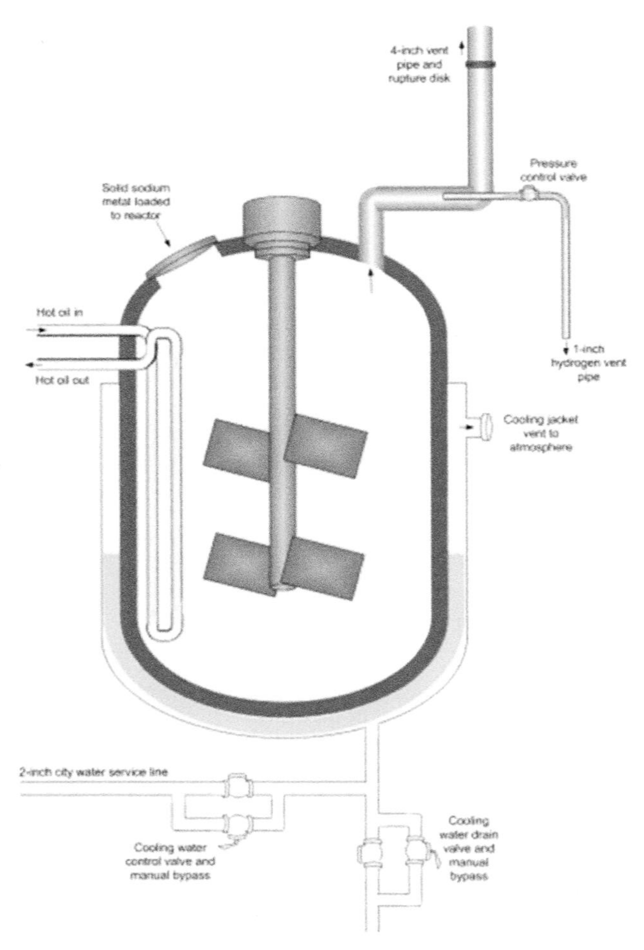

Figure 5.3. T2 Reactor cross-section
(CSB 2009)

Process Safety Competency. T2 was started by a chemical engineer and a chemist. Neither had experience designing and running processes involving chemical reactions. The chemist tested the chemistry in the lab and developed the process based on patent literature provided by a company called Advanced Fuel Development Technologies, who wanted T2 to manufacture MCMT for them. This lack of experience showed itself in several ways.

The chemist did the laboratory testing at a 1 liter scale and did not observe strong exothermic behavior. A fundamental concept that needs to be understood when scaling up an exothermic reaction is that the energy released increases as the cube of the reactor diameter, while the heat transfer area increases with the square of the diameter (without additional area from internal coils). Therefore, the rate and amount of heat generated increases faster than the ability to remove it. The need for cooling was discovered during process upsets in the first few batches, not during the laboratory tests.

The owners did not do any reaction testing such as adiabatic calorimetry (e.g., Accelerating Rate Calorimeter™ (ARC), Vent Sizing Package™ (VSP), Phi-Tec, or Automatic Pressure Tracking Adiabatic Calorimeter® (APTAC)), although this type of testing had been good engineering practice for years.

The CSB noted that process safety was not part of the chemical engineering curriculum in almost 90% of universities at the time of the incident. In its report, the CSB recommended to the AIChE and the Accreditation Board for Engineering and Technology, Inc. (ABET) that awareness of reactive chemical hazards be part of the baccalaureate program (CSB 2009). This recommendation was implemented by the ABET, in fact, the CSB notes that the action exceeded the CSBs expectations.

Hazard Identification and Risk Analysis. Even though a design consultant recommended that T2 do a Hazard and Operability (HAZOP) study on the process, T2 apparently did not do one. If the MCMT process had been reviewed by a competent PHA team questions such as, "what happens if the temperature is too high?" or "what if the cooling fails?" would have come up. These questions would lead to recommendations such as: determine what the safe operating temperature is, what happens if it is exceeded, how can we make the cooling system more reliable, or what other safeguards can be provided against high temperature and pressure?

Asking these questions could also have led to a better understanding of the emergency relief requirements. The emergency relief system (ERS) was based on the maximum rate of hydrogen generation in normal operation (CSB 2009). The ERS was inadequate for the reaction that occurred. After subsequent testing in a VSP, the CSB determined that the second exothermic reaction was so fast that the reactor could not have been successfully protected by a relief device. The only way to protect the reactor from overpressuring was to vent the reactor during the first reaction and allow the energy to be removed by boiling off the diglyme solvent and MCPD.

Management of Change (MOC). After one year of production the batch size was increased by one-third, without a safety review. However, without the needed competency to recognize reactive chemical hazards, an MOC would not have helped.

Emergency Management. T2 did not warn emergency responders of the presence of MCMT on site. MCMT is toxic by inhalation and skin contact.

Incident Investigation. Prior to the explosion, there had been unexpected exotherms in three of the first ten batches during the first reaction step when the process was scaled up to the main reactor. After the first exotherm (in Batch 1), the response was to adjust the batch recipe and to add cooling to the operating procedures. Uncontrolled exotherms also occurred in Batches 5 and 10. Nevertheless, after Batch 11, the process scale-up was considered successful. The owners did not recognize that the previous exotherms were actually near misses which could have had more severe consequences, and therefore failed to further investigate the causes of these exotherms.

A video about the T2 Laboratories explosion can be found on the CSB website at http://www.csb.gov/videos/.

5.3 Introduction to Chemical Reactivity

The CSB report *Improving Reactive Hazard Management* analyzed reactive chemistry incidents showing that they occur in a variety of equipment as shown in Figure 5.4 and result in severe consequences as shown in Figure 5.5. This focus on reactive chemical incidents highlighted that regulations did not address reactive chemical hazards as well as other chemical hazards.

The risks of a chemical reactivity incident result from the potential for an uncontrolled chemical release leading to the consequences shown in Figure 5.5.

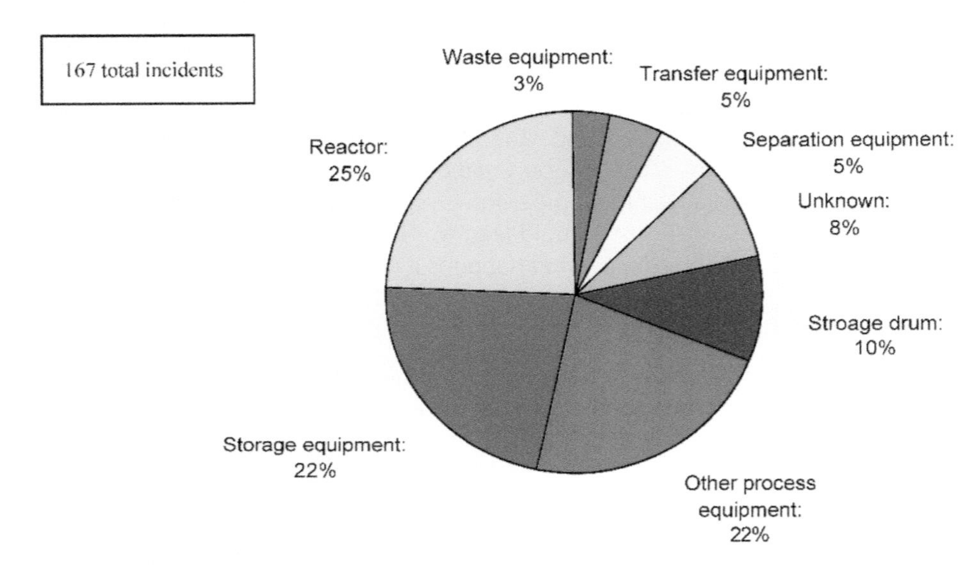

Figure 5.4. Equipment involved in reactive chemistry incidents (edited from CSB 2002)

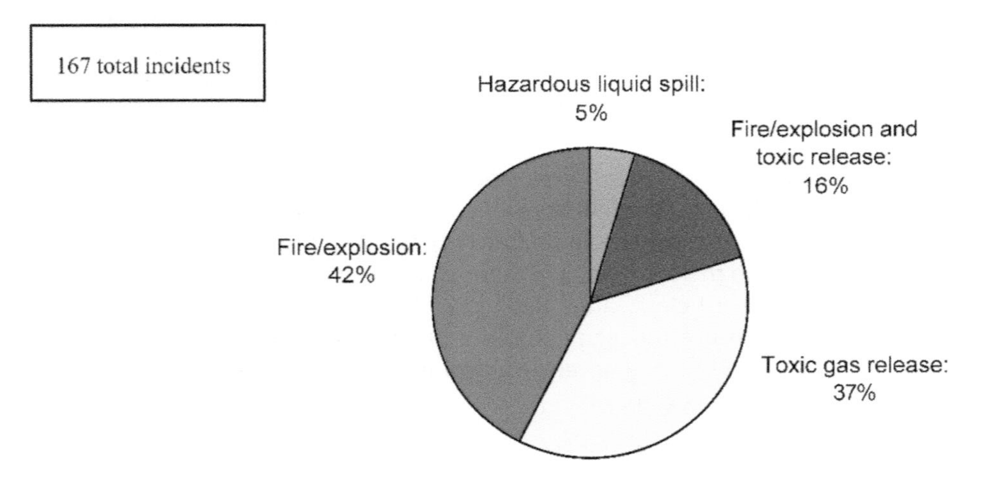

Figure 5.5. Consequences of reactive chemistry incidents (CSB 2002)

The CSB identified three reactive hazard types: chemical incompatibility, runaway reaction, and impact or thermally sensitive materials. (CBS 2002) Chemical reactivity can occur by design (e.g. in a chemical reactor) or accidentally (e.g. inadvertent mixing of incompatible materials, storage or handling).

Reactive chemical incidents typically involve:

- inadequate hazard identification & evaluation,
- inadequate procedures and training for storage and handling of reactive chemicals, and
- inadequate process design for reactive hazards.

Important definitions and terminology related to reactive chemistry include

Chemical Reactivity - The tendency of substances to undergo chemical change. (CCPS Glossary)

Exothermic Chemical Reaction - A reaction involving one or more chemicals resulting in one or more new chemical species and the evolution of heat. (CCPS Glossary)

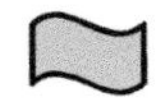

Endothermic Chemical Reaction - A reaction involving one or more chemicals resulting in one or more new chemical species and the absorption of heat. (CCPS Glossary)

Endothermic and exothermic reactions can generate gaseous or highly volatile products. Exothermic reactions have the potential for a runaway reaction leading to a dramatic increase in temperature, pressure (if the reaction is contained) and reaction rate.

Chemical reactions can occur throughout a process facility including in the storage of raw materials, in process streams, as the products, or in the process waste streams. Particular attention should be paid to waste streams where many chemicals can come together in ways that may not be identified unless specifically analyzed. Chemical reactivity types are listed in Table 5.1 and several reactive functional groups are listed in Table 5.2.

Table 5.1. Chemical Reactivity types and examples
(Crowl 2019)

Reactive Type	Example
Pyrophoric and spontaneously combustible	Readily reacts with oxygen, igniting and burning Aluminum alkyls, Raney nickel catalyst
Peroxide-forming	Reacts with oxygen to form unstable peroxides 1,3, butadiene
Water reactive chemical	Sodium, titanium tetrachloride, boron trifluoride
Oxidizer	Readily yields oxygen or other oxidizing gas to promote or initiate combustion Chlorine, hydrogen peroxide, nitric acid, HF
Self-reactive	Butadiene polymerization, acetylene decomposition, ethylene and propylene oxide, styrene, vinyl acetate
Chemical incompatibles	Caustic + muriatic acid
Impact sensitive or thermally sensitive	Trinitrotoluene (TNT)
Runaway reactions	Ethylene

Table 5.2. Some Reactive Functional Groups
(Crowl 2019)

Some Reactive Functional Groups	
Azide	N3
Diazo	-N=N-
Nitro	-NO2
Nitroso	-NO
Nitrite	-ONO
Nitrate	-ONO2
Fulminate	-ONC
Peroxide	-O-O-
Peracid	-CO3H
Hydroperoxide	-O-O-H
Ozonide	O3
Amine oxide	≡NO
Chlorates	ClO3

The key point is to either prevent the chemical reaction, or, if it is desired, to ensure that it can be safely contained in the equipment. The first step in managing chemical reactivity is to gather data on the reactive chemical properties of the materials being handled. These data are important process safety information (Refer to Section 2.5) Relevant data include:

- heat of reaction per mass, which is the change in the enthalpy of a chemical reaction that occurs at a constant pressure,
- maximum reaction temperature and pressure, which depends on starting conditions, additional heat inputs or cooling (such as vaporization),
- detected onset temperature, which is based on data from a specific calorimeter and may need to be "adjusted" for the specific equipment or vessel to be evaluated,
- temperature of no return, which is the temperature where the heat gain from reaction exceeds the heat losses from the equipment or vessel (the point at which control is lost), and
- identification of highly energetic reactions that may indicate shock sensitive material.

5.4 Reactive Chemicals Testing

In general, if insufficient data are available, then the materials should be subjected to screening evaluations or tests. Data can be gathered from existing reactive chemicals databases, from calorimetry and from the development of kinetic models. Calorimetry testing includes the following, amongst others.

- Differential Scanning Calorimetry (DSC) – DCS measures the difference in the amount of heat required to increase the temperature of a sample and reference as a function of temperature.
- Accelerating Rate Calorimetry (ARC) – ARC analyzes the properties of an exothermic reaction in a confined adiabatic environment, one in which a minimal amount of heat is lost.
- Advanced Reactive System Screening Tool (ARSST) for initial screening, and
- Vent Sizing Package (VSP) – VSP is a low thermal inertia adiabatic calorimeter used for process hazard characterization in which upset process conditions may be mimicked.

It is important to safely contain a chemical reaction in its equipment. For an exothermic reaction, this means maintaining a normal operating temperature below the temperature at which reaction heat gain exceeds heat loss to the surroundings. Beyond a certain temperature, the temperature of no return, if no action is taken, the reaction will proceed to a maximum rate. This may generate a pressure rise that can lead to loss of containment as illustrated in Figure 5.6 for the chemicals involved in the T2 Laboratories explosion.

Additional reactive chemicals testing might be warranted when the chemical reaction is highly energetic or results in significant pressure. Of key importance is the sizing of pressure relief systems to handle the potential overpressures..

These tests have limitations which should be understood when using the test data. The *Guidelines for Chemical Reactivity Evaluation and Application to Process Design* (CCPS 1995 a*)* and

Chemical Process Safety, Fundamentals with Applications, 4th Edition (Crowl 2019) are helpful references.

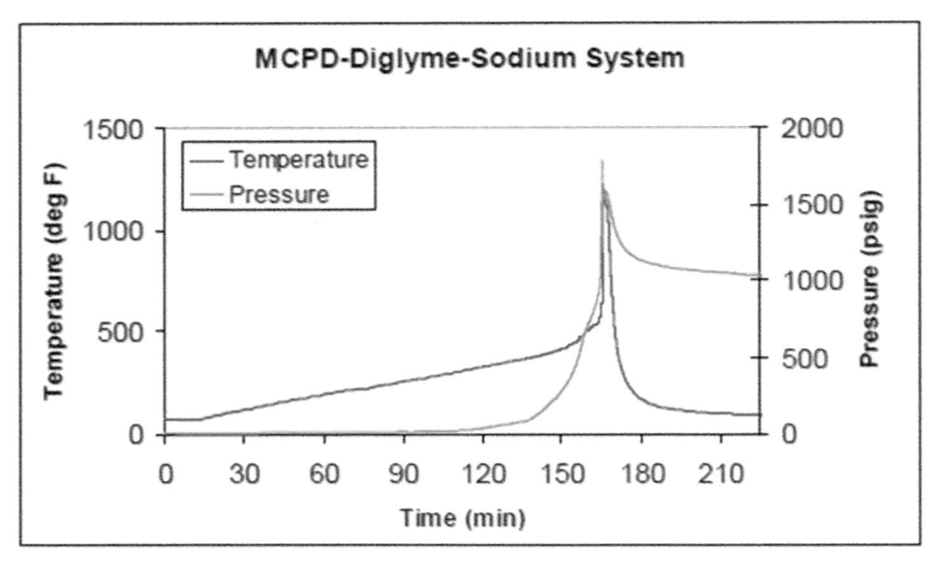

Figure 5.6. Temperature and pressure vs. time for T2 Laboratories explosion (CSB 2009)

5.5 Reactive Chemicals Hazard Screening and Evaluation

Many reactive chemical incidents occur because the reactivity of the chemical was not understood or because the interaction between chemicals was not understood. Reactive chemicals screening is an approach intended to easily identify potential reactive chemicals hazards even in those facilities that may not recognize they are handling reactive chemicals. A Reactive Hazard Screening Table and Flowchart is a tool provided in Section 5.7. The data required are simply a knowledge of the chemicals and how they are handled at a facility.

Where screening indicates a chemical reactivity hazard is expected, the next step is evaluation of the chemicals and their interaction to support design of the process and equipment. The first step is calculating the data as discussed in Section 5.4. Additional testing may be appropriate where this data are not readily available from databases and other sources.

The next step is to evaluate the chemicals for the conditions in which they are transported, stored, and processed. This includes both normal operating conditions and those during a process upset. Kinetic modeling can assist with this evaluation. Approaches and tools to support this are discussed in Section 5.8 including the Chemical Reactivity Worksheet that supports identification of chemical incompatibilities.

Once the chemical reactivity hazards are understood and characterized by data, then the process and equipment can be designed with the appropriate safeguards. The chemicals can be transported, stored, and handled so as to avoid inadvertent mixing. The temperatures can be managed to control reactivity. The mechanical design of equipment can provide containment for the temperatures and pressures required. Finally, effective overpressure protection can be designed, where appropriate.

5.6 Reactive Chemical Incident Prevention and Mitigation

Section 5.3 makes clear that understanding reactive chemical hazards and transporting, storing, and processing chemicals in a way to prevent a potential incident is the most important approach in managing this hazard. In addition to managing the temperature and pressure, the use of inhibitors is a means to control the rate of reaction. Some inhibitors require a minimum amount of oxygen to be present to work properly. Systems that are inerted with nitrogen to prevent the formation of a flammable atmosphere in a tank may need to be modified to include small amounts of oxygen. Periodic testing of the atmosphere is recommended to ensure adequate oxygen remains present over time.

Many things can be done to control reactive hazards: using less hazardous chemicals or a less energetic reaction pathway, separating incompatible materials, and training operators on relevant chemical reactivity hazards at their facility. (Crowl, 2019)

If, however, an incident does occur, mitigation techniques can be incorporated in the facility design. These may include:

- An emergency cooling system to cool the process below a temperature where a runaway reaction will not occur,
- A reaction quenching system that adds a diluent to reduce the concentration and stop the runaway reaction, and
- An overpressure relief system designed to relieve the pressure to a safe location.

5.7 What a New Engineer Might Do

The T2 Laboratories incident highlights what every engineer, new or experienced, should do. That is, understand the reactive chemical hazards of the materials being handled and processed at the facility. You may be asked to gather data on the chemicals, or the mixing or compatibility of chemicals. Analyze this data carefully. If it seems incomplete or incorrect, question it and dig a bit deeper. Request that reactive chemicals testing be conducted.

The Reactive Chemicals Hazard Screening and the Chemical Reactivity Worksheet are tools that can support your analysis. They are described in Section 5.8. These tools are free and easy to use.

Multiple resources that bring the chemical reactivity hazard to life are available from the CSB. A new engineer can benefit from reviewing the CSB investigations and videos relevant to this chapter as listed in Appendix G.

5.8 Tools

The following resources are available for helping to understand and assess chemical reactivity hazards.

5.8.1 Chemical Properties Data

Data on chemical properties are available through many sources.

Chemical properties data. Greater detail on these tools is provided in Chapter 7.

- **Safety Data Sheets (SDS).** The SDS includes chemical properties, hazards, and protective measures.
- **CAMEO Chemicals**. Data provided by the National Oceanic and Atmospheric Administration on thousands of chemicals. (NOAA)
- **Cole-Parmer Technical Compatibility Database.** Facilitates searches for chemical and material compatibility. (CP)
- **NIH WebWISER.** WISER is a system provided by the National Institutes of Health designed to assist emergency responders in hazardous material incidents. (NIH)

5.8.2 Chemical Reactivity Hazard Screening Table and Flowchart

The material in this section is adapted from *Essential Practices for Managing Chemical Reactivity Hazards* (CCPS 2003). Table 5.3 can be used as a form to document the screening questions described in this reference.

Table 5.3. Example form to document screening of chemical reactivity hazards

FACILITY:	COMPLETION DATE:	
COMPLETED BY:	APPROVED BY:	
Do the answers to the following questions indicate chemical reactivity hazard(s) are present?[1] ____		
AT THIS FACILITY:	***YES, NO, or NA***	**BASIS FOR ANSWER; COMMENTS**
1. Is intentional chemistry performed?		
2. Is there any mixing or combining of different substances?		
3. Does any other physical processing of substances occur?		
4. Are there any hazardous substances stored or handled?		
5. Is combustion with air the only chemistry intended?		
6. Is any heat generated during the mixing or physical processing of substances?		
7. Is any substance identified as spontaneously combustible?		
8. Is any substance identified as peroxide forming?		
9. Is any substance identified as water reactive?		
10. Is any substance identified as an oxidizer?		
11. Is any substance identified as self-reactive?		
12. Can incompatible materials coming into contact cause undesired consequences, based on the following analysis?		

SCENARIO	CONDITIONS NORMAL?[2]	R, NR or ?[3]	INFORMATION SOURCES; COMMENTS
1			
2			
3			

[1] Use Figure 5.7 with answers to Questions 1-12 to determine if answer is YES or NO

[2] Does the contact/mixing occur at ambient temperature, atmospheric pressure, 21% oxygen atmosphere, and unconfined? (IF NOT, DO NOT ASSUME THAT PUBLISHED DATA FOR AMBIENT CONDITIONS APPLY)

[3] **R** = Reactive (incompatible) under the stated scenario and conditions
NR = Non-reactive (compatible) under the stated scenario and conditions
? = Unknown; assume incompatible until further information is obtained

Figure 5.7 is a flowchart that shows how the questions in Table 5.3 are connected to determine whether a chemical reactivity hazard can be expected in your facility. The note about Chapter 3 and question numbers refers to the chapters in *Essential Practices for Managing Chemical Reactivity Hazards.*

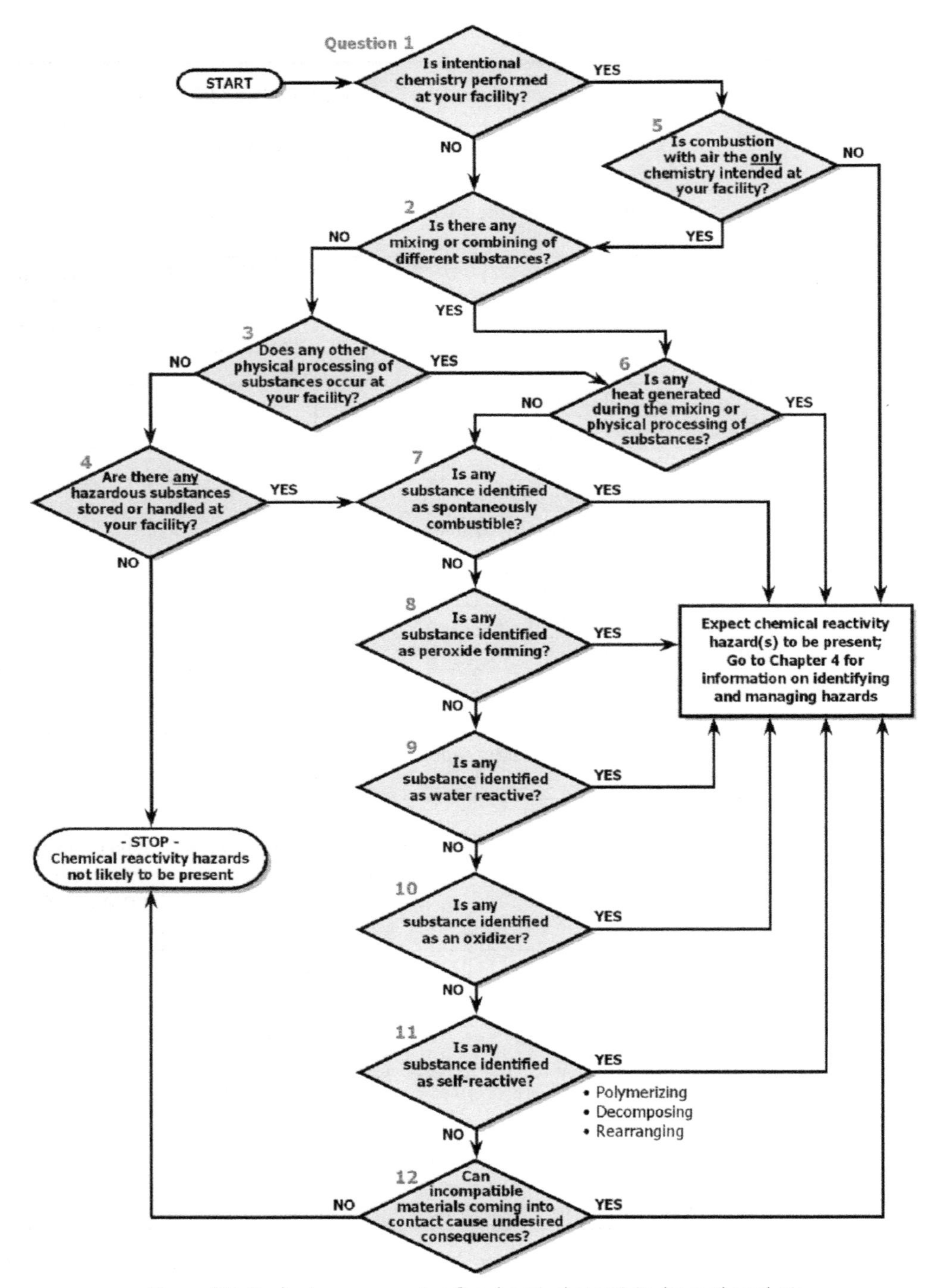

Figure 5.7. Preliminary screening for chemical reactivity hazard analysis (CCPS 2003)

5.8.3 Chemical Reactivity Worksheet (CRW)

The Chemical Reactivity Worksheet (CRW) is a free software program providing extensive process safety information (Refer to Section 2.5) The CRW includes data required to understand the hazards associated with the inadvertent and intentional mixing of reactive chemicals. This includes the chemical reactivity of thousands of common hazardous chemicals, compatibility of absorbents, and suitability of materials of construction in chemical processes. It is designed to be used by emergency responders and planners, as well as the chemical industry, to help prevent hazardous chemical incidents. It is available at https://www.aiche.org/ccps/resources/chemical-reactivity-worksheet. (CCPS)

Versions of the CRW were developed by the collaboration of several organizations including the Center for Chemical Process Safety, Environmental Protection Agency, NOAA's Office of Response and Restoration, The Materials Technology Institute, Dow Chemical Company, Dupont, and Phillips.

The CRW contains a database of chemical datasheets for thousands of chemicals. The chemical datasheets in the CRW database contain information about the intrinsic reactive hazards of each chemical, such as flammability, the ability to form peroxides, the ability to self-polymerize, explosivity, strong oxidizer or reducer capability, water or air reactivity, pyrophoricity, known catalytic activity, instability, and radioactivity. Datasheets also contain general descriptions of the chemicals, physical properties, and toxicity information. They also include case histories on specific chemical incidents, with references. The CRW also allows the creation of custom chemical datasheets, for example, to use in documenting properties of a proprietary chemical that is not in the CRW database.

The CRW uses chemical pairs. In order to fully understand inadvertent and intentional mixing, the reactivity of the entire mixture must be understood, not just the pairs.

The CRW includes a reactivity prediction worksheet to virtually "mix" chemicals to simulate accidental chemical mixtures, such as in the case of a train derailment, to learn what dangers could arise from the accidental mixing. For example, if the reaction is predicted to generate gases, the CRW will list the potential gaseous products, along with literature citations.

The CRW has two additional modules of particular interest to the chemical industry. One of them discusses known incompatibilities between certain chemicals and common absorbents which are used in the cleanup of small spills. The other module, new in CRW 4, contains information about known incompatibilities between certain chemicals and materials that are used in the construction of containers, pipes, and valving systems on industrial chemical sites.

The **Mixture Manager** screen provides a search for chemicals in the CRW's database, a preview of the information on the chemical datasheets, and the creation of virtual mixtures of chemicals. It also provides access to all the other features of the program from this screen, including the compatibility chart and hazards report for any mixture created, reference information about the reactive groups used in the CRW, and information about absorbent incompatibilities with certain chemicals.

The **Compatibility Chart** shows the predicted hazards of mixing the chemicals in a mixture in an easy-to-use graphical interface. The reactivity predictions are color-coded, and the cells on the chart can be clicked to find more information about specific predicted reactions. General hazard statements, predicted gas products, and literature documentation for the selected pair of chemicals are shown at the bottom of the chart. Other functionality to understand reactivity are: chemical information, physical properties, synonyms, a reactive groups tab, absorbent incompatibilities, and a place to start to look for compatible materials of construction. Finally, the Helphelp function is very useful to understand how to create mixtures and understand their reactivity.

The **Reactive Chemicals Checklist** contained in Appendix D is adapted from a CCPS Safety Alert *A Checklist for Inherently Safer Chemical Reaction Process Design and Operation*. (CCPS 2004) It provides a checklist addressing chemical reaction hazard identification and reaction process design.

Figure 5.8 is a chemical reactivity worksheet showing the reactivity of sulfuric acid and sodium hydroxide, a strong acid / strong base pair. The worksheet illustrates that the acid and base are incompatible and hazardous reactivity issues are expected. The hazard summary lists the hazards of mixing including corrosive reaction products and gas evolution that might result in pressurization. The potential gases tab lists acid and base fumes, and nitrogen oxides. This view of the CRW also summarizes the intrinsic chemical hazards and NFPA 704 ranking.

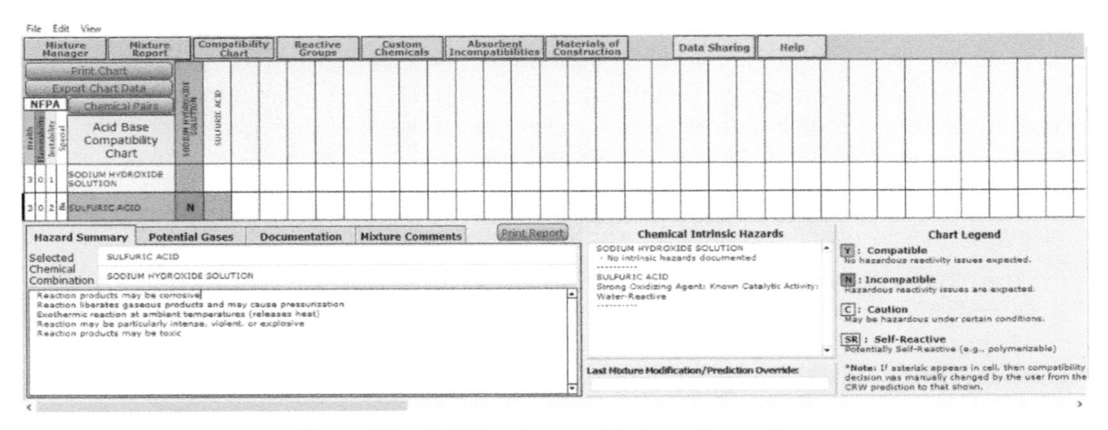

Figure 5.8. – CRW for strong acid strong base pair (CCPS)

5.8.4 *Documents*

The following books and resources are available for helping to understand reactive chemical hazards.

Bretherick's Handbook of Reactive Chemical Hazards, 7th Edition. Bretherick's is a 2-volume set of all reported risks such as explosion, fire, toxic or high-energy events that result from chemical reactions gone astray, with extensive referencing to the primary literature. (Bretherick & Urban 2006)

CCPS Essential Practices for Managing Chemical Reactivity Hazards. This book provides technical guidance to help small and large companies to identify, address, and manage chemical reactivity hazards. Appendix C has the flowchart developed for this book that guides the user through an analysis of the potential for chemical reactivity accidents. This book and flowchart are summarized in the article Screen Your Facilities for Chemical Reactivity Hazards (Johnson and Lodal 2003) (CCPS 2003).

CCPS Guidelines for Chemical Reactivity Evaluation and Application to Process Design. This book provides principles and strategies for the evaluation of chemical reactions, and for using this information in process design and management. (CCPS 1995 a)

CCPS Guidelines for Safe Storage and Handling of Reactive Materials. This book provides guidelines to manage the risks associated with storing and handling reactive materials. The book includes policies, design, operations and maintenance of facilities as well as human factors issues. (CCPS 1995 b)

CCPS Guidelines for Safe Warehousing of Chemicals. This book provides an understanding of the potential dangers inherent in warehousing chemicals. It offers a performance-based approach to such hazards as health effects, environmental pollution, fire, and explosion, and presents practical means to minimize the risk of these hazards to employees, the surrounding population, the environment, property, and business operations. These basic precepts can be used to evaluate the risks in initial or existing designs for warehousing facilities on a manufacturing site, for freestanding offsite buildings, and for strictly chemical or mixed-use storage. (CCPS 1998)

Chemical Reaction Hazards: A Guide to Safety. This book describes how to assess reactive chemical hazards before designing a plant. The book includes over 100 case studies. (Barton & Rogers 1997)

Chemical Handling Guides or Product Safety Bulletins. Chemical manufacturers may offer these documents on specific chemicals. They provide chemical properties information and also guidance on safe handling. Examples include the following.

- Nouryon Safety of Organic Peroxides
- LyondellBasell Propylene Oxide Product Safety Bulletin (LB)

CSB Improving Reactive Hazard Management. This report reviews reactive chemical incidents, considers how industry and regulatory bodies address them, and offers recommendations for reducing their occurrence. (CSB 2002)

Designing and Operating Safe Chemical Reaction Processes. This free document is for those responsible for the development, design and operation of chemical plant and processes. It provides information on the assessment of chemical reaction hazards for batch and semi-batch processes, and for the design, operation and modification of chemical reaction processes. (HSE 2009)

Essential Practices for Managing Chemical Reactivity Hazards. This book supports the analysis and management of chemical reactivity hazards. It provides specific guidance useful for the scale up of processes. (Johnson 2003)

National Fire Protection Association (NFPA) codes. The NFPA is a trade association that generates a large number of codes addressing fire and electrical hazards. The codes are often adopted by local authorities having jurisdiction thus making the code legally enforceable in that jurisdiction. These codes are a good source of knowledge addressing fire protection and suppression. Of note are the following. (NFPA)

- NFPA 491M Manual of Hazardous Chemical Reactions
- NFPA43 B Storage of Organic Peroxide Formulations
- NFPA 49 Hazardous Chemicals Data
- NFPA 325 Fire Hazard Properties of Flammable Liquids, Gases, and Volatile Solids
- NFPA 430 Storage of Liquid and Solid Oxidizers

5.9 Summary

Reactive chemical hazards can occur in a variety of industries and types of equipment. Incidents occur when the reactive hazard is unknown or when it is underestimated. Wherever chemicals are handled and processed, even where it is not understood that reactive chemistry can occur, the chemical properties should be understood, hazards screened, and safeguards implemented.

Do not assume that the chemicals can be controlled in all operating conditions at the facility. If the data are not available to verify this, then conduct testing to gain the data. Many sources are available to gather the chemical property data and useful tools such as the Chemical Reactivity Worksheet can be used to analyze the hazards.

5.10 Other incidents

Other incidents involving reactive chemicals include the following.

- Rohm & Haas Road Tanker Explosion, Teeside, U.K., 1976
- Arco Channelview Explosion, Texas, U.S., 1990
- Hickson Welsh Jet Fire, Yorkshire, U.K., 1992
- Hoechst Griesheim, Explosion, Frankfurt, Germany, 1993
- Port Neal AN Explosion, Sioux City, Iowa, U.S., 1994
- Napp Technologies Explosion, Lodi, New Jersey, U.S., 1995
- Bartlo Packaging Pesticide Explosion, West Helena, Arkansas, U.S., 1997
- Morton International Explosion, Paterson, New Jersey, U.S., 1998
- Concept Sciences Hydroxylamine Explosion, Allentown, Pennsylvania, U.S., 1999
- AZF AN Explosion, Toulouse, France, 2001
- Synthron Chemcial Explosion, Morganton, North Carolina, U.S., 2006
- Bayer CropScience Runaway Reaction and Pressure Vessel Explosion, Institute, West Virginia, U.S., 2008
- West Fertilizer AN Explosion, West, Texas, U.S., 2013
- Tianjin AN Explosion, China, 2015
- Arkema Fire, Crosby, Texas, U.S., 2017
- Seveso Disaster, Seveso, Italy, 1976
- Port of Beirut Ammonium Nitrate Explosion, Lebanon, 2020

5.11 Exercises

1. List 3 RBPS elements evident in the T-2 Laboratories reactive chemicals explosion incident summarized at the beginning of this chapter. Describe their shortcomings as related to this accident.
2. Considering the T-2 Laboratories reactive chemicals explosion incident, what actions could have been taken to reduce the risk of this incident?
3. The U.S. Chemical Safety and Hazards Investigation Board conducted an investigation of reactive chemical incidents. What were the 3 reactive hazards types they identified? Provide an example of each.
4. What can result if an organic peroxide is mixed with a hydrazine?
5. Can a metallic alloy pipe be used to safely transport sulfuric acid?
6. Using the CRW worksheet, develop an Inter-Reactivity Chart for the following chemicals: Styrene, Ethylbenzene, Aqueous Sodium Hydroxide, Aqueous Hydrogen Chloride. Submit a "screen shot" of your chart. What are the binary combinations of concern and why?
7. Describe the chemical reactivity that occurred in the Napp Technologies explosion in Lodi, New Jersey in 1995. What steps could have been taken to prevent this event?

5.12 References

ABET 2015, Criteria for accrediting engineering programs, Accreditation Board for Engineering and Technology, Baltimore, MD, https://www.abet.org/accreditation/accreditation-criteria/criteria-for-accrediting-engineering-programs-2021-2022/

Nouryon, "Safety of Organic Peroxides", https://www.nouryon.com/globalassets/inriver/resources/brochure-safety-of-organic-peroxides-lowres-en_us.pdf

Barton & Rogers 1997, *Chemical Reaction Hazards: A Guide to Safety*, Institute of Chemical Engineers, U.K.

Bretherick & Urban 2017, *Handbook of Reactive Chemical Hazards, 8th Edition*, Elsevier, Netherlands.

CCPS, "Chemical Reactivity Worksheet 4.0", https://www.aiche.org/ccps/resources/chemical-reactivity-worksheet.

CCPS Glossary, "CCPS Process Safety Glossary", Center for Chemical Process Safety, https://www.aiche.org/ccps/resources/glossary.

CCPS 1995 a, *Guidelines for Chemical Reactivity Evaluation and Application to Process Design*, Center for Chemical Process Safety, John Wiley & Sons, Hoboken, N.J.

CCPS 1995 b, *Guidelines for Safe Storage and Handling of Reactive Materials*, Center for Chemical Process Safety, John Wiley & Sons, Hoboken, N.J.

CCPS 1998. *Guidelines for Safe Warehousing of Chemicals*, Center for Chemical Process Safety, John Wiley & Sons, Hoboken, N.J.

CCPS 2003, *Essential Practices for Managing Chemical Reactivity Hazards*, Center for Chemical Process Safety, John Wiley & Sons, Hoboken, N.J.

CCPS 2004, "CCPS Safety Alert; A Checklist for Inherently Safer Chemical Reaction Process Design and Operation", March 1, Center for Chemical Process Safety, of the American Institute of Chemical Engineers, New York, NY.

Crowl 2019, Daniel A. Crowl and Joseph F. Louvar, *Chemical Process Safety, Fundamentals with Applications 4th Edition.*, Pearson, NY.

CP, Cole-Parmer Technical Compatibility Database, https://www.coleparmer.com/Chemical-Resistance.

CSB, U.S. Chemical Safety and Hazard Investigation Board, https://csb.gov/videos/reactive-hazards/.

CSB 2002, "Hazard Investigation Improving Reactive Hazard Management", Report No. 2001-01-H, U.S. Chemical Safety and Hazard Investigation Board, October.

CSB 2009, "T2 Laboratories, Inc. runaway reaction", Investigation Report, Report No. 2008-3-I-FL, U.S. Chemical Safety and Hazard Investigation Board, September.

HSE 2009, "Designing and Operating Safe Chemical Reaction Processes", https://www.hse.gov.uk/pubns/books/hsg143.htm.

Johnson and Lodal 2003, "Screen Your Facilities for Chemical Reactivity Hazards", *Chemical Engineering Progress*, American Institute of Chemical Engineers, New York, NY, August.

Johnson 2003, R. W., S. W. Rudy, and S. D. Unwin, *Essential Practices for Managing Chemical Reactivity Hazards*. Center for Chemical Process Safety, Hoboken, NJ USA: John Wiley & Sons.

LB, "Propylene Oxide Product Safety Bulletin", LyondellBasell, Houston, Texas, 2020.

NFPA, National Fire Protection Association, Quincy, MA, www.nfpa.org.

- NFPA 704, "Standard System for the Identification of the Hazards of Materials for Emergency Response"
- NFPA 491M "Manual of Hazardous Chemical Reactions"
- NFPA43 B "Storage of Organic Peroxide Formulations"
- NFPA 49 "Hazardous Chemicals Data"
- NFPA 325 "Fire Hazard Properties of Flammable Liquids, Gases, and Volatile Solids"
- NFPA 430 "Storage of Liquid and Solid Oxidizers"

NIH, https://webwiser.nlm.nih.gov/getHomeData

NOAA, CAMEO Chemicals, National Oceanic and Atmospheric Administration, https://cameochemicals.noaa.gov/.

6 Toxic Hazards

6.1 Learning Objectives

The learning objectives of this chapter:

- Explain chemical toxicity hazards,
- Identify the pathways for toxics to enter the human body,
- Explain exposure limits, and
- Identify where to find resources for chemical toxicity data.

6.2 Incident: Methyl Isocyanate Release Bhopal, India, 1984

6.2.1 Incident Summary

Just after midnight on December 3, 1984, a pesticide plant in Bhopal, India released approximately 40 metric tons of methyl isocyanate (MIC) into the atmosphere. The incident was a catastrophe; the exact numbers are in dispute; however, lower range estimates suggest at least 3,000 fatalities, and injuries estimates ranging from tens to hundreds of thousands. The impacted area is shown in Figure 6.1. The event occurred when water contaminated a storage tank of MIC which resulted in the release of a large toxic cloud.

Key Points:

Process Safety Culture – Culture is about what you do when no one is watching. When the culture degrades to the point that mechanical integrity and resources are in disrepair, it may be time to stop the operation.

Hazard Identification and Risk Analysis – As Trevor Kletz said, "what you don't have, can't leak". (Kletz) Do you really need that chemical in that quantity?

Management of Change – Multiple layers of protection only work if they are functional. Removal of any layer should be subject to management of change oversight.

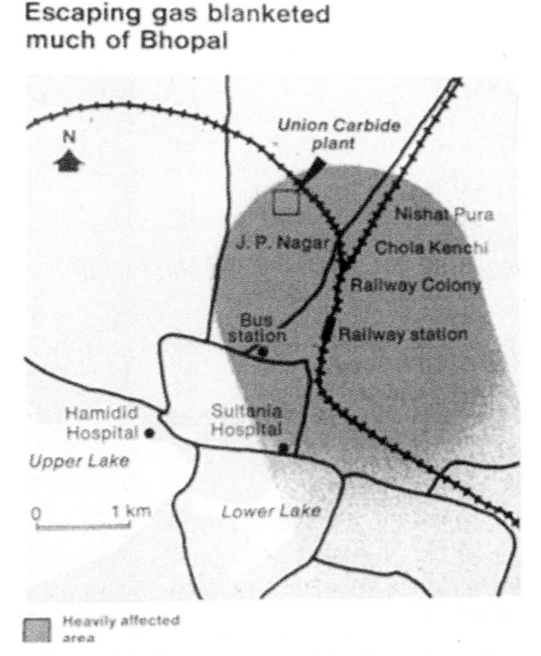

Figure 6.1. Overview map of the Bhopal vicinity (C&EN 1984)

6.2.2 *Detailed Description*

MIC is a flammable and highly toxic liquid. It is also water reactive, with the reaction being very exothermic. Water entered the MIC storage tank, and an exothermic reaction did occur. Investigators are not certain how the water got into the tank. Several theories have been developed about exactly what happened. One is that water entered the tank through a common vent line from a source over a hundred meters away. Another is that water was deliberately introduced by a disgruntled employee. Other theories include the idea that water entered from the scrubber over time because the tank was not pressurized or that the hose connections for water and nitrogen were confused with each other (Macleod 2014).

Whatever the initial source of the water contamination, several systems failures that could have mitigated the consequences of the event occurred. See Figure 6.2. In the figure RVVH is relief valve vent header and PVH is process vent header.

- Pressure gauges and a high temperature alarm which could have warned of an exothermic reaction initiation.
- A refrigeration system that cooled the liquid MIC was shut down and the refrigerant sold to save money. This could have removed heat from the reaction to prevent or reduce the amount of MIC that boiled up.
- The relief vents of the MIC tank were directed to a scrubber that could have absorbed a portion of the MIC; however, the vent gas scrubber was turned off.

- The scrubber was vented to a flare which could have burned the MIC; however, it was disconnected from the process while corroded pipework was being repaired.
- A fixed water curtain designed to absorb MIC vapors did not reach high enough to reach the gas cloud.

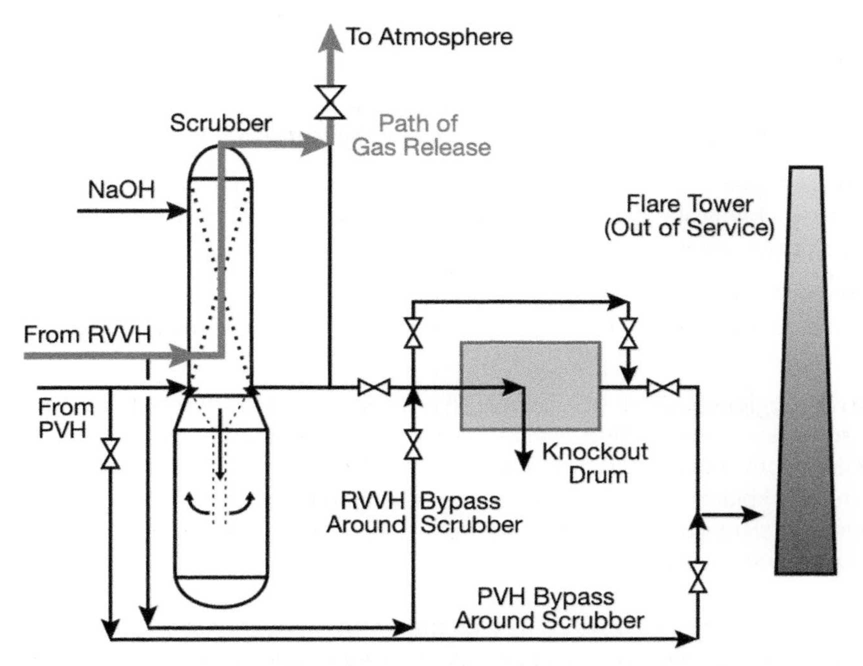

Figure 6.2. Emergency relief effluent treatment with scrubber and flare tower in series (CEP)

6.2.3 Lessons

Process Safety Culture. The plant in Bhopal was running under severe cost pressures. Not only were safety systems shut down to save money, but maintenance of the plant itself was cut and the plant was in disrepair. There plant had also cut staffing and training. These issues are evidence of a culture that prioritized cost over process safety.

Hazard Identification and Risk Analysis (HIRA). Inherent safety is an approach to process safety that emphasizes reducing or eliminating hazards as opposed to controlling them. Bhopal is an example of how the inherent safety strategy called minimization (the others are substitution, moderation, and simplification) could have reduced the consequences of an incident. The minimization strategy can be summed up by the saying "what you don't have, can't leak". MIC was an intermediate in the process to make the pesticide SEVIN™. The Bhopal plant had three 57,781 liter (15,000 gallon) MIC tanks. Normally, large intermediate tanks provide flexibility in a chemical process. In the case of MIC, however, a large inventory meant an increased hazard. The key lesson here is to assess the hazards from the inventories of hazardous chemicals in a HIRA before deciding how much, if any, to store.

Management of Change. Disabling of protective systems must be covered by a management of change review. The Bhopal plant was designed with several protection layers against an MIC release. No adjustments were made in the MIC storage tank protection strategy, such as increased monitoring or a reduction in the amount stored, as the layers were removed.

6.3 Toxins and Pathways

Many chemicals have toxic properties. The chemicals may be toxic to humans, to the environment, or both. Following are a few definitions key to discussion of toxic chemicals.

Toxicity. The quality, state, or degree to which a substance is poisonous and/or may chemically produce an injurious or deadly effect upon introduction into a living organism. (CCPS Glossary)

Toxicology. A science that deals with poisons and their effect and with the problems involved (such as clinical, industrial, or legal problems). (MW)

Industrial Hygiene: Industrial Hygiene (IH) is a science and art devoted to the anticipation, recognition, evaluation, control, and confirmation of protection from those environmental factors or stresses arising in or from the workplace which may cause sickness, impaired health and wellbeing, or significant discomfort among workers or among citizens of the community. (AIHA)

In Chapter 1 process safety was differentiated from occupational safety in terms of the potential impact on numbers of individuals, as opposed to one or two. Here, on the topic of toxicity, process safety is differentiated from occupational safety in that process safety events are acute whereas occupational safety is seen as chronic illness. Acute effects occur immediately on exposure or soon after. Chronic effects occur after repeated exposures over many months or years. Both acute and chronic effects can include mild health impacts, impaired function, irreversible health impacts, and death.

Toxins can impact people, and other living organisms, through 4 pathways as shown in Figure 6.3. Inhalation and absorption are the most common routes in industrial settings.

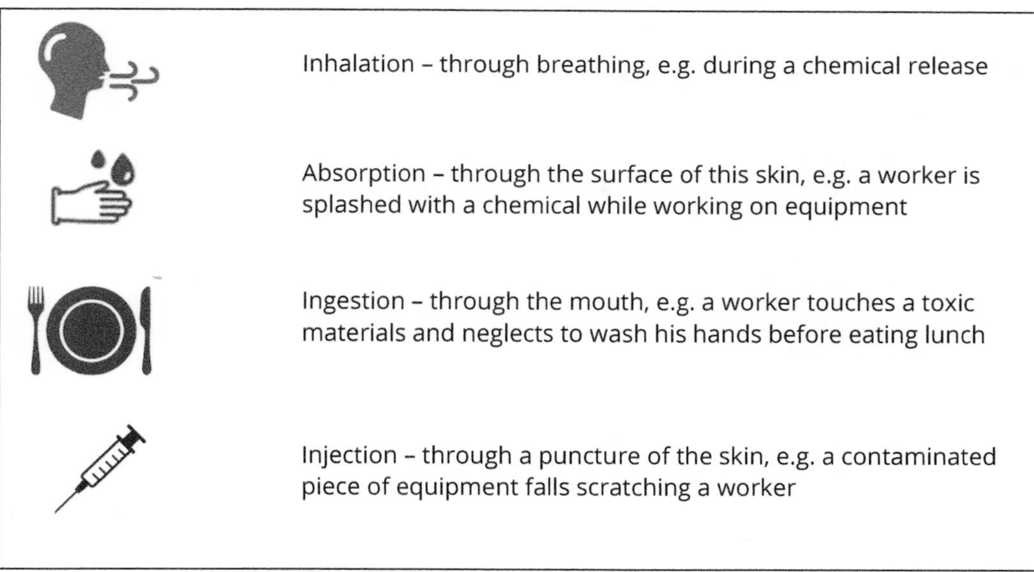

Figure 6.3. Toxic pathways

Toxins can have acute effects where symptoms develop rapidly to a critical level and chronic effects where symptoms develop slowly over a period of time. Example exposures are provided in Table 6.1.

Table 6.1. Example chemical exposure effects
(CDC)

Chemical	Acute Effect	Chronic Effect
Aniline	Impairs the blood's ability to transport oxygen which causes skin to turn blue	
Asbestos		Lung cancer and mesothelioma
Benzene	Irritating to skin, eyes, respiratory tract	May cause central nervous system damage, anemia, and leukemia
Hydrogen Cyanide	Eye irritation, headache, vomiting at lower levels to profound cardiovascular and respiratory effects at higher level, potentially fatal	
Lead		Memory loss, irritability, insomnia, depression, anorexia
Nitrogen gas	Lightheadedness at moderate levels, immediate asphyxiation fatality at high levels	
Phenol	Nausea, sweating, arrhythmia. Coma and seizures can occur up to 18 hours after exposure	
Phosgene	Irritant to skin, eyes, respiratory tract, pulmonary edema up to 24 hours after exposure, potentially fatal	

6.4 Exposure and concentration limits

The impact a toxin is dependent on both the toxic properties of the chemical and the duration of the exposure. A person can tolerate a certain amount of a toxin for a certain period of time and this varies for both the person and the toxin. For example, a small amount of alcoholic beverage can be tolerated with little effect; however, a large amount in a short time may cause impairment or alcohol poisoning. The same amount of alcohol consumed by a young healthy person and an older adult with preexisting health conditions can cause different effects. The difference in response can be due to age, weight, diet, health, and other factors.

Toxic hazards can be managed by identifying chemicals with toxic properties, understanding what concentration level can cause impacts, and managing potential exposure. Toxicologists have conducted testing to determine concentration levels at which health impacts occur. Exposure limits have been established by several organizations. They define a concentration level above which a human will have defined health impacts.

In the US, the Emergency Response Planning Guideline (ERPG) concentration values have become broadly accepted within industry and government. ERPG values have been developed for approximately 150 chemicals. The three ERPG levels are defined as follows.

- ERPG-1: the maximum airborne concentration below which nearly all individuals could be exposed for up to one hour without experiencing effects other than mild transient adverse health effects or perceiving an objectionable odor.
- ERPG-2: the maximum airborne concentration below which nearly all individuals could be exposed for up to one hour without developing serious health effects that could impair their ability to take protective action.
- ERPG-3: the maximum airborne concentration below which nearly all individuals could be exposed for up to one hour without developing serious health effects that could impair their ability to take protective action.

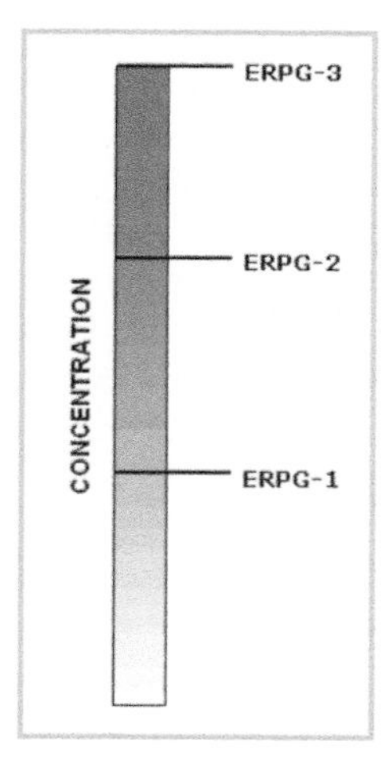

Figure 6.4 Emergency Response Planning Guideline (ERPG) concentration (NOAA)

Where ERPG criteria are not available, other criteria may be used. These may include the following: The ERPG, AEGL, and TEEL are all listed in the Preventive Action Criteria database listed in Section 6.7.

- Acute Exposure Guideline Levels (AEGL) values published by the U.S. Environmental Protection Agency (EPA). AEGLs are applicable to exposures ranging from 10 minutes to 8 hours.
- Temporary Emergency Exposure Limit (TEEL) values developed by U.S. Department of Energy.

ERPG and AEGL values focus on short term exposures relevant to process safety and emergency response. Other more appropriate parameters are used for longer term worker exposures, as might occur during spill cleanup operations including threshold limit value (TLV). These are set by the American Conference on Governmental Industrial Hygienists. Permissible exposure limits (PEL), set by OSHA, are used to evaluate long-term worker exposure to chemicals.

It should be noted that exposure levels mentioned in the previous paragraphs are based on defined time exposure time periods. In a workplace; however, it may be important to know at what level chemicals may have an immediate effect or at what levels no adverse effects are expected.

Immediately Dangerous to Life or Health (IDLH) means any condition that would interfere with an individual's ability to escape unaided from a permit space and that poses a threat to life or that would cause irreversible adverse health effects. (NIOSH) IDLH values are based on a 30-minute exposure.

Threshold Limit Value (TLV) - The exposure level of a chemical substance to which a worker can be exposed day after day for a working lifetime without adverse effects. (ACGIH)

A more recently developed system for classifying chemicals was developed by the United Nations. (UN) *The Globally Harmonized System of Classification and Labelling of Chemicals (GHS)* was developed over decades with the support of many countries and stakeholder organizations with expertise from toxicology to fire protection. The intent is to have a single, globally harmonized system to address classification of chemicals, labels, and safety data sheets. This harmonization supports hazard communication and facilitates international trade in chemicals. The GHS classifies materials by:

- physical hazards, including flammability and reactivity,
- health hazards, including toxicity and carcinogenicity, and
- environmental hazards, including to the aquatic environment.

To make it more complex, multiple toxins may be present in a workplace and thus may be involved in a single exposure incident. Each one may have different concentration criteria. It is not appropriate to assume that the lowest concentration criteria will apply to the chemical mixture. This may underestimate the hazard. The U.S. Department of Energy (Baskett 1999) and others have recommended an "additive" approach (which is similar to Le Chatelier's rule).

Chemicals can cause other health impacts. in addition to toxic properties, for example, by displacing air and thus reducing the oxygen level. Air is normally 21% oxygen. Effects of oxygen depletion below that level are listed in Table 6.2.

Table 6.2. Effects of oxygen depletion
(HSE)

Percent of Oxygen in Air	Symptoms
21-20	Normal
18	Night vision begins to be impaired
17	Respiration volume increase, muscular coordination diminishes, attention and thinking clearly requires more effort
12 to 15	Shortness of breath, headache, dizziness, quickened pulse, effort fatigues quickly, muscular coordination for skilled movement lost
10 to 12	Nausea and vomiting, exertion impossible, paralysis of motion
6 to 8	Collapse and unconsciousness occur
6 or below	Death in 6 to 8 minutes

6.5 Toxic Incident Prevention and Mitigation

The potential for a toxic exposure can be reduced or eliminated. Inherently safer design principles (refer to section 10.7.2) are the preferred way to prevent toxic incidents. The inventory of toxic material can be minimized or a less toxic material substituted. Material of construction can minimize leak points. Construction techniques (welded versus screwed piping) can reduce the likelihood of release. Double containment is often considered (e.g. annular pipe, double walled tanks, sealless pumps). Inspection, testing and preventive maintenance ensure equipment readiness for operation.

Where an exposure potential exists, the exposure can be minimized through applying proper industrial hygiene practices, the use of personal protective equipment, and the design of equipment. Equipment such as local exhaust or ventilation systems can prevent or reduce exposure concentrations to permissible levels. If a toxic release incident does occur, shelter-in-place systems are appropriate.

Shelter-in-place - A process for taking immediate shelter in a location readily accessible to the affected individual by sealing a single area (an example being a room) from outside contaminants and shutting off all HVAC systems. (CCPS Glossary)

Safe Haven - A building or enclosure that is designed to provide protection to its occupants from exposure to outside hazards. (CCPS Glossary)

It is necessary to design and install equipment, in advance of a toxic release, to support emergency response, including detection systems, alarm and communication systems, and potentially a shelter-in-place.

Detection systems detect the presence of or a specified concentration level a toxic chemical. Detection can include personal detectors worn by individuals working in areas handling specific hazardous chemicals such as hydrogen sulfide monitors or phosgene badges. Fixed detectors can be located at a potential leak source such as a pump moving a hazardous chemical. Fixed detectors can also be in the form of line-of-sight detectors designed to detect a gas between two, distant points, for example, along a side of a process unit.

Alarm or other notification systems are typically activated when a fixed detection system detects a predetermined chemical concentration level. Personnel should be trained to take the appropriate action when notified by the alarm. Communication systems may extend into the community to advise people to shelter in place.

A safe haven is needed for employees where toxic concentration levels may be dangerously high. A safe haven design includes an air-tight or pressurized facility and provides sufficient atmosphere for the number of occupants for the duration needed.

6.6 What a New Engineer Might Do

As with flammable, explosive, and reactive materials, an engineer should understand the toxic properties of materials being handled or processed. Tools are provided in the next section that can assist in identifying toxic properties and what exposure levels are of concern.

It is also important that an engineer follow the safety guidelines for toxics in terms of their own protection. This includes handling chemicals in ways defined by safe operating practices and using required personal protective equipment.

A new engineer may be asked to identify toxic hazards or to use a risk analysis, such as a What-If analysis (see section 12.3.3) to identify toxic release scenarios. They may apply inherently safer design strategies to minimize the risk due to toxics. They could be involved in the design of mitigation devices such as scrubbers, incinerators, and thermal oxidizers. They be asked to ship a material or manage an incoming shipment. In these cases, communication is important. The GHS provides classification for toxics such that there should not be any miscommunication during the shipment or in the labeling of chemicals. A new engineer could be asked to use toxic release modeling to understand a potential incident and create an emergency response plan.

A new engineer can benefit from reviewing the CSB investigations and videos relevant to this chapter as listed in Appendix G.

6.7 Tools

Resources necessary to understand toxic hazards include toxicological data resources and quantitative methods to determine the risk.

Toxicological data. Toxicological data can be found in many of the same tools and documents listed in Sections 5.8 for reactive chemicals data. These data sources provide the process safety information valuable in understanding the hazards of these chemicals. Chapter 13 provides guidance on determining the consequence effects and potential outcome of a toxic chemical release.

- Cameo Chemicals Database contains basic chemical descriptions, ID numbers, potential hazards, response recommendations, physical properties, and regulatory information. It is available at https://cameochemicals.noaa.gov. (NOAA)
- Emergency Response Planning Guideline (ERPG) values can be found at: https://www.aiha.org/get-involved/aiha-guideline-foundation/erpgs
- Protective Action Criteria - A database containing AEGL, ERPG, and TEEL values for more than 3000 chemicals can be found at: https://energy.gov/ehss/protective-action-criteria-pac-aegls-erpgs-teels-rev-29-chemicals-concern-may-2016. (PAC)
- BASF Medical Guideline provides medical care information for exposure to a large number of chemicals. It is intended for use by emergency responders and medical professionals and can be found at https://collaboration.basf.com/portal/basf/en/dt.jsp?setCursor=1_1032616.

Methods to address toxic impacts in risk assessment are described in the following.

Crowl and Louvar, *Chemical Process Safety Fundamentals with Applications, 4th Edition*. This book provides a compilation of many process safety topics. It links academic concepts to industrial process safety. (Crowl 2019)

CCPS *Guidelines for Chemical Process Quantitative Risk Analysis*. This guideline describes the application of quantitative risk analysis in process safety, including toxic risk analysis. It describes the identification of incident scenarios, evaluation of the risk by defining probability of failure and the consequence impacts. In understanding the risk, then risk reduction strategies may be evaluated. (CCPS 1999)

6.8 Summary

Chemicals can pose toxic hazards in addition to the fire and reactive hazards described in previous chapters. Various organizations have established lists of toxic chemicals and exposure limits for those chemicals. This assists in understanding what level concentration may impact people and the environment.

6.9 Other Incidents

This chapter began with a detailed description of the Bhopal incident. Other incidents involving toxic chemicals include the following.

- ICMESA Dioxin Release, Seveso, Italy, 1976
- Union Carbide MIC release, Bhopal, India, 1984
- Marathon Oil Refinery HF Release, Texas City, Texas, US, 1987
- Motiva Enterprises Sulfuric Acid Tank Failure, Delaware City, Delaware, US, 2001
- Georgia Pacific Hydrogen Sulfide, Pennington, Alabama, US, 2002
- CITGO HF Release, Corpus Christi, Texas, US, 2009
- DuPont Phosgene Release, Belle, West Virginia, US, 2010
- Millard Refrigeration Ammonia Release, Theodore, Alabama, US, 2010
- DPC Chlorine Release, Festus, Missouri, US, 2012

- Hube HF Release, Gumi, South Korea, 2012
- DuPont MMA Release, LaPorte, Texas, US, 2014
- Freedom Ind. Chemical Spill, Charleston, West Virginia, US, 2014
- Vizag Gas Leak, Visakhapatnam, India, 2020

6.10 Exercises

1. List 3 RBPS elements evident in the MIC release in Bhopal, India summarized at the beginning of this chapter. Describe their shortcomings as related to this accident.
2. Considering the MIC release in Bhopal, India, what actions could have been taken to reduce the risk of this incident?
3. List the four pathways through which toxic materials can impact people.
4. What is the exposure duration used for the ERPG concentration values? And for the IDLH concentration value?
5. What is the ERPG-2 and ERPG-3 for Chloroacetyl chloride? What is your reference?
6. What is an equivalent ERPG-3 for a vapor mixture of 0.2 mole fraction Chloroacetyl chloride and 0.8 mole fraction hydrogen chloride? Show your work.
7. What is the concentration yielding the same dose for a 15-minute exposure duration of methyl isocyanate as the ERPG-3 of 1.5 ppm by volume assuming the dose exponent, n, is 0.7? Show your work.
8. In the Freedom Industries incident in Charleston, West Virginia in 2014, what chemicals were released. How are they toxic to humans? How was the community at risk of being exposed to these chemicals?
9. What means might be used to protect humans from exposure to toxic materials?

6.11 References

ACGIH, American Conference of Governmental Industrial Hygienists, https://www.acgih.org/tlv-bei-guidelines/tlv-chemical-substances-introduction.

AIHA, American Institute of Industrial Hygienists, https://www.aiha.org/

BASF, Medical Guidelines, https://collaboration.basf.com/portal/basf/en

Baskett 1999, Craig, Baskett and et. al., "Recommended Default Methodology for Analysis of Airborne Exposures to Mixtures of Chemicals in Emergencies", *Applied Occupational and Environmental Hygiene,* vol. 14.

C&EN 1984, "India's Chemical Tragedy: Death toll at Bhopal still rising", *Chemical and Engineering News*, American Chemical Society, December.

CCPS Glossary, "CCPS Process Safety Glossary", Center for Chemical Process Safety, https://www.aiche.org/ccps/resources/glossary.

CCPS 1999, Guidelines for Chemical Process Quantitative Risk Analysis, Second Edition, Center for Chemical Process Safety, John Wiley & Sons, Hoboken, N.J.

CDC, Centers for Disease Control and Prevention, https://www.atsdr.cdc.gov/mmg/

CEP 2014, Willey, R.J. "Consider the Role of Safety Layers in the Bhopal Disaster", *Chemical Engineering Progress*, American Institute of Chemical Engineers, New York, NY December.

Crowl 2019, Daniel A. Crowl and Joseph F. Louvar, *Chemical Process Safety, Fundamentals with Applications 4th Edition*., Pearson, NY.

HSE, *"Methods of approximation and determination of human vulnerability for offshore major accident hazard assessment",* https://www.hse.gov.uk/foi/internalops/hid_circs/technical_osd/spc_tech_osd_30/spctecosd30.pdf.

Kletz 1978, What you Don't Have, Can't Leak", *Chemistry & Industry*, London, United Kingdom.

Macleod 2014, Macleod, Fiona, "Impressions of Bhopal", *Loss Prevention Bulletin*, Vol. 240, December.

MW, Merriam-Webster, https://www.merriam-webster.com/dictionary/toxicology.

NIOSH, National Institute of Occupational Safety and Health, https://www.cdc.gov/niosh/idlh/default.html.

NOAA, National Oceanic and Atmospheric Administration, https://cameochemicals.noaa.gov.

NOAA a, National Oceanic and Atmospheric Administration, https://response.restoration.noaa.gov/oil-and-chemical-spills/chemical-spills/resources/emergency-response-planning-guidelines-erpgs.html.

PAC, Protective Action Criteria, https://energy.gov/ehss/protective-action-criteria-pac-aegls-erpgs-teels-rev-29-chemicals-concern-may-2016.

UN, "Globally Harmonized System Of Classification And Labelling Of Chemicals (GHS) Sixth Edition", United Nations, New York and Geneva, 2011, https://www.un-ilibrary.org/transportation-and-public-safety/globally-harmonized-system-of-classification-and-labelling-of-chemicals-ghs-sixth-revised-edition_591dabf9-en.

Willey 2006, Willey, R. et. al., "The Accident in Bhopal: Observations 20 Years Later", prepared for American Institute of Chemical Engineers Spring Meeting, April 24 -26.

Shutterstock

- *Royalty-free* stock vector *ID: 1215841048 Breath icon.*
- *Royalty-free* stock vector *ID: 1478935706 Set of Medical icons*
- *Royalty-free* stock vector *ID: 240907861*
- *Royalty-free* stock vector *ID: 1546583303Syringe vector icon isolated on white*

7
Chemical Hazards Data Sources

7.1 Learning Objectives

The learning objective of this chapter is:

- Identify sources of chemical hazards data and understand the data provided.

7.2 Incident: Concept Sciences Explosion, Allentown, Pennsylvania, 1999

7.2.1 Incident Summary

"At 8:14 pm on February 19, 1999, a process vessel containing several hundred pounds of hydroxylamine (HA) exploded at the Concept Sciences, Inc. (CSI), production facility near Allentown, Pennsylvania. Employees were distilling an aqueous solution of HA and potassium sulfate, the first commercial batch to be processed at CSI's new facility. After the distillation process was shut down, the HA in the process tank and associated piping explosively decomposed, most likely due to high concentration and temperature.

Four fatalities resulted, including CSI employees and a manager of an adjacent business. Two CSI employees survived the blast with moderate-to-serious injuries. Four people in nearby buildings were injured. Six firefighters and two security guards suffered minor injuries during emergency response efforts.

The production facility was extensively damaged (Figure 7.1). The explosion also caused significant damage to other buildings in the Lehigh Valley Industrial Park and shattered windows in several nearby homes." (CSB 2002)

Key Point:

Hazard Identification and Risk Analysis - Hazard review methodologies need to be appropriate to the hazards being managed. A high hazard warrants a detailed review.

7.2.2 Detailed Description

Pure HA is a compound with the formula NH_2OH. Solid HA consists of colorless or white crystals that are unstable and susceptible to explosive decomposition and explodes when heated in air above 70°C (158°F). HA is usually sold as a 50 wt. % or less solution in water. The Chemical Safety Board Investigation report quoted CSI's safety data sheet (SDS) as stating "Danger of fire and explosion exists as water is removed or evaporated and HA concentration approaches levels in excess of about 70%". HA can be ignited by contact with metals and oxidants.

Figure 7.1. Damage to Concept Sciences Hanover facility
Courtesy Tom Volk, The Morning Call
(CSB 2002)

CSI developed a four step process to make 50% HA:

1. Reaction of HA sulfate and potassium hydroxide to produce a 30 wt.% HA and potassium sulfate aqueous slurry:
2. Filtration of the slurry to remove precipitated potassium sulfate solids.
3. Vacuum distillation of HA from the 30 wt. % solution to separate it from the dissolved potassium sulfate and produce a 50 wt. % HA distillate.
4. Purification of the distillate through ion exchange cylinders.

The distillation process is shown in Figure 7.2. The charge tank was a 2,500 gallon (9.5 m3) tank. In the first step of the distillation, a pump circulated 30 wt. % HA to the heating column, which is a vertical shell and tube heat exchanger.

The HA was heated under vacuum by 49°C (120°F) water. Vapor was drawn off to the condenser and collected in the forerun tank and concentrated HA was returned to the charge tank. When the concentration in the forerun tank reached 10 wt. %, it was then collected in the final product tank. At the end of the first step of the distillation, the HA in the charge tank was at 80 - 90 wt. % HA. At that point, the charge tank was supposed to be rinsed with 30 wt. % HA.

The first distillation done by CSI began on Monday afternoon, February 15, 1999. By Tuesday evening, the HA in the charge tank was approximately 48 wt. %. At that time, the process was shut down for maintenance after it was discovered that water had leaked into the charge tank through broken tubes in the heater column. The distillation restarted on Thursday afternoon and shut down at 11:30 PM. The distillation restarted late on Friday morning, after a feed line to the heater column was replaced. By about 7:00 PM, the concentration in the charge tank had reached 86 wt. % HA. The distillation was shut down at 7:15 PM. A manufacturing supervisor was called on Friday evening and arrived at the facility about 15 minutes before the explosion occurred, at 8:14 PM.

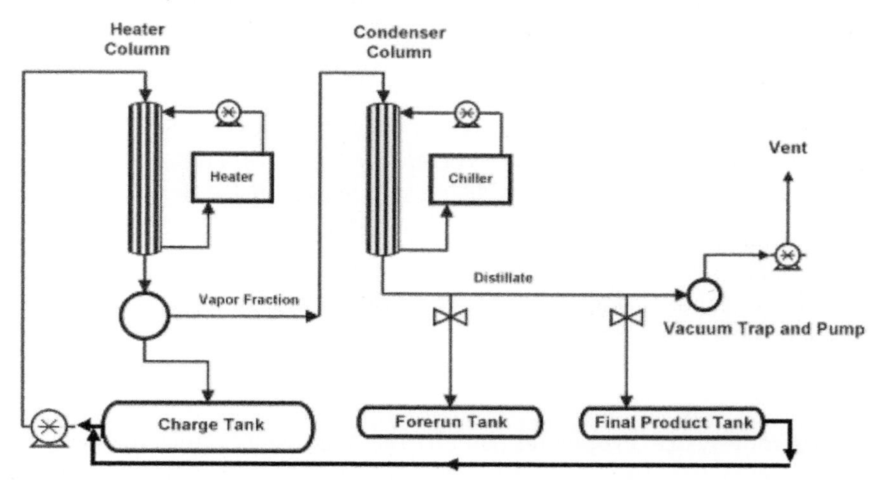

Figure 7.2. Simplified process flow diagram of the CSI HA vacuum distillation process (CSB 2002)

It is not known what initiated the explosion. Possibilities include: "addition of excessive heat to the distillation system, physical impacts from partial or total collapse of the glass equipment, or inadvertent introduction of impurities. Friction may have heated the mixture as it passed through the pump that supplied the heating column." (CSB, 2002)

CSI developed a process for making 50 wt. % HA that would normally cause HA to be concentrated to a level that was inherently unstable and subject to exothermic decomposition.

7.2.3 Lessons

Process Knowledge Management. Although management in CSI learned of the hazards of HA during pilot plant operation, the knowledge was not used in the design of the process, or in the hazard reviews that were conducted. CSI also did not review available literature about HA, which also would have shown that HA is subject to exothermic decomposition and had an explosive force equivalent to TNT. Nor did CSI attempt to do any testing to define the magnitude of the hazard of HA. Finally, CSI did not create engineering drawings or detailed operating procedures.

Hazard Identification and Risk Analysis. The hazard identification method used should be appropriate to the hazards being managed. The study should be conducted in a detailed manner by a competent team. CSI used a "What-If" review which was reported in a one-page document. The hazard review did not address the "prevention or consequences of events that could have triggered an explosion of high concentrations of HA". (CSB 2002). CSI did not implement any of the recommendations from the hazard review they did conduct.

7.3 Chemical Hazards Data

The previous chapters discussed chemical hazards including flammability, reactivity, and toxicity. Managing these hazards starts with identifying chemical properties and communicating them to those who are handling or processing them. This section covers some of the sources for chemical hazards data and a couple of commonly used systems for communication of chemical hazards.

7.3.1 Chemical Hazards Data Sources

Cameo Chemicals Database. This database contains basic chemical descriptions, ID numbers, potential hazards, response recommendations, physical properties, and regulatory information. It is available at https://cameochemicals.noaa.gov. (NOAA)

Chemical Reactivity Worksheet (CRW). This is a free software program providing extensive process safety information required to understand the hazards associated with the inadvertent and intentional mixing of reactive chemicals. This includes the chemical reactivity of thousands of common hazardous chemicals, compatibility of absorbents, and suitability of materials of construction in chemical processes. Refer to Section 5.8.3 for further details. (CCPS)

Design Institute for Physical Properties (DIPPR). DIPPR is an AIChE Technology Alliance. The DIPPR database includes thermo-physical properties. The database is being populated from an earlier DIPPR research program and from on-going experimental work. It is available at https://www.aiche.org/dippr. (AIChE) The AIChE offers a free version of the DIPPR database to students and faculty. Registration is : Register - DIPPR801.

NIOSH Pocket Guide to Chemical Hazards. This guide provides industrial hygiene data for hundreds of chemicals. This is available in print or online at https://www.cdc.gov/niosh/npg/default.html. It includes the following data. (NIOSH)

- Chemical names, synonyms, trade names, CAS, RTECS, and DOT ID and guide numbers
- Chemical structure/formula, conversion factors
- NIOSH recommended exposure limits (RELs)
- Occupational Safety and Health Administration (OSHA) permissible exposure limits (PELs)
- NIOSH immediately dangerous to life and health values (IDLHs)
- Physical description and chemical and physical properties of agents
- Measurement methods
- Personal protection and sanitation recommendations
- Respirator selection recommendations
- Incompatibilities and reactivities of agents
- Exposure routes, symptoms, target organs, and first aid information

NIH WebWISER. WISER is a system provided by the National Institutes of Health designed to assist emergency responders in hazardous material incidents. It provides data including physical characteristics, human health impact, and containment and suppression advice. It is available at https://webwiser.nlm.nih.gov. (NIH)

Phast Process hazard analysis software. Phast is a widely used commercial software from DNV GL. It includes an extensive chemicals library and facilitates the handling of chemical mixtures. Phast is discussed in Chapter 13 as it supports release, dispersion, and consequence modeling. It is available at https://www.dnv.com/software/services/phast/index.html. (DNV)

7.3.2 Chemical Hazards Communication

Safety Data Sheet (SDS). The Safety Data Sheet (SDS) came to wide use in the U.S. as a result of the OSHA Hazard Communication Standard (OSHA a) which required that a chemical manufacturer, distributor, or importer provide an SDS for each hazardous chemical. (OSHA b)

The SDS is used to communicate important information about a chemical including properties, health hazards, and emergency response procedures. An SDS for each chemical is to be available for workers where the chemicals are handled or processed. SDSs are often available in multiple languages. SDSs are a good source of process safety information but they are not always perfect and may not contain all the information needed, for example fire and explosion data for combustible powders. It is good to research and compare sources for any given data point.

A challenge has been that different regulations had different requirements for the content and organization of the SDS. With the volume of international chemical trade, the UN *Globally Harmonized System of Classification and Labeling of Chemicals (GHS)* was developed to create a single, consistent system. The length of an SDS is dependent on the chemical data. It should be sufficiently long to include all the required information. The SDS should include the contents, in the order shown, in Table 7.1. (UN GHS and OSHA c)

The transportation classification used in Section 14 is used as primary source in understanding the class of a chemical used to classify a process safety incident. (Refer to Chapter 9.)

Table 7.1. Safety data sheet sections and content

Section		Content
No.	Title	
1	Identification	product identifier; manufacturer or distributor name, address, phone number; emergency phone number; recommended use; restrictions on use
2	Hazard identification	all hazards regarding the chemical; required label elements
3	Composition/information on ingredients	information on chemical ingredients; trade secret claims
4	First-aid measures	important symptoms/effects, acute, delayed; required treatment
5	Fire-fighting measures	suitable extinguishing techniques, equipment; chemical hazards from fire
6	Accidental release measures	emergency procedures; protective equipment; proper methods of containment and cleanup
7	Handling and storage	precautions for safe handling and storage, including incompatibilities
8	Exposure controls/personal protection	OSHA's Permissible Exposure Limits (PELs); ACGIH Threshold Limit Values (TLVs); and any other exposure limit used or recommended by the chemical manufacturer, importer, or employer preparing the SDS where available as well as appropriate engineering controls; personal protective equipment (PPE)
9	Physical and chemical properties	the chemical's characteristics
10	Stability and reactivity	chemical stability and possibility of hazardous reactions
11	Toxicological information	routes of exposure; related symptoms, acute and chronic effects; numerical measures of toxicity
12	Ecological information	
13	Disposal considerations	
14	Transport information	
15	Regulatory information	
16	Other information	includes the date of preparation or last revision

NFPA 704 *Standard System for the Identification of the Hazards of Materials for Emergency Response*. This system is used by emergency responders to quickly identify the risks of hazardous materials involved in an emergency. This helps inform the response method, the materials used in the response, and the personal protective equipment that may be required. The NFPA 704 graphic uses a diamond as shown in Figure 7.3.

The four corners of the diamond represent health (blue), flammability (red), reactivity (yellow) and special hazards (white) with numbers or symbols in each corner indicating the severity or type of special hazard. A summary is provided in Table 7.2. Full details are contained in the NFPA 704 standard.

Figure 7.3. Example NFPA 704
(OSHA f)

Table 7.2. NFPA 704 hazards and rating

	Flammability (red)
0	Materials that will not burn under typical fire conditions
1	Materials that require considerable preheating, under all ambient temperature conditions, before ignition and combustion can occur
2	Materials must be moderately heated or exposed to relatively high ambient temperature before ignition can occur
3	Liquids and solids (including finely divided suspended solids) that can be ignited under almost all ambient temperature conditions
4	Material will rapidly or completely vaporize at normal atmospheric pressure and temperature or if readily dispersed in air and will burn readily
	Health (blue)
0	Poses no health hazard, no precautions necessary and would offer no hazard beyond that of ordinary combustible materials
1	Exposure would cause irritation with only minor residual injury
2	Intense or continued but not chronic exposure could cause temporary incapacitation or possible residual injury
3	Short exposure could cause serious temporary or moderate residual injury
4	Very short exposure could cause death or major residual injury
	Reactivity (yellow)
0	Normally stable, even under fire exposure conditions, and is not reactive with water
1	Normally stable, but can become unstable at elevated temperatures and pressures
2	Undergoes violent chemical change at elevated temperatures and pressures, reacts violently with water, or may form explosive mixtures with water
3	Capable of detonation or explosive decomposition but requires a strong initiating source, must be heated under confinement before initiation, reacts explosively with water, or will detonate if severely shocked
4	Readily capable of detonation or explosive decomposition at normal temperatures and pressures
	Special hazard (white)
OX	Oxidizer, allows chemicals to burn without an air supply
W	Reacts with water in an unusual or hazardous manner
SA	Simple asphyxiant gas
	Other symbols that are not included in NFPA 704 are sometimes used including for strong acids and bases, biohazards, radioactivity, and cryogenics

Labels and Pictograms. A key communication aspect of the UN *Globally Harmonized System of Classification and Labeling of Chemicals (GHS)* is the labelling and pictogram requirement. Labels shall include: (OSHA d)

- Name, Address and Telephone Number
- Product Identifier - e.g. chemical name
- Signal Word - e.g. "danger" or "warning"
- Hazard Statement(s) - nature of hazard e.g. flammable or toxic
- Precautionary Statement(s) - recommended measures to prevent adverse effects, and
- Pictogram(s)

Pictograms are graphic symbols used worldwide as shown in Figure 7.4. They quickly communicate chemical hazard information without the use of words which is helpful in international transportation.

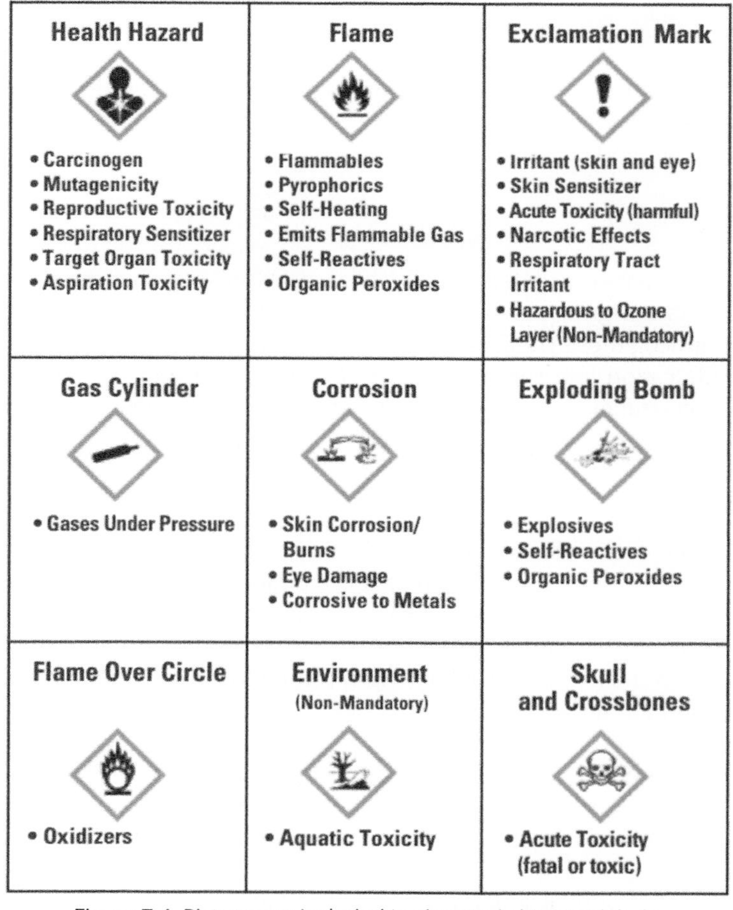

Figure 7.4. Pictograms included in chemical shipping labels (OSHA e)

Material Handling Guides. Organizations such as the American Chemistry Council and chemical manufactures publish guides on safe handling of chemicals. These are typically short documents that address the handling, storage, transportation, and compatibility with other chemicals and materials of construction. These are available through an internet search on the chemical name or manufacturer.

7.4 What a New Engineer Might Do

A new engineer may work closely with a chemist when dealing with chemicals to understand the hazards associated with the pure substances, and mixtures. They may also be involved in handling and processing chemicals or designing systems and procedures for others who handle and process chemicals. In either case, an engineer has a responsibility to understand and manage the hazards associated with chemicals. This includes researching chemical data and communication of the chemical hazards using the sources and systems identified in this chapter. Through this, the engineer will protect themselves, as well as others working with the chemicals.

7.5 Tools

The chemical hazards data sources and communication systems discussed in this chapter are themselves the tools that support the identification and understanding of chemical hazards.

7.6 Summary

It is imperative that engineers understand the hazards associated with the chemicals they are including in process designs. It is also imperative that they communicate these hazards to those who are handling these chemicals in the workplace. Many data sources are available to support the identification and communication of chemical hazards. Many of these are now aligning their categorizations and communications with the UN *Globally Harmonized System of Classification and Labelling of Chemicals (GHS).* (UN)

7.7 Exercises

1. List 3 RBPS elements evident in the Concept Sciences explosion summarized at the beginning of this chapter. Describe their shortcomings as related to this accident.
2. Considering the Concept Sciences explosion, what actions could have been taken to reduce the risk of this incident?
3. What pictogram is used in the Globally Harmonized System (GHS) for potassium permanganate? For anhydrous zinc chloride? What pictograms are used for acute toxicity?
4. Is anhydrous ammonia a fire hazard or a toxicity hazard? Draw the NFPA 704 diamond for it.
5. For a small spill of boron trifluoride at night, how far downwind should people be protected?
6. The MIC release in Bhopal, India was summarized in Chapter 6. For a large release of MIC at night, to what distance downwind should people be protected?

7. Draw the NFPA 704 diamond and list the ERPG values for sulfuric acid.
8. Draw the NFPA 704 diamond and list the ERPG values for hydrofluoric acid.
9. Refinery alkylation units use either a smaller volume of hydrofluoric acid or a greater volume of sulfuric acid to achieve the same amount of product. Both are supplied to the refinery by truck. Which poses a greater risk, a sulfuric acid alkylation unit or a hydrofluoric acid alkylation unit?
10. A facility laboratory tests for chlorine as a contaminant in its product. The laboratory building is a converted old residential building with copper piping. Can you identify any risks? Name the sources on which you base this view.

7.8 References

AIChE, Design Institute for Physical Properties, https://www.aiche.org/dippr.

CCPS, Chemical Reactivity Worksheet 4.0, http://www.aiche.org/ccps/resources/chemical-reactivity-worksheet.

CSB 2002, "The Explosion at Concept Sciences: Hazards of Hydroxylamine, Concept Sciences, Hanover Township, PA. February 19, 1999", Report No. 1999-13-C-PA, U.S. Chemical Safety and Hazard Investigation Board, March.

NFPA 704, "Standard System for the Identification of the Hazards of Materials for Emergency Response", National Fire Protection Association, Quincy, MA, www.nfpa.org.

NIH, https://webwiser.nlm.nih.gov/getHomeData

NIOSH, "Pocket Guide to Chemical Hazards", https://www.cdc.gov/niosh/npg/default.html.

NOAA, National Oceanic and Atmospheric Administration, https://cameochemicals.noaa.gov.

DNV, Phast Process hazard analysis software,
https://www.dnvgl.com/software/services/phast/index.html.

OSHA a, Hazard Communication, 29 CFR 1910.1200, Occupational Safety and Health Administration, https://www.osha.gov/laws-regs/regulations/standardnumber/1910/1910.1200

OSHA b, OSHABrief, "Hazard Communication Standard: Safety Data Sheets", Occupational Safety and Health Administration, https://www.osha.gov/Publications/OSHA3514.html.

OSHA c, Hazard Communication, Occupational Safety and Health Administration, https://www.osha.gov/dsg/hazcom/index.html.

OSHA d OSHABrief "Hazard Communication: Labels and Pictograms", Occupational Safety and Health Administration, https://www.osha.gov/Publications/OSHA3636.pdf.

OSHA e, Quick Card, "Hazard Communication Standard Pictogram", Occupational Safety and Health Administration,
https://www.osha.gov/Publications/OSHA3491QuickCardPictogram.pdf.

OSHA f, OSHA Quick Card, "Comparison of NFPA 704 and HazCom2012 labels", Occupational Safety and Health Administration, https://www.osha.gov/Publications/OSHA3678.pdf.

UN GHS, "Globally Harmonized System Of Classification And Labelling Of Chemicals (GHS) Sixth Edition", United Nations, New York and Geneva, 2011, https://www.un-ilibrary.org/transportation-and-public-safety/globally-harmonized-system-of-classification-and-labelling-of-chemicals-ghs-sixth-revised-edition_591dabf9-en.

8 Other Hazards

8.1 Learning Objectives

The learning objectives of this chapter are:

- Identify hazards related to energy and technology, and
- Identify where to find resources for this hazard data.

8.2 Incident: Fukushima Daiichi Nuclear Power Plant Release, Japan, 2011

8.2.1 Incident Summary

On March 11, 2011, one of the largest recorded earthquakes occurred off the coast of Japan. This caused a tsunami that led to 1,500 fatalities, 6,000 injuries, and many missing people. The tsunami waves flooded the Fukushima Daiichi nuclear power plant, impacting all six units on site. In the following days, the units overheated, and radioactive material was released in a melt-down event exposing surrounding communities and the environment. (IAEA 2015) People were evacuated within 20 km (12.4 mi) of the site for years. No human fatalities were attributed directly to the incident; however, since the accident there has been reporting of significant increases in thyroid cancer. The Fukushima Nuclear Accident Independent Investigation Commission (NAIIC) called for reforms in both the electric power industry and the related government and regulatory agencies. (NAIIC 2012)

Key Points:

Stakeholder Outreach – Make sure you are all working toward the same goal – safety. It is good to have a positive working relationship with other stakeholders. Remember, though, just because someone says it is 'OK' does not mean that it is in the best interest of safety.

Process Safety Competency – This underpins many elements in most management systems. If process safety is strong, so will be the business management. If the understanding of process safety aspects of the process technology is weak, then decisions over time will degrade overall risk management.

Hazard Identification and Risk Analysis – How unlikely is it, really? Potential emergency events can seem unrelated. They should be analyzed to consider whether one can prompt another. Simply deciding an event is unlikely may result in design, procedures, and emergency response falling short.

8.2.2 Detailed Description

Following the oil crisis of the 1970's, Japan moved to diversify its power sources. By 2010, nuclear power generation provided 29% of the total power generation in Japan. Five nuclear power plants were located on the northeastern coast of Japan. Fukushima Daiichi was operated by Tokyo Electric Power Company (TEPCO). Refer to Figure 8.1.

The Fukushima Daiichi design used boiling water reactors. The reactors were a closed loop system. Water boiled in the reactor producing steam that drove turbines to generate electric power. The steam was then condensed using cold water from the ocean, and then fed back to the reactor again.

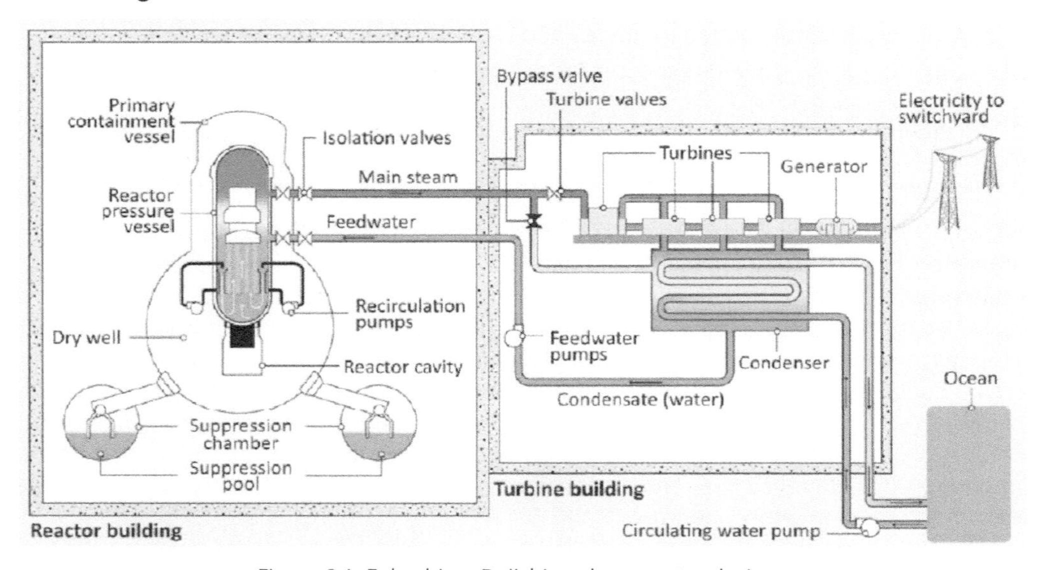

Figure 8.1. Fukushima Daiichi nuclear reactor design (IAEA 2015)

The Great East Japan Earthquake occurred at 4:46 PM. It was a magnitude 9.0 and lasted more than 2 minutes causing damage to structures and power infrastructure. Units 1, 2, and 3 were running at the time and shutdown automatically due to the earthquake seismic motion. A tsunami was created by the earthquake with the waves arriving 40 minutes after the initial shock. A wave of 14 to 15 m (46 to 49 ft) overwhelmed the Daiichi seawalls and flooded the site. This caused significant damage, loss of power, loss of control, and eventual loss of reactor containment.

Following the earthquake, TEPCO set up an emergency response center in Tokyo to manage the response and an on-site emergency response center at the Daiichi site. Evacuation and shelter-in-place orders were issued over the next three days.

After inserting the control rods (rods composed of chemical elements used to control the nuclear fission) to stop the reaction, heat continued to be generated. Cooling systems were powered or controlled by electrical power. The earthquake damaged off-site power supply resulting in a total loss of power supply to the plant. This loss of power isolated the units from their turbines resulting in increased temperature and pressure in the reactors. The operators followed appropriate procedures for the earthquake and loss of power in shutting down, isolating, and activating cooling systems. The incident progression is shown in Figure 8.2.

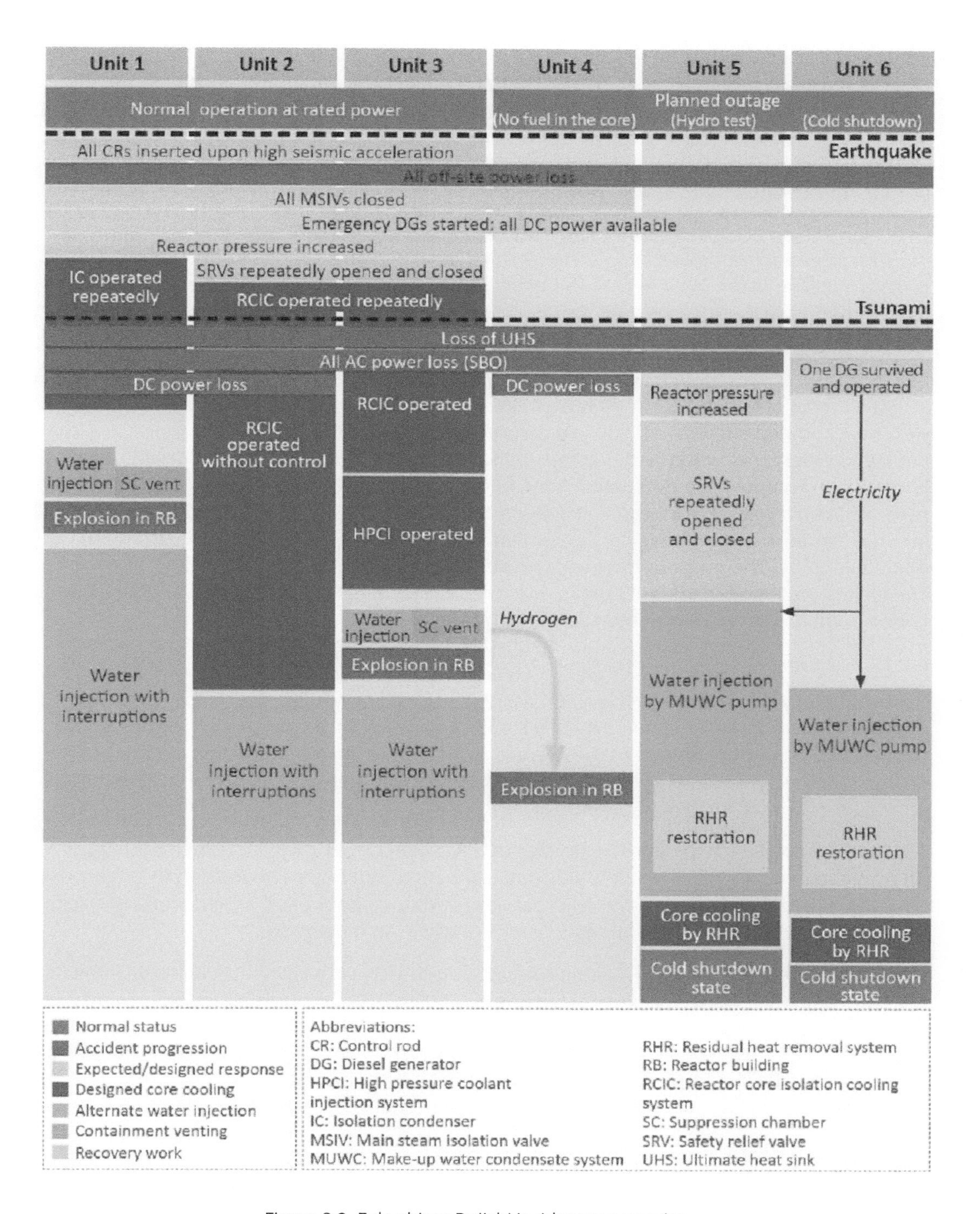

Figure 8.2. Fukushima Daiichi incident progression
(IAEA 2015)

The tsunami flooded the reactors and turbines resulting in loss of seawater intake for all units and thus loss of cooling. It also damaged the electrical equipment including the diesel generators, power distribution and switchgear which resulted in loss of the emergency diesel generators to provide cooling for all but one of the six units. DC power was provided as an additional emergency backup, but the batteries were flooded, and this power supply was lost to most of the units. With loss of power, the ability to monitor reactor pressure, water level and other aspects of core cooling was lost for three of the units.

The operators struggled with the loss of power and were taking various approaches to provide cooling water. With loss of the ability to monitor the process conditions, the worst-case scenario of a core overheating was assumed, and an evacuation and shelter-in-placeorder was issued at 9:23 pm on March 11. At 11:00 pm, radiation levels were detected outside the Unit.

Over most of March 12, efforts were made to restore cooling water and power to the units with no or limited success. At 3:30 pm on March 12, an explosion occurred in one unit that damaged emergency water and power supplies and an caused an abnormal rise in radiation levels. This prompted the evacuation zone being extended to 20 km (12.4 mi). On March 13, high radiation levels were detected at a second unit. On March 14, another explosion occurred injuring workers and damaging equipment. On March 15, explosions occurred in two additional units. The on-site emergency response center ordered the evacuation of all units. The highest radiation readings of the accident were recorded. Residents within a 30 km (19 mi) radius of the facility were ordered to shelter-in-place.

8.2.3 Lessons

Process Safety Competency. The NAIIC concluded that knowledge, training, inspection, and instruction were lacking. (NAIIC 2012) This points to a lack of process safety competency to support good practices in each of these areas. Without a deep understanding of both process safety and the process technology, the decisions made, and actions taken in these areas increased the risk of such an incident.

Process safety competency underpins many elements in most management systems. Without the mindset of being vulnerable and without considering each decision through a risk lens, the day-to-day decisions over the years can add up to poor asset integrity management, poor practices, and an inability to respond effectively in an emergency.

Stakeholder Outreach. The Japanese Fukushima NAIIC concluded that collusion between the government, regulators, and TEPCO was involved in the incident. (NAIIC 2012) The government agencies thought to be addressing public safety were found to be promoting nuclear power at the expense of safety. The events and the structural damage could have been foreseen. Structural improvements and improved emergency plans were not demanded by the regulator even though they were aware of the shortfalls.

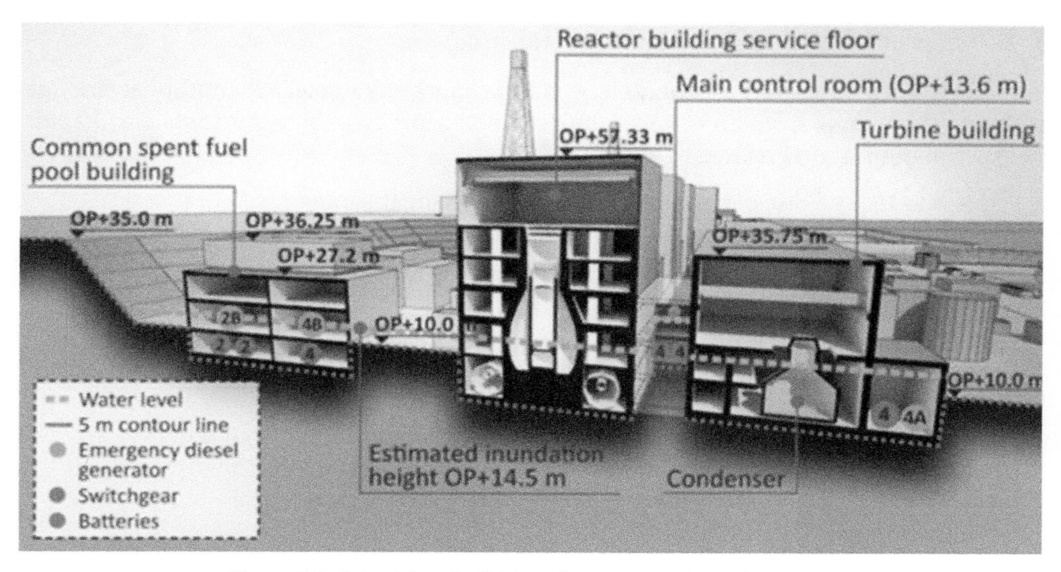

Figure 8.3. Fukushima Daiichi nuclear power plant elevations (IAEA 2015)

Hazard Identification and Risk Analysis. The nuclear industry is recognized for their use of probabilistic risk assessment. The Fukushima nuclear power plant was originally designed to withstand a magnitude 8 earthquake. Although the earthquake potential was recognized and addressed in design and procedures, the design basis was less than the magnitude 9 earthquakes that have occurred along the Pacific "ring of fire" and this was not clearly addressed in the risk assessments. Loss of externally supplied power was recognized and addressed in design and procedures. The tsunami potential was recognized but was also underestimated. However, the likelihood that these events could happen simultaneously was not well addressed. In hindsight, it is logical to see how one event can cause the next and thus their simultaneous occurrence is credible. Because the risk of a full loss of power was not recognized, the operators were not provided with appropriate procedures (loss of all power - main, diesel generator, and DC back-up). Refer to Figure 8.3. In the figure, OP refers to sea level at Onahama Port.

Emergency Management. The roles and responsibilities of the various regulators and agencies involved in the emergency response was not clear. This enabled the escalation of the emergency at the Fukushima nuclear power plant. Emergency preparedness and crisis management was lacking over the years which resulted in confusion and inefficient management of the situation during the emergency.

An effective emergency response is dependent on identification of the potential emergency, planning for it, including all those who may be impacted, and putting the systems in place to manage the event when, if, it occurs.

8.3 Types of Hazards (Beyond Chemical Hazards)

The hazards encountered in the workplace include more than solely chemical hazards (as discussed in Chapters 4 through 7). Recall the definition of process safety presented in Chapter 1. It also includes hazards related to energy.

> **Process Safety** - A disciplined framework for managing the integrity of operating systems and processes handling hazardous substances by applying good design principles, engineering, and operating practices.
>
> Note: Process safety focuses on efforts to reduce process safety risks associated with processes handling hazardous materials and ***energies***. Process safety efforts help reduce the frequency and consequences of potential incidents. These incidents include toxic or flammable material releases (loss events), resulting in toxic effects, fires, or explosions. The incident impact includes harm to people (injuries, fatalities), harm to the environment, property damage, production losses, and adverse business publicity. (CCPS Glossary)

Thinking more broadly, the list of potential hazards includes the following topics which will be discussed in this chapter.

- Kinetic energy
- Potential energy
- Electrical energy
- Meteorological and geological
- Health

Kinetic energy hazards are those due to the motion of equipment. Many workplaces include large pieces of rotating equipment such as turbines. If these fail, they often fail catastrophically throwing debris which can cause harm and destruction. A noteworthy incident is the failure of a turbine in the Sayano-Shushenskaya hydroelectric plant in Russia as shown in Figure 8.4 which resulted from heavy vibration. (IWP&D 2010) Kinetic energy hazards also include moving vehicles which can impact equipment and piping or trains that can derail.

Figure 8.4. Damage in the generator hall at the Sayano-Shusenskaya hydroelectric plant (AP)

Potential energy hazards result from the effects of gravity. For example, it might be helpful to locate feed tanks on a hill to use gravity feed as a means of simplifying the process and saving energy. The unintended consequences of this arrangement should be considered. A facility in Tacoa, Venzeula used such an arrangement. They had a large oil tank on a hill that caught fire and burned for eight hours. Topography played a key role in events that followed. The oil overflowed the tank dike resulting in a downhill flow of burning oil causing firemen and spectator fatalities. (CCPS 2003 and IFW 1982)

Objects being accidently dropped from a height is another example of a potential energy hazard. Workplaces have many requirements to lift heavy materials by crane during maintenance work or in transporting supplies. At a refinery, a section of a convection heater was dropped as it was being lifted near an anhydrous hydrogen fluoride tank. It sheared the truck loading line and the pressure relief line resulting in the tank contents of HF being released, forming a vapor cloud, and necessitating a shelter-in-place for neighboring residential areas. (HSE) This incident prompted a focus on safe lifting practices and an edict in many companies prohibiting lifting over live equipment. This same hazard is present on offshore platforms where most supplies arrive by boat and are lifted onto the platform. A dropped objects analysis is often conducted to determine where best to locate cranes and subsea equipment to avoid a dropped object resulting in damage to a pipeline or wellhead.

Electrical energy hazards typically result in occupational safety incidents harming a single person. Electricity can also be a source of ignition as discussed in Chapter 4. As relates to process safety, significant differences in potential can result in an arc-flash when conductors inadvertently contact each other. The rapid heating and expanding of gases during an arc-flash can result in an explosion that can throw debris and molten bits of metal. The best means of controlling this hazard is to deenergize equipment before working on it. (LANL)

Nature can pose **meteorological and geological hazards**. (CCPS 2019)

- Meteorological hazards are those that naturally occur due to the weather cycle or climactic cycles, and include flooding, temperature extremes, snow/ice storms, wildfire, tornado, tropical cyclones, hurricanes, storm surge, wind, lightning, hailstorms, drought, etc.
- Geological hazards are those occurring due to the movements of the earth and the internal earth forces, and include seismic events, earthquakes, landslides, sinkholes, tsunami, volcanic eruptions, and dam rupture.

Natural Hazards Triggering Technological Disasters (Natech) refers to the interaction between natural disasters and industrial accidents. More information on this area of research can be found at the UN Economics Commission for Europe. (UNECE) Identifying these natural hazards and including their potential impact in design and emergency response preparedness plans is important to prevent their resulting in a process incident. For example, natural hazards can result in loss of access and power to facilities which can, in turn, result in loss of cooling to reactive chemical storage. This was evident in the Arkema organic peroxide decomposition and fire following Hurricane Harvey flooding in 2017. (CSB 2018) Earthquakes can result in failure of equipment and piping resulting in fires.

Mining operations often create retention basins or dams to hold tailings. In Hungary, the Kolontar Tailings dam failure released a wave of bauxite into the surrounding area. (AGU) Tailing dam failures can result in both rapid water current and flooding hazards as well as longer term toxic exposures. Guidance is available such as the Mining Association of Canada's Guide to the Management of Tailings Facilities. (MAC)

Seismic hazards for process facilities include both potential toppling of tall structures and also the resulting escalation of the initial event. In seismic regions a seismic evaluation should be conducted of tall structures, such as distillation columns, to determine if actions should be taken to prevent damage due to a seismic event. Additional design considerations may need to be incorporated into facilities located in seismic zones.

Natural hazards can also result in secondary process safety impacts. Consider the eruption of the Iceland volcano that resulted in a shutdown of air travel in Europe as shown in Figure 8.5. (NASA) This meant that work to evaluate hazards, analyze risks, and implement systems to control process safety risks were put on hold until travel, and transportation of supplies, could resume.

Figure 8.5. Iceland volcano ash plume on May 2, 2010
(NASA)

Health hazards can also have a secondary impact on process safety. This is clearly seen in the COVID19 pandemic that changed the way work was conducted in 2020. By necessity, process safety experts had to create new ways to work. For example, HAZOPs (refer to Chapter 12) were conducted virtually, rather than in face-to-face meetings. Management of change reviews (refer to Chapter 18) were conducted with a single, socially isolated person in the plant with a camera providing a view of the equipment to the remote team.

8.4 What a New Engineer Might Do

Along with chemical hazards, a new engineer should include other hazards in their analyses and designs. When gathering data, an engineer should go beyond the chemical hazards data sources and research other data sources such as those defined in the CCPS Monograph: Assessment of and planning for Natural Hazards. (CCPS 2019)

Common activities for a new engineer include the following.

- Understanding meteorology and geology in the area in order to understand frequency & consequence of severe weather and natural disaster data.
- Creating emergency plans to address the meteorological and geological risk.
- Participating in emergency drills to prepare for these disasters.
- Ensuring equipment design is appropriate for the identified hazards (e.g., elevate pumps for flood, provide emergency power (diesel generators) for expected power outages, building design for the wind loads).
- Serving as an emergency responder.
- Participating in creating return to work plans.

New engineer activities on this topic will depend on their working location. Coastal locations may have exposure to hurricanes and associated flooding, while other areas may be concerned with wind and wildfires. Northern locations such as Canada and Alaska will be most concerned about very low temperatures. Desert locations may need to address sandstorms. Locations such as California and Turkey should address earthquake preparedness as demonstrated by the Fukushima incident.

8.5 Tools

Kinetic and potential energy hazards can be identified and managed through use of engineering tools learned in undergraduate engineering curriculum.

CCPS Monograph: *Assessment of and planning for Natural Hazards.* lists many data sources and approaches for identifying meteorological and geological hazards and addressing them in design and emergence response preparedness plans. (CCPS 2019) The monograph includes reference to the following data sources and design criteria.

Data sources.

- Federal Emergency Management Agency (FEMA) flood maps - https://msc.fema.gov/portal/home
- United States Geological Survey (USGS) seismic maps - https://earthquake.usgs.gov/hazards/designmaps/usdesign.php
- American Society of Civil Engineers (ASCE) seismic guide - https://hazards.atcouncil.org/#/
- National Oceanic and Atmospheric Administration (NOAA) tornado prediction - https://www.spc.noaa.gov/new/SVRclimo/climo.php?parm=allTorn
- Tornado Wind Prediction - https://hazards.atcouncil.org/#/ (ATC)
- National Hurricane Center (NHC) Storm surge maps - https://www.nhc.noaa.gov/nationalsurge/ (NHC and NOAA 2019 a)
- Snow load - https://hazards.atcouncil.org/#/ (ATC)
- NOAA Hurricane center - https://www.nhc.noaa.gov/climo/

Design guidance.

- ASCE Flood Resistant Design and Construction, ASCE 24 (ASCE 2014)
- ASCE Minimum Design Loads and Associated Criteria for Buildings and Other Structures, ASCE /SEI 7-16 (ASCE 2016)
- CCPS Guidelines for Safe Warehousing of Chemicals (CCPS 1998)
- CCPS Guidelines for Safe Storage and Handling of Reactive Materials (CCPS 1995)
- Guidelines for Siting and Layout of Facilities, 2nd Edition (CCPS 2018)
- FM Global Property Loss Prevention Data Sheets 1-2 Earthquakes (FM Global 2021)

- FM Global Property Loss Prevention Data Sheets 1-11 Fire Following Earthquake (FM Global 2016)
- FM Global Property Loss Prevention Data Sheets 1-29 Roof Deck Securement and Above-Deck Roof Components (FM Global 2020)
- FM Global Property Loss Prevention Data Sheets 1-34 Hail Damage (FM Global 2020 a)
- FM Global Property Loss Prevention Data Sheets 1-40 Flood (FM Global 2019)

CCPS Monograph: *Risk Based Process Safety During Disruptive Times.* addresses different types of disruptive events, including pandemics and provides insights for managing process safety during these events. It is organized by the RBPS elements and also addresses the human factors impact. (Refer to Chapter 16) The top three elements of highest importance are: Process Safety Culture, Asset Integrity & Reliability and Management of Change. (CCPS 2020)

8.6 Summary

It is easy to focus on the desired process, the safe handling of chemicals, and inadvertently ignore other hazards that can exist in the workplace. Kinetic and potential energy, electrical energy, geological and meteorological hazards, and health hazards can all lead to process safety events. Including these hazards in the scope of an analysis or the basis of a design can help prompt engineers to consider these hazards that may otherwise be overlooked.

8.7 Other Incidents

This chapter began with a detailed description of the Fukushima Daiichi incident. Other incidents involving kinetic and potential energy, electrical energy, geological and meteorological hazards, and health hazards include the following.

- Dust Explosion at Courrieres Mine, France, 1906
- Three Mile Island Nuclear Reactor Core Meltdown, Pennsylvania, U.S., 1979
- Chernobyl Nuclear Disaster, U.S.S.R., 1986
- NASA Challenger Disaster, Florida, U.S., 1986
- Loss of Space Shuttle Columbia, Texas, U.S., 2003
- Sewol Ferry Sinking, South Korea, 2014
- Arkema Organic Peroxide Decomposition and Fire, Texas, U.S., 2017

8.8 Exercises

1. List 3 RBPS elements evident in the Fukushima Daiichi Nuclear Power Plant release summarized at the beginning of this chapter. Describe their shortcomings as related to this accident.
2. Considering the Fukushima Daiichi Nuclear Power Plant release, what actions could have been taken to reduce the risk of this incident?
3. What chemical was involved in the Arkema release and fire in Crosby, Texas in 2017. What prevention measures did Arkema take to try to prevent this release?

4. Many refineries and chemical plants are located on the Gulf Coast and in the Los Angeles basin of the United States. In addition to the hazards associated with the chemicals they handle, what other hazards should they address?
5. Fertilizers are used to increase crop yield. The primary components in fertilizers are nitrogen, phosphorus, and potassium. Ammonia can be used as the nitrogen source. It is made from natural gas and air. Ammonia and nitric acid are used to make ammonium nitrate which is used as the fertilizer component. Phosphorus comes from phosphate rock which is surface mined and then treated with sulfuric acid and nitric acid. Potassium comes from potash which is also mined and includes a large tailings dam. (Madehow) What hazards should be addressed for the manufacture of these fertilizers?
6. Covid 19 had an impact on the way nearly everyone worked. How do you think it impacted process safety?
7. What hazard was involved in the Space Shuttle Columbia disaster? Describe the accident scenario.
8. Long pipelines are used to transport materials across regions, states, even countries. What hazards might threaten the pipelines?
9. Food processing plants such as ice cream factories and meat packing plants use refrigeration involving ammonia, propane, and nitrogen. What hazards and what risks might this pose?
10. Figure 8.6 shows a refinery during historic river flooding. What impacts might this have on process safety in the facility?

Figure 8.6. Coffeyville Refinery 2007 flood
(KDA)

8.9 References

AGU, American Geophysical Union, https://blogs.agu.org/landslideblog/2010/12/19/the-outcome-of-a-study-of-the-kolontar-tailings-dam-failure/.

AP Photo/Rossiiskaya Gazeta Newspaper, Sayano-Shusenskaya hydroelectric plant in southern Siberia.

ASCE, American Society of Civil Engineers, https://hazards.atcouncil.org/#/.

ASCE 2014, "Flood Resistant Design and Construction", ASCE/SEI 24-14, American Society of Civil Engineers.

ASCE 2016, "Minimum Design Loads and Associated Criteria for Buildings and Other Structures", ASCE /SEI 7-16, American Society of Civil Engineers.

ATC, Applied Technology Council, https://hazards.atcouncil.org/#/.

CCPS Glossary, "CCPS Process Safety Glossary", Center for Chemical Process Safety, https://www.aiche.org/ccps/resources/glossary.

CCPS 1995, *Guidelines for Safe Storage and Handling of Reactive Materials*, Center for Chemical Process Safety, John Wiley & Sons, Hoboken, N.J.

CCPS 1998, *Guidelines for Safe Warehousing of Chemicals*, Center for Chemical Process Safety, John Wiley & Sons, Hoboken, N.J.

CCPS 2003 *Guidelines for Fire Protection in Chemical, Petrochemical, and Hydrocarbon Processing Facilities*, Center for Chemical Process Safety, John Wiley & Sons, Hoboken, N.J.

CCPS 2018, *Guidelines for Siting and Layout of Facilities*, Center for Chemical Process Safety, John Wiley & Sons, Hoboken, N.J.

CCPS 2019, *Monograph: Assessment of and planning for Natural Hazards*, Center for Chemical Process Safety, John Wiley & Sons, Hoboken, N.J.

CCPS 2020, *Monograph: Risk Based Process Safety During Disruptive Times*, Center for Chemical Process Safety, John Wiley & Sons, Hoboken, N.J.

CSB, 2018, "Organic Peroxide Decomposition, Release, and Fire at Arkema Crosby Following Hurricane Harvey Flooding", Investigation Report, Report No. 2017-08-I-TX, U.S. Chemical Safety and Hazard Investigation Board.

FEMA, Federal Emergency Management Agency, https://www.fema.gov/flood-zones.

FM Global 2016, Property Loss Prevention Data Sheet 1-11 "Fire Following Earthquake", FM Global, Hartford, Connecticut.

FM Global 2020, Property Loss Prevention Data Sheet 1-29 "Roof Deck Securement and Above-Deck Roof Components", FM Global, Hartford, Connecticut.

FM Global 2020 a, Property Loss Prevention Data Sheet 1-34 "Hail Damage", FM Global, Hartford, Connecticut.

FM Global 2019, Property Loss Prevention Data Sheet FM 1-40 "Flood", FM Global, Hartford, Connecticut.

HSE, Health and Safety Executive, https://www.hse.gov.uk/comah/sragtech/casemarathon87.htm.

IAEA 2015. "The Fukushima Daiichi Accident", Technical Volume 1/5 Description and Context of the Accident, ISBN 978–92–0–107015–9 (set), International Atomic Energy Agency, Vienna, August. *Reproduced with permission*

IFW 1982, *Industrial Fire World*, https://www.industrialfireworld.com/540292/tacoa-venezuela-dec-19-1982.

IWP&D 2010, "Sayano Shushenskaya accident – presenting a possible direct cause", International Water Power & Dam Construction, 22 December 2010, https://www.waterpowermagazine.com/features/featuresayano-shushenskaya-accident-presenting-a-possible-direct-cause.

KDA, https://agriculture.ks.gov/divisions-programs/dwr/floodplain/resources/historical-flood-signs/lists/historical-flooding.

LANL, Los Alamos National Laboratory, *Electrical Safety Hazards Handbook*, https://www.lanl.gov/safety/electrical/docs/arc_flash_safety.pdf.

MAC, The Mining Association of Canada, https://mining.ca/our-focus/tailings-management/tailings-guide/

Madehow, http://www.madehow.com/Volume-3/Fertilizer.html

NAIIC 2012. "The official report of The Fukushima Nuclear Accident Independent Investigation Commission", The National Diet of Japan, 2012.

NASA, National Aeronautics and Space Administration, https://earthobservatory.nasa.gov/images/43883/eruption-of-eyjafjallajakull-volcano-iceland.

NHC, National Hurricane Center, https://www.nhc.noaa.gov/surge/.

NOAA 2019 a, National Oceanic and Atmospheric Administration, https://noaa.maps.arcgis.com/apps/MapSeries/index.html?appid=d9ed7904dbec441a9c4dd7b277935fad&entry=1.

NOAA, National Oceanic and Atmospheric Administration (NOAA), https://www.spc.noaa.gov/new/SVRclimo/climo.php?parm=allTorn.

UNECE, Natech, https://unece.org/industrial-accidents-convention-and-natural-disasters-natech.

USGS 2018, U.S. Geological Survey, https://www.usgs.gov/news/post-harvey-report-provides-inundation-maps-and-flood-details-largest-rainfall-event-recorded.

9
Process Safety Incident Classification

9.1 Learning Objectives

The learning objectives of this chapter are:

- Understand Process Safety Incident (PSI) classification using API RP 754 and IOGP 456,
- Explain the difference between and use of leading and lagging indicators,
- Discuss the relationship between PSI classification levels, and
- Classify a process safety incident.

9.2 Incident: Petrobras P-36 Sinking, Brazil, 2001

9.2.1 Incident Summary

A Petrobras offshore production platform, P-36 was located in Campos Basin off the coast of Brazil. Explosions occurred on the platform P-36 started on March 15, 2001 resulting in fatalities of eleven Petrobras employees. The platform stability was compromised by the explosions and it sank on March 20, 2001. Images of the P-36 are provided in Figures 9.1 and 9.2. The government investigation recommended "Review and application of a management system to ensure a strict compliance with standard procedures, including reviewing the definition of responsibilities with respect to maintenance, operation and safety." (ANP 2001)

Key Point:

Measurement and Metrics – "Efficiency and performance should not supersede the need and continuous pursuit of safe operations." (NASA 2008)

Figure 9.1. P 36 Platform shown during dry tow
(ANP 2001)

Figure 9.2. P 36 attempted salvage operations
(NASA 2008)

9.2.2 *Detailed Description*

The Petrobras P-36 was a semi-submersible production platform buoyantly supported by four columns and two pontoons.

An emergency drain tank located in one of the support columns had been shut down and isolated. The isolation valve leaked as it was not properly blinded off and hydrocarbons slowly leaked in and overpressured the tank causing the first physical explosion. A main cooling water pipe for the installation was located adjacent to the emergency drain tank and was ruptured in the explosion. More than one thousand alarms were triggered. Cooling water was flowing through the pipe for normal cooling and when ruptured it leaked water into the column and very quickly led to a noticeable tilt. The ruptured pipe also provided the firewater for the structure and the control system had been designed to make it difficult to shut-off the cooling water line as this would stop firewater supply.

As the emergency response team arrived to investigate the initial physical explosion in the column, they accidentally ignited flammable vapors released from the physical explosion of the tank and a second explosion occurred fatally injuring eleven firefighters. The column was normally a safe location with no hydrocarbons intended to be present, so the emergency response team did not test for flammable vapors before entering.

Flooding of the column short-circuited the seawater pump located at the bottom of the column in the pontoon. The sea chest valve to the ocean located at the bottom of the column is always open as it supplies the cooling water continuously. It is a manual valve and once the area flooded it could not be accessed.

P-36 continued tilting, and the remaining staff were safely evacuated. Salvage operations were attempted for five days but were unsuccessful. P-36, valued at $496 million, sunk in 1300 m (4265 ft) of seawater.

9.2.3 *Lessons*

Hazard Identification and Risk Analysis. The risk of a drain tank to contain hydrocarbons and thus violate the policy of no hydrocarbons in the columns was not identified.

Conduct of operations. Multiple doors designed to seal ballast compartments were left open. The union cited poor training of contractors. Additionally, the single isolation valve provided for the emergency drain tank failed to provide positive isolation (as a redundant valve would have). The valves to the ocean at the bottom of the column were designed to fail-open which resulted in flooding of the column and pontoon.

Measurement and Metrics. A Petrobras executive stated, in regard to positive financial performance, "the project successfully rejected ... prescriptive engineering, onerous quality requirements, and outdated concepts of inspection ...". It is clear that focusing on financial metrics distracted from the value of operational risk and process safety metrics.

9.3 Introduction to Metrics

Metrics are measurements. In 1970, the U.S. OSHA required employers to maintain a log of recordable occupational injuries and illnesses per a set of definitions they provided. This common method of measure led to industry comparing safety performance and motivation to continually improve this performance. Metrics such as the Recordable Incident Rate, and Days Away from Work Cases have become common safety performance metrics. The focus of these metrics, however, is solely on occupational safety.

Industry recognized the value of the occupational safety performance metrics and saw the need to improve process safety performance by creating process safety-specific metrics.

In 2007, CCPS published a document defining process safety metrics. API RP 754 *Process Safety Performance Indicators for the Refining and Petrochemical Industries* was originally published in 2010. (API RP 754)

The upstream oil and gas industry aligned with refining and petrochemical in creating a parallel document IOGP Report 456 – *Process safety – recommended practice on key performance indicators*. (IOGP 456) This allowed integrated oil and gas companies to collect consistent data statistics for upstream and downstream operations.

Since the original metrics documents were issued, CCPS, API, and IOGP have updated their relevant documents while keeping them aligned. An effort is underway to align the metrics with the UN, *Globally Harmonized System of Classification and Labelling of Chemicals (GHS)*. (UN) The metrics alignment has enabled companies to:

- Measure process safety performance, identify underperformance, and drive continuous improvement, and
- Compare company performance to industry performance, past company performance, or intra-company performance, and continuously improve process safety performance.

The system used in the CCPS, API, and IOGP documents is a four-tier system as shown in Figure 9.3. This approach is similar to Heinrich's incident pyramid that depicts a larger, bottom-level of minor incidents (a larger area of the triangle representing a larger number of incidents), a mid-level of incidents, and a top, small level of more serious accidents (the smaller area representing less incidents). In the process safety pyramid, Tier 1 and Tier 2 have been designated Process Safety Events (PSE) that have occurred. Tier 1 PSEs are of greater consequence; Tier 2 PSEs of lesser, but still serious, consequence.

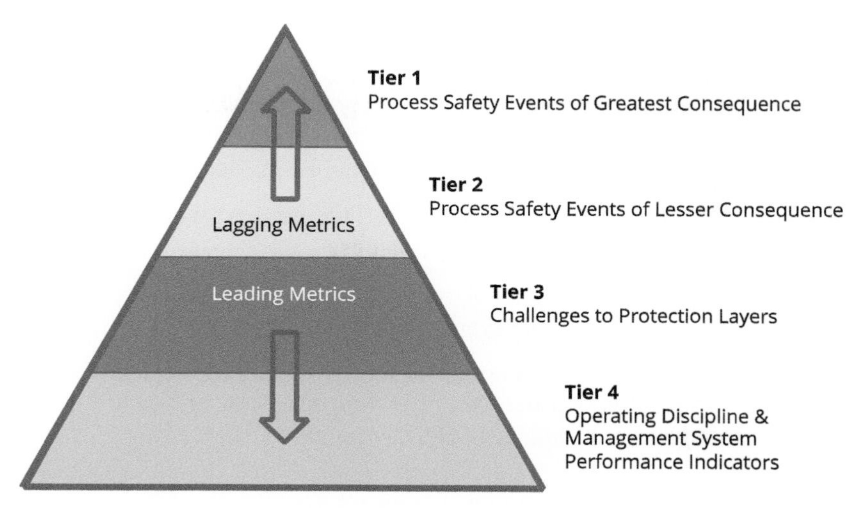

Figure 9.3. Process safety pyramid (adapted from CCPS 2019)

Process Safety Event (PSE) – An event that is potentially catastrophic, i.e., an event involving the release/loss of containment of hazardous materials that can result in large-scale health and environmental consequences. (CCPS 2019)

Tier 3 and Tier 4 are indicators of a potential degradation in process safety performance indicating that a PSE may occur but has not yet. These indicators focus on the more frequent, less severe incidents, as shown in the lower portions of the incident pyramid. A Tier 3 incident typically represents a challenge to the barrier or protection layer system that progressed along the path to harm but is stopped short of a Tier 1 or Tier 2 PSE consequence – designated as "challenges to protection layers". Indicators at this level provide an additional opportunity to identify and correct weaknesses within the barrier system. Tier 4 indicators are focused on operating discipline and management system performance. Indicators at this level provide an opportunity to identify and correct system-related weaknesses. Tier 4 indicators are indicative of process safety system weaknesses that may contribute to future Tier 3, Tier 2 PSEs, and possibly Tier 1 PSEs. In that sense, Tier 4 indicators help identify issues and opportunities for both learning and process safety system improvements.

A near miss process safety incident is defined as an event that under slightly different circumstances could have consequences that elevate the classification to Tier 1, 2, or 3. It is common industry practice to report and investigate all Tier 1, 2, and 3 events, and near miss events. For example, OSHA PSM regulation requires investigation of all process safety incidents and near misses that could have been a process safety incident.

Near Miss - An event in which an accident (that is, property damage, environmental impact, or human loss) or an operational interruption could have plausibly resulted if circumstances had been slightly different. (CCPS Glossary)

As stated, Tier 1 and Tier 2 PSEs are events that have occurred. These are termed lagging metrics. They usually involve multiple root causes, some of which involve degraded effectiveness or failure of management systems. It is good to measure the past and compare to it; however, it is even better to have an indication of a problem before an incident occurs. This is the intent of the Tier 3 and Tier 4 metrics which are considered leading metrics.

A combination of leading and lagging metrics is often the best way to provide a complete picture of process safety performance and effectiveness.

Lagging metric - A retrospective set of metrics based on incidents that meet an established threshold of severity. (CCPS Glossary)

Leading metric - A forward-looking set of metrics that indicate the performance of the key work processes, operating discipline, or layers of protection that prevent incidents. (CCPS Glossary).

Tier 1 and Tier 2 PSEs are described in significant detail in the API and IOGP practices. Classifying a PSE as either Tier 1 or Tier 2 is dependent on the amount of material released, referred to as a Loss of Primary Containment (LOPC). Further details on classifying LOPC incidents as Tier 1 or Tier 2 using API RP 754, 3rd edition are provided in Appendix E.

Loss of Primary Containment (LOPC) - An unplanned or uncontrolled release of material from primary containment, including non-toxic and non-flammable materials (e.g., steam, hot condensate, nitrogen, compressed CO_2 or compressed air). (CCPS Glossary)

Note: Steam, hot condensate, and compressed or liquefied air are only included in this definition if their release results in one of the consequences other than a threshold quantity release. However, other nontoxic, nonflammable gases with defined UN Dangerous Goods (UNDG) Division 2.2 thresholds (such as nitrogen, argon, compressed CO_2) are included in all consequences including, threshold release. (API RP 754).

Primary Containment - A tank, vessel, pipe, transport vessel or equipment intended to serve as the primary container for, or used for the transfer of, a material. Primary containers may be designed with secondary containment systems to contain or control a release from the primary containment. Secondary containment systems include, but are not limited to tank dikes, curbing around process equipment, drainage collection systems into segregated oily drain systems, the outer wall of double-walled tanks, etc. (CCPS Glossary)

Given the alignment of Tier 1 and Tier 2 definitions and the fact that they are lagging metrics, many companies publicly report these metrics. To facilitate comparison across companies of different size, rates are used. For example, the PSE Tier 1 rate (PSE1R) = (Total Tier 1 PSE Count / Total Work Hours) × 200,000.

API RP 754 also describes the severity weighting for Tier 1 indicators as shown in Figures 9.4 and 9.5. The weighting considers the consequence categories: safety/human health; direct cost from fire or explosion; release amount; community impact; and off-site environmental impact. Severity points are assigned according to the potential impact with larger consequences receiving more points. Like the PSE rate calculation, the severity weighting can be expressed as a point count or converted to a rate using work hours.

For leading metrics, each company should focus on areas of potential weakness. As this is company specific, and reflects company operational systems, the appropriate metrics are different. Thus, Tier 3 and Tier 4 metrics are too facility-specific for benchmarking or developing industry applicable criteria. They are intended for internal company use. Typical Tier 3 and Tier 4 metrics are listed in Table 9.1.

Table 9.1 Tier 1 Process Safety Event Severity Weighting

Severity Points	Consequence Categories				
	Safety/Human Health [c]	**Direct Cost from Fire or Explosion**	**Material Release Within Any 1-Hr Period** [a, d, e]	**Community Impact**	**Off-Site Environmental Impact** [b, c]
1 point	Injury requiring treatment beyond first aid to an employee, contractor, or subcontractor. (Meets the definition of a US OSHA recordable injury.)	Resulting in \$100,000 ≤ Direct Cost Damage <\$1,000,000.	Release volume 1x ≤ Tier 1 TQ < 3x outside of secondary containment.	— Officially declared shelter-in-place or public protective measures (e.g. road closure) for <3 hours, or — Officially declared evacuation <3 hours.	Resulting in \$100,000 ≤ Acute Environmental Cost <\$1,000,000.
3 points	— Days Away From Work injury to an employee, contractor, or subcontractor, or — Injury requiring treatment beyond first aid to a third party.	Resulting in \$1,000,000 ≤ Direct Cost Damage <\$10,000,000.	Release volume 3x ≤ Tier 1 TQ < 9x outside of secondary containment.	— Officially declared shelter-in-place or public protective measures (e.g. road closure) for > 3 hours, or — Officially declared evacuation > 3 hours < 24 hours.	— Resulting in \$1,000,000 ≤ Acute Environmental Cost <\$10,000,000, or — Small-scale injury or death of aquatic or land-based wildlife.

Table 9.1 continued

Severity Points	Consequence Categories				
	Safety/Human Health [c]	Direct Cost from Fire or Explosion	Material Release Within Any 1-Hr Period [a, d, e]	Community Impact	Off-Site Environmental Impact [b, c]
9 points	— A fatality of an employee, contractor, or subcontractor, or — A hospital admission of a third party.	Resulting in \$10,000,000 ≤ Direct Cost Damage <\$100,000,000	Release volume 9x ≤ Tier 1 TQ < 27xoutside of secondary containment.	Officially declared evacuation > 24 hours < 48 hours.	— Resulting in \$10,000,000 ≤ Acute Environmental Cost <\$100,000,000, or — Medium-
27 points	— Multiple fatalities of employees, contractors, or subcontractors, or — Multiple hospital admission of third parties, or	Resulting in ≥\$100,000,000 of direct cost damages.	Release volume ≥ 27x Tier 1 TQ outside of secondary containment.	Officially declared evacuation > 48 hours.	— Resulting in ≥ \$100,000,000 of Acute Environmental Costs, or — Large-scale injury or death of aquatic or land-based

[a] Where there is no secondary containment, the quantity of material released from primary containment is used. Where secondary containment is designed to only contain liquid, the quantity of the gas or vapor being released and any gas or vapor evolving from a liquid must be calculated to determine the amount released outside of secondary containment.

[b] Judging small, medium or large-scale injury or death of aquatic or land-based wildlife should be based on local regulations or Company guidelines.

[c] The severity weighting calculation includes a category for "Off-Site Environmental Impact" and injury beyond first aid (i.e. OSHA "recordable injury") level of Safety/Human Health impact that are not included in the Tier 1 PSE threshold criteria. However, the purpose of including both of these values is to achieve greater differentiation of severity points for events that result in any form of injury or environmental impact.

d For the purpose of Severity Weighting, general paving or concrete under process equipment, even when sloped to a collection system, is not credited as secondary containment.

e Material release is not tabulated for fires or explosions. These events severity will be determined by the other consequence categories in this table.

Table 9.2. Typical Tier 3 and Tier 4 process safety metrics
(derived from CCPS 2019)

Tier 3	
Challenges to Safety Systems	• Opening of a rupture disc, a pressure control valve to flare or atmospheric release, or a pressure safety valve when a pre-determined trigger point is reached • Activation of a safety instrumented system when an "out of acceptable range" process variable is detected, e.g. activation of high-pressure interlock
Process Deviations or Excursions	• Excursion of parameters such as pressure, temperature, flow outside of the standard operating limits (the operating "window" for quality control) but remaining within the process safety limits
Inspection, Testing and Preventive Maintenance (ITPM)	• Primary containment inspection or testing results outside acceptable limits • Discovery of a failed safety system upon testing, e.g. relief devices that fail bench tests at set points
LOPC not classified as Tier 1 or 2	• LOPC events that do not meet the Tier 1 or 2 criterion considered Tier 3 incidents
Other	• Dropping loads / falling objects within range of process equipment • Failure to remove line blanks in critical piping

Table 9.2 continued

Tier 4	
Process Safety Culture	• Process safety culture survey scores
Process hazard analysis	• Percentage of total PHAs documenting use of complete Process Safety Information (PSI) during the PHA • Number of PHA Recommendations
Facility Siting Risk Assessments	• Percentage of total PHAs documenting Facility Siting risk assessments
Operating Procedures and Maintenance Procedures	• Percentage of total number of operating or maintenance procedures reviewed/updated
Asset Integrity	• Percentage of total inspections of safety critical equipment completed on time • Percentage of time plant is in production with items of safety critical equipment in a failed state
Process Safety Training and Competency Assurance	• Percentage of individuals who completed required process safety competency training on time
Management of Change	• Percentage of MOCs that satisfied all aspects of the site's MOC procedure. • Percentage of identified changes that used the site's MOC procedure prior to making the change.
Action Item Follow-up	• Percentage of process safety action items that are past due
Fatigue Risk Management	• Amount of overtime • Number of extended shifts

9.4 What a New Engineer Might Do

New engineers are frequently involved in the collation of performance metrics. The calculations should be accurate and use the precise definitions provided to support comparison of performance as opposed to comparison of data anomalies. A new engineer should be very familiar with the relevant documents described in this chapter.

A common responsibility of early career engineers is to develop source models to calculate release amounts from various aperture releases, vessel overflows, and other loss of primary containment events. Once release amounts are known, the API RP 754 criterion for PSE is used to classify the incident as Tier 1, 2, 3, or near miss. Often, engineers face a short time frame to return classification due to company or regulatory requirements for reporting.

Leading and lagging indicator data are tracked and the data analyzed to identify trends and make suggestions for improvement.

New engineers, as any engineer, may notice things in the workplace that they suspect reflect on process safety performance. Perhaps this is a relief valve that is frequently lifting or a decrease in available training opportunities. Suggesting that these be measured and tracked may provide the data needed to focus attention on the topic and drive improvement.

9.5 Tools

The documents discussed in this chapter are the best tools in understanding, selecting, and enabling consistency in process safety metrics.

API RP 754 *Process Safety Performance Indicators for the Refining and Petrochemical Industries* addresses both leading and lagging metrics and includes detailed definitions and classifications. (API RP 754)

CCPS. Acknowledging that performance metrics continue to evolve, CCPS has created an evergreen webpage resource for process safety metrics. The CCPS webpage contains links to resources, reports, and research in multiple languages including those listed in the References in Section 9.8. The References also includes a link to a Process Safety Incident Evaluation Tool which assists the user in determining how to classify an incident. Consult the CCPS Metrics webpage at https://www.aiche.org/ccps/process-safety-metrics.

- Process Safety Metrics: Guide for Selecting Leading and Lagging Indicators
- Process Safety Leading and Lagging Metrics...You Can't Improve What You Don't Measure
- Process Safety Incident Evaluation Tool

CCPS Process Safety Incident Evaluation (PSIE) App has three main components: the Process Safety Incident Evaluation Questionnaire, the Severity Weighting Questionnaire, and the Chemical and Mixtures Database. The app leads a user, step-by-step, through the evaluation of an event leading to the determination of whether it meets Tier 1 or Tier 2 PSE Criteria. The app is available at app stores. (PSIE)

IOGP Report 456 – *Process safety – recommended practice on key performance indicators* addresses both leading and lagging metrics and includes detailed definitions and classifications. It focuses on the upstream oil and gas industry. (IOGP 456)

9.6 Summary

The phrase "what gets measured, gets done" sums up this chapter. Occupational injuries and illnesses were measured and have been continuously improving over the past 50 years. Without a common definition, process safety incidents were not measured. CCPS, API, and IOGP have created aligned systems to classify lagging process safety events and leading indicators. Companies are working to use these systems to drive improvements in process safety performance.

9.7 Exercises

1. List 3 RBPS elements evident in the Petrobras P 36 sinking summarized at the beginning of this chapter. Describe their shortcomings as related to this accident.
2. Considering the Petrobras P 36 sinking, what actions could have been taken to reduce the risk of this incident?
3. Considering the BP Texas City explosion described in Chapter 2, was this a Tier 1 or Tier 2 PSE? Suggest a Tier 3 metric and a Tier 4 metric that could have been measured to indicate an incident such as this was becoming more likely.
4. Considering the MIC release in Bhopal, India described in Chapter 6, was this a Tier 1 or Tier 2 PSE? Suggest a Tier 3 metric and a Tier 4 metric that could have been measured to indicate an incident such as this was becoming more likely.
5. A facility experiences an increasing number of "small" releases (less than 25 kg) of butane. Is this a process safety event? If so, what Tier is it?
6. An accidental release of 636 liters (4 barrels) of gasoline at a gasoline station results in a fire causing $50,000 damage. No one is injured. Is this a process safety event? If so, what Tier is it?
7. Propane is relieved through a pressure relief valve to the flare system. Estimates of the quantity put the release at 2268 kg (5000 lbs). Is this a process safety event? If so, what Tier is it?
8. A maintenance technician is working on the ammonia refrigeration system in the refinery methyl ethyl ketone unit and inadvertently causes a release of 4.5 kg (10 lb) of ammonia. He immediately begins coughing and his eyes are irritated and burning. He is kept in the hospital overnight for observation. Is this a process safety event? If so, what Tier is it?
9. An offshore drilling platform uses diethylamine as a corrosion inhibitor. A leak occurs in the diethylamine piping releasing 1500 kg (3307 lb) that flows into the sea. No one is injured. Is this a process safety event? If so, what Tier is it?
10. A blowdown drum overflows, and 28,769 liters (7600 gallons) of a pentane/hexane mixture rain out onto the ground in an area where no buildings or people are present. The overflow is stopped, and no one is injured. Is this a process safety event? If so, what Tier is it?

9.8 References

ANP 2001, "Analysis of the Accident with the Platform P-36, Report of the ANP / DPC Commission of Investigation", Agencia Nacional do Petroleo, Gas Natural e Biocombustiveis, July, http://www.anp.gov.br/images/EXPLORACAO_E_PRODUCAO_DE_OLEO_E_GAS/Seguranca_Operacional/Relat_incidentes/Analysis_of_the_Accident_with_the_Platform_P-36.pdf.

API RP 754, "Process Safety Performance Indicators for the Refining and Petrochemical Industries" 3rd Edition, American Petroleum Institute, Washington, D.C., US, 2021.

CCPS Glossary, "CCPS Process Safety Glossary", Center for Chemical Process Safety, https://www.aiche.org/ccps/resources/glossary.

CCPS 2011, "Process Safety Leading and Lagging Metrics...You Don't Improve What You Don't Measure", Center for Chemical Process Safety, New York.

CCPS 2019, "Process Safety Metrics Guide for Selecting Leading and Lagging Indicators Version 3.2", https://www.aiche.org/sites/default/files/docs/pages/ccps_process_safety_metrics_-_v3.2.pdf.

IOGP 456, "Process safety – recommended practice on key performance indicators", International Association of Oil and Gas Producers, London, U.K.

NASA 2008, "That Sinking Feeling: Total Loss of Petrobras P-36", https://sma.nasa.gov/docs/default-source/safety-messages/safetymessage-2008-10-01-lossofpetrobrasp36-vits.pdf?sfvrsn=c4a91ef8_4.

PSIE, CCPS Process Safety Incident Evaluation App, https://www.aiche.org/ccps/tools.

UN, "Globally Harmonized System Of Classification And Labelling Of Chemicals (GHS) Sixth Edition", United Nations, New York and Geneva, 2011, https://www.un-ilibrary.org/transportation-and-public-safety/globally-harmonized-system-of-classification-and-labelling-of-chemicals-ghs-sixth-revised-edition_591dabf9-en

10
Project Design Basics

10.1 Learning Objectives

The learning objectives of this chapter are:

- Identify and understand the purpose of basic process design documentation,
- Update basic process safety information, and
- Explain the concept of inherently safer design and the hierarchy of controls.

10.2 Incident: Mars Climate Orbiter lost contact, 1999

10.2.1 Incident Summary

The Mars Climate Orbiter (MCO) was launched on December 11, 1998 and contact was lost on September 23, 1999 as it entered into an orbit around Mars.

Key Points:

Stakeholder Outreach – Are you speaking the same language? In large projects and complex operations, it is important that people have the same understanding of relevant terminology and are using the same basis such that all the project/operation parts work safely together.

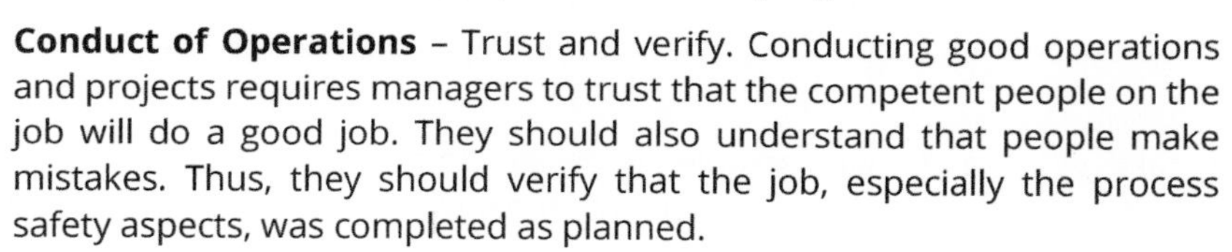

Conduct of Operations – Trust and verify. Conducting good operations and projects requires managers to trust that the competent people on the job will do a good job. They should also understand that people make mistakes. Thus, they should verify that the job, especially the process safety aspects, was completed as planned.

10.2.2 Detailed Description

The Mars Surveyor 1998 program included the Mars Climate Orbiter and the Mars Polar Lander which were launched separately with the intent to study the weather on Mars. The MCO would also serve as a communication relay for the Mars Polar Lander. (NASA 2018) The Mars Climate Orbiter included propulsion and equipment modules. The mass at launch was 629 kg (1387 lb) which includes 291 kg (642 lb) of propellant.

The spacecraft reached Mars. It passed behind Mars and contact was not re-established. Some of the spacecraft commands were in U.S. customary units instead of being converted to metric. A navigation error resulted from some spacecraft commands being sent in English units instead of being converted to metric. Due to this error, the MCO would have entered the Martian atmosphere at the incorrect altitude and would have been destroyed on entry.

A simple unit conversion error is why it happened. However, understanding why that unit conversion error happened gets into the root causes. Contributing causes listed in the NASA report are: (NASA 1999)

1. Undetected mismodeling of spacecraft velocity changes
2. Navigation Team unfamiliar with spacecraft
3. Trajectory correction maneuver not performed
4. System engineering process did not adequately address transition from development to operations
5. Inadequate communications between project elements
6. Inadequate operations navigation team staffing
7. Inadequate training
8. Verification and validation process did not adequately address ground software

10.2.3 Lessons

Stakeholder Outreach. NASA projects include a large array of contractors and subcontractors. Communication and project hand-offs can be challenging. Communication is key in large industrial projects that involve many engineering and construction contractors and subcontractors working around the globe to build a single installation. In both cases, keeping everyone communicating and working together well is required to deliver a successful project.

Conduct of Operations. Building on the large number of stakeholders, the way projects are managed must be controlled to support those communications and hand-offs. In this NASA case, ineffective communications between project elements and teams led to missed steps. The systems in place to verify that the project was proceeding as planned, did not address all areas.

Projects can take years and many people to design and construct. Often, business pressures or the desire to see the finished product push people to rush through verification steps. Despite these pressures, topics such as consistent language (units), should be included in verification processes. In a small project, this could be realized as a pre-startup safety review (PSSR). In a large project, it could be seen as a detailed verification and certification program that could take weeks to complete.

10.3 Introduction to Engineering Documentation

Project engineering is a collaboration between many engineers frequently in different companies and often in different locations around the world. It is clear from the Mars Orbiter mishap that engineering communication is imperative to a successful engineering project. The language of engineering communication is engineering documentation in the form of data, drawings, and specifications. Through clear and accurate engineering documentation, engineers can communicate complex details to one another and enable collaboration that takes a design from a concept to a detailed, ready-to-build project.

Engineering documentation describes the hazards, the process technology, the equipment, and the instrumentation in terms of the chemical, mechanical, and electrical engineering, respectively. It is the activity of the process knowledge management element of

RBPS. Many important pieces of process safety information are included in engineering documentation. (Refer to Section 2.5.)

Process Safety Information (PSI) – Physical, chemical, and toxicological information related to the chemicals, process, and equipment. It is used to document the configuration of a process, its characteristics, its limitations, and as data for process hazard analyses. (CCPS Glossary)

OSHA requires that employers complete a compilation of written process safety information before conducting any process hazard analysis and notes that this PSI also support management of change and incident investigation. The following are examples of PSI. (OSHA a)

- Hazards: physical, reactivity, and toxicity data (Refer to Chapters 4, 5, and 6)
- Process technology: process chemistry and safe upper and lower limits for parameters such as temperatures, pressures, flows or compositions
- Equipment: design codes and standards used, materials of construction, electrical area classification, and pressure relief system design
- Instrumentation and controls: distributed control systems and safety instrumented systems

It is challenging to maintain accurate data, especially over the life of an operating facility as changes are made. Engineering documentation should be checked and formally approved. Even with this, some data may not be updated, and mistakes may be made in recording changing data values. It is tempting to believe data when it is found in engineering documentation, but it is always important to verify that the data are correct. Check the data accuracy by confirming with other sources or going back to the original source.

10.4 Common Engineering Documentation

This section discusses common engineering documentation including the following.

- Block Flow Diagram
- Process Flow Diagram (PFD)
- Piping and Instrumentation Diagram (P&ID)
- Plot Plan
- Isometric

Projects include those that start with an empty site (green field) and those that are modifications to or additions to an existing process unit (brown field). The first step is to develop a block flow diagram. The block flow diagram is used at the conceptual phase when decisions are being made about the overall process. A block flow diagram can depict an entire facility as shown in Figure 10.1 or a single process unit.

Block Flow Diagram – A simplified drawing representing a process. It typically shows major equipment and piping and can include major valves. (CCPS Glossary)

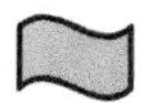

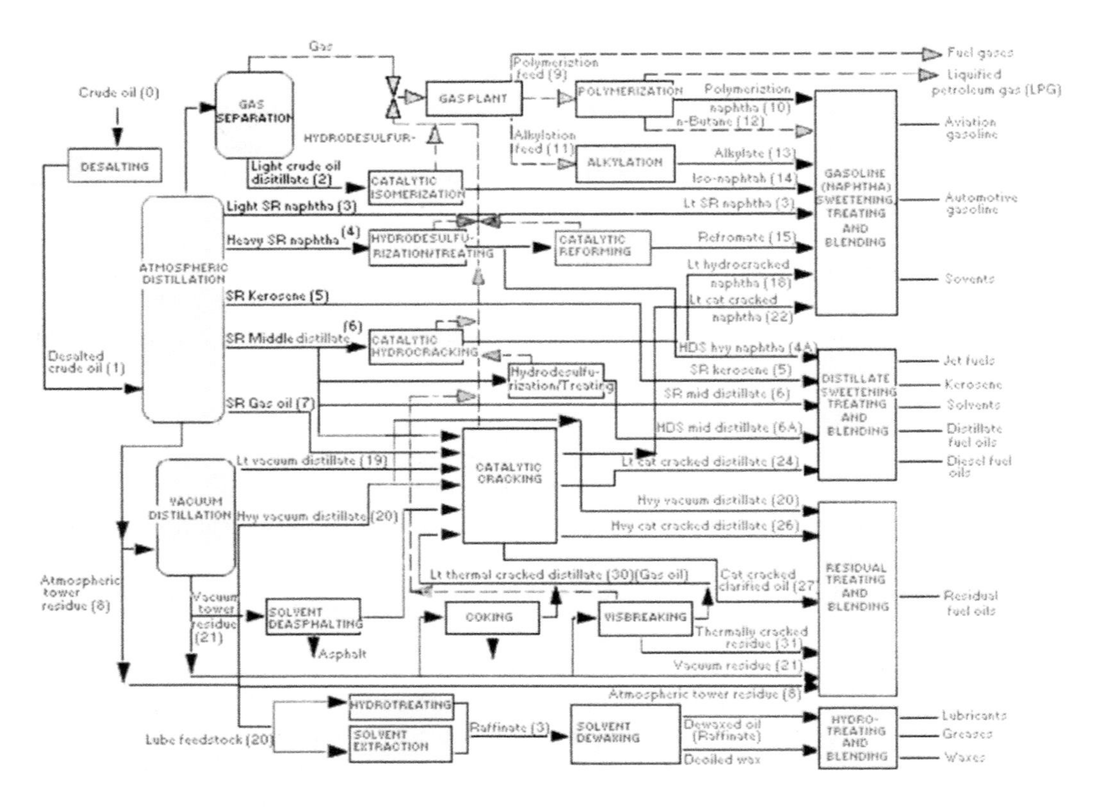

Figure 10.1. Example refinery block flow diagram
(OSHA b)

The block flow diagram is valuable in providing an overview of the process. As the conceptual design is progressed, more details are developed. The boxes and lines of a block flow diagram evolve to show greater detail including process flow interconnections, process controls, and operating parameters (e.g. temperature, pressure, level) as shown in Figure 10.2.

Process Flow Diagram – A diagram that shows the material flow from one piece of equipment to the other in a process. It usually provides information about the pressure, temperature, composition, and flow rate of the various streams, heat duties of exchangers, and other such information pertaining to understanding and conceptualizing the process. (CCPS Glossary)

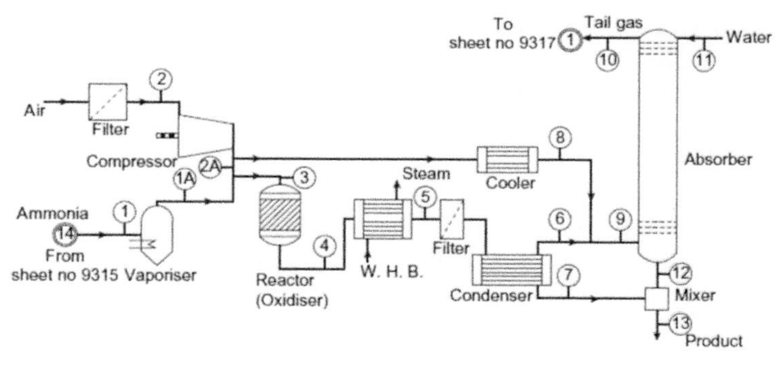

Flows kg/h pressures nominal

Line no. Stream component	1 Ammonia feed	1A Ammonia vapor	2 Filtered air	2A Oxidiser air	3 Oxidiser feed	4 Oxidiser outlet	5 W.H.B. outlet	6 Condenser gas	7 Condenser acid	8 Secondary air	9 Absorber feed	10 Tail(2) gas	11 Water feed	12 Absorber acid	13 Product acid	C & R Construction Inc
NH_3	731.0	731.0	—	—	731.0	Nil	—	—	—	—	—	—	—	—	—	Nitric acid 60 percent
O_2	—	—	3036.9	2628.2	2628.2	935.7	(935.7)(1)	275.2	Trace	408.7	683.9	371.5	—	Trace	Trace	100,000 t/y
N_2	—	—	9990.8	8644.7	8644.7	8668.8	8668.8	8668.8	Trace	1346.1	10,014.7	10,014.7	—	Trace	Trace	Client BOP chemicals
NO	—	—	—	—	—	1238.4	(1238.4)(1)	202.5	—	—	202.5	21.9	—	Trace	Trace	SLIGO
NO_2	—	—	—	—	—	Trace	(?)(1)	967.2	—	—	967.2	(Trace)(1)	—	Trace	Trace	Sheet no. 9316
HNO_3	—	—	—	—	—	Nil	Nil	—	850.6	—	—	—	—	1704.0	2554.6	
H_2O	—	—	Trace	—	—	1161.0	1161.0	29.4	1010.1	—	29.4	26.3	1376.9	1136.0	2146.0	
Total	731.0	731.0	13,027.7	11,272.9	12,003.9	12,003.9	12,003.9	10,143.1	1860.7	1754.8	11,897.7	10,434.4	1376.9	2840.0	4700.6	
Press bar	8	8	1	8	8	8	8	8	1	8	8	1	8	1	1	Dwg by Date
Temp. °C	15	20	15	230	204	907	234	40	40	40	40	25	25	40	43	Checked 25/7/1980

Figure 10.2. Example process flow diagram
(Towler and Sinnott)

The project is progressed further through the conduct of process modeling, development of heat and mass balances, and defining equipment specifications. The PFD evolves into a P&ID as shown in Figure 10.3. Typical P&ID symbols include those shown in Figure 10.4.

Piping and Instrumentation Diagram – A diagram that shows the details about the piping, vessels, and instrumentation. (CCPS Glossary)

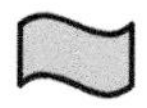

The P&ID includes details on the equipment, piping, and instrumentation. Details are included for normal operation as well as other operational phases such as start-up, catalyst regeneration, and emergency shutdown. Although a P&ID shows the relative location of process equipment, it is not to scale nor is it a model of equipment in the field.

Typically, multiple versions of P&IDs are issued as the engineering design is progressed and finalized. An "issued for construction" version documents the final design. Often an "as built" version documents the design as built capturing any final changes during the construction. In some cases, an "as operated" version is developed which may show disabled equipment and seasonally shutoff equipment.

Features of the P&ID include the following.

- Equipment and piping. All pieces of equipment (process and utility) and interconnecting piping, names and numbers, materials of construction, piping sizes, specification breaks (where materials or pressure specification changes).
- Process instrumentation and control. Temperature, pressure, and level instruments, their associated local or instrumented indications, and their use in the control of valves or other equipment.

- Valves. Control valves, emergency isolation valves, check valves, unit isolation valves and other types of valves. In addition, the failure position of valves on loss of instrument air or electrical power is shown.
- Safety systems. Pressure relief valves, depressuring systems, connections to flare systems.

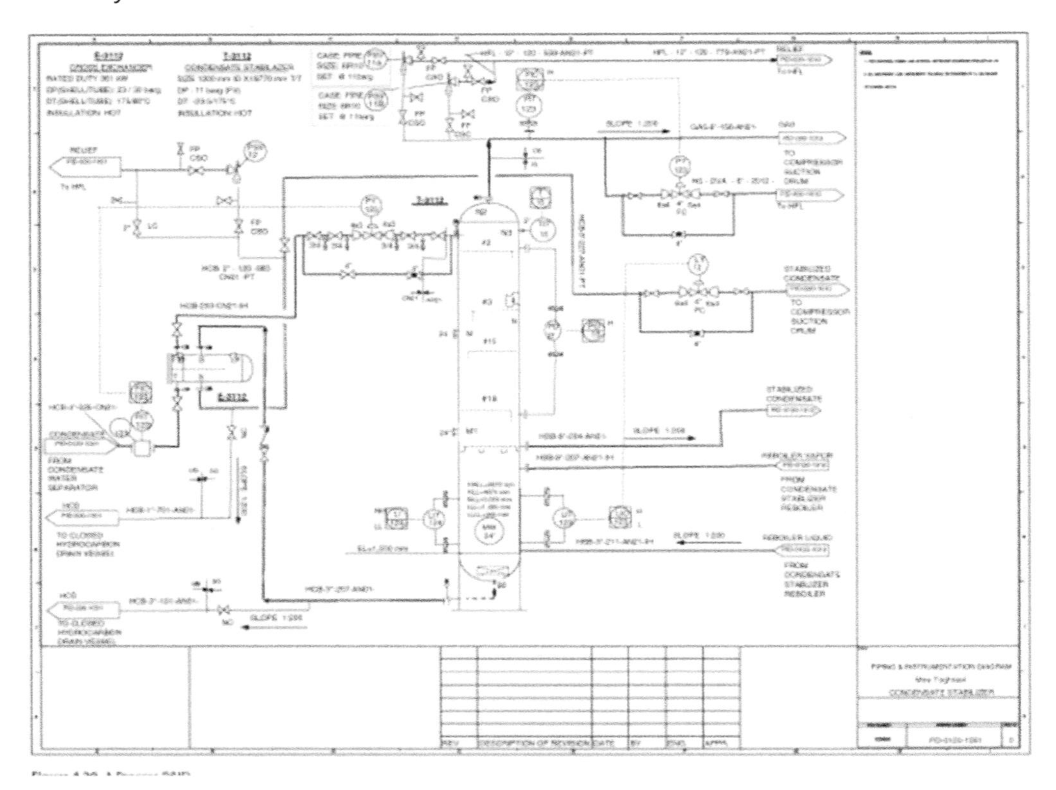

Figure 10.3. Example piping and instrumentation diagram (AIChE 2019)

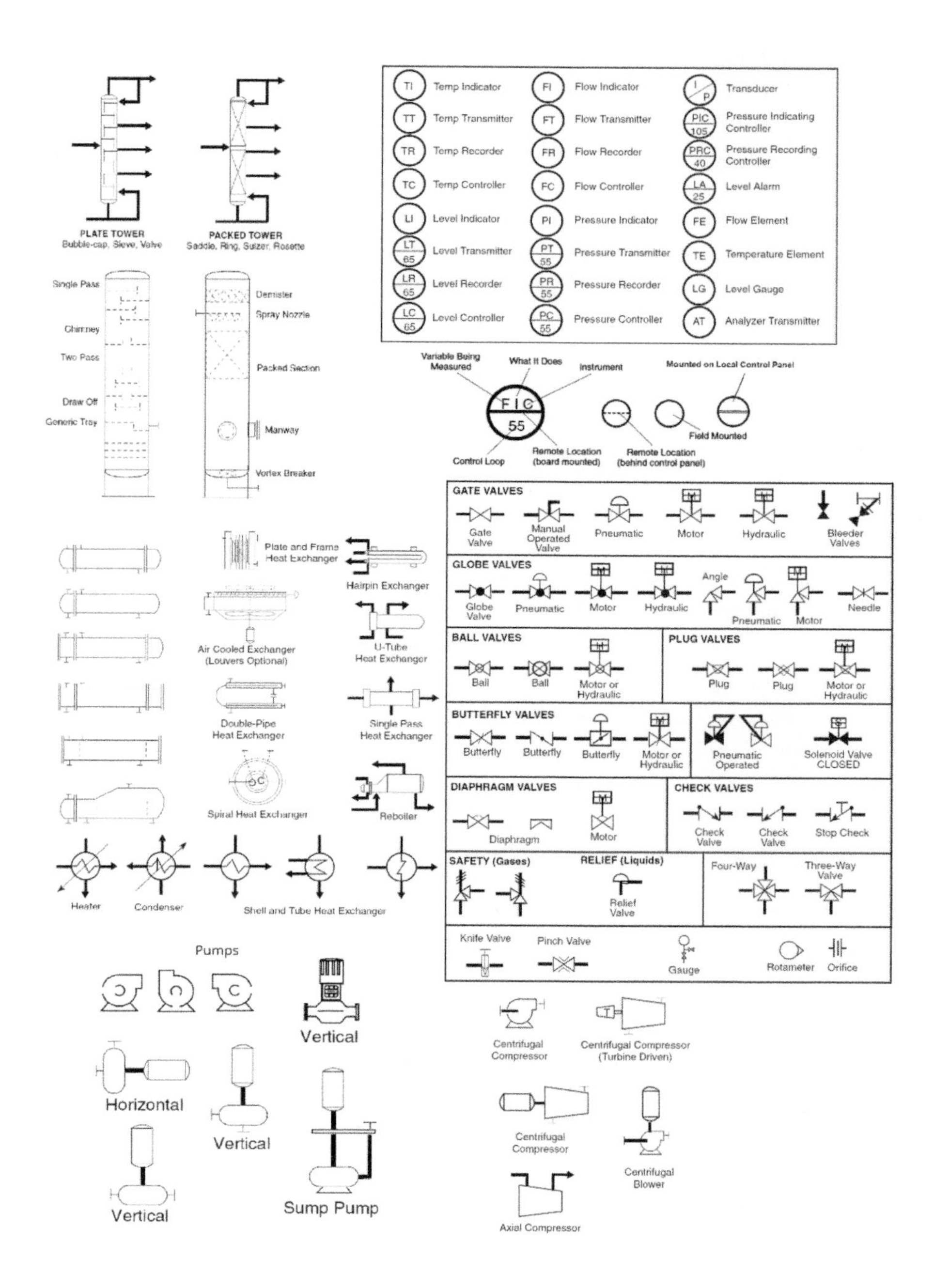

Figure 10.4. Typical P&ID symbols
(Patel)

The PFD and P&ID primarily describe the process. They do not include civil engineering or construction details.

Isometric Drawing – An isometric drawing depicts the process in three dimensions showing every turn and connection.

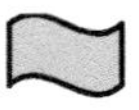

An isometric drawing, Figure 10.5, is developed to communicate to those fabricating the piping that will connect all process unit equipment. It includes line numbers, material type, size, weld locations and other details. A line designation table or line list is typically generated as an accompaniment to the P&ID and Isometrics drawing. This table usually lists process design conditions, pipe information, insulation requirements, and testing requirements.

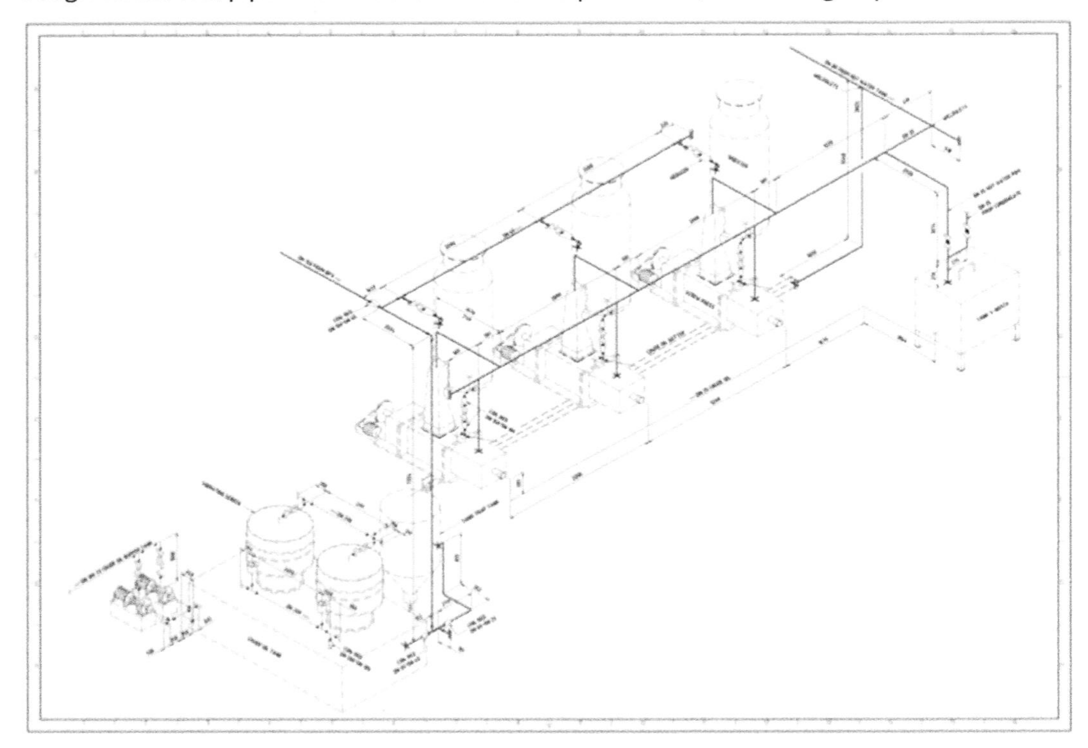

Figure 10.5. Example isometric drawing
(getdrawings)

Plot plans are developed for an overall site showing the relative location of offices, workshops, and process units, amongst other site features. They are also developed for a single process unit, as shown in Figure 10.6, showing the battery limits of the process unit and the location of process equipment as well as other details such as utilities and firefighting equipment. It is common for a basic plot plan to be annotated in several different versions to show important items or zones isolated (e.g. firefighting equipment, gas detectors, drainage, area classification, etc.) so that these can be seen clearly. Plot plan reviews are an important part of an engineering project as they are used to review maintenance access, equipment location with respect to occupied buildings and neighboring sites, and emergency response routes and assembly points, amongst other features.

Plot plan – A plot plan provides a top down, two-dimensional view of the layout

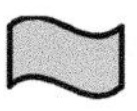

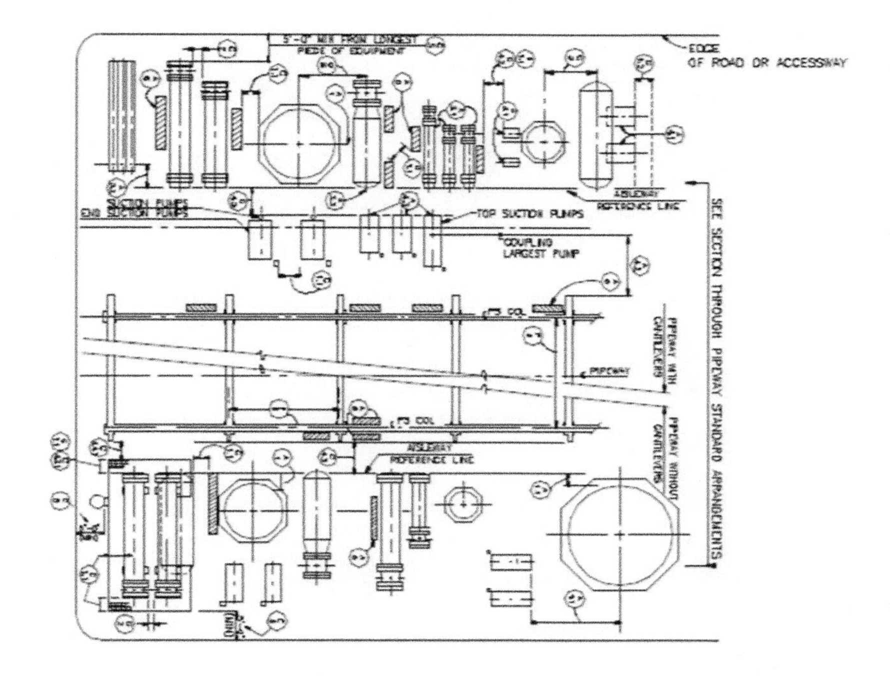

Figure 10.6. Example process unit plot plan (thepipingtalk)

10.5 Phases of a Project

The previous section on engineering documentation alluded to various phases of a project. Terminology and sequence may vary between companies; however, the project phases are typically formalized into the asset life cycle sequence shown in Figure 10.7. These project phases are described in Table 10.1. The hazard identification, consequence analysis, and risk assessment studies that are conducted during a project are described in Chapters 12, 13, and 14, respectively.

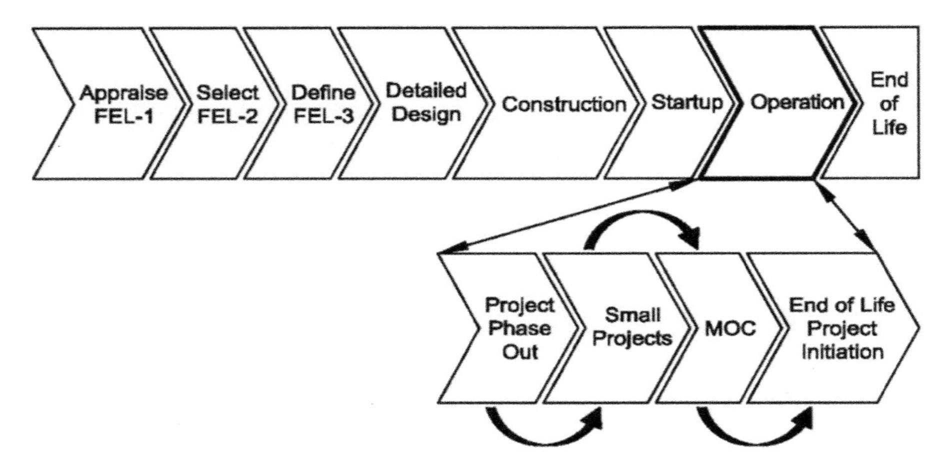

Figure 10.7. Asset life cycle stages including project phases (CCPS 2019 a)

Table 10.1. Asset life cycle stages including project phases (modified from CCPS 2020)

Asset Life Cycle Stages including Project Phases	Selected Activities and Process Safety Studies
Appraise phase	Develop and evaluate a broad range of project options, assess commercial viability, and rank feasible options to take forward. This stage is also referred to as Front End Loading (FEL) 1. Studies: Preliminary Hazard Analysis
Select phase	Evaluate concept options, maximizing opportunities and minimizing threats or uncertainties. A single concept to progress is normally chosen at this stage. This stage is also referred to as FEL 2. Evaluate inherently safer design options. Studies: Preliminary Hazard Analysis, What-If Analysis Selected PSI: chemical properties and composition
Define phase	Develop a basic design including plot plan, process flow diagrams, material and energy balances, and equipment data sheets. Schedule and cost are updated, and financial investment decisions may be made. This stage is also referred to as FEL 3. Studies: What-If Analysis Selected PSI: operating limits

Table 10.1 continued

Asset Life Cycle Stages including Project Phases	Selected Activities and Process Safety Studies
Detailed design phase	Detailed engineering of the chosen option is progressed. Applicable standards are applied. This includes P&ID drawings, detailed layout, equipment specification and data sheets, plot plans, fire protection, utilities etc. suitable to allow commercial bids. Typically, MOC is implemented starting after a HAZOP is completed. Studies: What-If Analysis, HAZOP, LOPA, consequence analysis, risk analysis, 3-D model review Selected PSI: refined chemical composition, MAWP, operating limits, pressure relief design, safety instrumented systems, hazardous area classification, risk register
Construction phase	This stage includes fabrication, construction, installation, and pre-commissioning.
Start-up phase	The commissioning and start-up phase to ensure that the completed facility meets its design specifications. Pre-Startup Safety Reviews (PSSR) or more detailed commissioning procedures occur in this phase. Emergency response plans are developed and response equipment operational. Selected PSI: refined documentation to "as built" status
Operation stage	This stage includes handover from projects to the operations and the years of operation. During operations, modifications may be made in the form of small projects and Management of Change (MOC) changes. Studies: Checklists, What-If Analysis, HAZOP, LOPA, consequence analysis, risk analysis Selected PSI: Keep all PSI current
Decommissioning (Abandonment) stage	At the end of the facility life, the facility is safely decommissioned. Abandonment should be treated like a new project. Studies: Checklists, What-If Analysis, HAZOP, consequence analysis, risk analysis

The engineering project phases include aspects that relate to many of the Risk Based Process Safety elements. *Compliance with standards* should be reflected in the engineering design. It should be compliant with regulations and based on recognized engineering codes and standards. As the project is progressed, *hazard identification and risk assessment* will drive the identification of hazards and their management through design. This is discussed further later in this Chapter and in Chapters 12 through 15. Projects may be designed with an intended life of many years. The *asset integrity and reliability* to support this lifespan is dependent on the materials of construction and corrosion management approaches specified in the engineering design. *Management of Change* ensures that design changes are reviewed to determine if any process safety or environmental barriers included in the design are compromised and if any new hazards are introduced. *Operational Readiness* verifies that the required changes have been successfully completed and that the facility is ready for operation. It checks that all the safety features agreed to in design have been implemented in the constructed facility and it is safe to start-up.

An engineering project involves more than just the engineers. *Workforce involvement* is important in that the operators and maintenance technicians know how the facility 'really' works and can aid in making design decisions to support operability and maintainability. An engineering project typically involves many contractors, both engineering specialists and construction, which makes *Contractor Management* important in understanding interface requirements both in terms of design and on the worksite. *Safe Work Practices* are essential to support the safe working on the project during construction, commissioning, startup, and into production. *Training and Performance Assurance* ensures that personnel are well trained for startup and for operations. This training should ensure understanding of *Safe Work Practices* and *Operating Procedures*.

All of these elements and actions in total support a healthy process safety culture. The project should remember the importance of Stakeholder Outreach in involving and informing external stakeholders about the project hazards and benefits.

10.6 Important Pieces of Process Safety Information

The previous sections described the phases of an engineering project, the engineering documentation that is developed and refined through the project phases, and the process safety information that this documentation contains.

The Mars Orbiter example illustrates how getting a seemingly small piece of data wrong can have a big impact. While all process safety information is important, some engineering data are fundamental to process engineering design and the safe operation of the process.

Maximum Allowable Working Pressure (MAWP). This is used in defining safe operating limits and is a factor used in determining pressure relief valve set pressures as shown in Figure 10.8.

Maximum Allowable Working Pressure (MAWP) - The maximum gauge pressure permissible at the top of a completed vessel in its normal operating position at the designated coincident temperature specified for that pressure. (API 520)

Operating Limits. Operating parameters may vary, and automated or manual process controls seek to keep the parameters within a safe range referred to as safe operating limits as shown in Figure 10.8. As the parameter strays near to or outside of the safe operating envelope, alarms are set to warn the operator. The operator will troubleshoot to determine the problem and take action to return the process to within the normal operating envelope. Where the safe operating limits are exceeded but the parameter remains within design limits, this range is referred to as the buffer zone. If the process parameter moves outside of the buffer zone by exceeding the safe design limit, then predetermined actions should be taken to return the process to a safe state.

Safe Operating Limits (SOL) – , Limits established for critical process parameters, such as temperature, pressure, level, flow, or concentration, based on a combination of equipment design limits and the dynamics of the process. (CCPS Glossary)

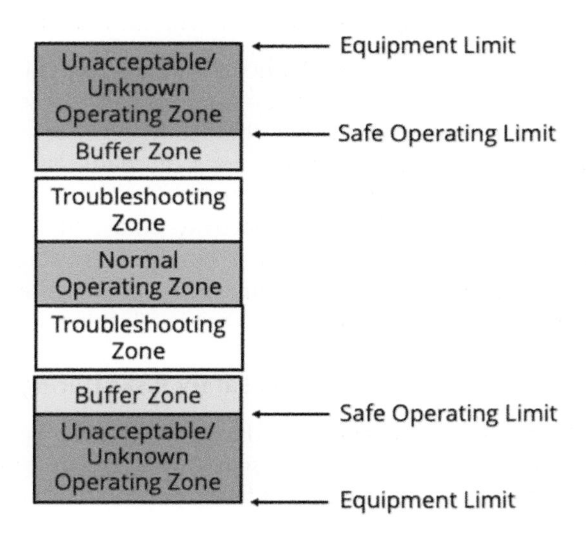

Figure 10.8. Relationship between operating zones and limits (redrawn from Forest 2018)

Chemical Composition. Composition may change over time or with different feedstock. Knowing the composition and managing changes in it can help to prevent corrosion, chemical incompatibilities, and poor process performance. Before an opportunity to change feedstock or change composition, a management of change should be conducted to ensure that the process unit materials of construction are suitable for the changed composition.

Valve Positions. It was previously noted that valves are shown on P&IDs. Their failure position is also indicated and warrants attention, e.g. FO for fails open, FC for fails closed and FL for fails last (position). When evaluating process flow and identifying process hazards, understanding which direction the process will, and won't, flow can be used to evaluate a potential problem.

Hazardous area classification. This was mentioned in Chapter 4 in the section on fire and explosion prevention. Classified areas are drawn on unit plot plans to indicate where protected electrical equipment, typically intrinsically safe, is required.

Maximum Intended Inventory. This is a list of all process chemicals by fixed and maximum controlled variable amount describing the total inventory by chemical expected in a process.

- Fixed inventory includes pumps, pipes, etc,
- Maximum controlled variable inventory (the maximum amount intended) includes columns, vessels, tanks, etc. controlled by passive means such as overflow, or active means with alarms, interlocks and safety instrumented systems,

- All process chemicals including raw materials, reactants, solvents, intermediate products, byproducts, finished goods, process aids, additives, lubricants, and process utilities, and
- Total inventory by chemical describes the inventory by equipment by chemical for pure components, and the weight percentage of chemicals in a mixture.

This is one of the most important pieces of process safety information helping to define the hazards of the process. It states the maximum amounts of each process chemical and the location of those chemicals in the process. This information helps to determine the "how bad can it be" portion of RBPS. The amounts and locations of chemicals be used in defining scenarios, as in PHAs (see chapter 12) and LOPA (see section 14.7) and conducting dispersion modeling (see chapter 13). In conjunction with reactivity data and the CRW, the relative amounts and locations of chemicals help us define hazards associated with inadvertent mixing. Understanding the maximum intended inventory provides a starting point for inherently safer design principles (see section 10.7.2) - especially minimization.

10.7 Methods to Prevent and Mitigate Process Safety Risks During Project Design

Section 10.5 described the staged project approach during which the engineering design evolves from a concept to a detailed engineering design. During this evolution, methods are available to prevent and mitigate process safety incidents that could occur over the life of the operating facility that is being designed.

10.7.1 Compliance with Standards

As discussed in Chapter 3, regulations and industry codes, and standards codify the learnings of industry. Company standards codify the learnings of a company and the specifics of how industry codes and standards should be applied. Process Safety Management systems also describe how process safety should be addressed in project engineering. Much attention is paid to learning from incidents and trying to prevent their future occurrence. Using regulations, codes, and standards as a foundation for engineering design is a key step in this direction.

10.7.2 Inherently Safer Design

Inherently safer design is a concept introduced by Trevor Kletz in an article titled "*What You Don't Have, Can't Leak*" and expanded subsequently. (Kletz 1978).

Inherently safer design (ISD) – A way of thinking about the design of chemical processes and plants that focuses on the elimination or reduction of hazards, rather than on their management and control. (CCPS Glossary)

Kletz's concept includes four principal techniques of achieving an inherently safer design as listed in Figure 10.9.

Minimize: Using or having smaller quantities of hazardous substances	**Substitute**: Replacing a chemical/material with a benign or less hazardous substance; or replace a process or processing technology with one that is benign or less hazardous	**Moderate**: Using less hazardous or energetic processing or storage conditions, a less hazardous form of a material, or facilities that minimize the impact of a release of hazardous material or energy	**Simplify**: Designing facilities that eliminate unnecessary complexity and make operating errors less likely, and which are forgiving of errors that are made.

Figure 10.9 Inherently safer design principles (CCPS 2019 b)

The "simplify" principle addresses human factors which will be discussed in Chapter 16. Humans make mistakes, even with the best knowledge and skills. Incorporating aspects in an engineering design to enable successful human performance will support good process safety performance.

These four techniques make ISD sound simple, but in reality, it is a bit more complex. "A technology can only be described as inherently saf***er*** when compared to a different technology...A technology may be inherently safer than another with respect to some hazards but inherently less safe with respect to others..." (DHS 2010) For example, minimizing the quantity of a hazardous material stored at a facility may reduce the facility risk; however, it may necessitate increased transportation of that hazardous material to supply the facility resulting in an increased risk along the transportation route.

ISD is applicable through the life cycle stages and includes manufacturing, transportation, storage, processing, and decommissioning. The greatest opportunity to implement ISD is during the design phases where the implementation of an ISD approach only involves a changing ink on an engineering drawing as opposed to modifying steel in a facility.

ISD is iterative during engineering design development. During conceptual phases, major decisions are made such as use of an inherently safer technology or choosing to export product by ship versus truck. As the design details are progressed, the size of vessels is determined, and the control systems are developed. In an operating facility, operating procedures can be provided in a clear and concise way with visual cues that makes them easy to understand and use.

10.7.3 Hierarchy of Controls

Hierarchy of controls is similar, in concept, to ISD and is used in many industries. It depicts hazard controls in order of decreasing effectiveness as shown in Figure 10.10. The first two levels in the figure are ISD principles and are typically more easily implemented during the early stages of a project prior to the detailed equipment design. The remaining levels reflect that "It is unlikely that any technology will be 'inherently safer' with respect to all hazards, and other approaches will be required to manage the full range of hazards and risks." (DHS 2010) When the engineering design is finalized or the project is completed, administrative controls are easier to implement.

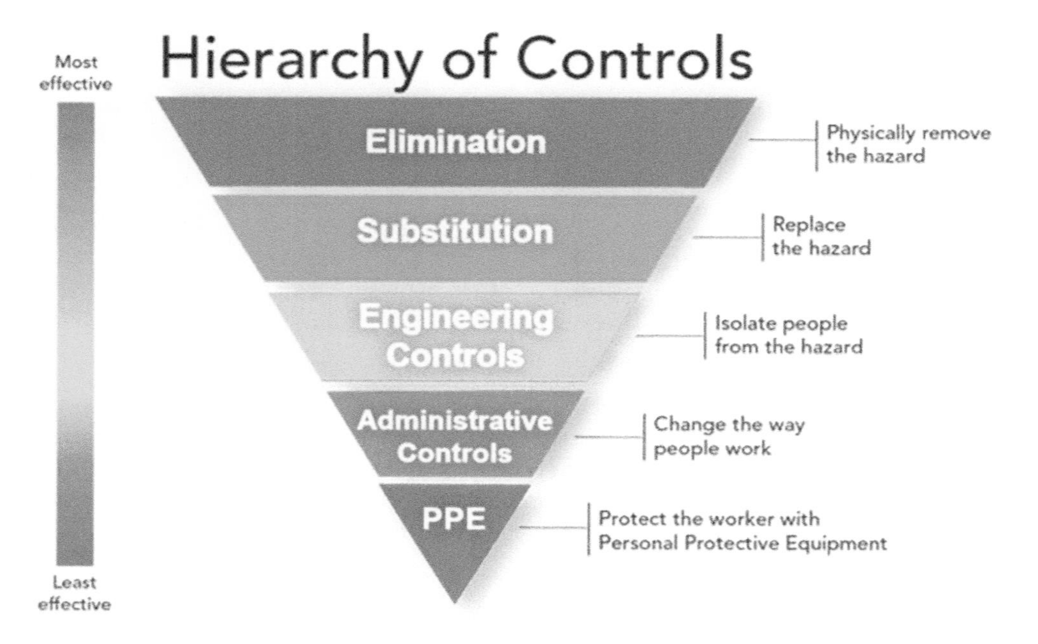

Figure 10.10. Hierarchy of controls
(CDC)

Considering the engineering and administrative control levels of the hierarchy of controls, the reliability of these controls can have a significant impact on process safety. In general,

PASSIVE engineering controls are more reliable than

ACTIVE engineering controls, which are more reliable than

ADMINISTRATIVE controls. (CCPS 2009)

Passive controls are those that function without an action required to enable them. For example, a dike around a storage tank and fireproofing are considered passive controls. Active controls require an action to enable them. For example, a shutdown system that requires a valve to close either by instrument control or by manual operation.

A combination of the concepts of ISD and hierarchy of controls is depicted in Figure 10.11. It describes how process safety may be incorporated into an engineering project design. The key first step is to identify and understand hazards. If hazards are not identified, they will not be managed. Next in ISD is eliminating the hazard or reducing the severity or likelihood through the four principles that Kletz presented. Then is segregating the hazard from the people or environment that could potentially be harmed through the use of the engineered and procedural safeguards lower in the hierarchy of controls that reduce the residual risks. This identifying, eliminating, reducing continues as the engineering details are defined over the project phases.

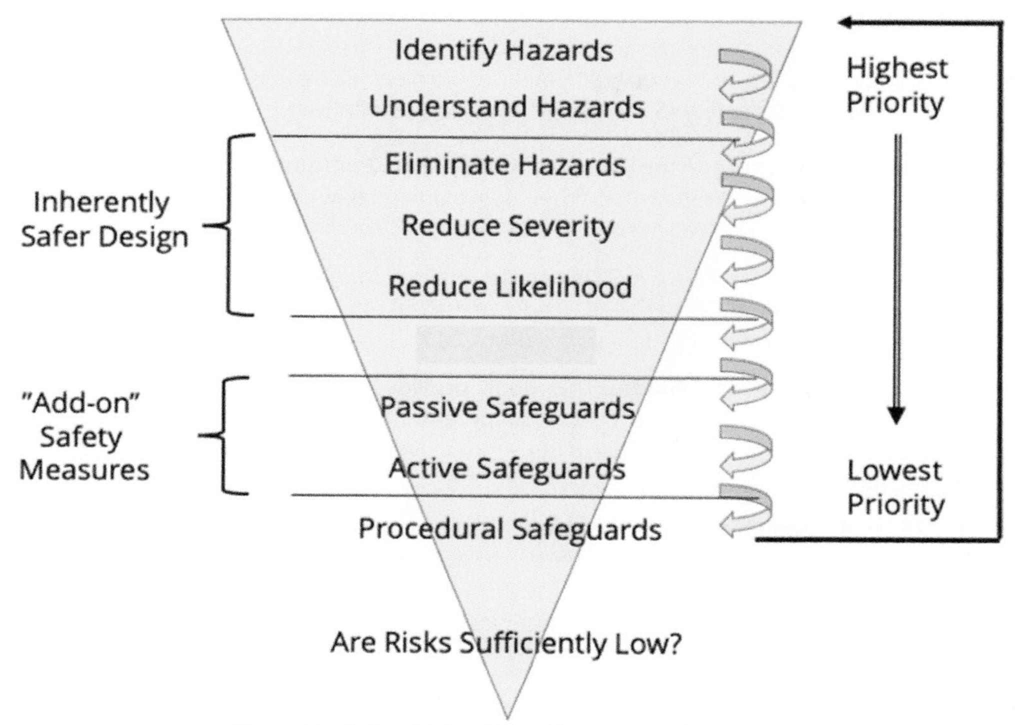

Figure 10.11. Combining ISD and hierarchy of controls
(adapted from Broadribb 2010)

10.8 What a New Engineer Might Do

When starting to work in a process plant, or doing engineering design or process development, a new engineer should learn what the process safety hazards are in the plant and process(es) they are assigned to and the systems in place to manage those hazards. Engineers should become familiar with the existing process safety information to understand the hazards and engineering controls of the process. Chemical, mechanical, instrumentation and electrical, and control engineers all contribute to developing and maintaining PSI.

Chemical engineers (often filling the role of the process engineer) are likely to be responsible for determining and keeping information on the process up to date. This information might include hazards of the chemicals, chemical reactions and reactivity hazards, heat and mass balances, relief device sizing basis and calculations, P&IDs, equipment specifications, and piping design detail. In research and development, and scale up, developing and organizing the PSI will be part of the chemical engineer's responsibility.

Engineers that are involved in making design decisions should keep in mind the concepts of inherently safer design and the hierarchy of controls. All approaches are not equal when it comes to reducing process safety risk. These concepts can be used to frame design decisions so that process safety is incorporated in the design process.

10.9 Tools

Tools that may be used to support engineering project design and the consideration of process safety in the design, include the following.

CCPS Guidelines for Integrating Process Safety into Engineering Projects. This book addresses all engineering projects, whether upstream or downstream, large or small. It is based on a project life cycle approach. (CCPS 2019 a)

The Project Management Institute **A Guide to the Project Management Body of Knowledge PMBOK Guide**, is a widely recognized resource for effective project management in any industry. It is available in multiple languages (PMI 2017)

The Oil & Gas Engineering Guide. This book provides a comprehensive description of engineering project activities, sequence, and documentation. It includes many illustrations and would support an engineer in any discipline to improve their understanding of engineering project execution. (Baron 2018)

ISO 17776:2016 Petroleum and natural gas industries — Offshore production installations — Major accident hazard management during the design of new installations. This document is directed toward larger projects but is applicable to smaller ones as well. It is also relevant to onshore facilities. (ISO 17776)

Other tools include process simulators and consequence modeling packages that allow design alternatives to be quickly evaluated for their contribution to risk reduction.

10.10 Summary

An engineering project is typically a collaboration of the efforts of many different engineers frequently in different locations. The clarity and accuracy of engineering documentation is crucial in making sure these engineers communicate effectively and avoid any errors in design. This documentation evolves as the project engineering details evolve. It continues into the operation phase of a facility's life and is used in management of change evaluating proposed modifications to the existing facility.

The best time to include process safety considerations in an engineering design is in the development stages. Inherently safer design and the hierarchy of controls are concepts that can be used to frame design decisions ensuring process safety is addressed. Building process safety studies into a project execution strategy and verifying that action items resolved before startup can support a successful project.

10.11 Other incidents

Other incidents involving problems in engineering design include the following.

- Union Carbide MIC release, Bhopal, India, 1984
- Space Shuttle Challenger, U.S., 1986
- Space Shuttle Columbia, U.S., 2003
- Air France Flight 447, Brazil, 2009

10.12 Exercises

1. List 3 RBPS elements evident in the Mars Climate Orbiter incident summarized at the beginning of this chapter. Describe their shortcomings as related to this accident.
2. Considering the Mars Climate Orbiter incident, what actions could have been taken to reduce the risk of this incident?
3. Name three things that are on a P&ID and not PFD.
4. In what project stage are the initial PFD and P&ID developed?
5. What engineering drawing shows how vessels and piping are located and connected in the field (in three dimensions)?
6. Give two examples of a passive control measure and two of an active measure.
7. Process safety events are categorized into 4 tiers. How are Tier 1 and 2 different from Tier 3 and 4?
8. The MAWP for a vessel is 100 psig. The pressure relief valve is set to relieve at a pressure of 100 psig. The process performs best at an operating pressure of 85 to 90 psig. Apply these values to the diagram shown in Figure 10.8.
9. State and explain the four principles of inherently safer design.
10. Which option is preferred in terms of process safety? Explain.
    - Locating the control room some distance from the operating unit?
    - Designing the operating unit such that high pressure and temperature excursions are sensed by the control system and corrective actions taken automatically by the control system to prevent an incident?
11. Providing pressure and temperature instrumentation read outs in the control room and training the operators on what is considered a high pressure and what action to take to address it?

10.13 References

AF, Automation Forum, https://automationforum.in/t/process-diagram-symbols/663.

AIChE 2019, "Piping and Instrumentation Diagram Development", American Institute of Chemical Engineers, John Wiley & Sons, Hoboken, N.J.

API 520, "Sizing, Selection, and Installation of Pressure-Relieving Devices in Refineries", American Petroleum Institute, Washington, D.C., 2000.

Baron 2018, *The Oil & Gas Engineering Guide*, Herve Baron, Editions Technip, Paris, France.

Broadribb 2010, "HAZOP/LOPA/SIL Be Careful What You Ask For!", American Institute of Chemical Engineers, 2010 Spring Meeting, 6th Global Congress on Process Safety, March 22-24.

CCPS Glossary, "CCPS Process Safety Glossary", Center for Chemical Process Safety, https://www.aiche.org/ccps/resources/glossary.

CCPS 2009. *Inherently Safer Chemical Processes: A Life Cycle Approach 2nd edition*, Center for Chemical Process Safety, John Wiley & Sons, Hoboken, N.J.

CCPS 2019 a, *Guidelines for Integrating Process Safety into Engineering Projects*, Center for Chemical Process Safety, John Wiley & Sons, Hoboken, N.J.

CCPS 2019b, *Guidelines for Inherently Safer Chemical Processes A Life Cycle Approach*, Center for Chemical Process Safety, John Wiley & Sons, Hoboken, N.J.

CCPS 2020, *Process Safety in Upstream Oil and Gas*, Center for Chemical Process Safety, John Wiley & Sons, Hoboken, N.J.

CDC, https://www.cdc.gov/niosh/topics/hierarchy/default.html#

DHS 2010, "Final Report: Definition for Inherently Safer Technology in Production, Transportation, Storage and Use", U.S. Department of Homeland Security.

Forest 2018, "Know Your Limits", *Process Safety Progress*, American Institute of Chemical Engineers, New York, NY, December.

ISO 17776, "Petroleum and natural gas industries — Offshore production installations — Major accident hazard management during the design of new installations", International Organization for Standardization, Geneva, Switzerland, 2016.

Getdrawings, http://getdrawings.com/isometric-pipe-drawing#isometric-pipe-drawing-29.jpg.

Kletz 1978, What you Don't Have, Can't Leak", *Chemistry & Industry*, London, United Kingdom.

NASA 1999, "Mars Climate Orbiter Mishap Investigation Board Phase I Report", National Aeronautics and Space Administration, November 10.

NASA 2018, "Mars Climate Orbiter", National Aeronautics and Space Administration, Space Science Data Coordinated Archive, https://nssdc.gsfc.nasa.gov/nmc/spacecraftDisplay.do?id=1998-073A

OSHA a, https://www.osha.gov/dts/osta/otm/otm_iv/otm_iv_2.html.

OSHA b, https://www.osha.gov/sites/default/files/publications/osha3132.pdf.

Patel, Varun, https://hardhatengineer.com/pid-pfd-pefs-pfs-drawing-symbols-legend-list/

PMI 2017, *A Guide to the Project Management Body of Knowledge PMBOK Guide*, Project Management Institute, Newtown Square, Pennsylvania.

Thepipingtalk, http://thepipingtalk.com/what-is-process-plant-plot-plan-basic-information-needed-to-start/.

Towler and Sinnott, Towler G, Sinnott R. Chemical Engineering Design: Principles, Practice and Economics of Plant and Process Design. 2nd ed. Boston: Elsevier; 2013.

11 Equipment Failure

11.1 Learning Objectives

The learning objectives of this chapter are:

- Understand how equipment may fail,
- Describe how to prevent equipment failure through asset integrity and reliability efforts, and
- Be knowledgeable of inspection, testing, and maintenance programs

11.2 Incident: Buncefield Storage Tank Overflow and Explosion, Hemel Hempstead, England, 2005

11.2.1 Incident Summary

A delivery of gasoline from a pipeline into a storage tank in the Buncefield depot in the U.K. began on Sunday morning, 11 December 2005. The level control and shutoff systems in place failed to operate and hence no alarm sounded. The tank overflowed and gasoline cascaded down the side of the tank. Up to 272 metric tonne (300 ton) of gasoline escaped from the tank (MIIB 2008). About 45 minutes after the release started, a series of explosions took place. These explosions generated overpressures higher than what would have been expected in a normal vapor cloud explosion. Some have speculated it was a Deflagration to Detonation Transition (DDT) event.

Forty-three people were injured and about 2,000 were evacuated from the area. If the incident had happened on a weekday during normal working hours, it could have resulted in more injuries, and even fatalities. The explosions caused the largest fire in peace time Europe that engulfed more than 20 large storage tanks over a large part of the Buncefield depot. The fire burned for five days, destroying most of the depot (Figure 11.1). In addition to destroying large parts of the depot, the incident caused widespread damage to surrounding property and disruption to local communities. Houses close to the depot were destroyed and others suffered severe structural damage. Buildings in the area, as far as 5 miles (8 km) from the depot, suffered lesser damage, such as broken windows, and damaged walls and ceilings. The Major Incident Investigation Board (MIIB) estimated the cost of the incident was £1 billion.

The occurrence of a detonation which consumes the entire flammable mass of the cloud (not just the mass inside the congested space) and produces much higher overpressures than a deflagration, was a surprise to experts. Prior to this event, it was not expected that a gasoline tank farm vapor cloud explosion could make the deflagration to detonation transition (DDT). The MIIB recommended that research be done to understand why the DDT occurred. The results of this research can be found in the HSE report titled "Buncefield Explosion Mechanism Phase 1". (HSE 2009 a)

Figure 11.1. Buncefield storage depot after the explosion and fires (HSE 2011)

Key Points:

Stakeholder Outreach – Process Safety Culture – A good process safety culture does not "live with" frequent instrument failures. Instead, it investigates to find out what is causing the failures and addresses the problem. Thus, the barriers against a process safety incident remains healthy.

Management of Change – A large change in throughput is a change that needs to be managed. This change may have implications on equipment, operations, and provision of adequate staffing.

Conduct of Operations – A process should be designed with layers of protection sufficient for the magnitude of the risk.

11.2.2 Detailed Description

The Buncefield depot is a large tank farm north of London near Hemel Hempstead in Britain. The Buncefield depot was constructed in 1968. At the time of the incident, three sites at the depot were operated by Hertfordshire Oil Storage (a joint venture between Total and Chevron), British Pipeline Agency (a joint venture between Shell and BP) and BP Oil.

The Buncefield depot or tank farm is a large site that stores gasoline, heating oil and aviation fuel in over 25 storage tanks (Figure 11.2). The fuels are received via two 0.25 m (10 in) and one 0.36 m (14 in) pipelines. Gasoline and heating oil from the tanks are offloaded into trucks for delivery, and the jet fuel is sent out by pipeline. The depot is about three miles away from the center of Hemel Hempstead, the nearest town.

The storage tank involved was Tank 912. Tank 912 was a 6,000 m^3 (1.6 million gal) floating roof tank with an automatic tank gauging (ATG) system that was monitored in the control room. Operators could operate the appropriate valves to shut off flow and/or divert it to other tanks. The tank had an alarm for high and high-high level that could be set by the supervisors. Tank 912 also had an independent high-level switch (IHLS) that would stop incoming flow at a high-high level by closing the inlet valves and provide an audible and visual alarm in the control room.

The tank started receiving about 550 m^3/hr (145,294 gal/hr) of gasoline (containing 10% isobutane) at about 7:00 PM on Saturday evening. The isobutane had been recently added to make a winter blend making the gasoline more volatile. At 3:00 AM Sunday the tank was about 2/3 full, but the level gauge stopped recording any further increase in level. The independent high-level switch (IHLS) shutdown did not work. At about 5:20 AM the tank began to overflow, but flow into the tank continued, even increasing in rate to about 890 m^3/hr (235,113 gal/hr).

As fuel continued to overflow from Tank 912, a dense vapor cloud up to 2 m (6.6 ft) tall and covering an area of about 500 x 350 m (1640 x 1148 ft) formed, engulfing a large portion of the facility (Figure 11.3) (HSE 2017). The final extent of vapor cloud explosion is marked in yellow. The first explosion occurred at 6:01 AM. Initially the ignition source was hard to determine, candidates include; a pump house, heaters in the emergency generator building, and car engines (witnesses stated their cars began to run erratically, (i.e. surging due to drawing in fugitive gasoline vapors). Subsequent analysis (see below) has settled on the pump house as the initial site of ignition. Further explosions occurred and the entire facility was engulfed in fire.

Figure 11.2. Buncefield storage depot before the explosion (HSE 2017)

Figure 11.3. Buncefield Terminal site and wider area after explosion (HSE 2017)

The automatic tank gauging level detector had a history of failing due to sticking and this had not been corrected. The IHLS did not function because a test lever for the switch was not locked in the neutral position. The lever enabled testing of the high-level function, and/or the function of the low-level function (if the low-level function was installed) of the IHLS. Failure to lock the lever in the middle position allowed the lever to slip into the low-level test position, disabling the high-level function.

Experts were surprised by the severe damage from the explosion, given the low level of congestion at the site. The extent of the damage was such that experts concluded that a DDT occurred. This surprise led to recommendations to do further study of the mechanism for the DDT.

Figure 11.4. Breakup of liquid into drops spilling from tank top (Buncefield 2008)

The following factors contributed to the DDT:

- Mist formation as the gasoline spilled over top of storage tank. Normally, a spill of a liquid from a storage tank would be modeled as evaporation from the pool created by the spill. As the gasoline spilled from the top of Tank 912, liquid droplets formed, and this enabled the transport of air into the vapor cloud (Figure 11.4). (Mists can also increase the hazard of a flammable release because they can ignite at temperatures below their flashpoint, although that was not the case in this incident.)
- Low or no wind causing little dispersion and dilution of the flammable cloud. The lack of wind meant the cloud did not disperse. When dispersion occurs the concentration of vapor in the cloud is reduced by entrainment of air. At Buncefield this lack of dispersion led to the large cloud with a large portion (or almost all) of it in the flammable zone.
- Strong ignition source from the pump house. The pump house was near Tank 912 and was completely submerged in the cloud. The ignition in the pump house led to an explosion inside the pump house itself, and this explosion created a strong ignition source that also created turbulence around the pump house, leading to a strong external explosion and the DDT.

- Containment by hedgerows and trees provided an elongated path for DDT. Hedgerows near the pump house served as more congestion for the vapor cloud. Also, a tree-lined street next to the facility caused further acceleration of the flame front leading to detonation. That vegetation could do this, in effect acting like a pipe rack or wall, came as a surprise.

Regarding detonations, the report, "Review of vapour cloud explosion incidents" (HSE 2017), has challenged the conclusion that the Buncefield explosion, and several others, were detonations, based on the nature of some of the physical damage at the explosion sites. It hypothesizes that a mechanism in between a VCE and a detonation is possible, and the HSE has called for more investigation of this phenomena. Interested readers can obtain and read the HSE report. For brevity, this book will continue to call the Buncefield and Jaipur explosions detonations. The important thing to remember is that with these types of events the potential damage can be much worse than the common consequence models would indicate.

11.2.3 Lessons

Process Safety Culture. The atmospheric tank gauging (ATG) system became stuck (not registering a level change) 14 times in the three months before 11 December. Each time it was fixed by either the operators or the maintenance crew. Sometimes the failure was not even logged. The willingness to continue to operate with such an unreliable level control is indicative of a poor safety culture and is an example of normalization of deviance.

Compliance with Standards. The land use planning standards in the U.K. assumed that facility operators were in compliance with appropriate requirements. The MIIB recommended using a risk-based approach to land use planning, requiring the operators to develop a risk management plan.

Hazard Identification and Risk Analysis. Prior to this incident, the scenario that occurred at Buncefield had not been considered credible. Since land use planning in the U.K. was based on the worst credible case, the scenario was not part of the Land Use Planning process. Subsequently, guidance was published to update and improve standards at gasoline storage depots.

Operating Procedures. The operating procedures were inadequate. The procedures were not detailed enough (e.g. no safe operating limits, were included), and the supervisors on each shift used the available level alarms differently.

Asset Integrity and Reliability. The IHLS failed to close the manual inlet valve because the test lever was not secured. A safety critical device such as this switch not only needs to be tested on a regular basis but needs to be put back into service properly. The staff did not have procedures for putting the switch back into operation. This led the HSE to issue an alert on now to test the switch. The MIIB recommended that these storage sites improve their maintenance systems and conduct regular proof testing.

Management of Change. In 2002 a large increase in throughput to the facility occurred when an adjacent facility was shutdown. No MOC was done to see if the control systems and staffing levels were adequate for the increased throughput. The IHLS was installed in 2004.

The design of the high-level switch allowed the failure mode to occur. The failure mode could have been eliminated if an MOC review had been performed when the switch was installed.

Conduct of Operations. The level control system was inadequate for the system. Only one computer was used to monitor the ATG system for all the tanks. The system was designed with no backup. The system could have been designed to provide an alarm if it detected an inconsistency between the level in a tank and the incoming flow, but this was not done. The site operators did not have access to independent information about flowrates, or even if flow was present for two of the three incoming lines. In a well-designed control system, the operators should have been able to see that even though the level indication was unchanged, and that flow was still coming into the tank.

The MIIB recommended that automatic, high integrity overflow prevention systems, independent of the tank level system be installed. In accordance with current best standards (IEC 61511). The MIIB also recommended that the receiving site have ultimate control of the storage site, not the transmitting site.

11.3 Typical Process Equipment

This section discusses typical equipment used in refining, chemical, oil and gas, and many other industries. The material presented here is an overview of design considerations contained in *Guidelines for Engineering Design for Process Safety*, 2nd Edition. (CCPS 2012)

11.3.1 Pumps, Compressors, and Fans

Overview. Pumps, compressors and fans are used to move fluids from one point to another. In doing so, they impart energy, in the form of pressure and temperature, to the fluid being moved. If they are run with the inlet and or outlet blocked, they can heat the contained fluid, which can create hazards that will depend on the properties of the fluid being moved. As rotating equipment items, they will have seals around rotating shafts, whose failure can lead to leaks. Again, the hazards from leaks will depend on the fluid being moved. Finally, they can just fail to run or run for too long, leading to potential hazardous consequences to other parts of the process.

Common failure modes for pumps and compressors include stopping, deadheading and isolation, cavitation/surging, reverse flow, seal leaks, casing failures, and motor failures. Overly rapid closing of valves on pumps, including emergency isolation valves, can cause a short-term high pressure (termed water hammer) that can damage or rupture piping. If pumps are run dry (i.e. the inlet is blocked), the bearings can be damaged, causing high temperature.

Knowledge of the properties of the fluid is necessary to assess the hazards of the potential failures of fluid transfer equipment. Deadheading or isolation of pumps and compressors can lead to uncontrolled reactions, exothermic decomposition, and explosion hazards when moving chemicals that are reactive, have thermal stability issues, or are shock sensitive because pumps and compressors impart energy to the fluid. An example of this would be pumping Ammonium Nitrate (AN) solutions which become more sensitive to deflagration and detonation at high temperatures. If loss of containment due to a seal failure occurs, then the release of materials that are flammable can lead to fire and explosion hazards and corrosive or toxic materials can create personnel hazards. A low boiling fluid can flash, so knowledge of the vapor pressure/temperature is needed.

Example 1. The pump in Figure 11.5 was destroyed because the mechanical seal failed. The light hydrocarbon being pumped was released; it ignited and burned – causing extensive local damage. No one was near the pump when the fire occurred, consequently no injuries resulted (CCPS 2002 a).

Example 2. A 75 HP centrifugal pump was operated with both suction and discharge valves closed for about 45 minutes. It was believed to be completely full of liquid. As mechanical energy from the motor was transferred to heat, the liquid in the pump slowly increased in temperature and pressure until finally the pump failed catastrophically (see Figure 11.6). One fragment weighing 2.2 kg (5 lb) was found over 120 m (400 ft) away. No one was in the area consequently no injuries resulted (CCPS 2002 b).

Figure 11.5. Damage from fire caused by mechanical seal failure (CCPS 2002 a)

Figure 11.6. Pump explosion from running isolated
(CCPS 2002 b)

Design considerations for process safety. Two different types of pumps and compressors include centrifugal and positive displacement. When pumps and other rotating equipment are running, the process fluid can leak from between the rotating shaft and the body of the pump. Leaks can result in fires or toxic releases if the fluids are flammable or toxic. Different types of seal configurations are available to prevent these leaks. The selection of pump and seal type is usually dependent on process considerations. Every type of pump and seal has process safety considerations.

With compressors, liquid entry into the compressor can cause catastrophic failure. Protection should be provided upstream of compressors to remove liquids and associated shutdown systems should also be provided.

Centrifugal pumps (Figure 11.7) are susceptible to leaks, deadheading, running isolated, cavitation and reverse flow. Design configurations that have two pumps in parallel can be especially vulnerable to these failure modes because the possibility of starting the wrong pump.

Centrifugal pumps, as with other rotating equipment, need shaft seals between the process fluid and the external environment. The simplest form of a seal is packing material. This can degrade with time and leak. Mechanical seals are the next type. In a mechanical seal pump, a seal face is kept in contact between the shaft and casing. These seals leak less than packing but do require a lubricating fluid that must be compatible with the process fluid. Mechanical seals can be single or double (Figure 11.8). In a double mechanical seal, two seals sit back to back inside a chamber external to the pump. The seal chamber is flushed with a fluid, and leaks are contained and can be detected in this fluid. Double mechanical seals are better at preventing leaks but require more complex maintenance.

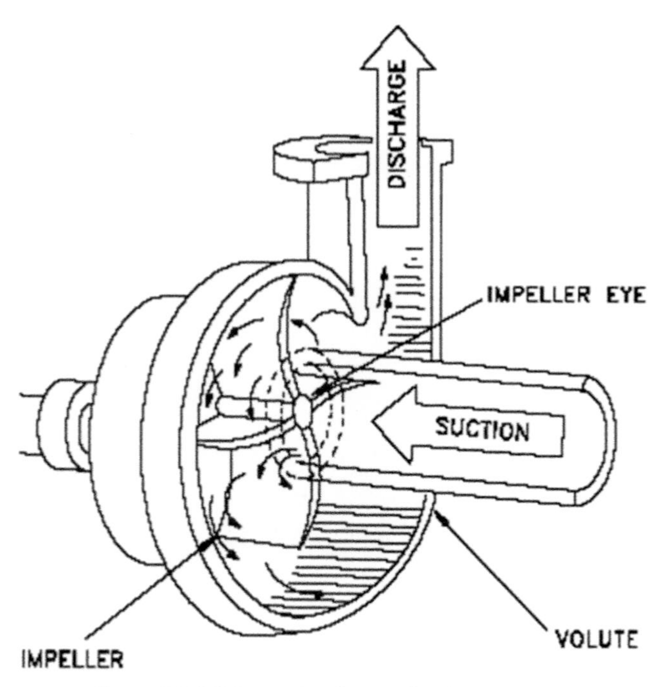

Figure 11.7. Schematic of centrifugal pump
(Kelley)

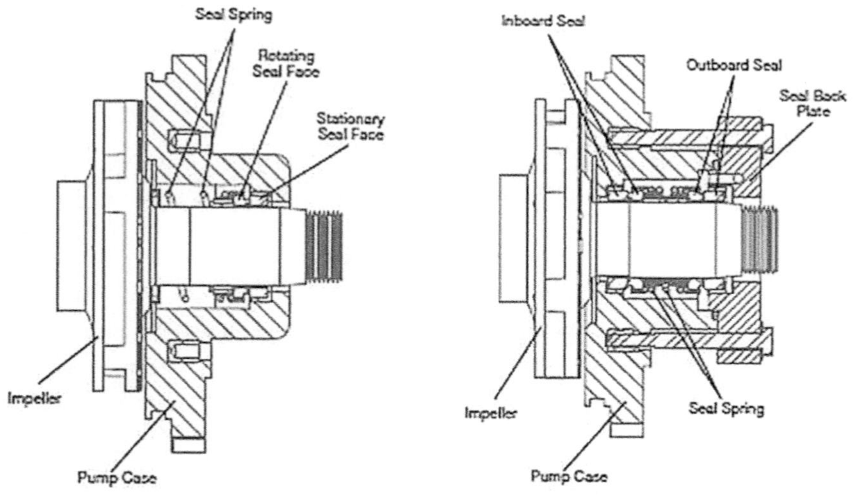

Figure 11.8. Single and double mechanical seals
(Berg)

Some pumps are sealless. Pumps with a magnetic drive with no direct connection between the motor and the pump shaft, are an example of these. If a very hazardous fluid is being pumped, sealless pumps can be the best choice.

Sealless pumps also have safety considerations. Selection of centrifugal sealed vs. sealless pumps carries a risk trade-off. Pumps with seals might fail more frequently with lower consequence, while sealless pumps might experience catastrophic, but less frequent, failures. For example, a sealless pump may be appropriate for a highly toxic fluid but not for a less hazardous one.

The hazards of running deadheaded or isolated were described in example 1 and 2 in this section. In a deadheaded pump, a blockage on the discharge side of the pump results in the flow reducing to zero and an increase in the discharge pressure. The energy input from the deadheaded pump increases the temperature and pressure of the fluid in the pump. For example, deadheading an ethylene oxide pump can add sufficient energy to cause the pump to explode. Designs should be considered to operate in a manner that prevents the pump from a deadhead operation if the pressure generated from deadheading the pump is more than the system overpressure protection provided. Return or kick-back lines are common and provide a return to allow continued operation and prevent excess temperature rise if the pump feed is stopped.

If a centrifugal pump stops while on-line, the fluid can flow in reverse, from the destination to the source if the piping and differential pressures allow it. Hazard Identification and Risk Analysis (HIRA) (Chapter 12) is needed to assess the consequences and protections needed for the reverse flow scenario. Generally centrifugal pumps have a non-return valve or check valve installed on the feed line to prevent reverse flow.

Many types of positive displacement (PD) pumps include the rotary screw pumps, gear pumps, as shown in Figures 11.9 and 11.10 and diaphragm pumps. Also included in this category are progressive cavity, piston pumps and peristaltic tubing pumps, some with large capacity. Diaphragm pumps are common in small scale applications and are vulnerable to failure due to excessive vibration and excessive pressure. Reverse flow is more difficult in positive displacement pumps, but positive displacement pumps can build up high pressures if deadheaded. Air driven diaphragm pumps can, however, be operated deadheaded. Because of the rapid buildup of pressure if many types of positive displacement pumps are deadheaded, some type of pressure relief or automatic shutoff device triggered by a pressure sensor is almost always included with the installation of positive displacement pumps. Many companies will also install a pressure relief device or high-pressure shutoff external to the pump.

Table 11.1 lists the common failure modes, consequences and design considerations for pumps, compressors, fans. No single table or source can list all the potential consequences of a given failure mode. This table is provided as a starting point only. While multiple failure modes are possible, the most important is typically that a blocked-in positive displacement pump can generate very high discharge pressures.

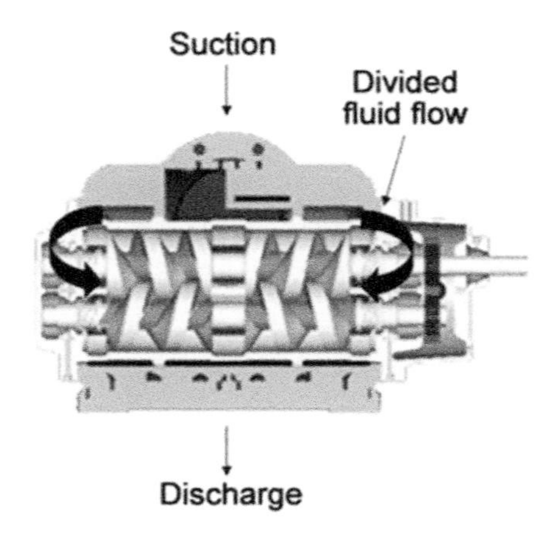

Figure 11.9. Two-screw type PD pump
(Colfax Fluid Handling)

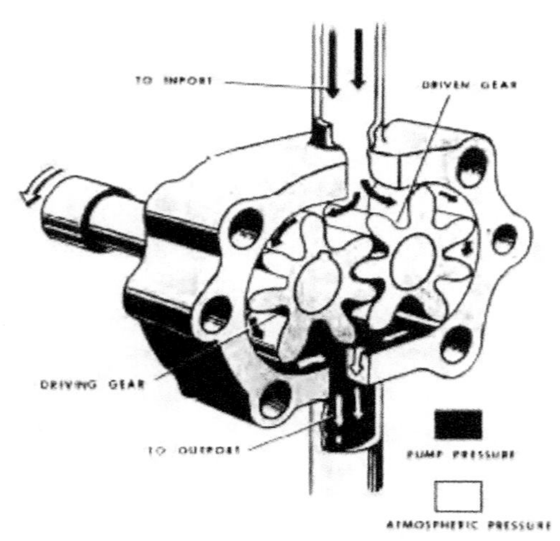

Figure 11.10. Rotary gear PD pump
(Tpub)

Table 11.1. Failure modes and design considerations for fluid transfer equipment

Failure mode	Causes	Consequences	Design considerations
Stopping	Power failure Mechanical failure Control system action (failure or intended)	Consequence to upstream or downstream equipment (HIRA needed) See Reverse Flow	Power indication on pump Low flow alarms/interlocks Level alarms and interlocks in other equipment
Deadheading or Isolation	Pump/compressor outlet blocked in by: Closed valves (manual, control, block) on discharge side, Plugged lines Blinds left in	Loss of containment due to, high temperature and pressure causing seal, gasket, expansion joint, pump or piping failure. Possible phase changes, reactions.	Overpressure protection. Minimum flow recirculation lines. Alarms/interlocks to shut down the pump or compressor on low flow or power Limit closing time for valves
Cavitation / Surging	Blocked suction by: Closed inlet valves Plugged filters/strainers	Loss of containment due to damage to seals or impellers	Low flow alarms/interlock to shut down the pump or compressor Vibration alarms/interlocks Differential pressure alarm on strainers
Reverse Flow	Pump or compressor stops	Loss of containment upstream Overpressure upstream Contamination upstream	Non-Return (Check) valves on discharge side (Check valves are difficult to count on; their dangerous failure modes are difficult to diagnose or test for until they are actually needed.) Automatic isolation valves Overpressure protection upstream Positive displacement pump
Seal Leaks	Particulates in feed Loss of seal fluids or flushes Small bore connections Age (wearing out)	Loss of containment due to damage to seals	Alarms or interlocks on seal fluid system to shutdown pump/compressor Double mechanical seals with alarm on loss of one seal Sealless pumps
Contamination / change of fluid	Liquid in compressor feed	Compressor damage See Seal leaks	Knock out pots before compressor

Figure 11.11 shows a Pump Application Data Sheet. The first block of information, Liquid Properties, specifically asks for safety information such as flammability, toxicity, regulatory coverage. Other properties that could be of interest could be thermal stability or reactivity of

the fluid. The next block, Materials of Construction, is important to safe processing. Use of the incorrect material of construction can lead to loss of containment.

Engineering standards that include design considerations for pumps are:

- API STD 610 "Centrifugal Pumps for Petroleum, Petrochemical and Natural Gas Industries, Eleventh Edition" (ISO 13709:2009 Identical Adoption)
- API STD 617 "Axial and Centrifugal Compressors and Expander-compressors"
- API STD 674 "Positive Displacement Pumps-Controlled Volume for Petroleum, Chemical, and Gas Industry Services"
- API STD 685 "Sealless Centrifugal Pumps for Petroleum, Petrochemical, and Gas Industry Process Service"

Pump Application Data Sheet

OEC FLUID HANDLING INC.

Equipment No.: ______ Date: ______
NAME ______ TITLE ______
COMPANY ______ PHONE ______
ADDRESS ______ FAX ______
CITY ______ STATE ______ ZIP ______ EMAIL ______

Send completed worksheets to:
OEC Fluid Handling Inc.
P. O. Box 2807
Spartanburg, SC 29304
Fax: 1-864-573-9299
Email: sales@oecfh.com

LIQUID PROPERTIES

Pump #: ______
Liquid: ______
Pump Temp °F: ______
SP. Gravity @ P.T.: ______
Viscosity: ______ ☐ SSU ☐ CPS ☐ Other: ______
PH: ______ % Solids: ______
Safety/Environmental:
☐ Flammable ☐ Explosive ☐ Carcinogenic ☐ Toxic
☐ Noxious ☐ Regulated ☐ FDA ☐ EHEDG
Comments: ______

SYSTEM

Discharge Pressure Required (PSIG): ______
Capacity (US GPM) Max: ______ Min: ______
Suction Lift: ______
Suction Conditions: ______
Duty Cycle: ☐ 24/7 ☐ 8-10 hrs ☐ Intermittent
Comments: ______

MOTOR/DRIVER REQUIREMENTS

☐ Electric Motor ☐ Engine ☐ Air ☐ Other: ______
Enclosure: ☐ ODP ☐ TEFC ☐ TENV ☐ EX. Proof ☐ Encap ☐ Inverter Duty ☐ Mag Drive ☐ DC Drive
Voltages: ☐ 3-60-230/460 ☐ 3-60-208 ☐ 3-50/60-208-220/440 ☐ 3-60-575 ☐ 1-60-115/230 ☐ 3-50-200/400 ☐ 3-50-220/380 ☐ 3-50-115/230 ☐ 3-50-220/440 ☐ 3-50-550
☐ Specify Voltages not listed above:
Phase: ___ Cycles: ___ Volts: ___
Additional Data: UL Label, fugitive emissions; tropical windings, motor heaters, special enclosures, etc.
(Specify): ______
Special Drives: ☐ V-Belt ☐ Inverter ☐ Air Motor
☐ Special (Specify): ______
Comments: ______

MATERIALS OF CONSTRUCTION

☐ Cast Iron ☐ Ductile Iron ☐ 316 Stainless Steel ☐ PVDF
☐ CPVC ☐ Hastalloy ☐ Alloy 20 ☐ Other: ______
Casing Connections: ☐ NPT ☐ Flanged ☐ Other: ______
Jacketing for cooling/heating: ☐ Yes ☐ No
O'ring Material: ☐ Buna ☐ TFE ☐ Viton ☐ Other: ______

STUFFING BOX

☐ Mechanical Seal
Preferred Seal Mfg.:
☐ Cartridge ☐ Double
☐ Single ☐ Lip
☐ Other: ______
☐ Packing
☐ Graphite
☐ Other: ______
Make: ______ Type: ______
Material: ______ Gland Type: ______
Comments: ______

BASE PLATE/MOUNTING

Pump Mounted: ☐ Horizontally ☐ Vertically
Base Plate: ☐ Fab Steel ☐ Chan Steel ☐ Cast Iron
Coupling: ☐ Jaw ☐ Spacer ☐ Other ______
Alignment Lugs: ☐ Yes ☐ No
Comments: ______
Painting: ☐ None ☐ Mfg. Std. ☐ Primed Only
Special Painting (Specify): ______

CUSTOMER REQUIREMENTS

Drawings: ______
Approval Dimensional Drawings: ☐ Yes ☐ No
Testing: ______
Hydro: ☐ None ☐ Witness ☐ Non-Witness
Performance: ☐ None ☐ Witness ☐ Non-Witness
Field: ☐ None ☐ Witness ☐ Non-Witness
Inspection prior to shipping: ☐ Yes ☐ No
Start-up Assistance: ☐ Yes ☐ No
Operator Training: ☐ Yes ☐ No
Maintenance Training: ☐ Yes ☐ No

Serving the Fluid Handling Process Needs of Industry | 1-800-500-9311 | 1-864-573-9200 | www.oecfh.com

Figure 11.11. Example application data sheet (OEC Fluid Handling)

11.3.2 Heat Exchange Equipment

Overview. Heat exchange equipment is used to control temperature by transferring heat from one fluid to another. Heat transfer equipment includes heat exchangers, vaporizers, reboilers, process heat recovery boilers, condensers, coolers and chillers. Much of what is stated in this section will also apply to heating/cooling coils in a vessel such as a reactor or storage tank.

Failures in heat transfer equipment can lead to loss of temperature control, contamination of the fluids, or loss of containment. Temperature is frequently a critical process variable, so failure of this equipment due to fouling, plugging, or loss of the heat transfer fluid supply can lead to undesired consequences. A HIRA is needed to assess the consequence. Heat exchange equipment can see thermal stress due to temperature gradients. This can lead to loss of containment. The Longford fire and explosion in Chapter 12 is an example of this failure mode.

Heat exchanger failure modes include the following.

- Fouling due to cooling water quality, low velocity, and microbiological fouling (both aerobic and anaerobic)
- Erosion
- Stress corrosion cracking
- Weld failures
- Tube to tube sheet failures (roll failure, seal weld failure)
- Poor fluid distribution at the baffles
- Heating or cooling media velocity not designed properly
- Non-condensable material accumulation

Hazards associated with heat exchange equipment include:

- Overpressure. due to blocked inlet/outlet streams
- Inadvertent mixing of chemicals
- Accelerated corrosion of downstream equipment, e.g.
- acid gas leaks to cooling water can form acid which rapidly corrodes carbon steel
- lack of oxidant in cooling water system can cause rapid algae fouling
- lack of oxygen control in a steam system can cause steel cracking
- Chemical release from cooling towers
- Water reactivity for heat exchangers using steam or water for heat transfer

The consequence of heat exchanger leaks depends on the nature of the process, the direction of the leak (process side to utility or vice versa), and the fluids involved. Failure to keep the fluids separate due to tube leaks can result in reactive chemical incidents (see example 1), or release of a toxic or flammable material into the low-pressure side where it can escape elsewhere, such as at a cooling water tower. A tube rupture can result in a shell rupture if the tube side is operating at significantly higher pressure than the shell. This risk is typically mitigated with shell side pressure protection that takes into account the tube side pressure.

Example 1. A plant had an explosion in the outlet piping of an oxidation reactor which ruptured a 0.9 m (36 in) pipe (see Figure 11.12). The explosion was caused by the reaction of molten nitrate salt, used to remove heat from the reactor, leaking into the piping where carbonaceous deposits had been trapped in a short dead-leg. Reactive chemical testing indicated that the reaction resembled closely the decomposition of TNT explosive. Fortunately, nobody was injured. The incident showed that it was critical to avoid leaks of the nitrate salt, to detect leaks if they did occur and to have a safe shutdown procedure if a leak occured (CCPS 2011).

Figure 11.12. Ruptured pipe from reaction with heat transfer fluid (CCPS h)

Example 2. Chapter 12 describes the explosion in the Longford gas plant in Australia. Process upsets led to the lean oil pumps being tripped off (stopped by an interlock). The loss of lean oil flow caused temperature drops in heat exchangers (resulting in ice formation on the exchangers), eventually resulting in a full bore exchanger shell brittle heat failure when hot oil was reintroduced to the cold section of the heat exchanger. This led to a gas release and explosion. A HIRA is needed to detect consequences such as these.

Design considerations for process safety. Several different types of heat exchangers include shell and tube, plate and frame, spiral, and air cooled are examples. See Figures 11.13 - 11.15.

One of the biggest concerns is inadvertent mixing of fluids due to tube leaks. Design considerations to prevent or mitigate this include the following.

- Where a toxic fluid is involved, put it on the tube side so tube leaks go into the shell side and can be detected in the cooling tower or piping at low non-hazardous concentrations. Also leaks from a shell failure are the utility fluid and not the toxic fluid.
- Carefully select materials of construction to resist corrosion on both sides.

- Design for drainage to reduce corrosion by installing exchanger in a sloped orientation (consider the baffle design to allow fluids to drain).
- Use double tube sheets for heat exchangers handling toxic chemicals or for materials where mixing must be avoided (see Figure 11.16).
- Consider fluid velocities, fluid properties, contaminants (solids and dissolved materials), and impingement.
- Ensure the vapor pressure of the process fluid at the maximum heating media temperature is less than the equipment maximum allowable working pressure.

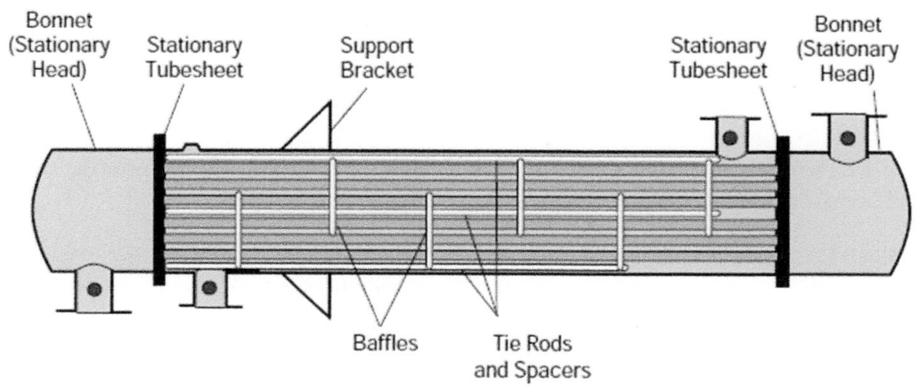

Figure 11.13. Shell and tube heat exchanger (Mukherjee)

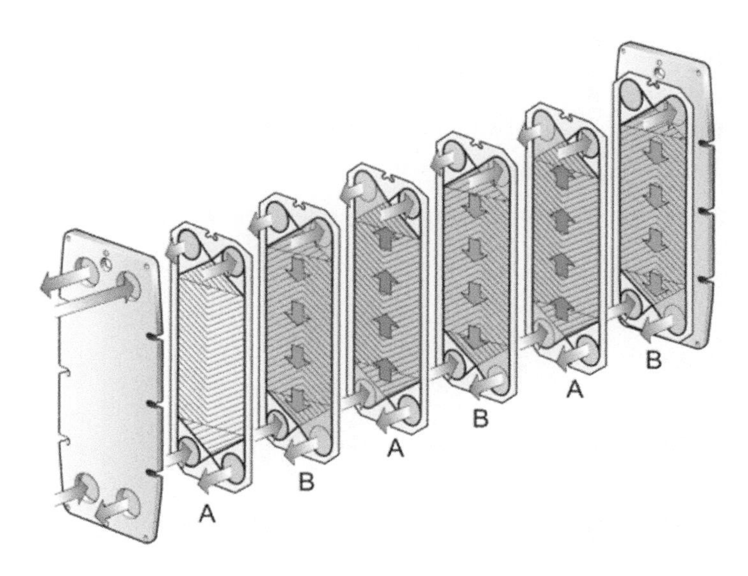

Figure 11.14. Principle of plate pack arrangement, gaskets facing the frame plate (Alfa Laval)

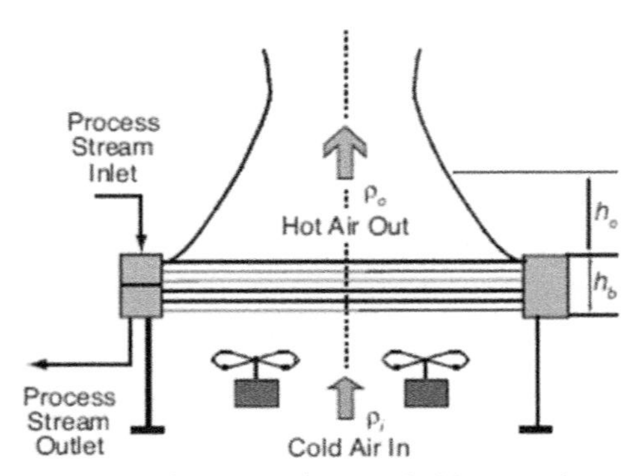

Figure 11.15. Schematic of air-cooled heat exchanger (Chu)

Control of temperature in a process is usually an important, if not critical, process parameter. Some design considerations to deal with this are as follows.

- Design for periodic cleaning to remove fouling.
- Provide a tube sheet vent nozzle or other a means to vent non-condensable gases from the process system.
- Tube pitch and spacing, flow distribution, fluid velocity and ΔT should be considered to prevent fouling.

Engineering standards that include design considerations for heat exchangers are as follows.

- ASME "Boiler and Pressure Vessel Code" (ASME)
- API STD 660 "Shell and Tube Heat Exchangers"
- Tubular Exchanger Manufacturers Association (TEMA)
- Heat Exchange Institute standards (HEI)

Table 11.2 presents some common failure modes and design considerations for heat exchangers.

Table 11.2. Common failure modes and design considerations for heat exchangers

Failure Mode	Causes	Consequences	Design Considerations
Leak from heat transfer surface	Corrosion from contaminants in the process fluids, and cooling fluids, and/or loss of treatment chemicals. Anaerobic attack under sediments and scale. Thermal stress (e.g. extreme heat/cold)	Loss of containment Inadvertent mixing and contamination of low pressure side, potential reactions, (HIRA needed)	Periodic inspection Choice of materials of construction Choice of heat transfer fluid Shell expansion joints Non shell and tube design Control of introduction of process fluids during startup and shutdown Monitoring of low pressure side fluid Toxic fluids in tubes, monitor shell side. Treatment chemicals
Rupture from heat transfer surface	Corrosion Thermal stress (e.g. extreme heat/cold) Operation out of design temperature range resulting in stress cracking, improvement, weakening of tubes or tubesheet (see loss of cooling or heating load) Blocking in one fluid side during operation	Potential rupture of heat exchanger Loss of containment	Emergency relief device Control of introduction of process fluids during startup and shutdown
Loss of cooling or heating fluid	Loss from supply Control system malfunction Pluggage or Misalignment of valves	Loss of process control (HIRA needed) High pressure	Alarms / interlocks on low flow or pressure of heat transfer medium High or low temperature alarms on process side
Inadequate heat transfer	Fouling Accumulation of non-condensable gases (mostly condensers)	Loss of process control (HIRA needed) High pressure	Ability to clean High or low temperature alarms on process side

11.3.3 Mass Transfer: Distillation, Leaching and Extraction, Absorption

Overview. Mass transfer operations are used to separate materials, purify products, and detoxify waste streams. Knowledge of the properties of the materials being handled is necessary to assess the hazards of the potential failures of mass transfer equipment.

Distillation (see Figure 11.16), stripping, and absorption frequently involve flammable materials; therefore, loss of containment can result in fires and explosions. High temperatures are used, especially in the reboilers, to drive the distillation/stripping; therefore, the thermal stability of the materials being handled should be understood. Loss of cooling to a reflux condenser can affect the composition of materials in a distillation, which again leads to the need to understand the effect of composition on the thermal stability characteristics of the material being handled. The Concept Sciences explosion in Chapter 7 is an example of a failure that led to a higher, and more hazardous, concentration than expected or intended. High levels of liquid in columns can lead to plugging of internals, high pressure, and loss of containment. The Texas City explosion in Section 3.1 is an example of this. Higher liquid loading on trays can result to damage to trays and result in more serious temperature upsets. An electrical power failure will stop all cooling water pumps but not necessarily any fired heaters. Thus, heat continues to enter the process, but no cooling occurs. This is often the design basis for the largest flare case for a process.

Packing material fires can occur where hydrocarbon residue that remains on column packing can self-ignite at elevated temperatures when exposed to the atmosphere. Iron sulfide, which is pyrophoric, can form from sulfur found in crude oil. Corrosion of carbon steel components can settle on packing and can ignite when exposed to the air or oxygen (Ender and Mannan).

Adsorption processes are exothermic. Carbon bed adsorbers are subject to fires due to this overheating. For certain classes of chemicals (e.g. organic sulfur compounds [mercaptans], ketones, aldehydes, and some organic acids) reaction or adsorption on the carbon surface is accompanied by release of a heat that may cause hot spots in the carbon bed. Adsorption of high vapor concentrations of organic compounds also can create hot spots. If a flammable mixture of fuel and oxygen are present, the heat released by adsorption or reaction on the surface of the carbon may pose a fire hazard (e.g., a fire may start if the temperature reaches the autoignition temperature of the vapor and oxygen is present to support ignition) (OSHA a and Naujokus). Figure 11.17 is a schematic of a carbon bed system, In Figure 11.17, the top bed is in absorption mode and the bottom bed is in recovery mode.

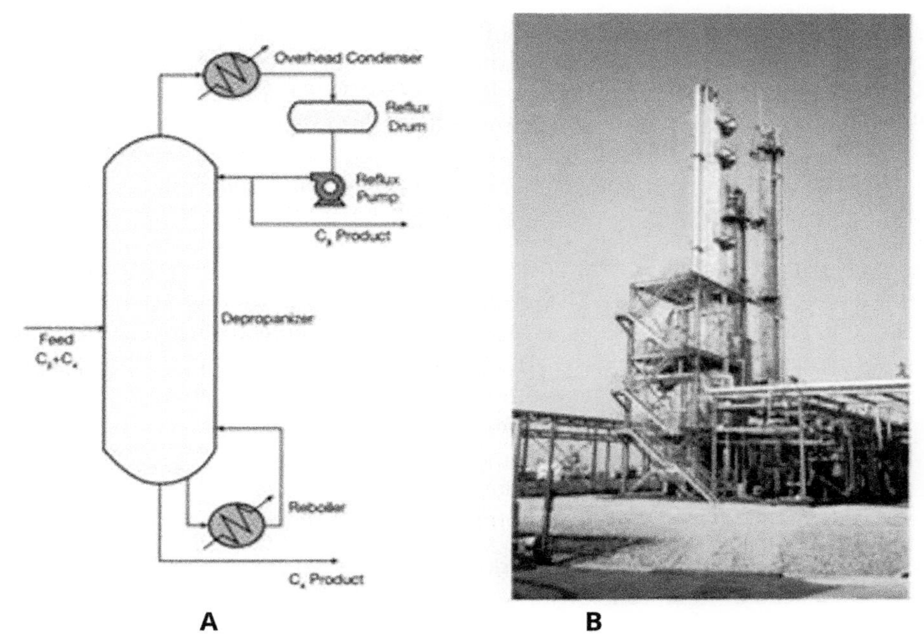

Figure 11.16.A. Example distillation column schematic (Bouck)

Figure 11.16.B. Typical industrial distillation column (©Sulzer Chemtech Ltd.)

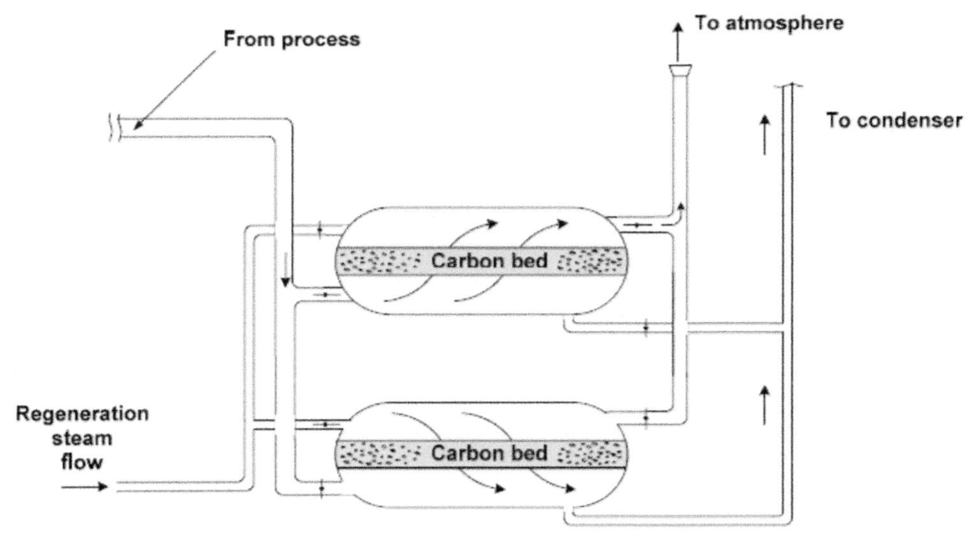

Figure 11.17. Schematic of carbon bed adsorber system (OSHA a)

Extractors will contain two immiscible fluids plus some materials being transferred from one phase to another. Loss of containment can result in flammable or toxic releases, depending on the nature of the materials. Failure of level control in extractors can result in the wrong material being sent to downstream equipment, leading to high levels or pressure in downstream equipment. As extractors are often run liquid full, loss of flow or level control can also lead to overpressure and damage to the unit if the maximum feed pump pressure excedes the maximum allowable working pressure.

Example 1. Distillation Column Incident. In 1969, an explosion occurred in a butadiene recovery unit in Texas City, Texas. The location of the center of the explosion was found to be the lower tray section of the butadiene refining (final purification) column. The butadiene unit recovered byproduct butadiene from a crude C4 stream. The overhead product of the refining column was a high-purity butadiene product. The heavy components of the feed stream, including vinyl acetylene (VA), were removed as a bottoms product. The bottoms vinyl acetylene concentration was normally maintained at about 35%. Explosibility tests had indicated that VA concentrations as high as 50% were stable at operating conditions. Highly concentrated VA, on the other hand, decomposes rapidly on exposure to high temperature.

When the butadiene unit was shut down to undertake necessary repairs, the refining column was placed on total reflux. The refining column explosion occurred approximately 9 hours after it was placed on total reflux. This operation had been performed many times in the past without incident. The operators did not observe anything unusual about this particular switch over to total reflux. Subsequent examination of the records indicated that the column had been slowly losing material through a closed, but leaking, valve in the column overhead line.

Loss of butadiene through the leaking valve resulted in substantial changes in tray composition in the lower section of the column. The concentration of vinyl acetylene in the tray liquid in the vicinity of the tenth tray apparently doubled to an estimated 60%. The loss of liquid level in the base of the column uncovered the reboiler tubes, allowing the tube wall temperature to approach the temperature of the steam supply. The combination of increased vinyl acetylene concentration and high tube wall temperature led to the decomposition of VA and set the stage for the explosion that followed. (PSLP a, b, c)

Example 2. Carbon Bed Incident. A bio-solids thickener tank in a refinery was overpressured by an internal explosion. The tank roof traveled about 200 feet. The tank was vented to a 208 l (55 gal) activated carbon adsorber to remove organics and control odors. After two years of operation, seal covers had been installed on the tank as part of an emissions control program. On the day of the event, fresh carbon was installed in the drum. The heat of adsorption increased the temperature in the drum to about 177°C (350°F). The hydrocarbons adsorbed on the carbon oxidized and a local hot spot was formed. Over several hours, the temperature increased to the point where the carbon started burning at about 427°C (800°F). This ignited the inlet vapors to the adsorber, and flames propagated back to the tank, igniting the vapor space.

Prior to the tank being sealed, hydrocarbons were able to escape to the atmosphere. When the tank was sealed better, this created a higher organic load to the carbon unit which generated more heat and provided fuel for the flames to propagate back to the tank. (Sherman)

Design considerations for process safety. Distillation is temperature, pressure, and composition dependent; special care must be taken to fully understand any potential thermal decomposition hazards of the chemicals involved.

Columns need adequate instrumentation for monitoring and controlling pressure, temperature, level, and composition. The location of sensing elements in relation to column internals must be considered so that they provide accurate and timely information and are in direct contact with the process streams.

A design feature of some columns is to provide a tall base (e.g. 3 m (10 ft)) to provide adequate Net Positive Suction Head (NPSH) to ensure that the bottoms pumps do not cavitate and fail. This also reduces the wetted area exposed to a ground-level fire.

Leaks from where piping or instrumentation is connected to these vessels is a common failure. Where the material is flammable, a fire can occur that can impact surrounding equipment. Column support structures and skirts should be fireproofed, as they are not cooled by internal fluid flow and a ground fire can lead to the column collapsing.

Overpressurization can result from freezing, plugging, or flooding of condensers, or blocked vapor outlets, if the heat input to the system is not stopped. API RP 520 *Sizing, Selection, and Installation of Pressure-Relieving Devices* and API RP 521 *Pressure-Relieving and Depressurizing Systems* provide extensive guidance on the placement and sizing of pressure relief valves and other overpressure protection systems. (API RP 520, API RP 521)

Emphasis should be placed upon the use of inherently safer design alternatives using concepts such as the following.

- Limiting the maximum heating medium temperature to safe levels
- Selecting solvents which do not require removal prior to the next process step
- Using a heat transfer medium that prevents freezing in the condenser
- Locating the vessel temperature probe on the bottom head to ensure accurate measurement of temperatures, even at a low liquid level
- Minimizing column internal inventory
- Avoiding dead legs that can corrode, plug or freeze

To prevent packing fires:

- Cool columns to ambient temperature before opening
- Wash the column thoroughly to remove residues and deposits
- Use chemical neutralization to remove pyrophoric material
- Purge columns with nitrogen
- Monitor temperatures of the packing and column as it is opened
- Minimize the number of open manways to reduce air circulation

To protect against carbon bed fires:

- Test the impact of the vapors on the carbon for potential heat release before putting the carbon adsorption system into service; if possible, identify reactions that are not already known.
- Measure the offgas for carbon monoxide and CO_2 to warn of hot spots
- Measure bed temperatures at a large number of places
- Provide fire control systems such as water sprays, nitrogen or steam
- Include flame arrestors to prevent the spread of fire from the carbon containers to the flammable chemical containers

11.3.4 Mechanical Separation / Solid-Fluid Separation

Overview. Mechanical separators are most often used to separate solids from liquids or gases. Typical equipment includes centrifuges, filters, and dust collectors.

Common failure modes for centrifuges include mechanical friction from bearings, vibration, leaking seals, static electricity, and overspeed. Vibration is both a cause of problems and an effect from other sources. Static charges can occur from the flow of the slurry and liquor into the unit, and the high speed of centrifuges. Static charges can accumulate due to the use of synthetic, non-conductive filter media. Both mechanical friction and static can ignite flammable liquids if used.

A concern for filters is exposure or loss of containment during opening and closing. Plate and frame filters have a high potential for leaks and should be avoided when flammable or toxic solvents are being filtered. The filter media can be overpressured. during the end of a cycle, which can cause further leaks.

Studies have shown that dust collectors are the equipment item most frequently involved in dust explosions. (CCPS 2005) Frequently, dust collectors are at the end of a process and collect the smallest particle size dusts, which are going to be the most hazardous if the dust is combustible. Common failure modes for dust collectors include loss of containment due to failure of the filter media (usually bags or cartridges), plugging of the filter media, and loss of grounding of filter bags. The filter media is often purged with air to prevent plugging. Failure of the purge system followed by reactivation can create a much worse dust cloud. Dust collectors with filter bags (often called baghouses) can have over a hundred bags, each one is a point where the bag cage can be isolated from ground (a metal bag cage, if electrically isolated, can develop a significant electrical charge). See Example 2 case history.

Example 1. A crystalline finished product was spinning in a batch centrifuge when an explosion occurred. The product had been cooled to -7°C (19°F) before it was separated from a methanol-isopropanol mixture in the centrifuge. It was subsequently washed with isopropanol precooled to -9°C (16°F). The mixture was spinning for about 5 minutes when the explosion occurred in the centrifuge. The lid of the centrifuge was blown off by the force of the explosion. The overpressure shattered nearby glass pipelines and windows inside the process area (up to 20 m (65 ft) away), but nearby plants were not damaged. No nitrogen inerting was used and sufficient air was drawn into the centrifuge to create a flammable atmosphere. Sufficient heat could also have been generated by friction to raise the temperature of the precooled solvent medium above its flash point. Because the Teflon® coating on the centrifuge basket had been worn away, ignition of the flammable mixture could also have been due to metal-to-metal contact between the basket and the bottom outlet chute of the centrifuge, leading to a friction spark. A static discharge might also have been responsible for the ignition. Since the incident, the company has required use of nitrogen inerting when centrifuging flammable liquids at all temperatures. (Drogaris)

Lessons learned include monitoring the oxygen concentration in conjunction with inerting and sealing the bottom outlet to minimize air entry. Because the ignition source was uncertain (static discharge, frictional heat), this incident illustrates why it often is prudent to assume an ignition source when designing for flammable materials. In his book, *Lessons from Disaster: How Organizations have No Memory and Accidents Recur*, Trevor Kletz is quoted as saying "Ignition source is always free." (Kletz 1993)

Example 2. An explosion occurred in a dust collector connected to the process vents for polyester plastic extrusion equipment. The explosion was safely vented. The bags were off their cages and charred, as shown in Figure 11.18. The dust collector housed 144 cages. During the investigation it was found that one of them was not grounded. A check of an identical unit revealed problems with the grounding between the cages and the tube sheet, as shown in Figure 11.19. It was also found that the type of bag used had been changed, but the need for a grounding strap was not understood, and that grounding/bonding checks were not part of the asset integrity program.

The type of bag used was changed to ones with an improved grounding strap to ensure grounding of the metal cage support inside the bag, and the procedures were fixed to include conductivity checks (Garland).

Figure 11.18. Damage to dust collector bags
(Garland)

Figure 11.19. Tube sheet of dust collector
(Garland)

Design considerations for process safety. If a centrifuge or filter is used for flammable or toxic materials, fully enclosed units with automatic filtration, washing and discharge cycles should be considered to avoid exposure or loss of containment. Inerting or use of an inert gas sweep or blanket should be strongly considered if flammable liquids are being used. Pressure or vacuum filters can be used in place of centrifuges to reduce the problems due to bearing failure or vibration. Some typical centrifuges and filters are shown in Figures 11.20 and 11.21.

When designing dust collectors, it is essential to know whether the dust involved is combustible and this can be understood using a dust risk hazards analysis as described in NFPA 652 "Standard on the Fundamentals of Combustible Dust". (NFPA 652) Most organic and metal dusts less than about 500 microns in diameter are not only combustible, but under certain circumstances can, if ignited, lead to flash fires and/or explosions. The explosion severity and minimum ignition energy of the dust should be measured. Tests should be done on the most representative materials possible. Commonly used protection measures for dust collectors are explosion vents, explosion suppression systems, and inerting. Figure 11.22 shows a schematic of a typical baghouse. Figure 11.23 shows a picture of a vented explosion from a dust collector.

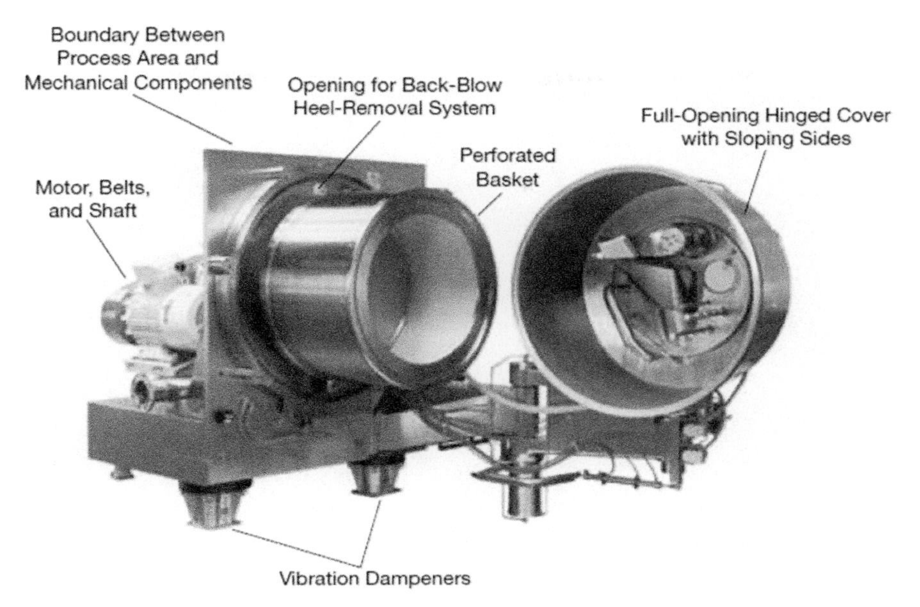

Figure 11.20. Horizontal peeler centrifuge with clean-in-place system and discharge chute (Patnaik)

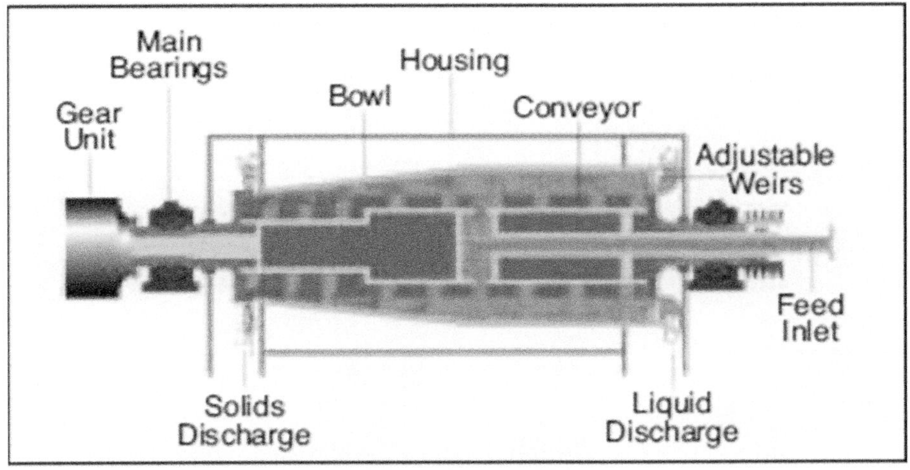

Figure 11.21. Cross sectional view of a continuous pusher centrifuge (Patnaik)

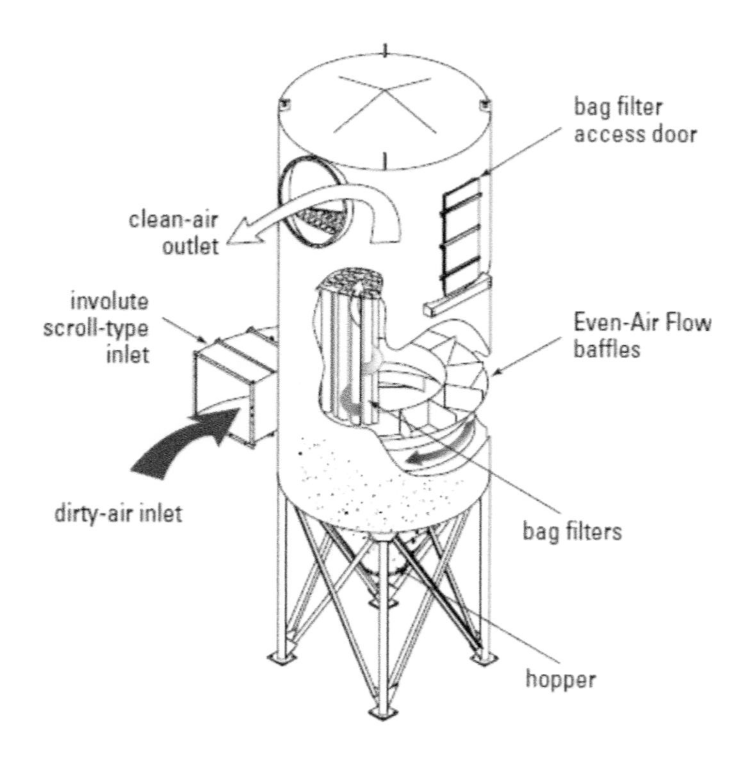

Figure 11.22. Schematic of baghouse
(Donaldson-Torit)

Figure 11.23. Dust collector explosion venting
(Fike)

11.3.5 Reactors and Reactive Hazards

Overview. The key process safety concern in the design of reactors is *runaway reactions*. Runaway reactions occur when the heat generation rate from an exothermic reaction exceeds the rate at which heat can be removed, causing an uncontrolled, often exponential, rise in temperature and usually an accompanying pressure rise. If the heat released by the reaction exceeds the cooling capacity, the reaction rate will accelerate (runaway) and may result in an excessive gas evolution or a vapor pressure increase that, in the absence of adequate overpressure relief protection, can rupture the reactor. If overpressure relief protection is adequate, then there will be loss of containment through the relief device either to a flare or to the atmosphere, depending on the design

During runaway reactions, the temperature can rise significantly, which may favor additional exothermic reactions. If this occurs, the composition may shift to produce a more toxic off-gas, as occurred in Seveso, Italy (see Example 1 below). If the potential for a runaway reaction exists, the characteristics and composition of offgases should be understood. Appropriate downstream systems to capture hazardous materials should be provided.

Common failure causes leading to runaway reactions in reactors include the following.

- cooling system failure
- mischarges (too much or too little of a reactant charged)
- wrong reactant charged
- reactants charged in the wrong order
- reactant quality (wrong concentration, reactant beyond shelf life)
- contamination with a catalytic impurity

The failure mode is then a loss of primary containment caused by overpressure. resulting in a physical explosion. agitator seal failure, or weld fatigue failure.

During the reaction, the reactants and solvents need to be well mixed for the reaction to proceed as planned and for efficient input or removal of heat. Thus, loss of agitation can be a cause of a runaway reaction. See the description of the Seveso incident below. A subset of agitator failure is starting an agitator too late. This allows a buildup of reactants that then suddenly are brought into contact with each other. Loss of cooling or insufficient cooling can likewise be a cause of a runaway reaction, as in the T2 Industries incident detailed below.

Examples of mischarges leading to runaway reactions would be an undercharge of a solvent meant to absorb some of the heat of reaction or overcharging a material that could result in a more exothermic reaction than the system was designed for. Charging a reactant that is at a higher concentration than expected is an example of this. Many reactions involve a catalyst. Using a catalyst that is past its recommended shelf life or undercharge of a catalyst can lead to the buildup of unreacted material that can then react and liberate more heat than the reactor was designed for.

Another category of reactive hazards is when reactions occur where you don't want them due to inadvertent mixing. The methyl isocyanate release in Bhopal (Section 3.15) is an example of this.

Example 1. This incident occurred in Seveso, Italy on July 10, 1976. A batch reactor was used to make 2,4,5-trichlorphenol (TCP) in two stages. Stage 1 was the reaction of 1,2,4,5-tetrachlorobenzene, and sodium hydroxide at 170 - 180°C (338 - 356°F) in the solvents ethylene glycol and xylene. Normally, at the end of the reaction, half of the ethylene glycol was removed by distillation and the batch cooled to 40 - 50°C (104 - 122°F). Steam, normally at 190°C (374°F), was used to heat the batch and for the distillation step. It was known that a runaway reaction could occur at 230°C (446°F).

A batch was started on a Friday, but the plant had to be shut down for the weekend. The distillation was in progress but not completed when the reactor had to be shut down. As other parts of the plant were being shut down, the steam temperature to the reactor rose to 300°C (572°F). At about 5 AM Saturday the reactor was shut down, and the agitator was shut off, but the reactor was not cooled down. The walls of the reactor were at 300°C (572°F). About 7.5 hours later, a runaway reaction occurred, bursting the rupture disk and releasing about 2 kg (4.4 lb) of dioxin, a very toxic chemical, into the atmosphere. The dioxin reached nearby residential areas. Many people developed Chloracne, a skin disease. A 17 km^2 (6.6 mi^2) area was made uninhabitable, with thousands of farm animal fatalities and contaminated soil.

The residual heat in the upper section of the reactor raised the temperature of the upper section of the liquid to about 200-220°C (392 - 428°F) (Figure 11.24). Investigations after the event showed that a slow exotherm begins at 185°C (365°F), which could cause a 57°C (135°F) adiabatic temperature rise, and another exotherm could start at 225°C (437°F) causing a 114 °C (237°F) temperature rise. Therefore, the residual heat was more than enough to raise the batch temperature to above the exotherm onsets. (Lees)

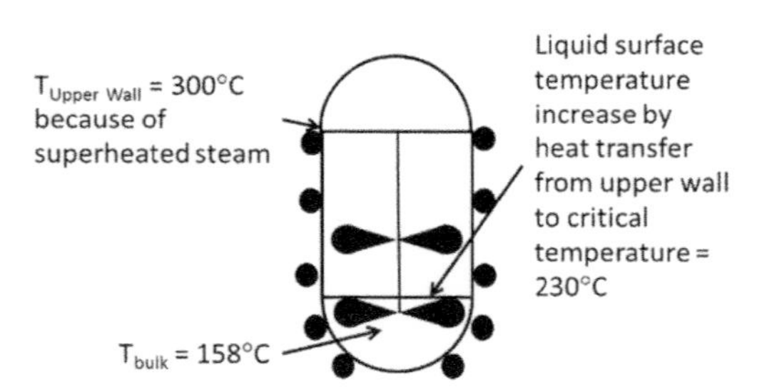

Figure 11.24. Seveso reactor
(adapted from SAChE presentation by Willey)

Example 2. The T2 Laboratory explosion described in Chapter 5 is an example of a runaway exothermic reaction. Figures 11.25 shows the T2 Laboratories before and after the explosion. A loss of sufficient cooling during the process likely resulted in the runaway reaction, leading to an uncontrollable pressure and temperature rise in the reactor. The pressure burst the reactor; the reactor's contents ignited, creating an explosion equivalent to 635 kg (1,400 lb) of TNT.

Figure 11.25. T2 Laboratories site before and after the explosion (CSB 2009)

Design considerations for process safety. The surface area to volume ratio drops as reactor size increases. The volume, and hence mass, in a reactor increases with the cube of the diameter: however, the surface area, through which heat transfer occurs, increases with the square of the diameter. Therefore, as a reactor gets bigger, the amount of potential heat released increases faster than the ability to remove that heat. Strategies to cope with this include adding coils for heat transfer or recirculating the reactants through an external cooler.

The least desirable way to run an exothermic reaction is in a *batch* reactor where all the reactants are added at one time and the reaction started. This is how the T2 Laboratories process described earlier was run. A better way to run an exothermic reaction is in a *semi batch* mode. In this mode, one or more reactants are added to the reactor gradually over the batch cycle. This enables the operator to stop the feeds if they see indication of anything going wrong, such as loss of cooling or agitation. The best way to run an exothermic reaction is in a continuous reactor. For the same production capacity, a continuous reactor will be smaller, and hence have better heat removal capabilities. In some processes, reactions are run in tubular reactors, which are essentially a continuous heat exchanger.

Reactors will almost always need an Emergency Relief System (ERS) to relieve the pressure from a potential overpressure scenario such as a runaway reaction. The Design Institute for Emergency Relief Studies (DIERS) has done research on the reactive relief design. This is a complex topic and a Subject Matter Expert (SME) is usually needed to do the actual ERS sizing. A hazard assessment is needed to determine appropriate relief design scenarios and chemical reactivity testing is needed to design these systems.

Alternatives or supplements to an ERS are adding chemicals such as inhibitors to stop the reaction (known as shortstopping), or rapidly emptying the reactor contents to dump (quench) tanks which are filled with water or other chemicals to stop the reaction.

Inadvertent mixing can be prevented by using dedicated charging lines and through operator training and written procedures.

Table 11.3 lists some common failure modes and design considerations for reactors. Injection of a reaction kill compound to stop any reaction occurring is a mitigation measure that is applicable to many of these failure modes.

Table11.3. Common failure modes and design considerations for reactors

Failure mode	Causes	Consequences	Design considerations
Loss of cooling	Loss of heat transfer medium from supply Control system failure	Potential runaway reaction	Emergency relief system Dual cooling modes, e.g. overhead condenser and reactor jacket Automatic actuation of secondary cooling medium on detection of low coolant flow, or high pressure, or high reactor temperature Automatic stopping of feeds of reactants or catalyst (with semi-batch or continuous reactors)
Loss of agitation	Loss of power Motor failure Agitator blades become loose/fall off	Potential runaway reaction	Emergency relief system Uninterrupted power supply backup to motor Agitator power consumption or rotation indication interlocked to stop the feed of reactants or catalyst or activate emergency cooling
Overcharge of reactant or catalyst	Error in measurement Control system failure	Potential runaway reaction Overflow of reactor	Emergency relief system Dedicated charge tanks sized to hold only the amount of reactant/catalyst needed Quantity of reactant/catalyst added limited by flow totalizer Redundant flow totalizers High level interlock / permissive to limit quantity of reactant/catalyst
Wrong reactant / catalyst	Misidentification Mix-up during product change	Potential runaway reaction	Emergency relief system Dedicated feed tank and reactor train for production of one product Control software preventing charge valve or pump operation until correct material bar code has been scanned
Step done out of sequence	Poor instructions/ training Human Performance issue	Potential runaway reaction	Controllers that verify a step has been done before advancing to the next step

11.3.6 Fired Equipment

Overview. Fired equipment, such as furnaces, flares, incinerators, thermal oxidizers, or heat transfer fluid heaters, are used to provide heat to processes, and dispose of combustible waste streams. For example, natural gas is converted to hydrogen in a reformer, which is a series of catalyst-packed tubes heated to several hundred degrees centigrade by a burner. Other uses of fired equipment include steam generation and heating of distillation column reboilers.

A common failure mode is accumulation of unburnt fuel due to loss of flame, too much fuel being fed, or insufficient air (oxygen) as examples. The unburnt fuel can then ignite and cause a fire or explosion. The second most common failure mode is tube failure, which can be caused by overheating, flame impingement, improper firing, thermal cycling, thermal shock, or corrosion. This can also result in fires and explosions. Liquid carry-over into flares and incinerators can also cause explosions as experienced in the Texaco Milford Haven Explosion. (HSE 1994)

Example 1. A heater was heavily damaged during startup as a result of a firebox explosion (Figure 11.26). The operator had some difficulty with the instrumentation and decided to complete the start up by bypassing the safety interlocks. This allowed the fuel line to be commissioned with the pilots out. The main gas valve was opened, and gas filled the heater. The heater exploded destroying the casing and several tubes. Fortunately, no one was injured. (CCPS a)

Figure 11.26. Damaged heater

(CCPS c)

Example 2. An explosion destroyed the furnace and adjacent column at a NOVA Chemical Bayport, TX plant (Figure 11.27). Before the explosion, an operator noticed flame stability problems with the low NO_x burners and began to manually adjust the airflow. During the few minutes that adjustments were being made to manage the burners, a loud puff was heard followed by a major explosion in the furnace. It appears that the explosion was caused by clogging in the nozzles on the burners resulting in an unstable flame. (CCPS 2004)

Figure 11.27. Heater and adjacent column at NOVA Bayport plant (CCPS d)

Example 3. After a shutdown for maintenance, a hydrogen reformer in an ammonia plant was being restarted. In the normal start-up procedure at the plant, nitrogen gas is passed through the primary reformer and a heating rate of 50°C (122°F) per hour is maintained at reformer outlet. This nitrogen flows in a closed loop, that is, it is recycled back into the reformer. This cycle continues until the temperature of 350°C (662°F) is obtained at the reformer outlet. To increase reformer outlet temperature, more burners are ignited.

Because of an emergency shutdown, sufficient nitrogen inventory was not available at site for startup. At least 8 to 10 more hours were required for nitrogen inventory makeup. To save production loss, the startup procedure was initiated. Furnace firing was started in the absence of nitrogen gas, and reformer outlet temperatures were monitored for a 50 °C (122°F) per hour heating rate. Reformer outlet temperatures were not increasing, so the firing rate was increased. During this period, many alarms appeared on the control system for convection zone temperatures. The alarms were inhibited to avoid any inconvenience to the control panel operator, because he was busy with the steam drum level control. Since no changes occurred in these outlet temperatures, the firing rate was further increased, and 56 of 72 burners were fired. This represents about 70% of the heat input, without any fluid flow through the reformer. The board operator instructed the plant operator to have a physical check of the reformer. The operator found that the reformer tubes were melting down inside the furnace.

The furnace was being fired and reformer outlet temperatures were being monitored without introduction of any nitrogen through the reformer. Because of the absence of any flow through the reformer, its outlet temperature did not increase and the increase in heat with no process flows resulted in high-tube temperatures and finally melting of the tubes. (Ramzan)

Design considerations for process safety. Process controls and process safety controls are handled by two control systems. Process safety considerations are usually handled by the Burner Management Systems (BMS). The BMS monitors temperatures, pressures, and the burner flames. Interlock trips and permissive interlocks are part of the BMS controls that include the ignition sequence, fuel shutoff, flame scanning, and purging, i.e., making sure excess fuel is purged before relighting occurs. A BMS is a critical safety system; it should either never be bypassed, or, if an organization believes it necessary, bypassed only after a management of change review where safety controls to be used in its place are established.

Combustion control systems control fuel to air ratios, firing rates, etc. accepting operator input, and adjusting to system demands. Combustion control systems may include process interlocks as well.

Tube rupture may be prevented by monitoring the tube skin temperatures (preferred) or monitoring the flow through the tubes.

In boilers, loss of the boiler water level supply could be catastrophic. Reliable level monitoring and control is paramount. Reliable level monitoring and control includes the design of a continuous supply of boiler feed water.

The hazards of fired equipment are so well known that many countries have specific industry fired equipment standards which define the design features and management systems required of a BMS. Standards that cover fired equipment include the following. These standards are valuable resources that should be referenced when doing a PHA or an MOC on fired equipment.

- NFPA 85 "Boiler and Combustion Systems Hazards Code"
- NFPA 86 "Standard for Ovens and Furnaces"
- NFPA RP 87 "Recommended Practice for Fluid Heaters"
- API RP 556 "Instrumentation, Control, and Protective Systems for Gas Fired Heaters"
- API RP 560 "Fired Heaters for General Refinery Service"

11.3.7 Storage

Overview. Storage of raw materials, intermediates and final product is necessary in a processing plant. Storage vessels include pressurized storage tanks, atmospheric storage tanks and silos/hoppers (for solids). Knowledge of the properties of the material is necessary to assess the hazards of a storage tank.

Common failure modes for storage tanks include the following.

- Loss of containment due to overfilling, mechanical failure, overpressurization, vacuum failures
- Internal fires or explosions caused by static electricity
- Failure due to addition of a hot material causing tank contents to react violently
- Uncontrolled reactions caused by loading the wrong material into a tank
- Boilover, the expulsion of contents caused by heat from the surface burning at the top of the tank reaching and vaporizing the water stratum at the bottom of the tank

- Rollover, the spontaneous and sudden movement of a large mass of liquid from the bottom to the top of a storage tank due to heat from the fire changing the fluid density gradient

Example 1. The Buncefield explosion and fire described at the start of this Chapter is an example of a gasoline storage tank overfill incident. The ensuing fire engulfed more than 20 large storage tanks over a large part of the Buncefield depot.

Example 2. The Cleveland LNG Tank failure in 1944 was the worst LNG accident in the USA. Wartime shortage of nickel led to use of a lower nickel stainless steel material than the now standard 9% Ni steel. This lower nickel steel was subject to brittle fracture, and this occurred leading to a catastrophic tank failure and total loss of the LNG tank contents. The LNG spilled into the tank yard and boiled off to methane. This cold methane, as liquid and cold vapor, was not lighter than the surrounding air. Thus, it flowed into the nearby neighborhood down streets and in sewer lines resulting in130 fatalities.

Example 3. In 1919 a 8,700 cubic meters (2.3 million gallon) tank of molasses suddenly broke apart, releasing its contents into the City of Boston. A wave of molasses over 5 m (15 ft) high and 50 m (1600 ft) wide surged through the streets at an estimated speed of 60 kph (35 mph) for more than 2 city blocks (Figure 11.28). The incident led to 21 fatalities and over 150 injuries. The tank was not properly inspected during construction and not hydrotested before filling it. Leaks between the welds had been observed, but no action was taken. (CCPS 2007)

Figure 11.28. Molasses tank failure; before and after
(CCPS e)

Example 4. A delivery truck arrived at a plant with a solution of nickel nitrate and phosphoric acid named "Chemfos 700" by the supplier. A plant employee directed the truck driver to the unloading location and sent a pipefitter to help unload. The pipefitter opened a panel containing 6 pipe connections, each of which fed to a different storage tank. Each unloading connection was labeled with the plant's name for the material stored in the tank. The driver told the pipefitter he was delivering Chemfos 700.

Unfortunately, the pipefitter connected the truck unloading hose to the pipe adjacent to the Chemfos 700 pipe, labeled "Chemfos Liq. Add." (Figure 11.29). This is similar to the human factor issue in the Formosa Plastics explosion (Chapter 16). The "Chemfos Liq. Add." tank contained a solution of sodium nitrite. Sodium nitrite reacts with Chemfos 700 to produce nitric oxide and nitrogen dioxide, both toxic gases. Minutes after unloading began, an orange cloud was seen near the storage tank (Figure 11.30). Unloading was stopped immediately, but gas continued to be released. 2,400 people were evacuated, and 600 residents were told to shelter in place. (CCPS d)

Figure 11.29 1 - Pipe connections in panel 2 and Chemfos 700; 2 - Liq. Add lines (CCPS f)

Figure 11.30. Cloud of nitric oxide and nitrogen dioxide (CCPS f)

Example 5. During painting, a tank's vacuum relief valve was covered with plastic to prevent potential contamination of the contents. When liquid was pumped out the covering prevented air/nitrogen from replacing the liquid volume. A vacuum developed, which led to the partial collapse of the tank, as shown in Figure 11.31. (CCPS 2002)

Overfill protection can consist of overflow lines to a safe place or instrumentation that automatically shuts off flow into the tank if the material is toxic or flammable. The overflow control system should have multiple devices to provide both redundancy and independence.

Corrosion is a major cause of structural failure. Improper manufacture or a change in service can lead to structural failure. Age and exposure to the humid environment can cause corrosion to the point of failure over time. If the tank is insulated, corrosion under insulation (CUI) can also lead to failure. Proper choice of the construction and coating materials; following the correct construction codes and standards; and performing ongoing inspection, and maintenance and testing are the main safeguards against these failures. New engineers may be asked to visit vendor shops to inspect and verify storage vessels meet the design specifications.

Filling or emptying a tank too quickly can cause overpressure or vacuum. Rapid cooling, for example after steam cleaning or filling with a hot material, can also cause a vacuum collapse. Pressure and vacuum protection can be installed. NFPA 30, "Flammable and Combustible Liquids Code" (NFPA 30), outlines the sizing of vents and emergency vents and API RP 2000 "Venting Atmospheric and Low-Pressure Storage Tanks" and API STD 2350 "Overfill Prevention for Storage Tanks in Petroleum Facilities" provide further guidance. The operator needs to know what the design rates are for filling and emptying so as not to exceed them. High- and low-pressure interlocks can be used to stop filling or emptying. Frangible or weak-seam roofs can also be provided for fixed roof atmospheric tanks which allow this seam to fail as opposed to the wall of the tank.

The overflow line for an atmospheric tank provides the overpressure protection for most atmospheric tanks and must be sized accordingly. Typically, atmospheric tanks are not designed for any significant pressure beyond a few inches of liquid height above the overflow. Vent lines should be separate from overflow lines on atmospheric tanks since they are normally sized for gas flow (breathing) and are too small for liquid flow. Combining the vent and overflow lines into one large line could result in two phase flow and subsequent back pressure resulting in over pressurization of the tank.

Clogging or blocking of vents can defeat pressure/vacuum protection systems (see the vacuum collapse example). Watch out for bird nests for externally located storage tanks. Polymerizing materials can plug vents. Cold weather can result in the freezing of vents and overflow seal pots which can also result in defeating pressure/vacuum protection. Regular inspection and maintenance of pressure/vacuum protection devices is necessary. (Figure 11.31)

Rollover can result if the material can stratify in the tank. One example occurs in LNG tanks. Depending on the source of the LNG it can be of different densities. The different-density LNG can layer in unstable strata within the tank. The layers may spontaneously roll over bringing bottom material near to the surface where it can flash-off due to the loss of hydrostatic pressure. LNG has a 600:1 volume change from gas phase to liquid phase. Thus, a small flash can generate a very large volume of vapor. Pressure relief systems may not be adequate for rollover. The force of the shifting mass can result in cracks or other structural failures in the tank. A control system of distributed temperature sensors and a pump-around mixing system can be used to provide rollover protection.

Figure 11.31. Tank collapsed by vacuum
(CCPS g)

An internal deflagration is possible if a flammable material is being stored. Static electricity is a common form of ignition. Static can be generated by the flow of fluid through pipes, or free fall of a liquid, or by the mixing of different phases in a tank, especially if one of the phases is non-conductive. The design for flammable liquids should avoid free fall of liquid by using a dip pipe or bottom feeding. Fill rates should be kept below certain levels until a dip pipe is covered. Guidance on fill rates to minimize generation of static electricity is provided in *Avoiding Static Ignition Hazards in Chemical Operations* (Britton). Inerting of the vapor space is another possible ignition control method. Tanks should be properly grounded to allow dissipation of static charges from all sources. Attached equipment should be bonded to the tank. NFPA 77, "Recommended Practice on Static Electricity", (NFPA 77) contains information about the generation and control of static charges. The CSB video "Static Sparks Explosion in Kansas" describes an example of an explosion in a storage tank caused by static electricity. (CSB) Another important safeguard needed on flammable storage tanks is a flame arrestor to prevent the flames of an external fire from propagating into the tank through the atmospheric vent. A flame arrestor is a device that allows the gas to pass through it but stops a flame.

Lightning strikes are another common cause of ignition in storage tanks. NFPA 780, "Standard for the Installation of Lightning Protection Systems", (NFPA 780) provides guidance for protection of structures containing flammable liquids. A lightning strike could ignite vapors in the vicinity of the seal of a floating roof atmospheric storage tank. These are termed rim seal fires and are usually not catastrophic unless the roof fails simultaneously (e.g. roof sinks) resulting in a full-surface fire. Usually, floating roof tanks are fitted with a foam dam around the circumference and firefighting foam chambers to add foam just to the dam and not the whole roof area. An internal floating roof tank is less susceptible because the vapor space is not flammable.

Addition of incompatible materials can cause reactions, see Example 4 in this section. The first step in prevention is identification of potentially incompatible materials that could be unloaded in the storage tank. The SDS of the material is the first place to look. Other sources include the following.

- Bretherick's Handbook of Reactive Chemical Hazards (Urban)
- Chemical Reactivity Worksheet, described in Chapter 5 (CCPS CRW)

If incompatible materials that could be unloaded are identified, design measures can include: positive identification of materials by sampling before unloading, locating storage tanks of incompatible materials in separate dikes, use of dedicated unloading stations with special fittings, clear labeling of unloading lines and storage tank, and clear operating procedures with written checks for material identification. If storage tanks unload into manifolds where other materials can be, precautions against backflow into the tank include check valves, or block valves interlocked to close if backflow is detected.

Self-reacting materials, such as monomers, or water reactive materials are special cases. Temperature control, for example, cooling, may be necessary for some self-reactive materials in warm climates. Monomers are shipped with inhibitors and have a shelf life, so the tanks can be sized for rapid turnover. Monomers can also plug normal and emergency vents, so the frequency of inspection and cleaning may need to be increased. Water reactive materials can have inert gas padded atmospheres to prevent water ingress.

Design considerations for process safety. Options available to the designer when designing a storage tank include placement, roof type, and pressure design. Tanks can be located underground or aboveground. Aboveground tanks can be fixed roof or floating roof. Storage tanks can be atmospheric or pressurized and equipped with pressure/vacuum protection devices.

The advantage of Underground Storage Tanks (UST) is that they cannot be exposed to an external fire from, for example, loss of containment of a flammable material from a tank in the same dike. USTs are also sheltered from swings in external temperatures. Underground tanks, however, have an increased risk of soil and/or groundwater contamination from leaks. Most underground tanks are now required to be double walled tanks or in a vault, with leak detection in the space between the tank walls or in the vault (Figure 11.32). Because of the risk of soil and groundwater contamination, the U.S. EPA and many states have strict regulations covering them. The U.S. EPA has a website with information about underground storage tanks at http://www.epa.gov/oust/index.htm.

A variant of the underground tank is the mounded tank design (Figure 11.33). This is an earth covered aboveground tank. The earth cover makes the tank almost immune to BLEVE. Earth covered tanks are frequently used for an LPG bullet tank. Mounded tanks have no groundwater contamination issue as they are above ground level and an impervious membrane or cathodic protection systems similar to those used for pipelines to address corrosion issues can be installed during construction.

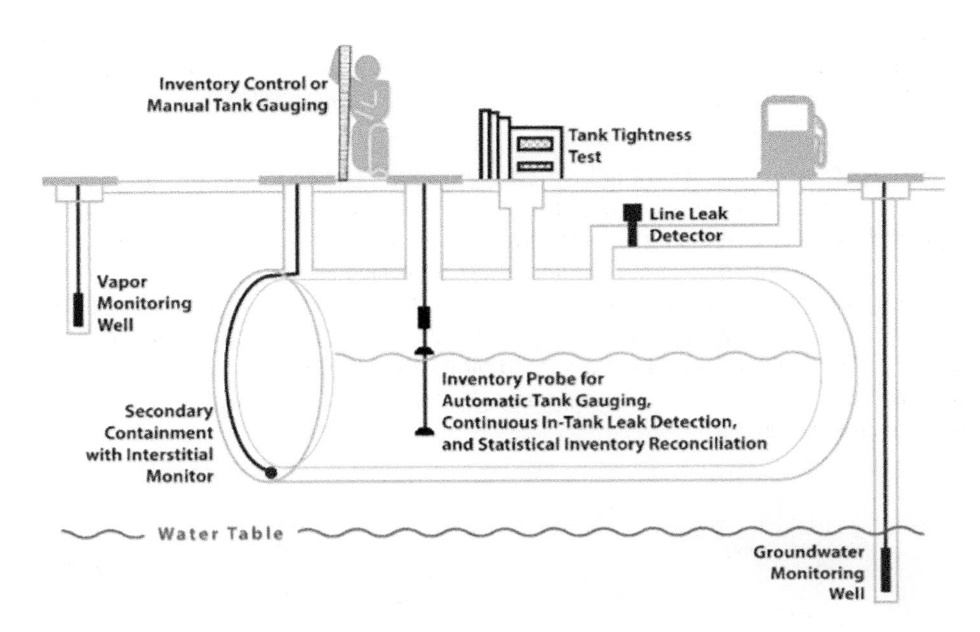

Figure 11.32. Schematic diagram of UST leak detection methods (EPA 2016)

Figure 11.33. Mounded underground tank

(BNH)

Fixed roof storage tanks are usually used for materials with a vapor pressure below 10.3 kPa (1.5 psia) at some specified temperature such as 20°C (68°F). Floating roof tanks can be used with materials with a vapor pressure up to 79.3 kPa (11.5 psia), at some specified temperature. Floating roof tanks can be open or have a secondary structural roof (Figure 11.34a. and 11.34b.). The advantage of a floating roof tank is that the roof has no vapor space above it; therefore, if the stored material is flammable, no flammable vapor can ignite in the headspace.

Floating roofs introduce other failure modes, however. The floating roof can tilt due to loss of flotation from one of its pontoons or get stuck on the anti-rotation pole and become wedged in one position. In that case filling or emptying the tank could lead to materials getting above the roof, which can result in loss of containment from tank collapse or a full surface fire from the exposed product. Also, leaks can occur between the tank wall and the floating roof seal, which, if flammable, can cause annular fires at the wall. The roof drain, the crooked pipe in the middle of the tank in Figure 11.34a, is a jointed pipe which is intended to drain rainwater into the tank dike. If it leaks, the entire contents of the tank can be released if no detection and response mechanism exists.

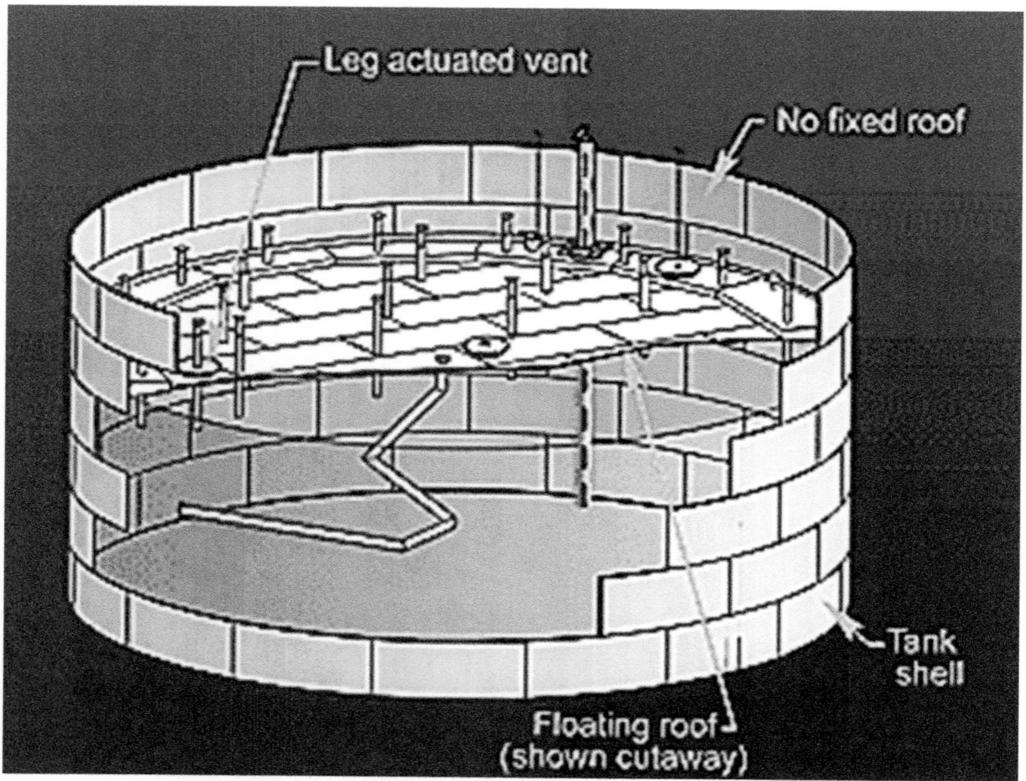

Figure 11.34 a. Schematics of external internal floating roof tank (Anson)

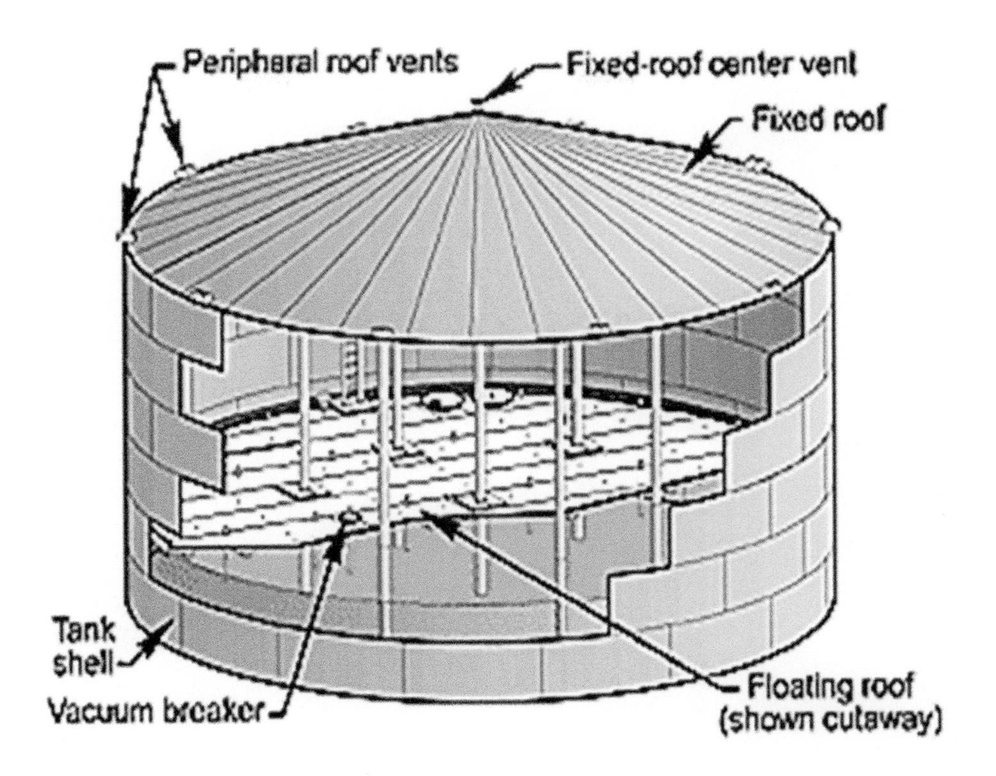

Figure 11.34 b. Schematics of internal floating roof tank (Anson)

The failure of the roof drain highlights another aspect of aboveground tanks; they must be inside dikes that provide secondary containment for leaks from the tank. The dikes need to be large enough to contain the tank volume plus some safety factor (an industry rule of thumb is 110% of the volume of the tank). Local and federal regulations usually require the dike to contain the entire contents of the largest tank within a dike. The dike needs to be maintained to prevent leaks.

Pressurized storage tanks are used for materials with higher vapor pressures, such as ammonia, butane or Liquefied Petroleum Gas (LPG) (Figure 11.35). Pressurized storage tanks are susceptible to a phenomenon called boiling liquid expanding vapor explosion (BLEVE). The most common cause of a BLEVE is exposure to external fire. A BLEVE occurs when a vessel containing liquid above its normal boiling point and under pressure fails catastrophically. The vessel may fail below its design pressure if the steel shell in the vapor space (which is not cooled by liquid contact) is exposed to the flames. This is because the ultimate tensile strength of the steel reduces to 50% of its original strength at 550°C (1022°F). Hydrocarbon flames are around 1150°C (2102°F). Therefore, the tank can fail catastrophically when the vapor space, an unwetted portion of the tank, exceeds 550°C (1022°F). Upon failure, the hot liquid instantly flashes, generating a large amount of vapor and a pressure blast. Since the initiating event is an external fire, if the vapor in the vessel is flammable it will typically ignite and create a large fireball. Many of the explosions in the Mexico City event in 1984 were BLEVEs.

Figure 11.35. Pressurized gas storage tank
(shutterstock)

A fixed water spray, deluge system, or firewater monitor nozzles can keep vessels cool enough, maintaining asset integrity when exposed to a fire. Use of a mounded storage tank, mentioned earlier in this section, is also an option to protect pressurized tanks from exposure to fire.

Codes and Standards that cover storage tanks include the following.

- API STD 620 "Design and Construction of Large, Welded, Low-pressure Storage Tanks"
- API STD 650 "Welded Steel Tanks for Oil Storage"
- API STD 651 "Cathodic Protection for Aboveground Petroleum Storage Tanks"
- API STD 2000 "Venting Atmospheric and Low-pressure Storage Tanks"
- API STD 2350 "Overfill Prevention for Storage Tanks in Petroleum Facilities"
- ASME "Boiler and Pressure Vessel Code"
- NFPA 30 "Flammable and Combustible Liquids Code"
- NFPA 58 "Liquefied Petroleum Gas Code"

11.3.8 Piping

Most facilities have miles of piping. This piping supports the flow of feed, product, and everything in between to and from the equipment previously discussed in this chapter. Piping comes in all sizes and materials. The piping material, the chemicals it is transporting, how it is protected, and its routing are factors in how it may be damaged.

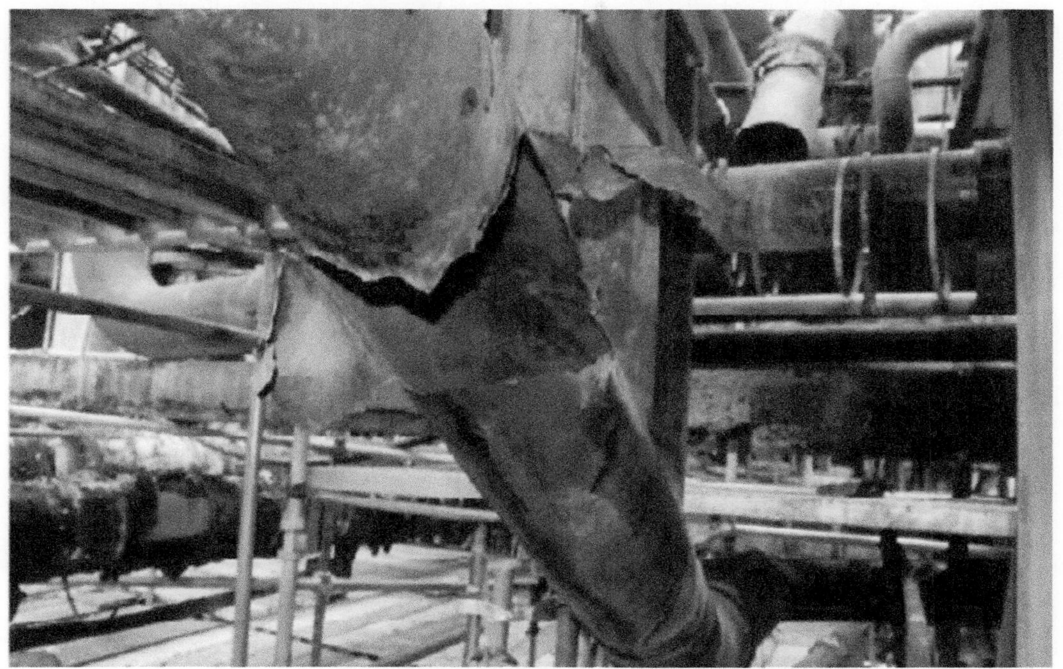

Figure 11.36. Piping rupture
(CSB 2015)

Example 1. Piping may corrode over time depending on the piping material and the composition of the chemicals flowing through the piping. An example of this is the Chevron Refinery in Richmond, California suffered a fire in 2012 (Figure 11.36). The source of the fire was a rupture of unit piping due to sulfidation corrosion applicable at high temperature. Other metallurgy and other chemical combinations can cause different types of corrosion.

Example 2. Through science and research, an improved understanding may be gained regarding the appropriate type of metallurgy for use with specific chemicals. Facilities that were constructed years ago used the understanding and materials of that time period. A hydrofluoric (HF) acid alkylation unit with a piping elbow installed in about 1973 contained more nickel and copper than would be recommended today in API RP 751 "Safe Operation of Hydrofluoric Acid Alkylation Units" or in a National Association of Corrosion Engineers (NACE a) paper. The elbow failed resulting in a fire, explosion, and release of toxic HF acid. Images of the HF unit before and after the explosion and fire are shown in Figure 11.37.

Figure 11.37. Comparison of HF Unit incident scene pre-and post-incident (CSB 2019)

Example 3. Piping may be vulnerable to impact due to its location. Section 8.3 described an incident involving a piece of equipment that was dropped on an HF Alkylation storage tank transfer line. Another form of impact is vehicular impact. Because piping is virtually everywhere in a facility, it is frequently located adjacent to roadways. Protecting piping at intersections and where it crosses overhead may be appropriate to prevent pipe damage and loss of containment.

Processes handling saturated gases (the carbon atoms are fully saturated with hydrogen), are subject to auto refrigeration as experienced in the Esso Longford explosion described in Chapter 12. Auto-refrigeration can occur on adiabatic expansion of gasses. The resulting low temperature can bring metals like carbon steel below their ductile-brittle transition temperature resulting in metal embrittlement. This has resulted in complete rupture of vessels and pipelines with loss of containment and gas explosions. In addition to LPGs, gases such as ammonia, chlorine, and hydrogen chloride can cause auto-refrigeration.

Failure of process equipment, including pumps and control valves, can lead to overpressure of process piping resulting in failure. This can lead to large flammable releases. Pressure relief valves routed to flare systems are typically installed to prevent overpressure. An additional source of overpressure is blocked in piping segments exposed to thermal radiation. Thermal relief should be provided in such situations.

Corrosion of piping and equipment is a common problem and can cause of loss of containment. This makes asset integrity and reliability a key PSM element for operations handling hazardous materials. NACE lists over 40 types of corrosion. (NACE b) A couple are highlighted as follows.

Sulfidation corrosion is due to the reaction between sulfur compounds, especially H_2S, and iron at temperatures of 230 – 430°C (450 – 800°F). This causes the thinning of materials such as steel, leading to failure of piping if not monitored and controlled. This can occur in processes handling materials that contain sulfur, such as crude oil. The hazard can be reduced by the use of steel with higher chromium content. Such steels are inherently safer than carbon steel with respect to sulfidation corrosion. A good asset integrity program is still required to manage the corrosion hazards. API RP 939-C "Guidelines for Avoiding Sulfidation (Sulfidic) Corrosion Failures in Oil Refineries" is a good reference (API RP 939 C).

In processes using hydrogen at high temperature, High Temperature Hydrogen Attack (HTHA) can occur. In HTHA, hydrogen diffuses into the steel walls of equipment at high temperatures and reacts with carbon in the steel, producing methane. This causes a local high pressure in the steel grains. The methane causes fissures to form on the steel, weakening it. HTHA is difficult to identify in its early stages, as the fissures are very small. By the time it can be detected, the equipment already has a higher likelihood of failure. High chromium steel is more resistant to HTHA and is, therefore, a safer material of construction.

API RP 941, "Steels for Hydrogen Service at Elevated Temperatures and Pressures in Petroleum Refineries and Petrochemical Plants" provides a curve (called Nelson curve) that shows the temperatures and pressures at which HTHA can occur for various metals.

Corrosion can also occur on the outside of the pipe wall. A common problem is Corrosion Under Insulation (CUI). This occurs when piping is insulated with fireproofing or thermal insulation. If a crack occurs in the insulation and water seeps into the space between the piping and the insulation, corrosion may occur undetected for some time. API RP 571, "Damage Mechanisms Affecting Fixed Equipment in the Refining Industry", address CUI.

Different from corrosion, erosion is also a potential threat to piping integrity. Erosion occurs as a material mechanically damages or thins the pipe wall. This can be caused by flow of the catalyst, such as in catalytic cracking, flow through the piping. In the upstream industry, it can be caused by sand that is entrained in the crude oil. Although it is a different mechanism than corrosion, the threat of loss of containment is the same.

Erosion and corrosion can team up to accelerate piping thickness loss. Some materials corrode, and the corrosion product forms a passive protection layer which limits corrosive damage. This protective layer can be stripped away by erosive forces, which exposes the underlying material to renewed corrosive attack.

An additional threat, primarily to small bore piping, is vibration. Small bore piping failures occur more frequently than for larger piping. Many releases have occurred due to vibration of small-bore piping such as instrument piping, which led to piping failure. Vibration can be caused by induced vibration, pulsating equipment such as reciprocating pumps, equipment subject to ocean waves, and fluid shock or 'hammer' caused by rapidly stopping or starting flow. (CCPS 2020)

Design considerations for process safety. Recognizing the potential for piping corrosion is the first step. The piping material may be selected that will not corrode in the service conditions, the piping may be designed to withstand the corrosion for many years by providing appropriate wall thickness, or the process may include a chemical injection to prevent or minimize the corrosive impact. Considering corrosion of materials in the design stage is only the first step; monitoring it throughout the life cycle is required. This is addressed in the following section.

With respect to vibration, a challenge is that frequently only larger piping is shown on piping isometrics with a note that small-bore piping is field installed. The result is that the installation is dependent on the skill of the installer and may not be subject to the engineering review that other piping and equipment receives. The length of unsupported or unrestrained piping should be reviewed. To isolate vibration, flexible connections may be used, but they are also weaker components that can fail. Both the amplitude (amount of movement) and the frequency (rate of movement) can affect how quickly vibration can cause equipment to fail. Technology exists to test and analyze vibration to determine the exact source. (CCPS 2020)

In addition to piping, flexible hose assemblies may be used, typically in loading and unloading operations, to transfer materials. "Guidelines for the management of flexible hose assemblies" provides information on maintaining the integrity of these systems. (EI)

11.4 Asset Integrity and Reliability

Asset integrity and reliability is the RBPS element that helps ensure that equipment is properly designed, installed in accordance with specifications, and remains fit for use until it is retired. The previous sections addressed how equipment can fail and provided design considerations for process safety Even with the best design, integrity issues can occur during operations. Putting in place a system to manage asset integrity and reliability is important to production and process safety.

The objective of the asset integrity element is to help ensure reliable performance of equipment. This equipment not only supports production, but it also serves as the primary envelope containing potentially hazardous materials. Increasing reliability maximizes run-time and thus production. It also prevents equipment failures, losses of containment, and process safety events – which can cause major production interruptions.

Asset Integrity – The condition of an asset that is properly designed and installed in accordance with specifications and remains fit for purpose. (CCPS Glossary)

Inspection, Testing, and Preventive Maintenance (ITPM) – Scheduled proactive maintenance activities intended to (1) assess the current condition and/or rate of degradation of equipment, (2) test the operation/functionality of equipment, and/or (3) prevent equipment failure by restoring equipment condition. (CCPS Glossary)

Asset integrity involves conducting damage mechanism reviews, inspections, tests, preventive maintenance; predictive maintenance, and repair activities that are performed by maintenance and contractor personnel at operating facilities. It also includes quality assurance processes, such as procedures and training, that underpin these activities.

Asset integrity activities occur at many places throughout an organization and extend throughout the life of the facility. At an operating facility, the asset integrity activities are an integral part of day-to-day operation involving operators, maintenance employees, inspectors, contractors, engineers, and others involved in designing, specifying, installing, operating, or maintaining equipment. Asset integrity activities range from technical meetings involving experts seeking to advance the state-of-the art in equipment design, inspection, testing, or reliability, to a plant operator on routine rounds spotting leaks, unusual noises or odors, or detecting other abnormal conditions.

Companies often establish asset integrity centers of excellence, establish corporate standards, and promote efforts to continuously improve the safety and reliability of process equipment. Industry sponsored technical committees and organizations are continuously working to advance the state of knowledge regarding proper design and ITPM practices to help ensure that equipment is fit for service at commissioning and remains fit for service throughout its life.

Fitness for Service (FFS) – A systematic approach for evaluating the current condition of a piece of equipment in order to determine if the equipment item is capable of operating at defined operating conditions (e.g., temperature, pressure). (CCPS Glossary)

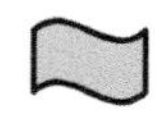

An effective asset integrity program focuses on the following.

- Equipment is designed, fabricated, and installed in accordance with design specifications and is fit for service at startup.
- A quality assurance program helps prevent equipment failures that could result from use of faulty parts/materials or improper materials of construction, improper

fabrication, installation, or repair methods. This includes Positive Material Identification (PMI) to verify the materials of construction of equipment. It also includes a system is used to manage repair parts and maintenance materials to prevent unintended use of inappropriate materials of construction.

- Processes are operated within the design limits of the equipment.
- Damage mechanisms are assessed, and results incorporated in asset integrity plans.
- ITPM tasks are conducted by trained and qualified individuals using procedures that conform to generally accepted standards and are completed as scheduled.
- Maintenance work conforms to design codes, engineering standards, and manufacturer's recommendations to help ensure that equipment remains fit for service.
- The performance of the asset integrity program is tracked, and actions are taken to address deficiencies. (CCPS 2020)

Damage Mechanism Reviews (DMRs) are conducted to determine credible degradation modes and susceptibilities of processing equipment. DMR identifies corrosion and damage mechanisms such as high temperature hydrogen attack and chloride stress corrosion cracking. The results of these studies are used in formulating inspection plans included in asset integrity plans. DMRs are the subject of API RP 571 "Damage Mechanisms Affecting Fixed Equipment in the Refining Industry, Second Edition". (API RP 571) In addition, California OSHA's recently revised "Process Safety Management for Petroleum Refineries" now requires that DMRs be performed and process hazard analyses address DMR reports that are applicable to the process unit being analyzed.

A Risk-Based Inspection program (RBI), as per API RP 580 and API RP 581 Risk-Based Inspection and Risk-Based Inspection Methodology are often used in industry. The RBI program provides a risk-based method to prioritize piping and vessels inspections accordingly.

Along with RBI, Integrity Operating Windows (IOW) can support mechanical integrity. IOWs are process variable limits beyond which the equipment integrity may be adversely impacted. Monitoring process variables such that IOWs are not exceeded prevent associated mechanical integrity issues.

11.5 What a New Engineer Might Do

Process engineers should understand how equipment fails, and design equipment to minimize the failure potential thereby reducing process risk. By reducing failure frequency, the process engineer can improve the reliability of equipment, lower operating costs, and create safer chemical process operations. A large part of equipment design is understanding material compatibility for material of construction selection.

Mechanical Engineers contribute significantly to asset integrity and reliability. They are responsible for developing test procedures, overseeing inspection programs, and analyzing the data from inspection programs that help predict and prevent equipment failure. To this end, a new mechanical engineer should become familiar with codes and standards related to asset integrity and ITPM.

Mechanical engineers often have the lead role in identifying failure rates in determining reliability. Failure rates can be used in reliability-centered maintenance and can be translated into frequencies of failure and probability of failure on demand which are key values used in semi-quantitative and quantitative risk analysis.

A technique for inspection, borrowed from process safety, is the use of risk-based inspection (API RP 580). This combines corrosion mechanisms predicting corrosion rates and consequence models predicting potential outcomes to prioritize sections of piping or process vessels for greater or less inspection attention. RBI has been shown to enhance safety and reduce inspection costs by focusing attention on the most important areas.

Reliability-Centered Maintenance (RCM) - A systematic analysis approach for evaluating equipment failure impacts on system performance and determining specific strategies for managing the identified equipment failures. The failure management strategies may include preventive maintenance, predictive maintenance, inspections, testing, and/or one-time changes (e.g., design improvements, operational changes). (CCPS Glossary)

Instrument, Electrical and Control engineers will often be tasked with designing Safety Instrumented Systems (SIS) and ensuring that the required reliability and probability of failure on demand is achieved. They may often be asked to develop procedures for safety instrument system test protocols, instrument calibrations and testing, control loop response capabilities etc. Similar to the mechanical engineer's role in asset reliability, instrumentation failure rates may be used to support semi quantitative and quantitative risk analysis.

All engineers can support asset integrity by being observant when walking through the facility. Watch and listen for vibrating equipment and report concerns to your supervisor. You may see or hear something that is not being monitored by maintenance inspections.

A new engineer can benefit from reviewing the CSB investigations and videos relevant to this chapter as listed in Appendix G.

11.6 Tools

Tools that may be used in understanding how equipment fails and supporting asset integrity to prevent that failure include the following.

API Standards and Practices. API has created many standards and practices that address the design and life cycle integrity of various types of equipment. These are available at www.API.org.

CCPS *Guidelines for Asset Integrity Management.* This book includes details on failure modes and mechanisms and testing an inspection programs for various types of equipment. (CCPS 2016)

CCPS *Guidelines for Improving Plant Reliability through Data Collection and Analysis.* This book provides guidance on how to collect, and use with confidence, process equipment reliability data for risk-based decisions. It provides the techniques to gather plant performance, maintenance, and repair data that can be used for in reliability/availability assessments; preventive maintenance programs; and risk analysis. (CCPS 1998)

CCPS *Guidelines for Mechanical Integrity Systems*. This book provides a basic familiarity of mechanical/asset integrity concepts and best practices and recommends approaches for establishing a successful mechanical/asset integrity program. (CCPS 2006)

CCPS *Guidelines for Safe Operations and Maintenance*. This book addresses process safety management program execution in the operations and maintenance for the entire life cycle of a facility. (CCPS 1995)

Chemical Reactivity Worksheet (CRW). This is a free software program providing extensive process safety information required to understand the hazards associated with the inadvertent and intentional mixing of reactive chemicals. It includes sections on compatibility of material, and suitability of materials of construction. Refer to Section 5.8.3 for further details. (CCPS CRW)

Equipment failure statistics. It can be a challenge to find equipment failure statistics for the process industry. The following are good sources.

- **Offshore Release Database (OREDA).** This is maintained by the U.K. HSE for North Sea oil and gas installations. It was implemented after the Piper Alpha disaster. (OREDA)
- **International Association of Oil & Gas Producers (IOGP) Risk Assessment Data Directory.** This resource focuses on equipment types found offshore. While not applicable to chemical plants or reactors, it does cover many of equipment types discussed in this chapter. (IOGP 2019)

OSHA Technical Manual – Section IV: Chapter 2 – Petroleum Refining Process. This provides information on basic refinery processes. Additionally, it provides integrity threats to common types of equipment which is relevant beyond the refining industry. This manual can be accessed at https://www.osha.gov/dts/osta/otm/otm_iv/otm_iv_2.html.

11.7 Summary

In the Buncefield incident a simple error during maintenance of a level control switch contributed to a major explosion and fire. It would be naïve to think that well designed equipment will retain its integrity over its life cycle. Understand how equipment can fail and address this in asset integrity programs. Asset integrity should aim to ensure that appropriate materials of construction are specified, equipment is installed as specified, and protected over its operating life against threats such as corrosion, erosion, and impact. Inspection, testing, and preventive maintenance programs can monitor asset integrity and identify opportunities to address integrity issues before a process safety event occurs.

11.8 Other Incidents

This chapter began with a description of the Buncefield storage tank overflow and explosion. Other incidents involving equipment failure and asset integrity include the following.

- Toxic Gas Release, Seveso, Italy, 1976

- Mexico City Fire and BLEVEs, Mexico City, Mexico, 1984
- Space Shuttle Challenger, U.S., 1986
- BP Hydrocracker Explosion, Grangemouth, U.K., 1987
- Arco Chemical Explosion, Channelview, Texas, U.S., 1990
- Hickson Welsh Jet Fire, Castleford, U.K., 1992
- Ammonium Nitrate Explosion, Port Neal, Iowa, U.S., 1994
- Texaco Refinery Explosion, Milford Haven, U.K., 1994
- Esso Gas Plant Explosion, Longford, Australia, 1998
- Equilon Refinery Coking Unit Explosion and Fire, Anacortes, Washington, U.S., 1998
- Motiva Sulfuric Acid Tank Explosion, Refinery, Delaware City, Delaware, 2001
- Hayes Lammerz Dust Explosion, Huntington, Indiana, U.S., 2003
- DPC Chlorine Release, Glendale, Arizona, U.S., 2003
- Space Shuttle Columbia, U.S., 2003
- BP Isomerization Explosion, Texas City, Texas, U.S., 2005
- Olive Oil Multiple Storage Tank Explosions, Spoleto, Italy, 2006
- Imperial Sugar Dust Explosion, Port Wentworth, Georgia, U.S., 2008
- Varanus Island Pipeline, Western Australia, 2008
- Deepwater Horizon Well Blowout, Gulf of Mexico, U.S., 2010
- DuPont Phosgene Release, Belle, West Virginia, U.S., 2010
- PG&E Pipeline Explosion, San Bruno, California, U.S., 2010
- Williams Olefins Heat Exchanger Explosion, Geismer, Louisiana, U.S., 2013
- Freedom Industries Chemical Spill, Charleston, West Virginia, U.S., 2014
- Philadelphia Energy Solutions Refinery HF Acid Alklyation Unit Fire and Explosion, Philadelphia, Pennsylvania, U.S., 2019

11.9 Exercises

1. List 3 RBPS elements evident in the Buncefield storage tank overflow incident summarized at the beginning of this chapter. Describe their shortcomings as related to this accident.
2. Considering the Buncefield storage tank overflow incident, what actions could have been taken to reduce the risk of this incident?
3. What equipment failed in the BP Texas City explosion incident described in Chapter 2?
4. What equipment failed in the Imperial Sugar Dust explosion incident described in Chapter 4?
5. What equipment failed in the Fukushima Daiichi Nuclear Power Plant incident described in Chapter 8?
6. Describe three common failure mechanisms for a centrifugal pump. What controls might be added to prevent this equipment failure?
7. Describe three common failure mechanisms for compressors. What controls might be added to prevent this equipment failure?

8. Describe three common failure mechanisms for shell and tube heat exchangers. What controls might be added to prevent this equipment failure?
9. Describe three common sources of loss of primary containment for storage tanks. What controls might be added to prevent this equipment failure?
10. What can be expected to happen if a high pressure pump is run with the inlet valve closed? With the outlet valve closed?
11. A refinery process stream includes sulfur compounds. The process is operated at 260°C (500°F). The piping is carbon steel. Is there a risk to the integrity of the piping? If so, what is it and how could it be managed?
12. A refinery naphtha hydrotreater uses hydrogen at more than 260°C (500°F) to treat the naphtha. What type of corrosion might the unit be subject to?
13. Piping is often insulated either to protect people from touching hot pipe walls, to reduce heat loss, or to protect the piping from flame impingement in the event of a fire. This insulation can be in place for many years and can deteriorate and crack. How might this affect the integrity of the pipe itself?
14. How might you determine how often a piece of equipment should be inspected?

11.10 References

Alfa Laval, https://www.alfalaval.com/globalassets/documents/products/heat-transfer/plate-heat-exchangers/gasketed-plate-and-frame-heat-exchangers/baseline/instruction-manual-baseline-en.pdf.

Anson, http://www.ansonindustry.com/floating-roof-tank.html.

ASME, "Boiler and Pressure Vessel Code", Section VIII, Division 1: Rules for Construction of Pressure Vessels, American Society of Mechanical Engineers, New York, NY.

API RP 520, "Sizing, Selection, and Installation of Pressure-Relieving Devices", American Petroleum Institute, Washington, D.C.

API RP 521, "Pressure-Relieving and Depressurizing Systems", American Petroleum Institute, Washington, D.C.

API RP 556, "Instrumentation, Control, and Protective Systems for Gas Fired Heaters", American Petroleum Institute, Washington, D.C.

API RP 560, "Fired Heaters for General Refinery Service", American Petroleum Institute, Washington, D.C.

API RP 571, "Damage Mechanisms Affecting Fixed Equipment in the Refining Industry", American Petroleum Institute, Washington, D.C.

API STD 610, "Centrifugal Pumps for Petroleum, Petrochemical and Natural Gas Industries, Eleventh Edition", American Petroleum Institute, Washington, D.C.

API STD 660, "Shell and Tube Heat Exchangers", American Petroleum Institute, Washington, D.C.

API RP 751, "Safe Operation of Hydrofluoric Acid Alkylation Units", American Petroleum Institute, Washington, D.C.

API RP 580, "Risk-Based Inspection", American Petroleum Institute, Washington, D.C.

API RP 581, "Risk-Based Inspection Methodology", American Petroleum Institute, Washington, D.C.

API RP 939-C, "Guidelines for Avoiding Sulfidation (Sulfidic) Corrosion Failures in Oil Refineries", American Petroleum Institute, Washington, D.C.

API RP 941, "Steels for Hydrogen Service at Elevated Temperatures and Pressures in Petroleum Refineries and Petrochemical Plants", American Petroleum Institute, Washington, D.C.

API STD 617, "Axial and Centrifugal Compressors and Expander-Compressors", American Petroleum Institute, Washington D.C.

API STD 620, "Design and Construction of Large, Welded, Low-pressure Storage Tanks", American Petroleum Institute, Washington, D.C.

API STD 650, "Welded Steel Tanks for Oil Storage", American Petroleum Institute. Washington, D.C.

API STD 651,"Cathodic Protection for Aboveground Petroleum Storage Tanks", American Petroleum Institute, Washington, D.C.

API STD 660, "Shell-and-Tube Heat Exchangers", American Petroleum Institute, Washington, D.C.

API STD 674, "Positive Displacement Pumps-Controlled Volume for Petroleum, Chemical, and Gas Industry Services", American Petroleum Institute, Washington, D.C.

API STD 685, "Sealless Centrifugal Pumps for Petroleum, Petrochemical, and Gas Industry Process Service", American Petroleum Institute, Washington, D.C.

API STD 2000, "Venting Atmospheric and Low-Pressure Storage Tanks", American Petroleum Institute, Washington, D.C.

API STD 2350 "Overfill Prevention for Storage Tanks in Petroleum Facilities", American Petroleum Institute, Washington, D.C.

Berg, J. "The Case for Double Mechanical Seals", *Chemical Engineering Progress*, June 2009.

BHN, https://www.bnhgastank.com/mounded-bullets-lpg.html.

Bouck, Doug, "Distillation Revamp Pitfalls to Avoid", *Chemical Engineering Progress*, February 2014.

Britton, L.G., *Avoiding static ignition hazards in chemical operations*, Center for Chemical Process Safety, John Wiley & Sons, Hoboken, N.J.,1999.

Buncefield 2008, "The Buncefield incident, 11 December 2005: the final report of the major incident investigation board", U.K. Health and Safety Executive. (http://www.hse.gov.uk/comah/buncefield/miib-final-volume1.pdf)

CCPS CRW, Chemical Reactivity Worksheet 4.0, https://www.aiche.org/ccps/resources/chemical-reactivity-worksheet.

CCPS a, Process Safety Beacon, "The Seal that Didn't Perform", July 2002, http://sache.org/beacon/files/2002/07/en/read/2002-07%20Beacon-s.pdf.

CCPS b, Process Safety Beacon, "It's a Bird, It's a Plane, It's a Pump", October 2002, http://sache.org/beacon/files/2002/10/en/read/2002-10%20Beacon-s.pdf.

CCPS c, CCPS Process Safety Beacon, "Interlocked for a Reason", June 2003, http://sache.org/beacon/files/2003/06/en/read/2003-06%20Beacon-s.pdf.

CCPS d, CCPS Process Safety Beacon, "Avoid Improper Fuel to Air Mixtures", January 2004, http://sache.org/beacon/files/2004/01/en/read/2004-01%20Beacon-s.pdf.

CCPS e, CCPS 2007, Process Safety Beacon, "The Great Boston Molasses Flood of 1919", May 2007, http://sache.org/beacon/files/2007/05/en/read/2007-05-Beacon-s.pdf.

CCPS f, CCPS Process Safety Beacon, "What if You Load the Wrong Material Into a Tank?", April 2012, http://sache.org/beacon/files/2012/04/en/read/2012-04-Beacon-s.pdf.

CCPS g, CCPS Process Safety Beacon, "Vacuum is a Powerful Force!", February 2002, http://sache.org/beacon/files/2002/02/en/read/2002-02-Beacon-s.pdf.

CCPS h, CCPS 2011, Process Safety Beacon, "Understand the Reactivity of Your Heat Transfer Fluid", February 2011, http://sache.org/beacon/files/2011/02/en/read/2011-02-Beacon-s.pdf.

CCPS Glossary, "CCPS Process Safety Glossary", Center for Chemical Process Safety, https://www.aiche.org/ccps/resources/glossary.

CCPS 1995, *Guidelines for Safe Operations and Maintenance*, Center for Chemical Process Safety, John Wiley & Sons, Hoboken, N.J.

CCPS 1998, *Guidelines for Improving Plant Reliability through Data Collection and Analysis*, Center for Chemical Process Safety, John Wiley & Sons, Hoboken, N.J.

CCPS 2005, *Guidelines for Safe Handling of Powders and Bulks Solids*, Center for Chemical Process Safety, John Wiley & Sons, Hoboken, N.J.

CCPS 2006, *Guidelines for Mechanical Integrity Systems*, Center for Chemical Process Safety, John Wiley & Sons, Hoboken, N.J.

CCPS 2012, *Guidelines for Engineering Design for Process Safety 2nd Edition*, Center for Chemical Process Safety, John Wiley & Sons, Hoboken, N.J.

CCPS 2016, *Guidelines for Asset Integrity Management*, Center for Chemical Process Safety, John Wiley & Sons, Hoboken, N.J.

CCPS 2020, Process Safety Beacon, "Not all vibrations in process equipment are 'good vibrations'", November, www.aiche.org/ccps/process-safety-beacon.

Colfax Fluid Handling, https://www.colfaxcorp.com/

CSB, "Static Sparks Explosion in Kansas", Case Study No. 2007-06-I-KS, U.S. Chemical Safety and Hazard Investigation Board, Washington, D.C.

CSB 2009, "Investigation Report T2 Laboratories, Inc. Runaway Reaction", Report No. 2008-3-I-FL, U.S. Chemical Safety and Hazard Investigation Board, Washington, D.C.

CSB 2015, "Final Investigation Report Chevron Richmond Refinery Pipe Rupture and Fire", Report No. 2012-03-I-CA, U.S. Chemical Safety and Hazard Investigation Board, Washington, D.C.

CSB 2019, "Fire and Explosion at Philadelphia Energy Solutions Refinery Hydrofluoric Acid Alkylation Unit, Factual Update", No. 2019-06-I-PA. U.S. Chemical Safety and Hazard Investigation Board, Washington, D.C.

Chu, "Improved Heat Transfer Predictions for Air-Cooled Heat Exchangers", *Chemical Engineering Progress*, November, 2005.

Donaldson-Torit, www.donaldson.com/et-us

Drogaris, G. *Major Accident Reporting System: Lessons Learned from Accidents Notified,* Elsevier, Amsterdam, 1993.

EI, "Guidelines for the management of flexible hose assemblies", Energy Institute, https://publishing.energyinst.org/topics/asset-integrity/guidelines-for-the-management-of-flexible-hose-assemblies.

Ender, Christophe and Laird, Dana, "Minimize the Risk of Fire During Column Maintenance", *Chemical Engineering Progress*, p. 54-56, September 2003.

EPA 2016, "Operating and Maintaining Underground Storage Tank Systems", EPA 510-K-16-001, February, Environmental Protection Agency, Washington, D.C.

Fike, www.fike.com

Garland, R. Wayne, "Root Cause Analysis of Dust Collector Deflagration Incident", *Process Safety Progress*, Vol. 29, No. 4, December 2010.

Haslego, Haslego and Polley, "Designing Plate-and Frame Heat Exchangers", *Chemical Engineering Progress*, September 2002.

HEI, Heat Exchange Institute, www.heatexchange.org

HSE 1994 The explosion and fires at the Texaco Refinery, Milford Haven. 24th July, Health and Safety Executive, https://www.hse.gov.uk/comah/sragtech/casetexaco94.htm.

HSE 2009a, "Buncefield Explosion Mechanism Phase 1, Vols. 1 and 2", U.K. Health and Safety Executive.

HSE 2011, "Buncefield: Why did it happen?", U.K. Health and Safety Executive.

HSE 2017, "Review of vapour cloud explosion incidents", U.K. Health and Safety Executive.

IEC 61511, "Standard for Safety Instrumented Systems", International Electrotechnical Commission, Geneva, Switzerland, https://www.iec.ch/.

IOGP 2019, "Risk Assessment Data Directory - Report 434-01 Process release frequencies", International Association of Oil & Gas Producers, https://www.iogp.org/bookstore/product/434-00-risk-assessment-data-directory-overview/.

Kelley, J. Howard, "Understand the Fundamentals of Centrifugal Pumps", *Chemical Engineering Progress*, October 2010.

Kletz 1993, *Lessons from Disaster: How Organization have no Memory and Accidents Recur*, Gulf Professional Publishing, Houston, Texas.

Lees, *Loss Prevention in the Process Industries, Volume. 3*, Elsevier, Netherlands, 2012.

Mannan, Sam, "Best Practices in Prevention and Suppression of Metal Packing Fires", Mary Kay O'Connor Process Safety Center, August 2003.

MIIB 2008, "The Buncefield incident, Volume 1", Major Incident Investigation Board, https://www.icheme.org/media/13707/buncefield-miib-final-report-volume-1.pdf.

Mukherjee, R., "Effectively Design Shell-and-Tube Heat Exchangers", *Chemical Engineering Progress*, February 1998.

NACE, , https://www.nace.org/home.

NACE, "Specification for Carbon Steel Materials for Hydrofluoric Acid Alkylation Units", Conference Pater 03651, National Association of Corrosion Engineers, Houston, Texas.

Naujokas, A.A., "Spontaneous Combustion of Carbon Beds", *Plant/Operations Progress*, April 1995.

NFPA 30, "Flammable and Combustible Liquid Storage Code", National Fire Prevention Association, Quincy, Massachusetts, 2015.

NFPA 58. "Liquefied Petroleum Gas Code", National Fire Protection Association. Quincy, Massachusetts.

NFPA 77, "Recommended Practice on Static Electricity", National Fire Prevention Association, Quincy, Massachusetts.

NFPA 85, "Boiler and Combustion Systems Hazards Code", National Fire Prevention Association, Quincy, Massachusetts.

NFPA 86, "Standard for Ovens and Furnaces", National Fire Prevention Association, Quincy, Massachusetts.

NFPA RP 87, "Recommended Practice for Fluid Heaters", National Fire Prevention Association, Quincy, Massachusetts.

NFPA 652, "Standard on the Fundamentals of Combustible Dust", National Fire Prevention Association, Quincy, Massachusetts.

NFPA 780, "Standard for the Installation of Lightning Protection Systems", National Fire Prevention Association, Quincy, Massachusetts.

OEC Fluid Handling, www.oecfh.com

OSHA, Safety Hazard Information Bulletin "Fire Hazard from Carbon Adsorption Deodorizing Systems", August 17, 1992, https://www.osha.gov/dts/hib/hib_data/hib19970730.html.

OSHA, *OSHA Technical Manual* – Section IV: Chapter 2 – Petroleum Refining Process, https://www.osha.gov/dts/osta/otm/otm_iv/otm_iv_2.html.

OREDA, Offshore and Onshore Reliability Database, https://www.oreda.com.

Patnaik, T., "Solid-Liquid Separation: A Guide to Centrifuge Collection", *Chemical Engineering Progress*, July 2012.

PSLP a, Jarvis, H.C. "Butadiene Explosion at Texas City-2", *Plant Safety & Loss Prevention*, Vol. 5, 1971.

PSLP b, "Butadiene Explosion at Texas City-1", *Plant Safety & Loss Prevention*, Vol. 5, 1971.

PSLP c, Keister, R.G., et al. "Butadiene Explosion at Texas City-3", *Plant Safety & Loss Prevention*, Vol. 5, 1971.

Ramzan, Naveeed, et al, "Root Cause Analysis of Primary Reformer Catastrophic Failure: A Case Study", *Process Safety Progress*, Vol. 30, No. 1, March 2011.

Sherman, R.E., "Carbon-Initiated Effluent Tank Overpressure Incident", *Process Safety Progress*, Vol. 15, No. 3, Fall 1996.

Shutterstock, Royalty-free stock photo ID: 1340068283

Sulzer Chemtech Ltd., www.sulzer.com/en/

TEMA, Tubular Exchanger Manufacturers Association, http://kbcdco.tema.org/

Urban, P.G., *Bretherick's Handbook of Reactive Chemical Hazards 7th Edition*, Academic Press, New York, NYISBN:978-0-12-372563-9, 2006.

12
Hazard Identification

12.1 Learning Objectives

The learning objectives of this chapter are as follows. Having completed the chapter, the reader should be able to:

- Understand hazard identification methods.
- Participate in hazard identification studies, and
- Create a potential process safety incident scenario.

12.2 Incident: Esso Longford Gas Plant Explosion, Victoria, Australia, 1998

12.2.1 Incident Summary

A major explosion and fire occurred at Esso's Longford gas processing site in Victoria, Australia in 1998. Two employees were fatally injured, and eight others injured. The incident caused the destruction of Plant 1 and shutdown of Plants 2 and 3 at the site. This shutdown resulted in total loss of gas supply to Victoria and consequential business interruption and economic impact.

A process upset in a set of absorbers eventually caused temperature decreases and loss of flow of a "lean oil" stream. This allowed a metal heat exchanger to become very cold and brittle. When operators restarted flow of the lean oil to the heat exchanger, it ruptured, releasing a cloud of gas and oil. When the cloud reached an ignition source, the fire flashed back to the release point resulting in additional equipment ruptures and an escalating fire.

Figure 12.1. Photograph of the failed end of GP905 reboiler (LRC)

Key Points:

Hazard Identification and Risk Analysis. If you don't identify a hazard, you won't manage it. The gas plant #1 hazard identification study had been planned, but never carried out. Operators were not aware of the potential hazard of the heat exchanger failing due to brittle fracture and did not know how to respond appropriately.

Management of Change. MOC is not just about equipment. Managing the changes in process safety tasks in job descriptions is key. The plant's process safety engineering staff was relocated and the role that they filled in Management of Change review was not managed and not replaced. The Supervisors and operators were not prepared for the increased troubleshooting responsibilities.

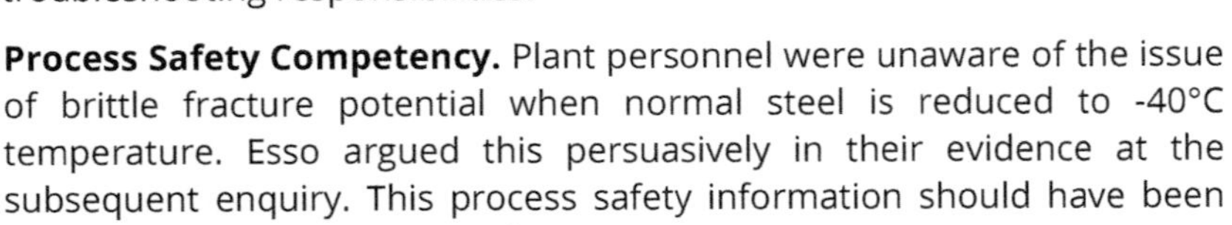

Process Safety Competency. Plant personnel were unaware of the issue of brittle fracture potential when normal steel is reduced to -40°C temperature. Esso argued this persuasively in their evidence at the subsequent enquiry. This process safety information should have been understood by plant personnel.

12.2.2 Detailed Description

The plant involved, Plant No. 1, was a lean oil absorption plant, which separated methane from LPG by stripping the incoming gas with a hydrocarbon stream called "lean oil". Methane rises to the top of the towers, with heavier hydrocarbons dissolving in the liquid hydrocarbon condensate, see Figure 12.2.

Plant No. 1 had a pair of absorbers operating in parallel. Each absorber had a gas/liquid disengaging region at the base where a mixture of gas and liquid hydrocarbons entered the absorbers. During the previous night shift, the hydrocarbon condensate level had started to increase in the base of Absorber B. As the normal disposal of condensate to Gas Plant No. 2 was not available, the alternative condensate disposal route was to a Condensate Flash Tank, see Figure 12.3. Under this set of circumstances, it was normal to increase the temperature at the base of the absorber, but this was not done. The inlet to the Condensate Flash Tank was protected against excessively low temperatures by an override on the absorber level controllers. The consequence; therefore, was that the disposal rate of condensate from the absorber became less than the inlet flow, resulting in a buildup of liquid condensate in the absorber base.

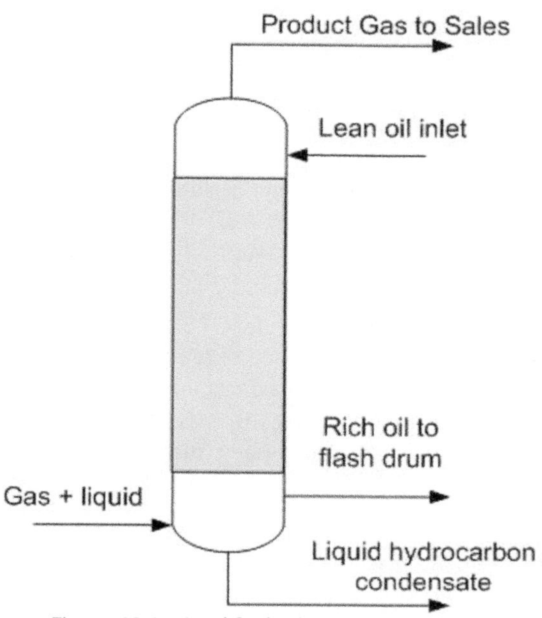

Figure 12.2. Simplified schematic of absorber
(CCPS 2008 a)

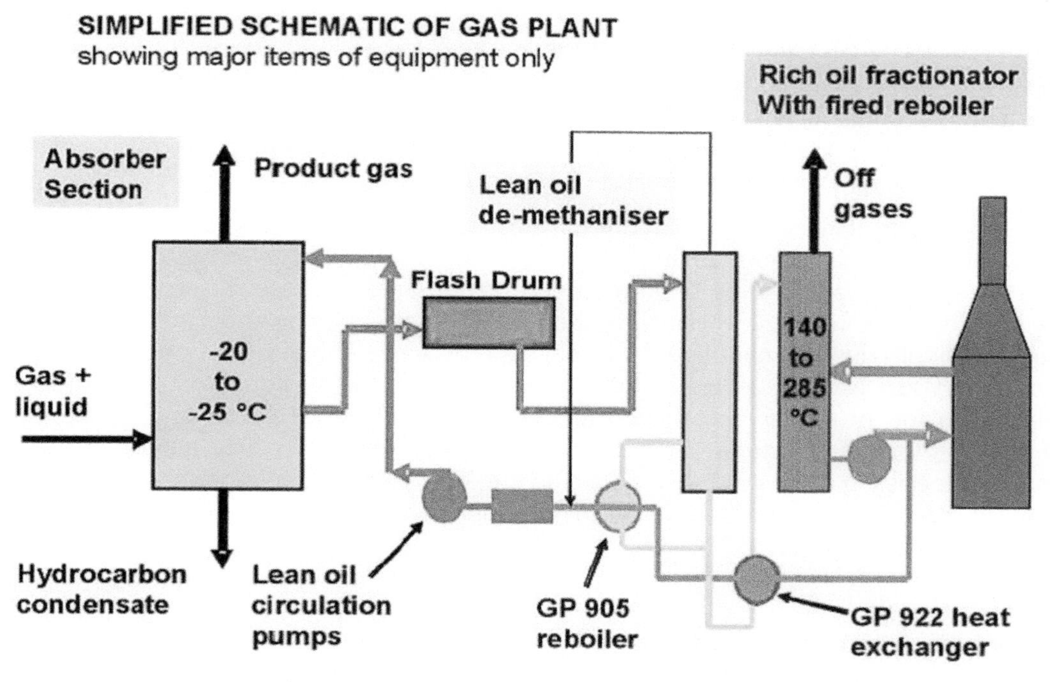

Figure 12.3. Simplified schematic of the gas plant
(CCPS 2008 a)

The condensate level rose in the absorber to a point where it mixed with the exiting rich stripping oil stream. Condensate mixed with rich oil flashed over the rich oil level control valve resulting in a much-reduced temperature in the downstream Rich Oil Flash Tank. This caused temperatures to drop across the plant as rich oil flowed through the recovery process where hydrocarbons were stripped from the rich oil before returning it to the absorbers as lean oil. Eventually, the lean oil pumps tripped out, causing major thermal excursions on a plant with a high degree of process and thermal integration. Loss of lean oil was a critical event but was not communicated to the supervisor until he returned from the morning production meeting one hour after the pumps had tripped.

Temperatures in parts of the plant fell to -48°C (-54°F). At 08:30 AM, a condensate leak occurred on heat exchanger GP922. The absence of lean oil flow meant that the condensate flowing through the rich oil system was not warmed as it entered the recovery section. The reason for the leak was probably due to a strong thermal gradient created while attempts were being made to re-establish the process. Other parts of the process showed signs of intense cold with ice forming on uninsulated parts of heat exchangers and pipework.

At 10:50 AM, the leak from GP922 was getting worse, and the Supervisor decided to shut down Gas Plant No: 1. By 12:15 PM, two maintenance technicians had completed retightening of the bolts on GP922 without making any appreciable difference to the leak. It was decided that the only way to stop the leak was to slowly warm GP922 by starting a flow of warm lean oil through it. However, initial attempts to restart the lean oil pumps were unsuccessful. Ten minutes later, after operating a hand switch to minimize flow through another heat exchanger, GP905, that heat exchanger ruptured, releasing a cloud of gas and oil.

It is estimated that the cloud traveled 170 m (558 ft) before reaching fired heaters where ignition occurred. After flashing back to the point of release flames impinged on piping, which started to fail within minutes. A large fireball was created when a major pressure vessel failed one hour after the fire had started. It took more than two days to isolate all hydrocarbon streams and finally extinguish the fire (CCPS 2008 a).

The investigation concluded that the immediate cause of the incident was loss of lean oil flow leading to a major reduction in temperature of GP905, resulting in embrittlement of the steel shell. This was followed by introduction of hot lean oil in an attempt to stop the hydrocarbon leak in GP922 which led to excess thermal stress in the end plate which failed catastrophically due to embrittlement. Throughout the whole sequence of events, operators and supervisors had not understood the consequences of their actions to re-establish the plant. Esso and the Government were desperate not to shut down the plant, as it supplied all the gas to the State of Victoria. They found their drawings were out of date and they needed to walk the lines to discover what to isolate. In the end they had to shut down the plant and that left the state without gas for between nine and nineteen days, causing major industrial disruption and job losses.

12.2.3 Lessons

Hazard Identification and Risk Analysis. Gas plant #1 had not been subject to a hazard identification study as had been done for the other two gas plants at the site. A Hazard and Operability study, HAZOP, had been planned in 1995, but never carried out. Flow and temperature deviations, like those that occurred at Longford Plant No. 1, are typically systematically reviewed as part of a HAZOP study. Because the HAZOP was not performed, the hazardous consequences of these deviations were never identified. This leads to other management safety issues. Procedures and training may be incomplete or inadequate; hence operators will have no knowledge of the seriousness of the deviation. They will not know what to do, and, as in this case, can take the wrong action.

Management of Change. Many of the plant's engineers and all of the process safety engineers were relocated to the head office in Melbourne, Australia in 1992. The plant had no Management of Change review about the effect of removing the process safety tasks that the engineers had previously fulfilled. As a result, their critical roles with respect to process safety were not replaced. The subject of Management of Organizational Change (CCPS 2013) has been frequently overlooked in the past.

Process Safety Competency. Supervisors and operators were given greater responsibility for operating the plant, including troubleshooting, for which they were not properly prepared. They were not competent to perform the functions the engineers served.

12.3 Hazard Identification Introduction

Hazard Identification and Risk Analysis (HIRA) encompasses the entire spectrum of hazard identification and risk analyses, from simple and qualitative to detailed and quantitative. This Chapter focuses on hazard identification. Identifying a hazard is fundamental to managing it. If a hazard is unrecognized, then it won't be addressed in engineering design, operating procedures, training or other elements of process safety management.

Hazard Identification - Part of the Hazard Identification and Risk Analysis (HIRA) method in which the material and energy hazards of the process, along with the siting and layout of the facility, are identified so that a risk analysis can be performed on potential incident scenarios. (CCPS Glossary)

It can sometimes be challenging for engineers or engineering leaders that design a project to understand the value of a hazard identification study. This is not malicious, it's simply human nature. Having spent much time and energy on developing a detailed engineering design for a process to work as planned, they find it hard to believe that there might be an unrecognized hazard in the design. It requires a purposeful mindset shift to move from designing it to work to finding out how it might not work as intended. It is for this reason that designers should participate in, but not lead, a HAZOP study. An independent leader is better able to ensure that all potential hazards are identified.

Hazard identification studies can find anything from errors in drawings to scenarios where the process could exceed design conditions. These hazards may be identified through the inclusion of operators, who know how the plant really works; the senior operator, who remembers that piping change that isn't reflected on the drawing; and a process licensor who shares specific engineering details.

Engineering design for process safety should consistently and systematically identify hazards. This enables the evaluation of the hazards and incorporation of prevention and mitigation measures in the design to reduce the associated risk. Process hazards come from many sources, including the following:

- Material and chemistry used (e.g., flammability, toxicity, reactivity)
- Process variables - the way the chemistry works in the process (e.g., pressure, temperature, concentration)
- Equipment failures

Recall from Chapter 1 that a hazard is a chemical or physical characteristic that has the potential for harming people, property, or the environment. Hazard identification involves analyzing a process and thinking about what scenario could exist where a hazard might result in undesired consequences. Specifically, hazard analyses are used to identify weaknesses in design and operation of facilities that could lead to a hazardous material release (loss of containment).

Process hazard analysis (PHA) - An organized effort to identify and evaluate hazards associated with processes and operations to enable their control. This review normally involves the use of qualitative techniques to identify and assess the significance of hazards. Conclusions and appropriate recommendations are developed. Occasionally, quantitative methods are used to help prioritized risk reduction. (CCPS Glossary)

This Chapter introduces a variety of hazard analysis techniques that can be used during various stages of the design and during the operation of a facility. Many regulations require process hazard analysis be conducted for new facilities and revalidated, in the case of OSHA PSM every five years. (OSHA) The revalidation may take the form of redoing the study, confirming the previous study is valid and up to date, or a blend of these approaches. Further guidance on revalidation studies is provided in *Guidelines for Revalidating Process hazard analysis, 1st edition.* (CCPS 2001) The approach of confirming the continued validity of the previous study might include the following steps.

- evaluate that the previous hazard identification analysis used an appropriate analysis method
- evaluate that the previous hazard identification analysis was accurate and complete
- determine if the previous hazard identification analysis has been updated to reflect changes that went through the management of change process
- ensure that process safety information is current
- verify that action items from the previous hazard identification analysis have been resolved
- ensure that learnings from incident investigations have been implemented
- document the revalidation

The type of hazards analysis to be used is often specified as one of the following techniques described in this chapter. Detailed information on these techniques can be found in *Guidelines for Hazard Evaluation Procedures*. (CCPS 2008 b)

- Preliminary Hazard Identification (HAZID) Analysis
- Checklist Analysis
- What-If Analysis
- Hazard and Operability Study (HAZOP)
- Failure Modes and Effects Analysis (FMEA)
- Fault Tree Analysis
- Event Tree Analysis

Hazard analyses are typically qualitative studies although the last three in this list can be quantified. The qualitative and quantitative data and from hazard analyses are often used in consequence and risk assessment which are discussed in Chapters 13 and 14.

A hazard identification study is only as good as its follow-up. Hazard identification studies typically include a coarse estimation of consequence severity and likelihood which can be plotted on a risk matrix to prioritize action. (Refer to Section 14.5.1.) Action items should be tracked to closure and implemented changes should be verified to ensure that they mitigated the hazard as intended.

12.3.1 Preliminary Hazard Identification (HAZID) Analysis

Preliminary Hazard Identification (HAZID) Analysis is sometimes referred to as Hazard Identification or Preliminary Hazard Analysis. HAZID focuses in a general way on the hazardous materials and major process areas of a plant. It is typically conducted by a small, multidisciplinary team. It is most often conducted during the research and development or early in the development of a process when the design is conceptual, and few design details are available. It is often followed by other hazard analyses when greater design details are available. It can be a cost-effective way to identify hazards early in a facility's life when it is easy to make changes. Identifying hazards and considering consequences early in a project can be very useful when making site selection decisions. It is also commonly used as a design review tool before a process P&ID is developed.

A HAZID formulates a list of hazards and generic hazardous situations by considering the conceptual design and known process characteristics. As each hazardous situation is identified, the potential causes, consequences, and possible safeguards are listed in a worksheet. Table 12.1 provides an overview of Hazard Identification requirements and results.

HAZID is most beneficial early in a project concept phase to identify potential opportunities for inherently safer design options.

Table 12.1. Preliminary hazard identification study overview

Typically Used During	Resource Requirements	Type of Results	Advantages and Disadvantages
Research and development	Material, physical, and chemical data	Rough screening of general hazards	Provides a quick focus on big issues
Conceptual design Pilot plant operation	Basic process chemistry Process flow diagram	Ranking of hazardous areas or processes	Potential to miss something due to limited design details

12.3.2 Checklist Analysis

A Checklist Analysis uses a written list of items or procedural steps to verify the status of a system. They can be simple or very detailed. Checklists are frequently used to evaluate compliance with standards and practices. In some cases, analysts use a more general checklist in combination with another hazard evaluation method to discover common hazards that the checklist alone might miss. While developing a good checklist may be challenging, using the checklist is easy and can be applied at any stage of the process life cycle.

A checklist used in hazard identification can be focused on a single process. For example, a checklist for distillation units can be developed with input from distillation experts and process safety engineers and can include learnings from industry distillation unit incidents. A checklist could also be focused on a specific topic that is then applied to all hazard identification studies. For example, a human factors checklist can be applied to every hazard identification study conducted to ensure that human factors considerations are addressed.

Checklists may be available from various sources including the following.

- Chemical Process Safety, Fundamentals with Applications, 4th edition, Chapter 11 (Crowl 2019)
- Guidelines for Revalidating Process hazard analysis, 1st edition (CCPS 2001)
- Guidelines for Hazard Evaluation Procedures, 3rd Edition, (CCPS 2008 b)

A danger with checklist analysis is that it can constrain the thinking of the hazard identification team to the items on the checklist. This might fail to identify novel hazards. However, good leadership and team brainstorming can avoid this pitfall. An advantage of the checklist approach is that it guarantees that the team will address potentially obscure sequences that have occurred in the past and thus are included in the checklist. Table 12.2 provides an overview of Checklist Analysis requirements and results.

Checklist Analysis_is well suited for determining if regulatory requirements or industry/corporate guidance is being followed.

Table 12.2. Checklist analysis overview

Typically Used During	Resource Requirements	Type of Results	Advantages and Disadvantages
Conceptual design Pilot plant operation Detailed engineering Construction / startup Routine operation Decommissioning Expansion or modification During What-If or HAZOP studies to address facility siting, human factors, and other general issues	Material, physical, and chemical data Basic process chemistry Process flow diagram Operating procedures Piping and Instrumentation Diagrams	Response to pre-defined questions. Documentation of compliance.	Can be used with less experienced personnel if the experience is captured in the checklist. Quality of the analysis is only as good as the quality of the checklist. Checklists that are too long or don't relate specifically enough to the process being analyzed may have a tendency to be completed without thorough evaluation.

12.3.3 What-If Analysis

The What-If Analysis technique is a brainstorming approach in which a multidisciplinary team of experienced people familiar with the subject process ask questions such as:

- What if the wrong material is delivered?
- What if Pump A stops running during start-up?, and
- What if the operator opens valve B instead of valve A?

The purpose of a What-If Analysis is to identify hazards, hazardous situations, or scenarios that could produce undesirable consequences. The team identifies possible causes, their consequences, and existing safeguard and documents this in a worksheet. In some What-If analyses, the consequences can be risk ranked to facilitate prioritization of any recommendations. They then suggest recommendations for risk reduction where improvement opportunities are identified or where safeguards are judged to be inadequate. The method can involve examination of possible deviations from the design, construction, modification, or operating intent. It requires a basic understanding of the process intention, along with the ability to mentally envision possible deviations from the design intent that could result in an incident. This is a powerful technique if the staff is experienced; however, an inexperienced team may overlook potential causes and consequences. Table 12.3 provides an overview of What-If Analysis requirements and results.

What-If Analysis is well suited for addressing "what can go wrong?" by identifying cause-consequence pairs.

Table 12.3. What-If analysis overview

Typically Used During	Resource Requirements	Type of Results	Advantages and Disadvantages
Research and development Conceptual design Pilot plant operation Detailed engineering Construction / startup Routine operation Decommissioning Expansion or modification During HAZOP studies to address issues such as loss of utilities.	Material, physical, and chemical data Basic process chemistry Process flow diagram Piping and Instrumentation Diagrams	Scenario-based documentation of What-If questions, consequences, safeguards, risk ranking, and recommendations, if any.	Allows an experienced facilitator to efficiently address issues of concern Inexperienced facilitators may miss potential process deviations if they don't brainstorm all potential What-If questions.

12.3.4 Hazard and Operability Analysis

The Hazard and Operability (HAZOP technique systematically reviews a process or operation to determine whether deviations from the design or operation intent can lead to undesirable consequences. A HAZOP typically focuses on hazards and major operability issues. This technique can be used for continuous or batch processes and can be adapted to evaluate written procedures. A HAZOP is typically conducted by a multidisciplinary team consisting of the following.

- Leader
- Scribe
- Process design engineer
- Operator
- Instrument and control engineer
- Process safety engineer
- Other specialists as appropriate for the equipment under review

The HAZOP is dependent on detailed engineering design data. Documents typically used in a HAZOP analysis include the following.

- PFD
- P&ID
- Equipment datasheets
- Cause and effect charts
- Management of Change records
- Incident reports

In a HAZOP study, the team uses a structured, systematic approach to identify hazard and operability problems resulting from deviations from the process's design intent that could lead to undesired consequences. Due to the defined structure and systematic nature of the HAZOP, it is viewed as one of the most rigorous hazard identification techniques.

HAZOPs are often used in project engineering design. Selecting the best time for the HAZOP can be challenging. Conducting it earlier, for example when the engineering design is comprehensive, but not yet final, allows time in the project engineering schedule to incorporate any changes prompted as a result of HAZOP recommendations. Conducting it later, when the engineering design is final, allows the engineer to use the HAZOP to confirm that no unresolved hazards remain in the design.

The HAZOP is conducted following the flow shown in Figure 12.4. The first step is to divide the process into "nodes" which can be a single piece of equipment or a portion of a process unit such as a reactor or an overhead condensing system. The process parameters (i.e., level, temperature, pressure, flow) are defined for each node. An experienced team leader systematically guides the team through each node. A fixed set of words called "guidewords" (more, less, no, other than, reverse, as well as, and part of) are applied to each parameter. This results in a "deviation" such as no flow or high temperature. The team brainstorms potential causes for the deviation. Then consequences are identified along with existing safeguards. If the team believes the safeguards are insufficient, they can make recommendations. The cause and consequences can be risk ranked to facilitate prioritization of recommendations. The study is documented in a HAZOP worksheet as shown in Figure 12.5. Table 12.4 provides an overview of HAZOP requirements and results.

HAZOP is the best option when a thorough, structured, systematic process hazard analysis is needed.

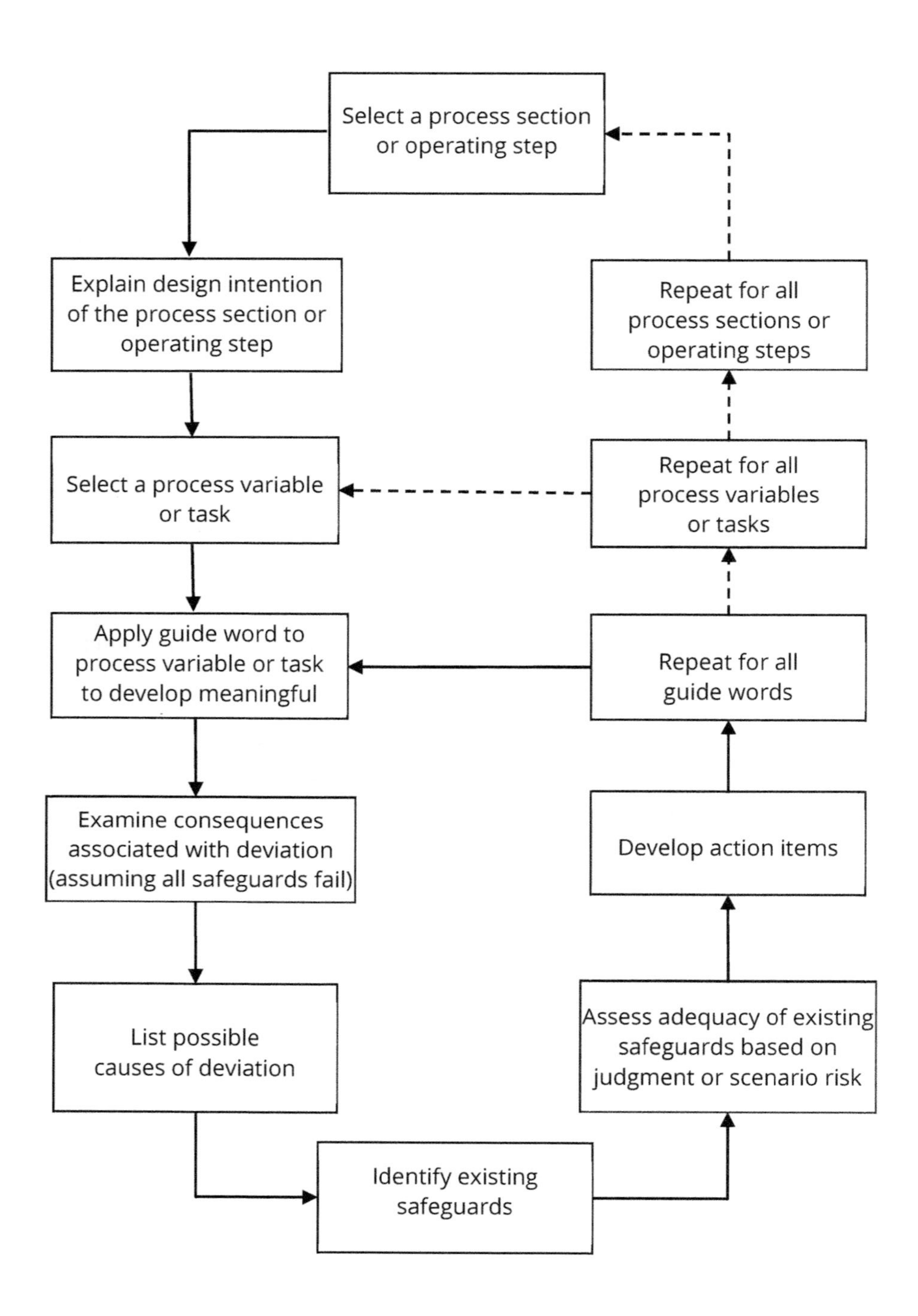

Figure 12.4. HAZOP analysis method flowchart
(CCPS 2008 b)

HAZOP WORKSHEET							
Area:							
Unit:							
Node:							
Drawings:							
Design Intent:							
No.	Guideword	Deviation	Causes	Consequences	Safeguards	Recommendations	Action by

Figure 12.5. Example HAZOP analysis worksheet (enggcyclopedia)

Table 12.4. HAZOP overview

Typically Used During	**Resource Requirements**	**Type of Results**	**Advantages and Disadvantages**
Pilot plant operation Detailed engineering Routine operation Expansion or modification	Material, physical, and chemical data Basic process chemistry Process flow diagram Piping and Instrumentation Diagrams	Scenario-based documentation of deviations, causes, consequences, safeguards, risk ranking, and recommendations, if any.	Provides a structured methodology to systematically and consistently analyze hazard scenarios. Provides input to Layer of Protection Analysis by identifying high consequence scenarios. Potential for redundancy in covering hazards.

HAZOP, like the other hazard identification methods, is a qualitative analysis. The higher risk scenarios from a HAZOP analysis are frequently used as the foundation for a Layer of Protection Analysis (LOPA). LOPA uses simplifying rules to evaluate initiating event frequency, independent layers of protection, and consequences to provide an order-of-magnitude estimate of risk. The primary purpose of LOPA is to determine if the scenario has sufficient layers of protection to prevent or mitigate the consequences. LOPA is discussed in Chapter 14 along with other risk assessment techniques.

12.3.5 Failure Modes and Effects Analysis (FMEA)

The FMEA method originated in the U.S. Military where it was used to assess potential equipment failures and reliability issues. The purpose of an FMEA is to identify single equipment and system failure modes and the consequences of these failures. This analysis typically generates recommendations for increasing equipment reliability, thus improving process safety. The failure mode describes how equipment fails to provide the function the user expects. Using a pump as an example, the pump can fail to start, stop, pump at expected head, fail to contain the process, or fail to run at expected intervals without maintenance. The effect of the failure mode is the evidence a failure has occurred (e.g., visible leak, low pressure, etc.). An FMEA identifies single failure modes that either directly result in or contribute significantly to an incident. Human operator error is usually not examined directly in an FMEA; however, the consequences of inadequate design, improper installation, lack of maintenance, or improper operation are usually manifested as an equipment failure mode.

During an FMEA, hazard analysts describe potential consequences and relate them only to equipment failures; they rarely investigate damage or injury that could arise if the system operated successfully. An FMEA is not as efficient as other methods such as HAZOP analyses in identifying an exhaustive list of combinations of equipment failures that lead to incidents, since it examines all failure modes that result in safe outcomes as well as those that can lead to or contribute to loss events. It also does not address human factors well. Each individual failure is considered as an independent occurrence, with no relation to other failures in the system, except for the subsequent effects that it might produce. However, in special circumstances, common cause failures of more than one system component may be considered.

The results of an FMEA are usually listed in tabular format, equipment item by equipment item. Generally, hazard analysts use FMEA as a qualitative technique, although it can be extended to give a priority ranking based on failure severity or criticality in which case it is termed an FMECA. Proactive tasks, put in place as a result of an FMEA, reduce the likelihood of an initiating event, and thus, lower the likelihood of a process safety incident. Table 12.5 provides an overview of FMEA requirements and results.

FMEA is well suited to provide details of a failure modes for individual pieces of equipment or systems.

Table 12.5. FMEA overview

Typically Used During	Resource Requirements	Type of Results	Advantages and Disadvantages
Conceptual engineering Detailed engineering Routine operation Expansion or modification	Material, physical, and chemical data Basic process chemistry Process flow diagram Piping and Instrumentation Diagrams	Identified failures and safeguards	Designed to analyze potential equipment failures Not a team approach Experience of analyst is essential

12.3.6 *Fault Tree Analysis (FTA)*

Fault Tree Analysis is a technique that focuses on one particular incident or main system failure and provides a method for determining causes of that event. FTA is well described in *Guidelines for Chemical Process Quantitative Risk Analysis*. (CCPS 1999) The purpose of an FTA is to identify combinations of equipment failures and human performance issues that can result in an incident. FTA is well suited for analyses of highly redundant systems. For systems particularly vulnerable to single failures that can lead to incidents, it is better to use a single-failure-oriented technique such as FMEA or HAZOP Study. FTA is often employed in situations where another hazard evaluation technique (e.g., HAZOP Study) has pinpointed an important incident of interest that requires more detailed analysis.

The fault tree is a graphical model, as shown in Figure 12.6, that displays the various combinations of equipment failures and human performance issues that can result in the main system failure of interest (called the Top Event). This allows the hazard analyst to focus preventive or mitigative measures on significant basic causes to reduce the likelihood of an incident.

Fault Tree Analysis is a deductive technique that uses Boolean logic symbols (i.e., AND gates shown as a flat arch, OR gates shown as an arrowhead in Figure 12.6) to break down the causes of a top event into basic equipment failures and human performance issues (called basic events). The analyst begins with an incident or undesirable event that is to be avoided and identifies the immediate causes of that event. Each of the immediate causes (called fault events) is further examined in the same manner until the analyst has identified the basic causes (shown as circles in Figure 12.6) of each fault event or reaches the boundary established for the analysis (shown as a diamond shape in Figure 12.6). The resulting fault tree model displays the logical relationships between basic events and the selected top event.

Top events are specific hazardous situations that are typically identified through the use of a more broad-brush hazard evaluation technique (e.g., What-If Analysis, HAZOP Study). A fault tree model can be used to generate a list of the failure combinations (failure modes) that can cause the top event. These failure modes are known as cut sets. An important qualitative outcome of an FTA is the minimal cut set (MCS) is a smallest combination of component failures which, if they all occur or exist simultaneously, will cause the top event to occur. For example, a car will not operate if the cut set "no fuel" and "broken windshield" occurs. However, the MCS is "no fuel" because it alone can cause the Top event; the broken windshield has no bearing on the car's ability to operate.

Fault tree analysis can be quantified. Where frequency data are available for the basic events, the resultant frequency of the top event can be calculated using Boolean algebra or arithmetical approximations. This data may be used in quantitative risk assessment.

Fault-Tree and Event Tree are suited to provide estimated of scenario likelihood or frequency.

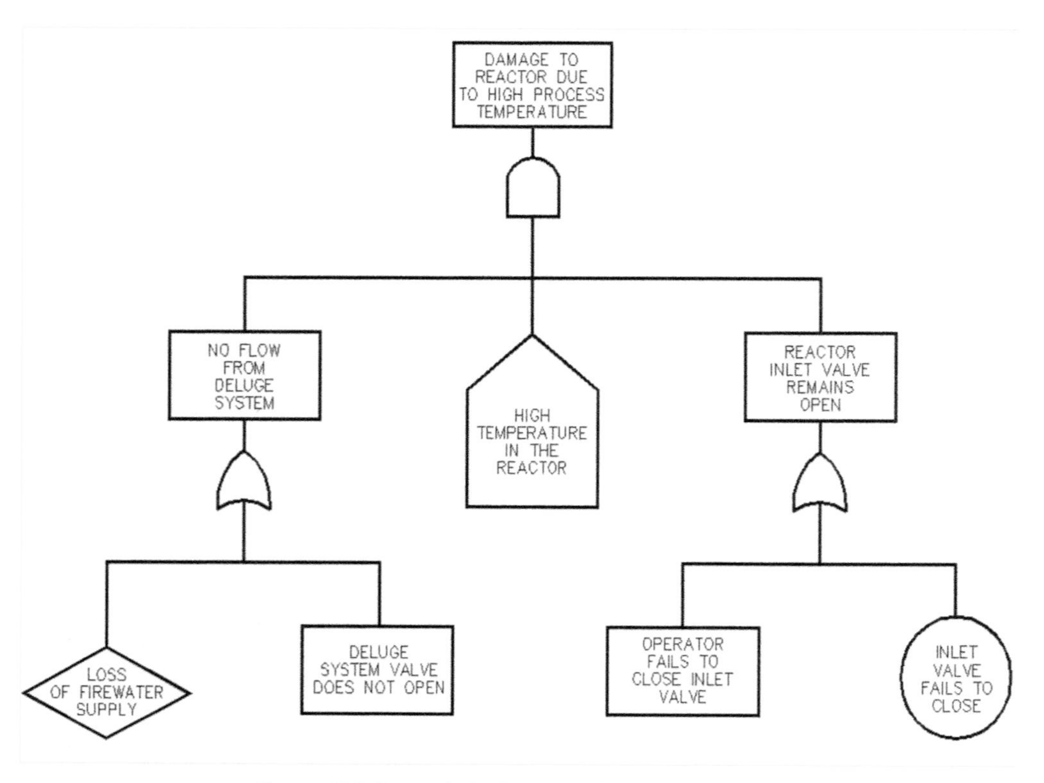

Figure 12.6. Example fault tree analysis diagram
(CCPS 2008 b)

12.3.7 Event Tree Analysis

Event trees are used to identify all the various outcomes that can result from a specific initiating event such as equipment failure. An event tree graphically shows the outcome each protective or mitigative system as a success branch and a failure branch. The branches can also represent wind or weather conditions that affect the hazardous material dispersion. The probability of the success plus the probability of the failure must equal one. Thus, when the frequency of the initiating event is known, the probabilities of success and failure can be applied to determine the frequency of the various outcomes. This data is also used in quantitative risk assessment.

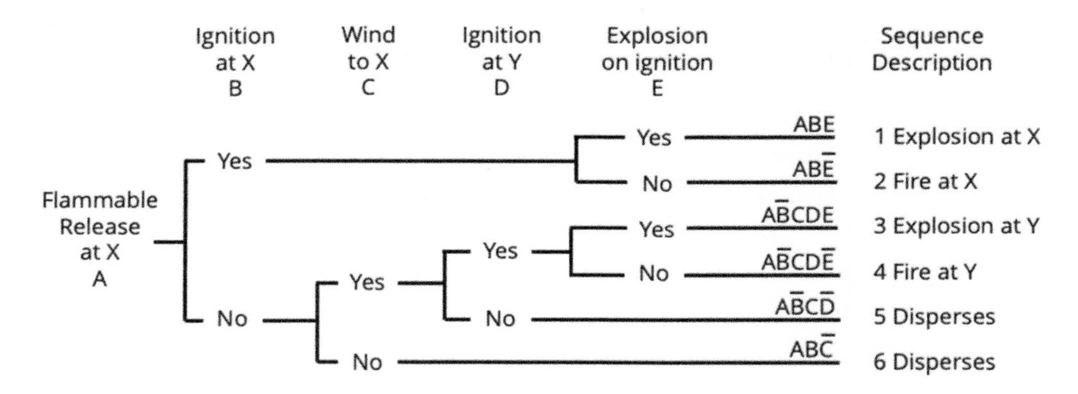

Figure 12.7. Example event tree diagram
(redrawn from EFCE 1985)

12.3.8 Hazard Identification Analyses and the Asset Life Cycle

Hazard identification analyses should be performed throughout the life of a process as an integral part of an organization's program to manage process safety. This is discussed in the *Guidelines on Integrating Process Safety into Engineering Projects.* (CCPS 2018) Hazard analyses can be performed to help manage process risks from the earliest stages of research and development; in detailed design and construction; periodically throughout the operating lifetime; and continuing until the process is decommissioned and dismantled. By using this life cycle approach along with other process safety activities, hazard analyses can point designers to inherently safer design opportunities and reveal deficiencies in design and operation before a unit is sited, built, operated, or decommissioned. Table 12.5 identifies typical hazard evaluation objectives and their appropriate process stages, as well as suggested hazard analysis techniques.

Table 12.5. Typical hazard evaluation objectives at different stages of a process life cycle

Process Phase	Example Objectives	Hazard Analysis Technique
Research and development	Identify chemical reactions or interactions that could cause runaway reactions, fires, explosions, or toxic gas releases. Identify process safety data needs.	Preliminary Hazard Identification (HAZID) Analysis What-If
Conceptual design	Identify opportunities for inherent safety. Compare the hazards of potential sites. Provide input to facility layout and buffer zones.	Checklist Preliminary Hazard Identification (HAZID) Analysis What-If What-If / Checklist

Table 12.5 continued

Process Phase	Example Objectives	Hazard Analysis Technique
Pilot plant	Identify ways for hazardous materials to be released to the environment. Identify ways to deactivate the catalyst. Identify potentially hazardous operator interfaces. Identify ways to minimize hazardous wastes.	Checklist Preliminary Hazard Identification (HAZID) Analysis What-If What-If / Checklist Hazard and Operability Study Failure Modes and Effects Analysis Fault Tree Analysis Event Tree Analysis
Detailed engineering	Identify ways for a flammable mixture to form inside process equipment. Identify how a loss of containment might occur. Identify which process control malfunctions will cause runaway reactions. Identify ways to reduce hazardous material inventories. Evaluate whether designed safeguards are adequate to control process risks to tolerable, required or as low as reasonably practical (ALARP) level. Identify safety-critical equipment that must be regularly tested, inspected, or maintained.	What-If What-If / Checklist Hazard and Operability Study Failure Modes and Effects Analysis Fault Tree Analysis Event Tree Analysis
Construction and start-up	Identify error-likely situations in the start-up and operating procedures. Verify that all issues from previous hazard evaluations were resolved satisfactorily and that no new issues were introduced. Identify hazards that adjacent units may create for construction and maintenance workers. Identify hazards associated with vessel cleaning procedures. Identify any discrepancies between as-built equipment and the design drawings.	Checklist What-If What-If / Checklist Critical Task Analysis

Table 12.5 continued

Process Phase	Example Objectives	Hazard Analysis Technique
Routine operation	Identify employee hazards associated with the operating procedures. Identify ways an overpressure transient might occur. Update previous hazard evaluation to account for operational experience. Identify hazards associated with out-of-service equipment.	Checklist What-If What-If / Checklist Hazard and Operability Study
Process modification or plant expansion	Identify whether changing the feedstock composition will create any new hazards or worsen any existing ones. Identify hazards associated with new equipment.	Checklist Preliminary Hazard Identification (HAZID) Analysis What-If What-If / Checklist Hazard and Operability Study Failure Modes and Effects Analysis Fault Tree Analysis Event Tree Analysis
Decommissioning	Identify how demolition work might affect adjacent units. Identify any fire, explosion, or toxic hazards associated with the residues left in the unit after shutdown.	Safety Review Checklist What-If What-If / Checklist

12.3.9 Performing a Good Quality Process Hazard Analysis

A good process hazard analysis will result in a comprehensive listing of potential incidents with causes and recommendations for managing the hazards. Additionally, the PHA can identify high hazard scenarios that warrant more detailed consequence and risk analysis (see chapters 13 and 14).

Just as with any analysis, an analyst can do a good job or a poor job on a process hazard analysis. In order to conduct a quality PHA, the analyst should plan the preparation, the analysis, and the follow-up.

The preparation for an analysis should include gathering the appropriate data and the appropriate expertise. The data can include that listed in Section 12.3.4 and more. The expertise should include people familiar with the process, the operation, the local environment, and others as needed. Preparation will also include selecting a location that will encourage the analysis participants to stay engaged in the analysis as opposed to being distracted by their other duties. The type of analysis should be identified in advance and will influence the data and expertise needed. The scope of the analysis should be determined in advance. This could be a single piece of equipment, an entire process unit, or a modification and the associated tie-in points in an existing facility.

Leading a good quality analysis requires good facilitation skills. Keeping the team on-task for hours, days, even weeks is no small challenge. Encourage all to participate and control those who tend to dominate the conversation.

The details of the analysis should be captured in worksheets. It should be recognized that these worksheets may be picked up in months, maybe years, and someone will try to understand what the analysis covered and what was found. Avoid using shorthand and abbreviations that might only be meaningful to the team members who participated. Documentation should be a complete listing of the cause, consequence, safeguards, and recommendations, as appropriate. Where no feasible cause can be determined or no adverse consequence found, this should be documented. Documenting only when a hazard is found will leave those trying to understand the analysis or respond to its recommendations wondering what the complete scenario was or if it was even addressed. Useful guidance for recording hazard identification studies is provided in *Guidelines for Hazard Evaluation Procedures*. (CCPS 2008 b)

12.4 What a New Engineer Might Do

As a new engineer, or an engineer new to process safety, it is very likely that you will participate in some form of process hazard analysis in your first few years in the process industries. Participation in a process hazard analysis such as a HAZOP is an excellent way to learn about a process and how it is actually operated and maintained.

New processes and substantially modified processes require the involvement of many engineering disciplines in addition to the role of chemical engineers, and thus all of these engineering disciplines can help support process hazard analysis either on a full time or part time basis. Mechanical engineers are often helpful in identifying vulnerabilities such as materials of construction, stress cracking, thermal cycling, and stress analysis that may contribute to hazardous events. Civil engineers may be needed to identify concerns/solutions to external events such as flooding, earthquakes, and high wind loading. Instrument and Control engineers are often crucial to identifying control reliability, control response, and control suitability for addressing consequences identified. Electrical engineers often provide insight into critical distribution system reliabilities, needs for redundancy, and issues involving electrical coordination. Nearly all process hazard analysis will require the involvement of a process safety engineer to either lead the analysis or support the critical thinking.

Section 12.3.9 discussed non-technical skills that new engineers must possess to support quality process hazard analysis including documenting process hazards analyses and participating in, scribing for, and leading analysis teams. These require writing skills, public speaking skills as well as human relation skills such as good listening, assertiveness, and respect for another person's opinion.

Process hazard analysis necessarily requires identification of scenarios that can lead to impacts such as environmental releases, injuries and fatalities, and property damage. Engineers should learn to state these in fact-based terms. For example, consider identifying that a large release of a flammable material inside a congested area can lead to an explosion if ignited. One should not write, "This will blow up the entire unit and kill everybody!" Instead, "if ignited, this can potentially lead to damage to the equipment or processing unit, and one or more fatalities, depending on occupancy levels." is a more precise statement. The more quantitative a statement can be made, the better, for example, "a 13.8 kPa (2 psig) overpressure zone extending 61 m (200 ft)" is preferable to a statement such as "a pretty big explosion".

One area that most new engineers must master is action item management. Action item management can be tedious, but it is critical. The recommendations from a PHA are intended to reduce risk. If these recommendations are not thoughtfully considered and actions taken where appropriate, then the hazardous situation will persist and may, eventually, result in a process safety incident. Conducting the hazard identification study does not reduce the risk, completing the action items reduces the risk. Each action item or recommendation assigned should have the following characteristics.

- A specific person or persons responsible for resolution
- A specific date is given for resolution
- Completion of an action item or recommendation should be fully documented to include: assumptions, engineering calculations if any, deviation and approval of abnormal closure, approval of extension of closure, a detailed description and proof of that the recommendation is complete, completion of any MOC or PSI (including SOP's and training) associated with closure, and final approval of closure.

A rule of thumb is, if it isn't documented, it isn't done. Each company will have different systems to achieve item resolution, but these characteristics are usually part of that system. Many companies have an item resolution database in which the engineer must become proficient.

12.5 Tools

Tools that may be used in preparing for and conducting process hazard analysis include the following. Documenting the process hazard analysis in the worksheets can be a big task. Commercially available software programs are available to facilitate this documentation.

CCPS Guidelines for Chemical Process Quantitative Risk Analysis. This book provides guidance on the quantitative methods discussed in this chapter such as fault tree analysis and event tree analysis. (CCPS 1999)

CCPS *Guidelines for Engineering Design for Process Safety.* This book focuses on process safety issues in the design of chemical, petrochemical, and hydrocarbon processing facilities. It includes a chapter providing an overview of process hazard analysis tools. (CCPS 2012)

CCPS *Guidelines for Hazard Evaluation Procedures.* This book provides a comprehensive overview of the topic. It includes worked examples, references for further reading, and charts and diagrams that reflect the latest views and information. (CCPS 2008 b)

CCPS *Guidelines for Revalidating Process hazard analysis, 1st edition.* This book provides guidance on conducting revalidation studies in an efficient and effective manner. (CCPS 2001)

Crawley and Tyler *HAZOP: Guide to Best Practice, 3rd Edition.* This book describes and illustrates the HAZOP study method, provides material on preparations, linkage to human factors and to LOPA. This edition will assist the reader in delivering optimum safety and efficiency of performance of the HAZOP team. (Crawley)

Hazard identification study documentation software. Many software packages are available to facilitate the documentation of hazard identification studies. A selection of these are listed below.

- **Spreadsheet software.** Although simple, spreadsheet software such as Excel can be used to document these studies.
- **Sphera PHA-Pro.** This is a popular vendor of hazard identification software that facilitates the leading and documentation of these studies. (Sphera)
- **Primatech PHAWorks.** This software facilitates the documentation of hazard identification study worksheets. It also includes generic checklists. (Primatech)
- **AE Solutions AEShield.** This software facilitates the documentation of hazard identification studies and the linkage of those studies to subsequent studies such as LOPA (discussed in Chapter 14) and to SIL verification (discussed in Chapter 15). (AE Solutions)

12.6 Summary

Hazard identification is fundamental to risk management. If the hazard is not identified, it cannot be managed. Process hazards analyses are used to identify hazards during the engineering design of a project and the operation of a facility. Many different process hazard analysis methods are available, and it is important to select the appropriate method for the purpose and for the engineering design data available. It is also important to have a competent study leader and a team with the appropriate technical expertise for the equipment being studied. Hazard identification analyses are typically qualitative although they may risk-rank scenarios for the purpose of prioritizing recommendations. Hazard Identification studies frequently are the source for data that are used in risk assessment or other process safety analysis. Hazard identification is only the first step. Having identified the hazard, actions should be taken to prevent its occurrence or minimize the impact. Having a process in place to ensure that recommendations or actions are considered, changes are made, and validating that the change addressed the hazard identified are all key steps in a successful hazard identification program.

12.7 Other Incidents

This chapter began with a description of the Esso Longford gas plant explosion. Other incidents involving hazard identification include the following.

- T2 Laboratories Reactive Chemicals Explosion, Jacksonville, Florida, U.S,. 2007
- Celanese Pampa Explosion, Texas, U.S., 1987
- Hickson Welsh Jet Fire, Yorkshire, U.K., 1992
- Port Neal AN Explosion, Sioux City, Iowa, U.S., 1994
- Georgia Pacific Hydrogen Sulfide, Pennington, Alabama, U.S., 2002
- Hayes Lammerz Dust Explosion, Indiana, U.S., 2003

- Buncefield Storage Tank Overflow and Explosion, U.K., 2005
- Olive Oil Storage Tank Explosion, Italy, 2006
- Fukushima Daiichi Nuclear Power Plant Release, Japan, 2007
- Valero-McKee LPG Fire, Amarillo, Texas, U.S., 2007
- Imperial Sugar Dust Explosion, Port Wentworth, Georgia, U.S., 2008
- Air France Flight 447, Brazil, 2009
- DuPont Phosgene Release, Belle, West Virginia, U.S., 2010
- Millard Refrigeration Ammonia Release, Theodore, Alabama, U.S., 2010
- PG&E Pipeline Explosion, San Bruno, California, U.S., 2010
- West Fertilizer AN Explosion, West, Texas, U.S., 2013
- DuPont MMA Release, LaPorte, Texas, U.S., 2014

12.8 Exercises

1. List 3 RBPS elements evident in the Esso Longford Gas Plant explosion incident summarized at the beginning of this chapter. Describe their shortcomings as related to this accident.
2. Considering the Esso Longford Gas Plant explosion incident, what actions could have been taken to reduce the risk of this incident?
3. Briefly describe three hazard identification techniques and when in the facility life cycle would the technique may be most appropriate?
4. Identify three HAZOP parameters commonly associated with a process tank or vessel. Identify two deviations for each of the three parameters that could lead to an undesired consequence and explain why.
5. A pump is used to unload a 18,927 l (5000 gal) methanol tank truck into a low-design pressure storage tank at ambient temperature. The elevation of the top of the storage tank and frictional pressure loss of the piping is such that the pump must deliver at least 30 psig discharge pressure. The unloading should take approximately one hour. Write a design intent statement for this pump.
6. Who typically conducts a hazard identification study?
7. A simple process starts with feed from a storage tank that is pumped to a heater. The outlet of the storage tank has a valve. The pump inlet and discharge also have valves. What "what-if?" questions might be appropriate?
8. Create a fault tree for a large house fire that started with a small grease fire in the kitchen where the house was equipped with a smoke detector in the hallway.
9. Draw an event tree where a pressurized release of propane is the initiating event. The propane could ignite immediately, it could form a cloud that travels downwind where it could be ignited, or it could disperse harmlessly.
10. A project is being initiated. It involves flammable material feedstocks that could be supplied by ship, rail, or truck. The site is to include process units, offices, warehousing, and maintenance facilities. The process itself is to be chosen from one of two licensors whose processes involve different temperatures and pressures. The site is on a hillside in a tropical environment that receives significant rainfall. What type of hazard identification study is appropriate in this instance? What questions might be asked?

12.9 References

AESolutions, AEShield, https://www.aeshield.com.

CCPS Glossary, "CCPS Process Safety Glossary", Center for Chemical Process Safety, https://www.aiche.org/ccps/resources/glossary.

CCPS 1999, *Guidelines for Chemical Process Quantitative Risk Analysis*, Center for Chemical Process Safety, John Wiley & Sons, Hoboken, N.J.

CCPS 2001, *Guidelines for Revalidating Process Hazard Analysis, 1st edition*, Center for Chemical Process Safety, John Wiley & Sons, Hoboken, N.J.

CCPS 2008 a, *Incidents That Define Process Safety*, Center for Chemical Process Safety, John Wiley & Sons, Hoboken, N.J.

CCPS 2008 b, *Guidelines for Hazard Evaluation Procedures*, Center for Chemical Process Safety, John Wiley & Sons, Hoboken, N.J.

CCPS 2012, *Guidelines for Engineering Design for Process Safety*, Center for Chemical Process Safety, John Wiley & Sons, Hoboken, N.J.

CCPS 2013, *Guidelines for Managing Process Safety Risks During Organizational Change*, Center for Chemical Process Safety, John Wiley & Sons, Hoboken, N.J.

CCSP 2018, *Guidelines on Integrating Process Safety into Engineering Projects*, Center for Chemical Process Safety, John Wiley & Sons, Hoboken, N.J.

Crawley, Crawley, F. and Tyler, B., *HAZOP: Guide to Best Practice, 3rd Edition*, Institution of Chemical Engineers, Elsevier, U.K., 2015.

Crowl 2019, Daniel A. Crowl and Joseph F. Louvar, *Chemical Process Safety, Fundamentals with Applications 4th Edition*., Pearson, NY.

Enggcyclopedia, www.enggcyclopedia.com/2012/05/hazop-study/

EFCE 1985, "Risk Analysis in the Process Industries, Report of the International Study Group on Risk Analysis", European Federation of Chemical Engineering, EFCE Publications, Series No. 45, Institute of Chemical Engineers, Rugby, England.

LRC. "The Esso Longford Gas Plant Accident Report of the Longford Royal Commission", Government Printer for the State of Victoria, No. 61 – Session 1998 -99. June 28,1999, https://www.parliament.vic.gov.au/papers/govpub/VPARL1998-99No61.pdf

OSHA, Hazard Communication, 29 CFR 1910.1200, Occupational Safety and Health Administration, https://www.osha.gov/laws-regs/regulations/standardnumber/1910/1910.1200

Primatech, PHAWorks https://www.primatech.com/software/phaworks

Sphera, https://sphera.com/pha-pro-software/

13 Consequence Analysis

13.1 Learning Objectives

The learning objectives of this chapter:

- Understand basic concepts of source, release, dispersion, consequence and impact modeling, and
- Know where to access tools to support consequence analysis.

13.2 Incident: DPC Enterprises L.P. Chlorine Release, Festus, Missouri, 2002

13.2.1 Incident Summary

On the morning of August 14, 2002, a chlorine transfer hose failed releasing 21,772 kg (48,000 lb) of chlorine over a 3-hour period during a railroad tank car unloading operation at DPC Enterprises, L.P., near Festus, Missouri. The facility repackages bulk dry liquid chlorine into 0.9 metric ton (1 ton) containers and 68 kg (150 lb) cylinders for commercial, industrial, and municipal use in the St. Louis metropolitan area.

Key Points:

Asset Integrity and Reliability – Did you get what you purchased? It is often difficult to simply visually determine if that pipe, hose, or valve is what you thought you were purchasing. Positive Material Identification (PMI) should be used to verify that materials are delivered as specified, especially where use of an incorrect material may lead to failure.

Emergency Management – We are in it together. Emergency response plans and drills should recognize and test the assets and limitations of the neighboring emergency response capabilities.

Asset Integrity and Reliability – Will your Emergency Shutdown (ESD) system work in an emergency? ESD system design should consider the operating and environmental conditions, including that of upstream equipment that might impact the system. ESD system testing should verify that the entire system works - from a sensor or button to the closing of a valve.

Chlorine is a toxic chemical. Concentrations as low as 10 parts per million are classified as "immediately dangerous to life or health". The wind direction on the day of the release carried the majority of the chlorine plume away from neighboring residential areas; however, some areas were evacuated. Sixty-three people from the surrounding community sought medical evaluation at the local hospital for respiratory distress, and three were admitted for overnight observation. The release affected hundreds of other nearby residents and employees, and the community was advised to shelter-in-place for 4 hours. Traffic was halted on Interstate 55 for 1.5 hours. Three DPC workers received minor skin exposure to chlorine during cleanup activities. (CSB 2003)

13.2.2 Detailed Description

DPC Enterprises bought the Festus repackaging facility in 1998 and added chlorine detectors and an ESD system to the chlorine repackaging area. The facility is part of the DX Distribution Group network of 18 repackaging and distribution companies.

DPC Festus is located on an 8-acre site in the Plattin Creek Valley of Jefferson County, at 1785 Highway 61. The facility receives bulk dry liquid chlorine in 82 metric ton (90 ton) tank cars and repackages it into 68 kg (150 lb) cylinders and 0.9 metric ton (1-ton) containers. DPC Festus employs 12 full-time personnel including four packaging operators (packagers), four truck drivers, two administrative staff, a sales representative, and an operations manager.

The chlorine repackaging process is a one-shift operation, typically running weekdays from 6:00 am to 4:00 pm. At the end of the day, all tank car valves are manually closed, chlorine in the piping system is directed to the bleach production process, a vacuum is pulled, and the ESD button is pressed to close all ESD valves. The chlorine transfer hoses remain connected to the tank car overnight.

Figure 13.1. Failed chlorine transfer hose and release (CSB 2003)

Chlorine is a toxic chemical. Chlorine exposure occurs through inhalation or through skin or eye contact. When inhaled in high concentrations, chlorine gas causes suffocation, constriction of the chest, tightness in the throat, and edema of the lungs. At around 1,000 parts per million (ppm), it is likely to be fatal after a few deep breaths. According to the National Institute for Occupational Safety and Health chlorine gas concentrations of 10 ppm are classified as "immediately dangerous to life or health" (IDLH). Depending on many factors—such as release volume, terrain, temperature, humidity, atmospheric stability, and wind direction and speed—a chlorine gas plume can travel several miles in a short time at concentrations well above IDLH.

At room temperature, chlorine is a greenish-yellow gas. It has a very pungent and irritating bleach-like odor provides warning of high concentrations. Chlorine gas can be detected by smell at concentrations well below 1 ppm.

The chlorine repackaging process operation involves the following:

- Connecting an 82 metric ton (180,000 lb) chlorine tank car to one of three unloading stations.
- Transferring liquid chlorine from the tank car through the process piping system to filling stations.
- Loading the filled 68 kg (150 lb) cylinders and 1 ton containers onto trucks for distribution.
- Cleaning and preparing empty cylinders and containers for reuse.
  In addition to repackaging chlorine, the Festus facility also runs a continuous bleach manufacturing process.

A chlorine tank car has four manually operated, one-inch valves and a pressure relief device mounted within a protective dome on top of the tank. Two valves are used for liquid chlorine discharge and two valves are connected to the vapor space; however, at DPC Festus, one of these valves supplied "pad air" to pressurize the tank car during chlorine unloading and the other was not in use. An excess flow valve that closes when the rate of flow exceeds 6804 kg/hr (15,000 lb/hr), is located beneath each liquid valve. Liquid chlorine is withdrawn from inside the tank car through dip pipes attached to the excess flow valves.

The facility operated one of the three unloading stations at a time. DPC specifications call for each chlorine transfer hose assembly to be constructed of a Teflon inner liner (plastic), a Hastelloy C-276 structural reinforcement braid layer (metal) for pressure containment, and a high-density polyethylene (HDPE) spiral guard for abrasion protection.

The DPC Festus ESD system is designed to shut off accidental releases of chlorine from the repackaging system. The ESD system is activated either automatically or manually by several ESD buttons located throughout the facility. At detection of 5 ppm chlorine, the system alarms with flashing lights and an audio alarm. At concentrations of 10 ppm, the ESD valves are automatically closed and a higher decibel audio alarm sounds. Each tank car station is equipped with five ESD valves with local indication of valve position. The ESD system is manually activated at the end of each day; however, the DPC standard operating procedures did not require verification that the ESD valves closed using the local indictors.

On August 12, 2002, a tank car containing 81,647 kg (180,000 lb) of chlorine was connected to station #3, which served all chlorine filling operations until the time of the release on August

14. The facility repackaging production records indicate that the car contained 36287 kg (80,000 lb) of chlorine at the time of the incident. It was later determined that 21,772 kg (48,000 lb) of chlorine had been released.

The chlorine repackaging system is on standby during morning and afternoon breaks, lunch, and cylinder change-outs. In both standby and shutdown modes, the chlorine transfer hoses remain connected to the tank car.

Early on August 14, four DPC packagers, a truck driver, and the operations manager started up the chlorine filling and container preparation operations for the day. Mid-morning, two of the packagers and the truck driver went to the designated smoking area outside the repackaging building; the others remained in the breakroom. Twenty minutes later, the three men outside heard a loud pop (rupture of the 2.5 cm (1 in) chlorine transfer hose) and observed a continuous release of chlorine at tank car station #3. They immediately evacuated the area.

The leak activated an area chlorine detection monitor audio alarm. The employees in the breakroom tried to identify the leak source but found chlorine entering the repackaging building and evacuated the building. The operations manager pushed the ESD button as he exited in an attempt to manually shut off the chlorine release. However, the release continued for nearly 3 hours until HAZMAT personnel closed the tank car valves.

DPC had four self-contained breathing apparatus (SCBA) units. The packagers were trained on use of the SCBA and on how to respond to a chlorine release; however, the SCBAs were not maintained and arranged for easy access so the packagers were not able to grab the SCBAs as they left the building.

The nine DPC personnel working evacuated within ten minutes. Seven followed the emergency plans to the assembly point, two did not but were contacted on the radio.

DPC Festus had no sirens or other community-wide alert systems to notify the estimated 1,500 people that live and work within a 1.6 km (1 mi) radius of the plant. A drive-through "bull horn" notification, followed by door-to-door evacuation, was conducted at a neighboring mobile home park and residential area. It took emergency response personnel over 1 hour to evacuate the areas. Sixty-three people from the surrounding community sought medical evaluation at the local hospital; three persons were admitted and released the following day. Three workers also received minor skin exposure to chlorine during cleanup activities after the release.

Hastelloy C-276 and 316L stainless-steel structural braiding are identical in appearance. DPC relied on information from the supplier to verify that the chlorine transfer hose met required specifications; the lack of an internal Quality Assurance (QA) management system, including verification of braid material, allowed the incorrect hose to be installed and left in operation until it failed.

Inspection of the ESD valves showed ferric chloride corrosion product on the valve balls that prevented the valves from closing properly. The valve balls were constructed of Monel, which is resistant to moisture-induced corrosion in chlorine service. The corrosion products came from upstream at the pad air supply and tank car assemblies, as well as from parts of the plant liquid and pad air carbon steel piping. The DPC personnel did not understand the causes and effects of moisture-induced corrosion in the chlorine repackaging system and so were not alerted to deteriorating equipment conditions.

According to the Chlorine Institute, the excess flow valve is designed to close automatically against the flow of liquid chlorine if the valve is broken off in transit. It may close if a catastrophic leak involving a broken connection occurs, but it is not designed to act as an emergency shut-off device during transfer. The tank car excess flow valves were designed to close only if the flow rate exceeds their set point of 6804 kg/hr (15,000 lb/hr). These valves remained open during the release.

13.2.3 Lessons

Asset Integrity and Reliability. The DPC QA management system did not ensure that chlorine transfer hoses met required specifications prior to installation and use. Companies should develop and implement a quality assurance management system, such as Positive Material Identification (PMI), to confirm that equipment is of the appropriate construction for its intended use. PMI is a chemical analysis that verifies the percentage of metals (e.g., iron, nickel) in various alloys, such as stainless steel and Hastelloy. A PMI program can be used to verify critical part components as a final check prior to shipping, receiving, and use.

The DPC testing and inspection program did not include procedures to ensure that the process emergency shutdown (ESD) system would operate as designed. The ESD testing procedures did not require verification that the valves closed. The mechanical integrity (MI) program failed to detect corrosion in the chlorine transfer and pad air systems before it caused operational and safety problems. Companies should implement procedures and practices to ensure the emergency shutdown (ESD) system operates properly including the verification that the ESD valves will close to shut down the flow.

Companies should implement a mechanical integrity (MI) program that ensures critical process equipment and components are designed, fabricated, installed, inspected, tested, and maintained in a manner that preserves the originally intended integrity of the equipment. Furthermore, management should provide adequate oversight to ensure that only trained and qualified personnel carry out these activities. Preventive maintenance; and inspection programs should include the various operating conditions that may be seen over the life cycle of the equipment. These operating conditions may include changes in environmental conditions, chemical composition or, in this case, exposure to corrosion products that migrated from other parts of the system.

Emergency Management. Lack of clear emergency response plans and supporting equipment resulted in additional exposure to neighboring residents and businesses. Companies should develop, communicate, test, and learn from the use of emergency response plans. The roles and responsibilities of facility emergency response personnel should be clearly described. These plans should include local emergency responders and should accurately reflect their capabilities and resources, including community notification systems. Drills should be coordinated to involve local emergency response authorities.

13.3 Consequence Analysis Overview

Techniques used to identify hazards were discussed in Chapter 12. This Chapter takes those hazards and considers what happens if they are realized as process safety incidents. Frequently, an incident involves a loss of containment of a material. Consequence analysis models a loss of containment by developing a source model, transporting the vapor downwind, estimating the consequences in terms of fire, explosion, and toxic effects, and determining the outcome to people, property, and the environment as shown in Figure 13.2.

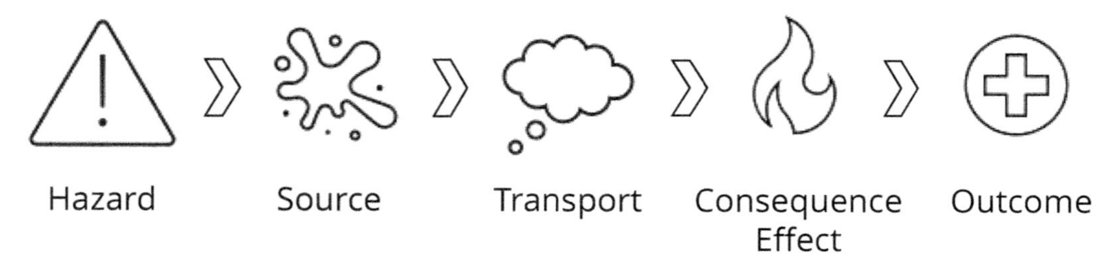

Figure 13.2. Consequence analysis stages

Consequence analysis is complex. As is apparent from the figure, estimating outcomes involves multiple individual models with fully integrated thermodynamics (mixture properties, enthalpies, entropies, etc.) and a wide array of assumptions. An analyst should understand all of these models to be able to assess potential consequences of loss of primary containment events.

This chapter will discuss each of these topics, from source onward. The hazard could be the material. The loss of containment could be the rupture or break of a pipeline, a hole in a tank or pipe, or due to a run-away reaction or a fire external to the vessel.

- A source model is selected to describe how materials are discharged from the process. The source model provides a description of the rate of discharge, the total quantity discharged (or total time of discharge), and the state of the discharge, that is, solid, liquid, vapor, or a combination.
- A transport model takes the source model and describes how a vapor is dispersed downwind. Additional transport models address hazardous materials spilled to rivers and waterways although this is beyond the scope of this book.
- Consequence effects are estimated based on the dispersion. For flammable releases, fire and explosion models estimate the thermal radiation or explosion overpressure based on the source model and dispersion model. For toxic releases, toxic concentration levels are estimated based on the vapor dispersion.
- The outcome is expressed in terms of the impact on humans, the environment, or property and is estimated based on the consequence models.

A more detailed description of the consequence analysis stages is illustrated in the logic diagram in Figure 13.3. In each of the stages several of different models, e.g. dispersion phenomena or fire types, may be relevant as illustrated in Figure 13.4.

The modeling in each of these stages is based on engineering principles and scientific data. This Chapter will not delve into the detailed engineering equations and instead will address the subject of consequence analysis at an introductory, conceptual level. Detailed treatment of the subject is available from the following texts.

- CCPS Guidelines for Consequence Analysis of Chemical Releases (CCPS 1999)
- CCPS Chemical Processes Quantitative Risk Analysis (CCPS 1999)
- Crowl and Louvar Chemical Process Safety Fundamentals with Applications, Fourth Edition (Crowl 2019)
- CCPS Guidelines for Use of Vapor Cloud Dispersion Models, Second Edition (CCPS 1996)

Although this chapter discussed the fundamentals of conducting a consequence analysis, in reality, an engineer will likely use one of the tools discussed in Section 13.10. Nonetheless, the fundamentals behind the software tools should be understood to ensure they are valid for use and that the right modeling options are selected for the particular scenario being modeled.

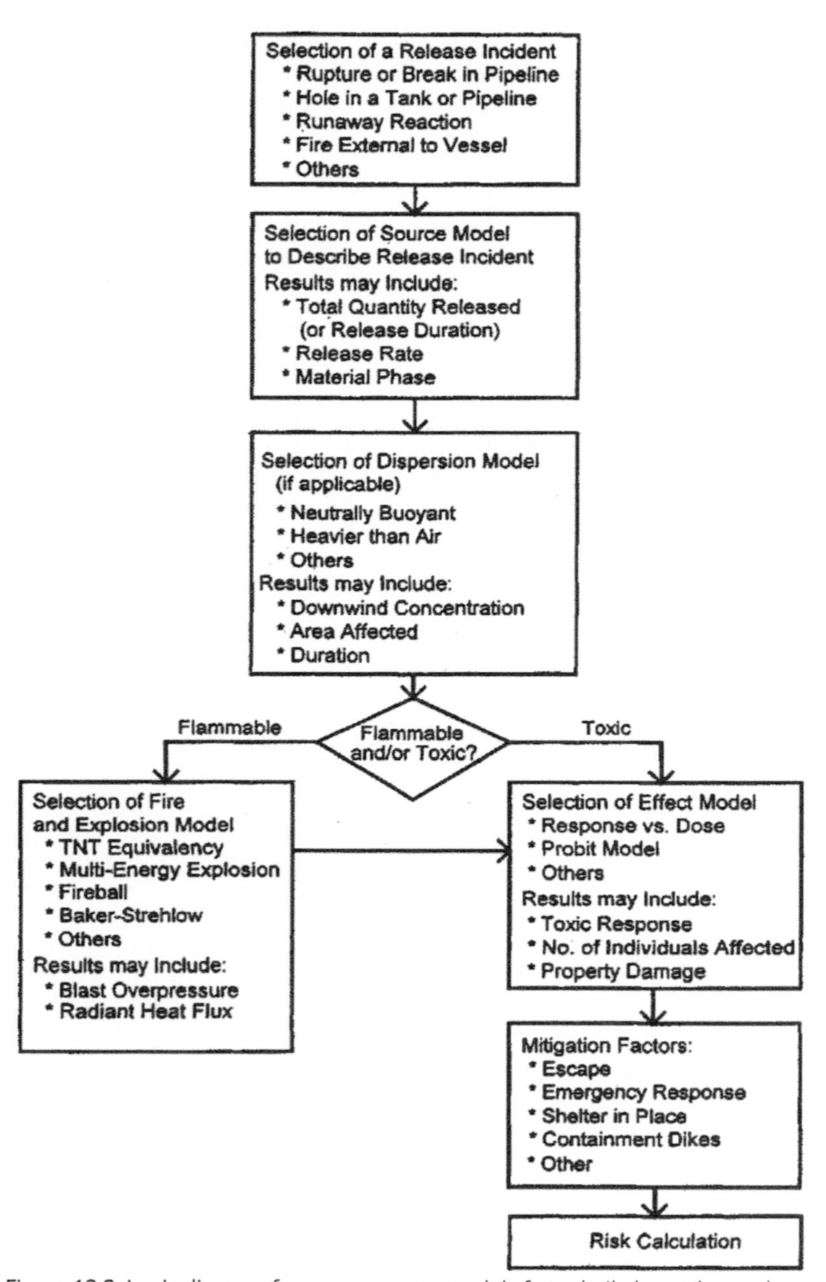

Figure 13.3. Logic diagram for consequence models for volatile hazardous releases (CCPS 1999)

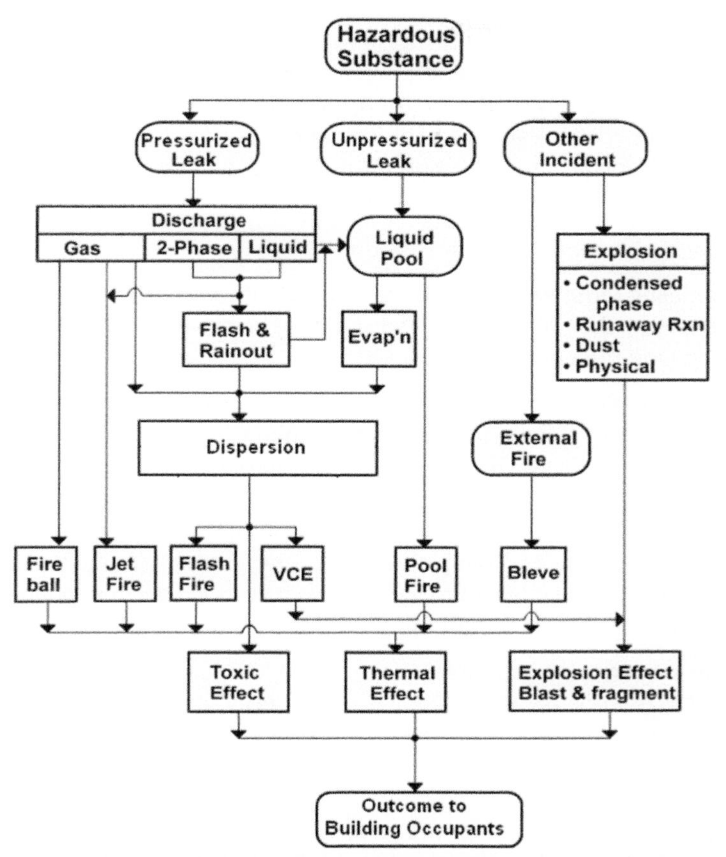

Figure 13.4. Block Diagram showing relationship between consequence models (CCPS 2021)

Figure 13.4 shows the progression of an incident from its initiation to its final outcome. Oval blocks are states or conditions and rectangular blocks are models. Pressurized leaks may be the most important category and, of these, flashing 2-phase liquids might be the most important subcategory. All routes are possible; however, and have occurred with serious outcomes.

Each rectangular block represents a model, but multiple alternative models ranging from simple to highly complex are possible. Each model will be initiated with a set of assumptions and may need rigorous thermodynamic inputs. Thus, good consequence modeling requires correct mathematical models, assumptions, and thermodynamics. As data flows between multiple models and interacts with thermodynamics, it is common to do consequence calculations throughout in SI units, reverting to local units if desired at the end.

The following sections explain each block in turn. Readers should return to this figure as they read this chapter to understand the context of the model under discussion.

13.4 Source Term Models

Source term models are used to quantitatively define the release scenario by estimating discharge rates, total quantity released (or total release duration), extent of flash and evaporation from a liquid pool and aerosol formation. Source term model outputs are the direct input to fireball and jet fire consequence models and to dispersion models that model the concentration fields downwind from the source.

13.4.1 Scenario Specification

Consequence modeling requires a detailed scenario specification. This includes details such as the material discharged, the hole location and size, the release temperature and pressure, the inventory released, and the operator response.

A process unit contains many vessels, pipes, and flanges. Where might the leak occur? It is first important to know where the hazardous chemical is present in the process. For example, it could be present as a vapor in the top of a vessel or it could exist as a liquid in the bottom of a vessel. A leak from a pressure vessel is less likely than from piping. The flanges connecting piping and vessels are likely to leak first in an overpressure. event. Given the number of vessels and lengths of piping in a process unit, it is still hard to know exactly where a leak might occur. It is often sufficient to identify sources as lying within an isolatable section and the precise location is not so important.

Many release scenarios are possible. For example, the release could be from a pressure relief valve in which case release details are easier to specify. The release could be from a large storage tank or from a ship. Whatever the case, specify as many of the scenario details as possible to improve the accuracy of the consequence analysis.

13.4.2 Isolatable Section

The next question is how much material is leaked. It is likely that the emergency procedure (see Chapter 19) addressing a leak in the process unit may specify operator actions such as shutting down the process unit and isolating the section that contains the leak by closing emergency isolation valves. This will limit the leak duration and volume. The leak duration can be determined based on the specific scenario. This may depend on the detection and reaction time for automatic isolation devices and response time of the operators for manual isolation. The rate of valve closure in longer pipes may require a longer closure time to avoid water hammer and hence can affect the duration of release.

An alternative is to use a conservative simplification, for example, the U.S. Department of Transportation (DOT 1980) LNG Federal Safety Standards specified a 10 minute leak duration. Other studies (Rijnmond Public Authority,1982) have used 3 minute for systems with leak detection combined with remotely actuated isolation valves. Other issues to consider when analyzing discharges include the following.

- The released volume can be no greater than the capacity of the isolated vessels and piping once isolation is achieved.
- Time dependence of transient releases: Decreasing release rates due to decreasing upstream pressure.

- Reduction in flow: Valves, pumps, or other restrictions in the piping that might reduce the flow rate below that estimated from the pressure drop and discharge area.
- Inventory in the pipe or process between the leak and any isolation device.

13.4.3 Hole Sizes

A primary input to source calculation is the leak hole size. Holes occur in process equipment due to corrosion, impact, fatigue, brittle fracture, and other mechanisms. The mechanism can influence whether a small hole or a full-bore pipe rupture is likely. No general consensus exists for appropriate hole sizes. Analysts use a variety of approaches for hole size including the following.

- World Bank (1985) suggests characteristic hole sizes for a range of process equipment (e.g., for pipes 20% and 100% of pipe diameter are proposed).
- Some analysts use 50 and 100 mm (2 and 4 in) holes, regardless of pipe size.
- Some analysts use a few hole sizes to represent the full range possible, such as 5, 25, 50, and 150 mm (0.2, 1, 4, and 6 in) and full-bore ruptures for pipes less than 152 mm (6 in) in diameter.
- IOGP data set provides a means to estimate leak frequencies spreading the total frequency across any number of hole sizes selected. (IOGP 2019)

13.4.4 Discharge Phase

Source models require careful consideration of the discharge phase. This is dependent on the release process. Standard texts on vapor-liquid equilibrium or commercial process simulators provide useful guidance on phase behavior. The starting point is defined by the initial condition of the process material before release. This may be normal process conditions, or an abnormal state reached by the process material prior to the release. The end point will normally be at a final pressure of one atmosphere.

Table 13.1 is a partial list of typical scenarios grouped according to the material discharge phase, i.e. liquid, gas, or two-phase. Different models are appropriate for each of these. Figure 13.5 shows selected discharge scenarios with the resulting effect on the material's release phase. Gasket failure, either full or partial, is often the cause of liquid or gas discharges from equipment leaks.

Table 13.1. Typical discharge scenarios

Liquid Discharges
• Hole in atmospheric storage tank or other atmospheric pressure vessel or pipe under liquid head • Hole in vessel or pipe containing pressurized liquid below its normal boiling point
Gas Discharges
• Hole in equipment (pipe, vessel) containing gas under pressure • Relief valve discharge (normally vapor only) • Boiling-off evaporation or volatile vapors from liquid pool • Generation of toxic combustion products as a result of fire • Vapor phase from a pressure vessel leak above the liquid level (subsequent boil-off will also contribute to the release
Two-Phase Discharges
• Hole in pressurized storage tank or pipe containing a liquid above its normal boiling point • Relief valve discharge (e.g. due to a runaway reaction or foaming liquid)

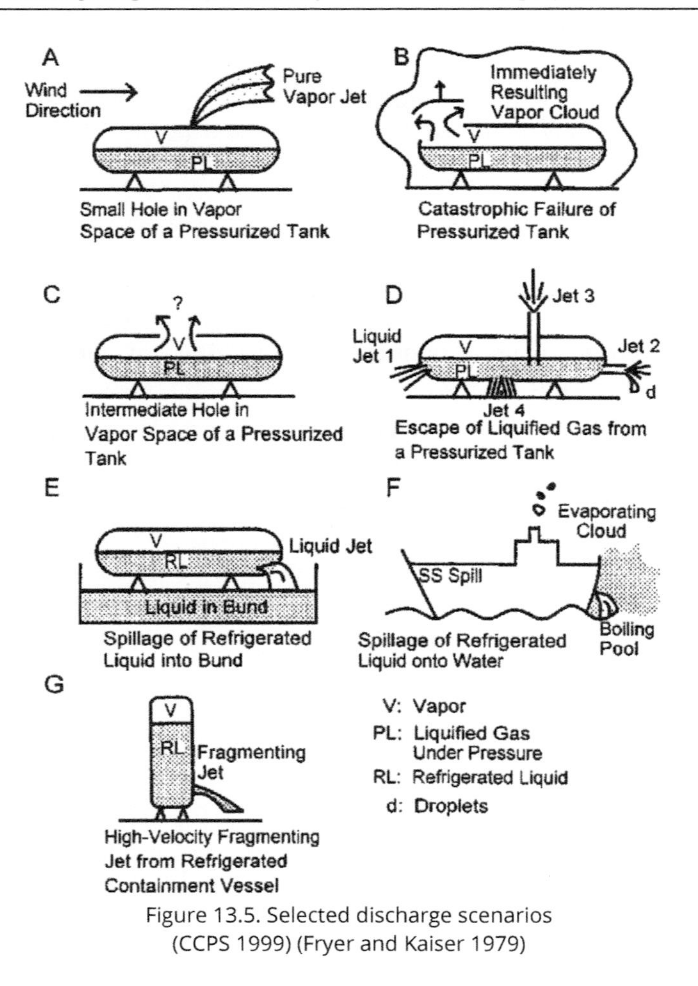

Figure 13.5. Selected discharge scenarios
(CCPS 1999) (Fryer and Kaiser 1979)

Liquid Discharge. For liquid discharges, the Bernoulli equation is used. The driving force for the discharge is normally pressure, with the pressure energy being converted to kinetic energy during the discharge. Liquid head can also contribute to the driving force. For pipe flow, the mass flux through the pipe is constant and, for pipes of constant cross-sectional area, the liquid velocity is constant along the pipe as well. In all cases, frictional losses occur due to the fluid flow.

Gas and liquid discharge equations contain a discharge coefficient which will affect the discharge rate. A discharge coefficient (often 0.6 – 1.0) is applied to account for irregular hole shapes compared to idealized circular sharp-edged holes. All discharge rates will be time-dependent due to changing composition, temperature, pressure, and level upstream of the hole. Average discharge rates are case-dependent, and intermediate calculations may be necessary to model a particular release. The mass flow rate of two-phase flashing discharges will always be bounded by pure vapor and liquid discharges calculations.

Gas Discharges. Gas discharges may arise from several sources: from a hole at or near a vessel, from a long pipeline, or from relief valves or process vents. Different calculation procedures apply for each of these sources. The majority of gas discharges from process plant leaks will initially be sonic or choked flow. The sonic discharge equation is used combined with an estimate of the discharge coefficient. For gas discharges, as the pressure drops through the discharge, the gas expands. For gas discharges through holes, the mechanical energy balance is integrated along an isentropic path to determine the mass discharge rate. A simple rule of thumb for many pure materials is that the gas mass discharge rate is 10% of the liquid mass discharge rate for the same conditions and hole size.

Two-Phase Discharge. When released to atmospheric pressure, any pressurized liquid above its normal boiling point will start to flash and two-phase flow will result. Two-phase flow is also likely to occur from depressurization of the vapor space above a volatile liquid, especially if the liquid is viscous or has a tendency to foam. For consequence modeling, the discharge models must be selected to maximize the mass flux.

13.4.5 Flash, Evaporation, Aerosol, and Pool Spread Models

A discharge can be in the form of a gas, two-phase, or a liquid as shown in Figure 13.4. Figure 13.6 illustrates this point further. Aerosol formation is also possible for case B if the release velocities are high.

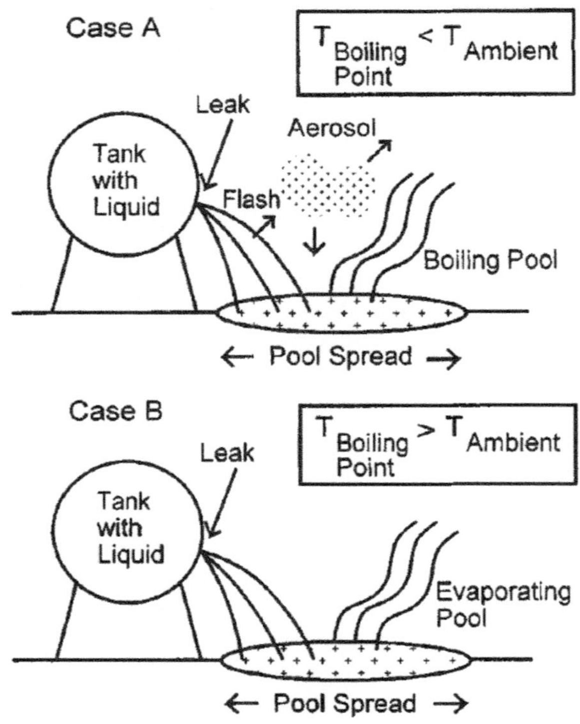

Figure 13.6. Two common liquid-release situations (CCPS 1999)

Flash and Evaporation Models. The purpose of flash and evaporation models is to estimate the total vapor or vapor rate that forms a cloud, for use as input to dispersion models.

If the liquid is stored under pressure at a temperature above its normal boiling point (superheated), it will flash partially to vapor when released to atmospheric pressure. The vapor produced may entrain a significant quantity of liquid as droplets. Some of this liquid may rainout onto the ground, and some may remain suspended as an aerosol with subsequent possible evaporation. The liquid remaining behind is likely to form a boiling pool which will continue to evaporate, resulting in additional vapor loading into the air. An example of a superheated release is a release of liquid chlorine or ammonia from a pressurized container stored at ambient temperature or release of LNG from a cryogenic tank at atmospheric pressure. Over time the entire spill will vaporize, but the rate will depend on the amount of spread of the liquid pool and the heat transfer from the ground or solar flux.

If the liquid is not superheated, but has a high vapor pressure (volatile), then vapor emissions will arise from surface evaporation from the resulting pools. The total emission rate may be high depending on the volatility of the liquid and the total surface area of the pool. Examples include the release of liquid toluene, gasoline, or alcohol. For liquids which exit a process as a jet, flow instabilities may cause the stream to break up into droplets before it impacts the ground. These droplets may rain out, evaporate, or remain in the aerosol/vapor cloud. The Buncefield incident described in Chapter 11 was caused by a gasoline spill due to a tank overfill and cascade of gasoline that acted like a waterfall with significant aerosol formation, which led to a large vapor cloud and subsequent explosion.

If the liquid released is superheated, during the flash a significant fraction of liquid may remain suspended as a fine aerosol. Some of this aerosol may eventually rain out, but the remainder will vaporize due to air entrained into the cloud. In some circumstances ground boiloff of the rainout may be so rapid that all the discharge may enter the cloud almost immediately. In other cases, the quantity of liquid may be so great that it cools the ground enough to sufficiently reduce surface vaporization from the pool.

The flash from a superheated liquid released to atmospheric pressure can be estimated in several ways. For pure materials, a pressure enthalpy diagram or a thermodynamic data table can be used. For liquids that are accelerated during the release, such as in a jet, a common approach is to assume an isentropic path. A standard equation for the prediction of the fraction of the liquid that flashes can be derived by assuming that the sensible heat contained within the superheated liquid due to its temperature above its normal boiling point is used to vaporize a fraction of the liquid.

For flashing mixtures, a commercial process simulator would normally be used as manual treatment of multicomponent mixtures is time consuming.

Table 13.2. Input and output for flash models

Input	Output
• Heat capacity • Latent heat of vaporization • Boiling point temperature • Initial temperature and pressure	The vapor-liquid split

Most evaporation models are based on the solution of time dependent heat and mass balances. Momentum transfer is typically ignored. These models rarely give atmospheric vapor concentrations or cloud dimensions over the pool, which may be required as input to dense gas or other vapor cloud dispersion models.

Evaporation from liquid spills onto land is typically well defined as many spills occur into a dike or other retention system that allows the pool size to be estimated. Spills onto water are unbounded and calculations are often empirical. Vaporization from a pool is determined using a total energy balance on the pool. The heat flux is the net total energy into the pool from radiation via the sun, from convection and conduction to the air, from conduction via the ground, and other possible energy sources, such as a fire.

The modeling approaches are divided into two classes: low and high volatility liquids. High volatility liquids are those with boiling points near to or less than ambient or ground temperatures. For highly volatile liquids, the vaporization rate of the pool is controlled by heat transfer from the ground (by conduction), the air (both conduction and convection), the sun (radiation), and other surrounding heat sources such as a fire or flare. The cooling of the liquid due to rapid vaporization is also important. This approach seems to work adequately for LNG, ethane, and ethylene. The higher hydrocarbons (C3 and above) require a more detailed heat transfer mechanism. This model also neglects possible water freezing effects in the ground, which can significantly alter the heat transfer behavior. For liquids having normal boiling points near or above ambient temperature, diffusional or mass transfer evaporation is the limiting mechanism. The vaporization rates for this situation are not as high as for flashing liquids or boiling pools.

A simplified approach for smaller releases of liquids with normal boiling points well below ambient temperature is to assume all the liquid enters the vapor cloud, either by immediate flash plus entrainment of aerosol, or by rapid evaporation of any rainout.

Table 13.3. Input and output for evaporation models

Input for boiling liquid pools	Input for nonboiling liquid pools
• leak rate	• leak rate
• pool area (for spills onto land)	• pool area (for spills onto land)
• wind velocity	• wind velocity
• ambient temperature	• ambient temperature
• pool temperature	• pool temperature
• ground density	• saturation vapor pressure
• specific heat	• mass transfer coefficient
• thermal conductivity	• solar thermal input
• solar radiation input parameters	
Output: The time-dependent mass rate of boiling or vaporization.	

Pool Spread Models. An important parameter in all of the evaporation models is the area of the pool. Pool spreading models are based primarily on the opposing forces of gravity and flow resistance and typically assume a smooth, horizontal surface. If the liquid is contained within a diked or other physically bounded area, then the area of the pool is determined from these physical bounds if the spill has a large enough volume to fill the area. If the pool is unbounded, then the pool can be expected to spread out and grow in area as a function of time. The size of the pool and its spread is highly dependent on the level and roughness of the terrain surface -most models assume a level and smooth surface. One approach is to assume a constant liquid thickness throughout the pool. The pool area is then determined directly from the total volume of material. This approach produces a conservative result, assuming the spill is on a flat surface, the pool growth is not constrained, and the pool growth will be radial and uniform from the point of the spill. More complex models include consideration of gravity spread and flow resistance terms for both laminar and turbulent flow but does not include evaporation or boiling effects. The approach is significantly different if the pool is on water versus land.

A pool spread model solves the simultaneous, time-dependent, heat, mass, and momentum balances. Factors important to this pool spread modeling include the following.

Table 13.4. Input and output for pool spread models

Input to determine spill rate	Input for materials	Input for physical characteristics
• Tank pressure • Liquid height • Hole diameter • Discharge coefficient • Density	• VLE data • Heat capacity • Heat of vaporization • Liquid density • Emissivity • Viscosity	• Ground density and thermal conductivity • Ambient temperature • Wind speed • Solar radiation
Output: The radius or radial spread velocity of the pool from which the total pool area and depth is determined.		

Aerosol Models. The fraction of released liquid vaporized is a poor predictor of the total mass of material in the vapor cloud, because of the possible presence of entrained liquid as droplets (aerosol). Aerosol and rainout models provide estimates of the fractions of the liquid that remain suspended within the cloud and the fraction reaching the ground. Aerosols may form through two mechanisms: mechanical and thermal. The mechanical mechanism assumes that the liquid release occurs at high enough speeds to result in surface stress which causes the liquid phase to breakup into small droplets. The thermal mechanism assumes that breakup is caused by the flashing of the liquid to vapor. At low degrees of superheat, mechanical formation of aerosols dominates; at higher degrees of superheat, a flashing mechanism dominates. Several methods exist to calculate aerosol formation and rainout and ongoing research projects are studying these, but this is still an area of significant uncertainty.

Aerosol entrainment has very significant effects on cloud dispersion that include the following.

- The cloud will have a larger total mass.
- There will be an aerosol component (contributing to a higher cloud density).
- Evaporating aerosol can reduce the temperature below the ambient atmospheric temperature (contributing to a higher cloud density).
- The colder cloud temperature may cause additional condensation of atmospheric moisture (contributing to a higher cloud density).

Taken together, these effects tend to significantly increase the actual density of vapor clouds formed from flashing releases. The prediction of these effects is necessary to properly initialize the dispersion models. Otherwise, the cloud's hazard potential may be grossly misrepresented.

Several different approaches can be used to address rainout. One approach is based on the elevation and orientation of the release and the jet velocity, the amount of rainout of aerosol and the resultant mass of material in the cloud can be estimated using the settling velocity of the droplets. The amount of moisture in the ambient air should be included in these considerations. All these steps were shown in Figure 13.4 in simplified form.

13.5 Transport Models

The next step in consequence analysis is taking the material from the source term model and modeling where it goes dependent on the physical properties and weather. This step uses transport, sometimes called dispersion, models. Three kinds of vapor cloud behavior can be defined: positively buoyant, neutrally buoyant, and negatively buoyant (dense gas).

Most releases of flammables occur as high pressure or liquefied gas releases. For these types of releases, the primary dilution mechanism is due to entrainment of air by shear as the release jets into the surrounding air. The initial dilution with air by the jet may reduce the concentration to below the Lower Flammable Limit within a short distance from the release location. Also, the concentration following jet mixing is often used as the initial starting point for dense gas dispersion. Further discussion on jet mixing is included in the section on momentum in section 13.5.2.

Many releases in the process industry lead to dense gas plumes due to molecular weight or initial temperature. For example, releases of LNG which has a molecular weight of 16 and hence is normally buoyant, is dense at the low temperature of -160°C (-260°F) required for it to be liquefied. Dispersion modeling is not a simple topic. Care is required in selecting the right equations for the dispersion parameters and for predicting concentration. Collection and analysis of meteorological and other input data may be time consuming.

13.5.1 Positively and Neutrally Buoyant

Historically dispersion modeling was developed to predict dispersion from power station chimney plumes as shown in Figure 13.7.

The Pasquill-Gifford models are used in dispersion modeling of positively and neutrally buoyant releases. These models are estimations based on the concept of Gaussian dispersion and can be solved algebraically or by using graphs. The dispersion is closely tied to the meteorological data, especially wind speed and atmospheric stability. Due to mixing at the edge of the plume, there may be local concentrations that deviate above and below the average concentrations estimated by models. Additionally, major wind direction shifts may cause a dispersing plume to change direction or meander. While such changes do not have a major effect on the hazard area of the plume relative to its centerline, they do matter with respect to the specific area impacted. For toxic or flammable clouds, it may be desired to plot a particular isopleth corresponding to a concentration of interest (e.g., fixed by toxic load or flammable concentration). This isopleth usually takes the form of a skewed ellipse as shown in Figure 13.8. The puff model describes near instantaneous releases of material. The plume model describes a continuous release of material.

A U.S. EPA document titled Workbook of Atmospheric Dispersion Estimates is a helpful resource on this topic. (EPA 1970)

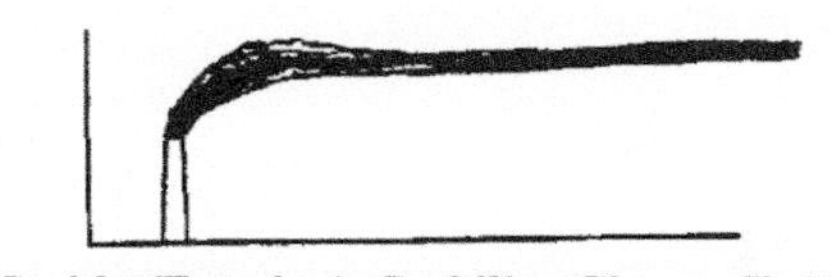

Stable (Fanning), Stability Classes E, F

Neutral Below, Stable Above (Fumigation)

Unstable (Looping), Stability Classes A, B

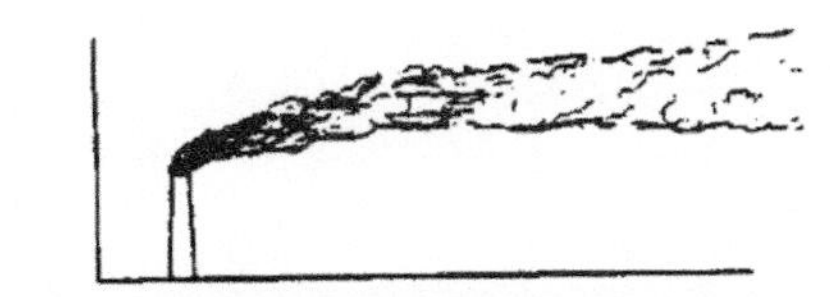

Neutral (Coning), Stability Class D

Stable Below, Neutral Aloft (Lofting)

Figure 13.7. Atmospheric stability effects
(CCPS 1999) (Slade 1968)

Table 13.5. Input and output for neutral and positively buoyant plume and puff models

<table>
<tr><th>Input for puff model</th><th>Input for plume model</th></tr>
<tr><td>• total quantity of material released
• atmospheric conditions
• height of the release above ground
• distance from the release</td><td>• rate of release
• atmospheric conditions
• height of the release above ground,
• distance from the release</td></tr>
<tr><th colspan="2">Output</th></tr>
<tr><td colspan="2">Time averaged concentration at specific locations (in the three spatial coordinates: x, y, z) downwind of the source</td></tr>
</table>

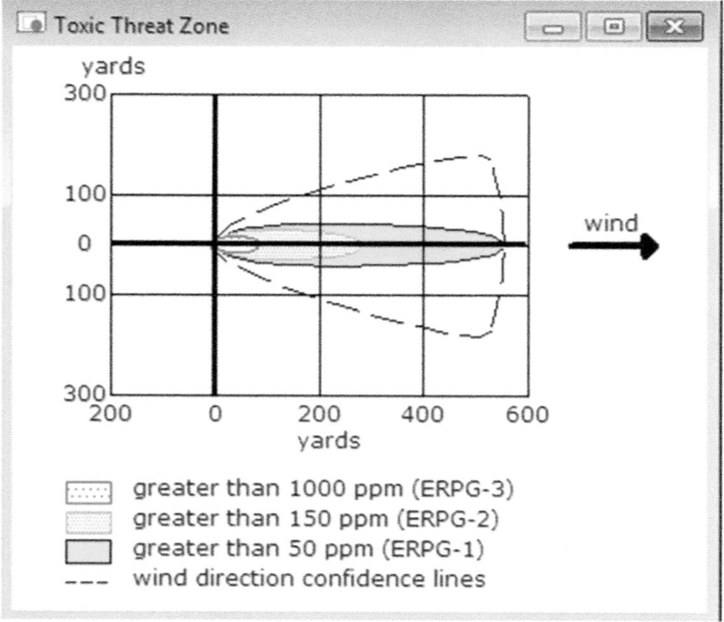

Figure 13.8. Typical dispersion model output
(EPA ALOHA)

13.5.2 Weather and Other Effects

As a vapor cloud is transported, it mixes with the atmosphere. The topics discussed in this section are important to the atmospheric mixing and dispersion.

Table 13.6. Input and output for dispersion models

Input	
• release rate of the gas (or quantity) • atmospheric stability • wind speed • surface roughness • temperature • pressure • release diameter	• height of the release above the ground • release geometry (point, line, or area source) • momentum of the material released • buoyancy of the material released
Output	
The time-dependent mass rate of boiling or vaporization	

Atmospheric Stability. Atmospheric stability relates to turbulent vertical motion in the atmosphere. Stable atmospheric conditions lead to the least amount of atmospheric mixing and unstable conditions to the most. These effects are shown in Figure 13.7 where the behavior of the plume changes markedly depending on the stability of the atmosphere. The atmospheric conditions are normally classified according to six Pasquill stability classes; A, the least stable, through F, the most stable. The stability classes are correlated to wind speed and the intensity of sunlight which warms the earth's surface. During the day, solar energy input and increased wind speed results in unstable atmospheric conditions, while at night the reverse is true. Cloudy conditions with moderate wind speeds lead to neutral (D) stability. Wind speed data can be gathered from meteorological sources such as a local airport. Stability classes can be estimated based on wind speed, cloudiness, and solar elevation as described in CCPS Guidelines for Consequence Analysis of Chemical Releases. (CCPS 1999)

Wind Speed. Wind speed is significant as gas will be diluted as it flows downwind. As the wind speed is increased, the material is carried downwind faster, but the material is also diluted faster by a larger quantity of air. Significant local variations in wind speed and direction are possible due to terrain effects even over distances of only a few miles. Wind speed and direction are often presented in the form of a wind rose presented in compass point form. Note a wind rose shows the likelihood of wind from, not towards, a specified direction (i.e. a west wind is from the west). Wind data are normally quoted on the basis of 10 m (33 ft) height. Wind speeds are reduced substantially within a few meters of ground due to frictional effects. As many smaller discharges of dense materials remain near ground level, wind data should be corrected from 10 m (33 ft) to that relevant for the actual release.

In the absence of detailed meteorological data for a particular site, two common weather combinations (stability and wind speed) used in many consequence analysis studies are D at 5 m/s (11 mph) and F at 2 m/s (4.5 mph). (CCPS 1999) The first is typical for windy cloudy situations and the latter for still nighttime conditions.

Local Terrain Effects. Terrain characteristics affect the mechanical mixing of the air as it flows over the ground. Thus, the dispersion over a lake is considerably different from the dispersion over a forest or a city of tall buildings. Most dispersion field data and tests are in flat, rural terrains. Most models require a surface roughness as an input, and this should be selected with care. It is not the average height of surface features although it is related to this.

Height of the Release above the Ground. The height of a release effects the downwind concentrations. As the release height increases, the ground concentration decreases since the resulting plume has more distance to mix with fresh air prior to contacting the ground.

Release Geometry. An ideal release for Gaussian dispersion models would be from a fixed-point source. Real releases are more likely to occur as a line source (from an escaping jet of material), or as an area source (from a boiling pool of liquid).

Momentum of the Material Released and Buoyancy. A typical dense gas plume exhibits a combination of dense and Gaussian behavior with initial plume rise due to momentum, followed by plume bendover and sinking due to dense gas effects. Far downwind from the release, due to mixing with fresh air, the plume will behave as a neutrally buoyant cloud. This is shown in Figure 13.9.

Since most releases are in the form of a jet rather than a plume, assess the effects of initial momentum and air entrainment on the behavior of a jet. Near its release point where the jet velocity differs greatly from the wind velocity, a jet entrains ambient air due to shear (velocity difference), grows in size, and becomes diluted. For a simple jet (neutral buoyancy), its upward momentum remains constant while its mass increases. Therefore, if vertically released, the drag forces increase as the surface area increases and eventually horizontal momentum dominates. The result is that the jet becomes bent over at a certain distance and is dominated by the wind momentum. If the jet has positive buoyancy (buoyant jet), the upward momentum will increase, and the initial momentum will become negligible compared to the momentum gained due to the buoyancy. Then, the jet will behave like a plume. For a dense or negatively buoyant jet, upward momentum will decrease as it travels. Finally, it will reach a maximum height where the upward momentum disappears and then will start to descend. This descending phase is like an inverted plume.

No simple rule defines a particular combination of stability and wind speed as worst case. Often F2 gives the longest plume, but it is also the narrowest. If a plume passes over a small densely populated area the width of the plume can be more important than its length. This is why most analysts run several combinations of stability and wind speed.

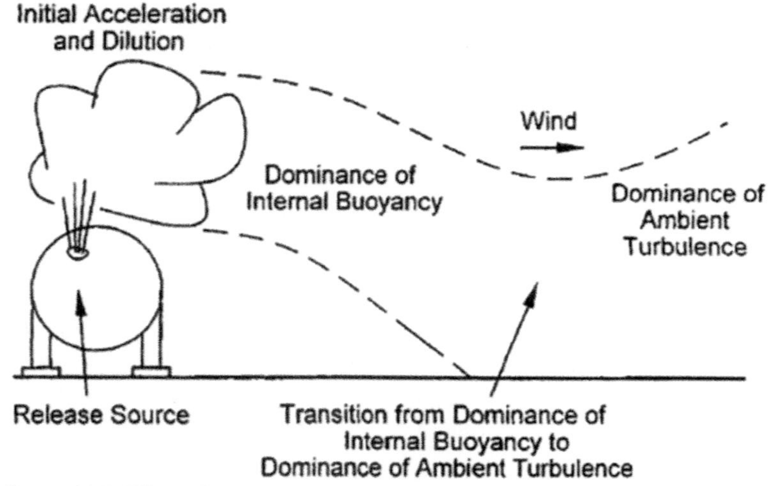

Figure 13.9. Effect of initial acceleration and buoyancy on a dense gas release (CCPS 1999)

13.5.3 *Dense Gas Dispersion*

Many materials of concern in process safety are denser than air and require dense gas modeling that factor in density mixing effects. A dense gas is defined as any gas whose density is greater than the density of the ambient air through which it is being dispersed. This result can be due to a gas with a molecular weight greater than that of air, or a gas with a low temperature due to auto-refrigeration during release, or other processes. The behavior of dense gas dispersion differs markedly from neutrally buoyant clouds. The major mechanisms include gravity slumping, air entrainment, and thermodynamic processes. When dense gases are initially released, these gases slump toward the ground and move both upwind and downwind as initially the dense spreading effect upwind may exceed the wind speed effect downwind. Dense gases spread more widely and often not as far as in neutrally buoyant dispersion and thus are not well modeled by gaussian simplifications. (CCPS 1996)

Distinct modeling approaches have been attempted for dense gas dispersion: mathematical and physical.

Mathematical. The most common mathematical approach has been the box model (also known as top-hat or slab model), which estimates overall features of the cloud such as mean radius, mean height, and mean cloud temperature without calculating detailed features of the cloud in any spatial dimension. Gaussian (Pasquil-Gifford) models can be used but the mechanisms are incorrect and thus models with dense gas mechanisms are preferred. The other form of mathematical model is the more rigorous computational fluid dynamics (CFD) approach that solves the complete three-dimensional conservation equations. The CFD model is typically used to predict the wind velocity fields, with the results coupled to a more traditional dense gas model to obtain the concentration profiles (Lee 1995). An advantage of CFD models is the ability to account for changes in terrain, buildings, and other irregularities. A disadvantage is they require substantial definition of the problem including a representation of the release environment in three-dimensional space in order to start the CFD computation. The method requires moderate computer resources. CFD modeling is typically used to analyze offshore installations where this level of detail is required close to the potential release source and for forensic investigations where the analyst tries to link observed outcomes to several possible release scenarios.

Physical. Physical (scale) models employing wind tunnels or water channels have been used for dense gas dispersion simulation. An advantage is the ability to model specific situations with obstructions or irregular terrain. A disadvantage is the inability to achieve exact similarity in all scales and re-creation of atmospheric stability and velocity distributions. The use of scale models is not a common risk assessment tool in consequence analysis although it has been applied successfully for specific problems such as dispersion of LNG plumes resulting from spills into containment.

Table 13.7. Input and output for dense gas dispersion models

Input (varies depending on model)	
Source Term • Cloud mass or volume • Temperature • Concentration • Gas density • Cloud dimensions (height, width) • Local information • Wind speed • Atmospheric stability • Surface roughness • Ground heat capacity, thermal conductivity	Physical/Chemical Information • Molecular weight • Atmospheric boiling temperature • Latent heat of vaporization • LFL • Toxic concentration or toxic dose To estimate cloud size or plume generation rate • Hole size • Release phase (gas, liquid, two-phase) • Flash fraction • Aerosol and rainout fractions • Release duration • Pool boiloff (from rainout fraction) • Cloud initial dilution • Cloud geometry
Output (varies depending on the model):	
• source term summary (if calculated by model): jet discharge or pool boiloff rate, temperature, aerosol fraction, rainout, initial density, initial cloud dimensions, time variance. • cloud dispersion information: cloud radius and height (or other dimensions as appropriate), density, temperature, concentration, time history at a particular location, concentrations and width at to specified distances. • special information: terrain effects, chemical reaction or deposition, toxic load at particular location, mass of flammable material in cloud.	

13.6 Consequence Effect Modeling

Many types of consequences are possible for a release including fires, explosions, and toxic cloud dispersion. The potential types of fire and explosions are described in Chapter 4 and summarized in Table 13.8.

Table 13.8. Types of fires and explosions

Fire	Pool Fire
	Flash Fire
	Jet Fire
Explosion	Physical: vessel rupture, BLEVE and fireball, rapid phase transition
	Chemical: thermal or runaway reaction, propagating reaction (confined and unconfined vapor cloud)

The consequences of concern for fires and fireballs are typically thermal radiation effects and for explosions are typically blast overpressure. effects and projectile effects.

13.6.1 Pool Fires

Pool fires tend to be localized, affecting a facility rather than a community risk. The primary effects of pool fires are due to thermal radiation. Grading and drainage are important considerations in the prevention of pool fires as a pool will form around a drain or in a natural low point. The determination of the thermal effects depends on flame geometry, including height, tilt and drag and geometric view factor with respect to the receiving source are calculated. Typical thermal flux from a hydrocarbon pool fire on ground is 160 kW/m^2 (70,754 $BTU/hr/ft^2$) and clean burning LNG on a water surface can be over 300 kW/m^2 (95,163 $BTU/hr/ft^2$).

Table 13.9. Input and output for pool fire models

Input	Output
• the type of fuel • the duration of the fire • burning rate • pool size and smokiness • flame surface emitted power • atmospheric transmissivity	The received thermal radiation at target locations

Flame temperatures vary with fuel type. Thus, the thermal radiation from the fuel is different as is the impact on equipment in the radiation zone. For example, the flame temperatures used for establishing firewalls are 850°C (1562°F) for wood/cellulosic fires and 1150°C (2102°F) for hydrocarbon fires. Firewalls for cellulosic material fires are A-rated followed by the time in minutes (e.g. A-60 for a 60 minute fire rating). Firewalls for hydrocarbon fires are H-rated followed by the time in minutes (e.g. H-30). Fireproofing designed for residential use where wood and other cellulosic materials may be the predominant fuel would not be effective for hydrocarbon fires. Hydrocarbon fires tend to get hotter, faster. Fireproofing for hydrocarbon fires is specifically designed and tested to withstand hydrocarbon fire radiation levels and also survive the impact of firefighting hose streams. Relevant fire test standards include the following.

- EN 1363 "Fire Resistance Tests"
- ASTM E1529 "Standard Test Methods for Determining effects of Large Hydrocarbon Pool Fires on Structural Members and Assemblies"

13.6.2 Flash fires

A flash fire is the nonexplosive (low flame speed) combustion of a vapor cloud. Major hazards from flash fires are from thermal radiation and direct flame contact for people or objects inside the fire. The thermal radiation is less significant because the thermal radiation and exposure time are low and most of the thermal energy is contained in the burning gases. As with VCE, the flammable mass is determined by identifying the cloud volume between the upper and lower flammable limits. Identify the dimensions of the flammable cloud (from dispersion modeling) and use this to determine consequence zone. Dense gas models are important for this estimate as they tend to be wider and shorter and show the potential impact inside the process facility. Gaussian models can be misleading.

13.6.3 Jet Fires

Jet fires typically result from the combustion of a material as it is being released from pressurized process equipment. Jet fires are highly turbulent and are often two-phase burning mixtures. The main concerns are direct impingement and local radiation effects. One method, the API (1996) method was originally developed for flare thermal analysis but is now applied to jet fires arising from accidental releases. Flare models apply to gas releases from nozzles with vertical flames. For accidental releases, the release hole is typically not a nozzle, and the resulting flame is not always vertical. The fraction of energy converted to radiant energy versus convective energy is determined empirically based on limited experimental data. The view factors and atmospheric transmissivity are determined using published correlations. The pressurized jet focuses the fire also yielding high heating where direct contact occurs. It also imparts a mechanical force on surfaces it contacts. Jet fire thermal flux (outside the flame envelope) can be over 300 kW/m^2 (95,163 BTU/hr/ft^2). Firewalls suitable for jet fires are rated J followed by minutes (e.g. J-15). This heat and force from direct impact can be challenging for firewalls. ISO 22899 "Determination of the resistance to jet fire fires of passive fire protection materials" specifically addresses this topic.

Table 13.10. Input and output for jet fire models

Input for jet flame model	Input for point source model
• Input for jet flame model • flame height (based on reaction stoichiometry and molecular weight) • view factor formulas • humidity data	• total energy generation rate (based on mass flow rate of combustible material) • view factor formulas • humidity data
Output: The received thermal radiation at target locations	

13.6.4 Physical Explosion

When a vessel containing a pressurized gas ruptures, the resulting stored energy is released. A steam drum burst is a well-known example. This can also apply to pressurized hydrocarbon vessels where the relief system has not functioned correctly. A BLEVE event, discussed in the next section, is another example. This energy can produce a shock wave and throw a small number of large vessel fragments for hundreds of meters. If the contents are flammable it is possible that ignition of the released gas could result in additional thermal consequence affects. The total energy from the bursting vessel is distributed to the fracturing of the vessel, generation a shock wave, and throwing of the fragments. Exactly what proportion of available energy goes to which is difficult to determine. Several methods use the calculated equivalent amount of TNT energy to estimate shock wave effects. In general, vessels of pressurized gas do not have sufficient stored energy to represent a threat from shock wave beyond the plant boundaries. However, these types of incidents can result in domino effects particularly from the effects of the projectiles produced.

Several different methods can be used to estimate projectile size and trajectory, but these have a high uncertainty as the specific way in which a vessel will fail is not known. These methods are more suited for accident investigations, where the number, size and location of the fragments is known. Very few Chemical Process Quantitative Risk Assessment (CPQRA) studies have incorporated projectile effects on a quantitative basis.

13.6.5 BLEVE and Fireball

A BLEVE is a sudden release of a large mass of pressurized superheated liquid to the atmosphere and was discussed in Chapter 4. A BLEVE occurs when an external fire, either through thermal radiation or direct flame impingement, weakens the vessel above the liquid level as the vapor space provides less internal cooling and the vessel wall fails, typically when it reaches 550°C (1022°F). As hydrocarbon fires burn at 1150°C (2102°F), there is only a short period, often only 15 minutes, before a BLEVE may occur. Note at 550°C (1022°F) the ultimate tensile strength of steel is reduced by half and this fully exhausts the design safety factor in shell thickness. A special type of BLEVE involves flammable materials, such as LPG. At the beginning of a BLEVE, a fireball is formed quickly due to the rapid ejection of flammable material as it flashes due to depressurization of the vessel. Ignition occurs as the cause of the failure is an external fire. This is followed by a much slower rise in the fireball due to buoyancy of the heated gases. Methods to determine consequences from a BLEVE are discussed in CCPS *Guidelines for Chemical Processes Quantitative Risk Assessment* and CCPS *Vapor Cloud Explosion, Pressure Vessel Burst, BLEVE and Flash Fire Hazards*. (CCPS 1999 and CCPS 2010)

BLEVE models are a blend of empirical correlations (for size, duration, and radiant fraction) and more fundamental relationships (for view factor and transmissivity). BLEVE models require the material properties (heat of combustion and vapor pressure), the mass of material, and atmospheric humidity. Fragment models are fairly simplistic and require vessel volume and vapor pressure. The output of a BLEVE model is usually the radiant flux level and duration. BLEVE models require some care in application, as errors in surface flux, view factor, or transmissivity can lead to significant error.

A BLEVE and fireball are significant threats to firefighters as they approach an emergency scene. Understanding when potential for a BLEVE exists and planning an appropriate response are important to the safety of the firefighters. Water spray can be used to cool the area of flame impingement – if the water can be applied without putting firefighters at risk. Protecting vessels that could be exposed to external flame impingement with fireproofing is also a means to reduce the vessel wall heating and delay or prevent a BLEVE.

13.6.6 Vapor Cloud Explosions (VCE)

Dispersion analysis can be used to define the extent of the flammable portion of a vapor cloud. If the vapor cloud is ignited before it is diluted below its lower flammability limit, a VCE or flash fire will occur.

Vapor clouds are normally ignited at the edge as they drift to an ignition source such as a fired heater or a vehicle. The effect of ignition is to terminate further spread of the cloud in that direction. Thus, a site with many ignition sources on or around it would tend to prevent clouds from reaching their full hazard extent, as most such clouds would find an ignition source before this occurs. Early ignition, before the cloud becomes fully formed, might result in a flash fire or an explosion of smaller size. Late ignition could result in an explosion of the maximum possible effect.

The main consequence of a VCE is blast overpressure.. The blast effects produced depend on whether a deflagration (flame front less than sonic velocity) or detonation (flame front greater than sonic velocity) results. (Refer to Chapter 4 for definitions.) Thermal expansion occurs as the fuel is combusted and this drives flame acceleration. Flame acceleration is influenced by congestion within the facility and confinement of the vapor cloud.

A deflagration event requires congestion and only the flammable material in the congested space contributes to the explosion. Once the flame passes through the congested space, its velocity drops and it becomes a flash fire event. A detonation event, once initiated, is self-sustaining and the entire flammable mass contributes, regardless of whether the whole cloud is in congested space or not.

Important parameters in explosion analysis are the properties of the material: lower and upper flammable limits (LFL and UFL), flash point, autoignition temperature, heat of combustion, molecular weight, and combustion stoichiometry. The upper and lower flammable limits are used to determine the flammable mass. Some analysts use the mass between the upper and lower flammable limits, some use the mass between half the lower flammable limit and the upper flammable limit. Using half of the lower limit, as opposed to the full lower limit, is conservative as it includes more material in the flammable range. The impulse, (the area under the explosion pressure-time curve) is necessary to determine the dynamic loading effects on a structure.

The three common VCEs models are explained in greater detail in *Vapor Cloud Explosion, Pressure Vessel Burst, BLEVE and Flash Fire Hazards*. (CCPS 2010) They are

- TNT equivalency model
- TNO multi-energy model
- Baker-Strehlow-Tang model

TNT Equivalency Model. The TNT equivalency model represents a VCE as a TNT detonation of equivalent energy. This was one of the first models developed; however, case studies and experimental data have shown that the results are not representative of a VCE. It is now understood that VCE blast effects are determined not only by the explosion energy, but more importantly, by the combustion rate. Therefore, use of the TNT equivalency model is no longer recommended for vapor cloud explosion analysis.

TNO Multi-Energy Method. The multi-energy model recognizes that only the confined portion of the cloud, not the entire volume of a vapor cloud, contributes to the blast effects. This method uses blast curves plotting overpressure . vs. distance and impulse vs. distance. The initial blast strength is represented in a series of 10 curves, representing levels of congestion. The curve used significantly impacts the results. Limited guidance is available on which curve to use although most analysts use curves 5, 6, or 7. The TNO method is based on interpretations of actual VCE incidents.

Baker-Strehlow-Tang (BST) Method. This method is also a set of blast curves. The curve selected based on flame speed which is correlated to the combined effects of fuel reactivity, confinement, and obstacle density. The Baker-Strehlow-Tang method is based on interpretations of actual VCE incidents.

Reactivity is classified as low, (such as methane and carbon monoxide), average, and high (hydrogen, acetylene, ethylene, ethylene oxide, and propylene oxide). All other fuels are classified as average reactivity.

Confinement is based on 2D and 3D symmetries as shown in Figure 13.10. In 3D the flame is free to expand spherically from an ignition source, such as a release from a tank in an open field. In 2D, the expansion is in 2 dimensions as if constrained between two plates, such as a in a multi-level process unit. In 1D, the flame front is only free to expand in 1 dimension, as if in a pipe. 1D is not included "because the maximum flame speed achieved in true one-dimensional expansion conditions (i.e., a pipe) is a function of the length-to-diameter ratio of the pipe in addition to pipe geometry (elbows, tees, etc.), fuel reactivity, and congestion level. Many fuels are able to undergo a deflagration to detonation transition in a 1-D geometry if the combination of length-to-diameter ratio and obstacle density are sufficiently high. As a result, the use of a single number to represent all length-to diameter ratios is overly simplified and a more detailed analysis is recommended for all such [1-D] cases". (BakerRisk 2005)

Type	Dimension	Description	Geometry
Point Symmetry	3-D	"Unconfined volume," almost completely free expansion.	
Planar Symmetry	2-D	Platforms carrying process equipment; space beneath cars; open-sided multi-story buildings	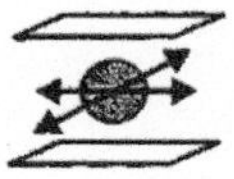
Line Symmetry	1-D	Tunnels, corridors, or sewage systems	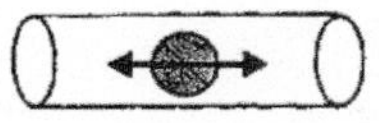

Figure 13.10. Degrees of confinement (CCPS 1999)

Type	Obstacle blockage ratio per plane	Pitch for obstacle layers	Geometry
Low	Less than 10%	One or two layers of obstacles	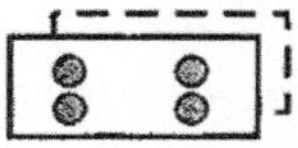
Medium	Between 10% and 40%	Two or three layers of obstacles	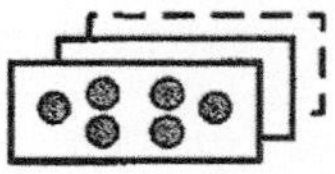
High	Greater than 40%	Three or more fairly closely spaced obstacle layers	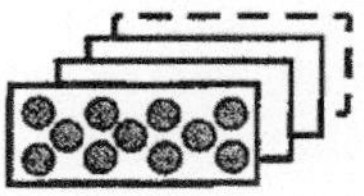

Figure 13.11. Levels of obstacle density (CCPS 1999)

Obstacle density is classified as low, medium and high, shown in Table 13.11, as a function of the blockages and the distances between blockages. Low density assumes few obstacles in the flame's path. A high obstacle density may occur in a process unit which contains many closely spaced structural members, pipes, pumps, heat exchangers and other turbulence generators. Medium density falls between the two categories.

Table 13.11. Input and output for VCE models

Input for TNT Equivalence Model	Input for TNO Multi-energy Model	Input for Baker-Strehlow-Tang Model	CFD Models
• mass of flammable material in the vapor cloud • lower heat of combustion for the vapor	• mass of flammable material in the vapor cloud • lower heat of combustion for the vapor • degree of confinement • relative blast strength	• mass of flammable material in the vapor cloud • lower heat of combustion for the vapor • chemical reactivity • obstacle density • degree of confinement	• source term for release model • Wind stability • full local geometry • grid to be used, including symmetries • physical and chemical properties • time step to be used
Output			
The overpressure and impulse with distance			

In addition to the three VCE models described in this section, computational fluid dynamics (CFD) models can be used. CFD models are a combination of transport model and blast model. They are time dependent and give more refined results. They are better suited for complex and limited geometries, such as an offshore platform and for accident investigation as they are easier to link to differentiated blast damage. A disadvantage as compared to the simpler models described here are that they require a detailed model of the area geometry and can be time consuming to set up.

13.6.7 Confined Explosions

Confined explosions are deflagrations which are primarily due to thermal expansion constrained within vessels and buildings. Thermal effects can generate a six-fold volume increase thus often a confined explosion is limited to 6 bar. Confined explosions can include dust explosions, combustion reactions, thermal decompositions, or runaway reactions within process vessels and equipment. In general, a deflagration occurring within a building or low strength structure such as a silo is less likely to impact the surrounding community and is more of an in-plant threat because of the relatively small quantities of fuel and energy involved. Shock waves and projectiles are the major threats from confined explosions.

The technique is based on the determination of the peak pressure. If the values of peak pressure calculated exceed the burst pressure of the vessel, then the consequences of the resulting explosion should be determined. Consequences include shock wave, fragments, and a burning cloud. For most pressure vessels designed to the ASME Code, the minimum bursting pressure is at least four times the "stamped" maximum allowable working pressure (MAWP). Specialist help is desirable for the container strength and calculations.

This analysis provides overpressure versus distance effects and also projectile effects. Using NFPA 68, "Standard on Explosion Protection by Deflagration Venting", overpressures can be estimated for vented vessels and buildings, which allows estimates to be made of the expected damage levels.

13.7 Outcome Models

The final step in Figure 13.4 is to establish outcomes. The consequence models described in this chapter generate a variety of outcomes from the consequence effects. These outcomes are based on chemical concentrations with distance, thermal radiation levels, and blast effects with distance. In order to understand the impact on people, the environment, and property, these outcomes are translated to human vulnerability, environmental impacts, and structural damage.

One method of assessing the consequence of an incident outcome is the direct effect model, which predicts effects on people or structures based on predetermined criteria (e.g., death is assumed to result if an individual is exposed to a certain concentration of toxic gas). In reality, the consequences may conform to probability distribution functions. A statistical method of assessing a consequence is the dose-response method. This is coupled with a probit equation to linearize the response. The probit equation is simply the single-tailed normal distribution with a mean of 5 and standard deviation of 1. It was developed as a simplification to avoid the need to use negative numbers in most circumstances.

The probit (probability unit) method described by Finney (1971) reflects a generalized time-dependent relationship for any variable that has a probabilistic outcome that can be defined by a normal distribution. The probit method can be applied to toxic, thermal, and explosion effects.

13.7.1 Dose-Response Functions

Most toxicological considerations are based on the dose-response function. Toxicological data is obtained by conducting experiments where a fixed dose is administered to a group of test organisms until effects are obtained. A range of responses is expected for a fixed exposure due to variability in response of living organisms. Some of the organisms will show a high level of response while some will show a low level. A typical plot of the results is modeled as a "bell-shaped" curve.

The experiment is repeated for different doses and Gaussian curves are drawn for each dose. The mean response and standard deviation are determined at each dose. A complete dose-response curve is produced by plotting the cumulative mean response at each dose. For convenience, the response is plotted versus the logarithm of the dose. The logarithm form arises because in most organisms some subjects can tolerate high levels of exposure while others are sensitive. Simpler forms of dose response are useful for emergency response planning and these use a standard time of exposure (30 or 60 minutes) and provide different levels of impact from mild impact to serious injury (e.g. ERPG discussed in Section 13.7.2).

For most engineering computations, an equation is more useful than the dose-response curve. For single exposures, the probit (probability unit) method provides a transformation method to convert the dose-response curve into a straight line. Probit equations are available for a variety of exposures, including exposures to toxic materials, heat, pressure, radiation, and impact. For toxic exposures, the causative variable is based on the concentration; for explosions, the causative variable is based on the explosive overpressure or impulse, , depending on the type of injury or damage. For fire exposure, the causative variable is based on the duration and intensity of the radiative exposure. The probit method is generally the preferred method of choice for consequence analysis studies. A limitation is the restricted set of chemicals for which probit coefficients are published.

13.7.2 Toxic Outcomes

Toxic impact models are employed to assess the consequences to human health as a result of exposure to a known concentration of toxic gas for a known period of time. For consequence analysis, the toxic effects are due to short-term exposures, primarily due to vapors. Chronic exposures are not considered here. Many releases involve several chemical components or multiple effects. The cumulative effects of simultaneous exposure to more than one material are not well understood.

Predictions of gas cloud concentrations and durations at specific locations are available from neutral and dense gas dispersion models.

Toxic concentration criteria and methods include the following which were discussed in Chapter 6.

- Emergency Response Planning Guidelines for Air Contaminants (ERPGs) issued by the American Industrial Hygiene Association (AIHA)
- Acute Exposure Guideline Levels (AEGL) maintained by the U.S. EPA in cooperation with the National Academies. (EPA AEGL)
- Immediately Dangerous to Life or Health (IDLH) levels established by the National Institute for Occupational Safety and Health (NIOSH)
- Emergency Exposure Guidance Levels (EEGLS) and Short-Term Public Emergency Guidance Levels (SPEGLs) issued by the National Academy of Sciences/National Research Council (NAS)
- Probit Functions

Once the concentration-time information is determined, the next step is to determine the toxic dose. Toxic dose is usually defined in terms of concentration per unit time of exposure raised to a power multiplied by duration of exposure.

Table 13.12. Input and output for toxic impact models

Input	Output
• Toxic gas concentrations and durations of exposure at all relevant locations • Toxic criteria for specific health effects	• The identification of populations at risk of death or serious harm • The percentage of the population that may be affected by a given toxic gas exposure

13.7.3 Thermal Radiation Exposure Outcomes

API Standard 521 provides a short review of the effects of thermal radiation on people. (API STD 521) The effect of thermal radiation on structures depends on whether they are combustible or not and the nature and duration of the exposure. Thus, wooden materials will fail due to combustion, whereas steel will fail due to thermal lowering of the yield stress. Many steel structures under normal load will fail rapidly when raised to a temperature of 932-1112°F (500-600°C) as the ultimate tensile strength is halved at that temperature. This was evident in the collapse of the World Trade Center towers on September 11, 2001. (NIST) Concrete will survive for much longer. Flame impingement on a structure is more severe than thermal radiation. Selected criteria for thermal radiation damage are shown in Table 13.13. For comparison, solar flux arriving at earth is 1.3 kW/m2 (412 BTU/hr/ft^2); however, after attenuation in the atmosphere and absorption by water vapor, a local maximum of 1.0 kW/m2 (317 BTU/hr/ft^2) is typical.

Table 13.13. Effects of thermal radiation
(CCPS 2012)

Radiation Intensity		Observed effect
BTU/hr/ft^2	kW/m^2	
11,900	37.5	Sufficient to cause damage to process equipment
8,000	25	Minimum energy required to ignite wood at indefinitely long exposures (nonpiloted)
4,000	12.5	Minimum energy required for piloted ignition of wood, melting of plastic tubing
3,000	9.5	Pain threshold reached after 8 sec.; second degree burns after 20 sec.
1,200	4	Sufficient to cause pain to personnel if unable to reach cover within 20 sec. however blistering of the skin (second degree burns) is likely; 0% lethality
500	1.6	Will cause no discomfort for long exposure

13.7.4 Overpressure Outcomes

Explosion overpressure can cause harm to people inside and outside of buildings. People can survive a higher overpressure than typical structures can. The overpressure can cause objects inside the building to be thrown, damage to the building, and building collapse, all of which can result in harm to the building occupants. Analyzing structural response is a specialist topic requiring expert civil engineering assistance.

Overpressure duration is important for determining effects on structures. The same overpressure level can have markedly different effects depending on the duration. Therefore, some caution should be exercised in application of simple overpressure criteria for buildings or structures. These criteria can, in many cases, cause overestimation of structural damage.

An object struck by a blast wave experiences loading. When the blast wave hits the front wall of a building, it is reflected off the wall and builds up a local, reflected, overpressure which can be approximately twice the incident (side-on) overpressure. This is illustrated in Figure 13.12 and fully described in *Guidelines for Vapor Cloud Explosion, Overpressure Vessel Burst, BLEVE and Flash Fire Hazards*. (CCPS 2010) Damage levels observed at selected overpressures are noted in Table 13.14.

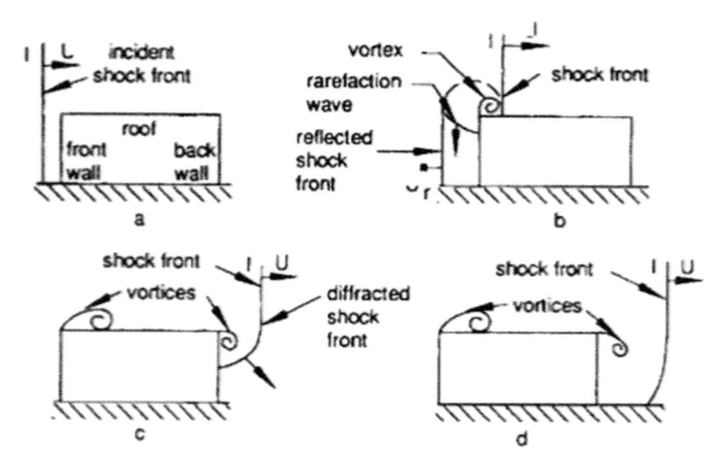

Figure 13.12. Interaction of a blast wave with a rigid structure (Baker 1973) (CCPS 2010)

Table 13.14. Selected overpressure levels and damage (CCPS 1999) (Clancy 1972)

Pressure		Damage
psig	kPa	
0.15	1.03	Typical pressure for glass breakage
1.0	6.9	Partial demolition of houses, made uninhabitable
2.5	17.2	50% destruction of brickwork houses
3	20.7	Heavy machines (3000 lb) in industrial building suffered little damage; steel frame building distorted and pulled away from foundation
10	68.9	Probable total destruction of buildings; heavy machine tools (7000 lb) moved and badly damaged; very heavy machine tools (12,000 lb) survive.

Building Damage Levels (BDL) are used as a criterion for the evaluation of existing buildings to limit the hazards to which occupants are potentially exposed. New buildings are designed for their intended location. BDLs are defined for various types of building construction including masonry buildings and pre-engineered metal buildings. Selection of a BDL includes an implied estimate of the occupant vulnerability. Building damage increases as the severity of the blast load increases. BDLs are categorized into damage states ranging from minimal damage to collapse as listed in Table 13.15 and illustrated in Figures 13.13 and 13.14.

Table 13.15. Typical industry building damage level descriptions (Baker 2002) (CCPS 2012)

Building Damage Level (BDL)	BDL Name	Damage Description
1	Minor	Onset of visible damage to reflected wall of building
2A	Light	Reflected wall components sustain permanent damage requiring replacement, other walls and roof have visible damage that is generally repairable
2B	Moderate	Reflected wall components are collapsed or very severely damaged. Other walls and roof have permanent damage requiring replacement
3	Major	Reflected wall has collapsed. Other walls and roof have substantial plastic deformation that may be approaching incipient collapse.
4	Collapse	Complete failure of the building roof and a substantial area of walls

Figure 13.13. Pre-engineered metal building BDL1
(CCPS 2012) (Explosion Research Cooperative)

Figure 13.14. Pre-engineered metal building BDL3
(CCPS 2012) (Explosion Research Cooperative)

People outside of buildings or structures are susceptible to injury from the blast (e.g. ear drum rupture) and injuries indirectly from the blast such as being knocked down or being impacted by projectiles.

13.7.5 Environmental Outcome

Environmental outcomes from fires, explosions and toxic releases are much harder to measure. Environmental impacts can include effects to air, water, and land.

Air impacts. For a fire there will most likely be a smoke plume that will extend offsite. The plume may have unburned toxic materials and particles that could impact both onsite employees as well as the neighboring communities. A toxic vapor cloud can also impact both on and offsite. Either of these can prompt a shelter-in-place to protect people from being exposed to the airborne contaminants.

Water impacts. Chemicals can be spilled directly into water, e.g. from river pipeline crossings or from ships. They can also be spilled indirectly, e.g. from a tank, and flow downhill or into drainage systems and find their way to water courses. Also, water runoff from firefighting could contain toxic materials harmful to the environment. Water collection systems should be designed to collect, and process water used during emergencies. Toxic impacts can extend for large distances downriver if spilled into a watercourse as was the case in the Freedom Industries Chemical Spill in West Virginia in 2014.

Land impacts. As with water impacts, impacts on land can occur directly or indirectly. The land where a release occurs can be polluted by the contaminants. Pollutants can settle out of a vapor cloud or smoke plume as they disperse downwind thus impacting animals and vegetation. Effects can be most harmful to farm animals or wildlife grazing on impacted land.

13.8 Data and Uncertainties

Consequence modeling involves a large quantity of data. This data describes the chemical, the release characteristics, and the environmental characteristics. A significant portion of the analysis time can and should be spent on gathering the most accurate data possible. This helps to support an accurate estimation of the consequence and impacts. Some tools have built-in data sets for thermodynamics and common parameters (e.g. surface roughness).

All models have inherent uncertainties. For consequence models, these may be due to:

- an incomplete understanding of the geometry of the source term, i.e. hole size, phase, duration and orientation
- unknown or poorly characterized physical properties, i.e. the ground condition, atmospheric wind speed stability and temperature
- a poor understanding of the chemical or release process, and
- unknown or poorly understood mixture behavior.

Considering that consequence analysis data will be used in understanding and designing systems to minimize risk to humans and the environment, typically best estimate values are selected for these characteristics and modeling parameters. In some cases, conservative values are used; however, consider the result of all the conservative assumptions in total.

Conservatism on top of conservatism can result in an unrealistic estimation of consequence. Selecting values that are best estimates for all parameters avoids undue conservatism in the final result.

To illustrate conservative modeling, consider a problem requiring an estimate of the gas discharge rate from a hole in a storage tank. This discharge rate will be used to estimate the downwind concentrations of the gas, with the intent on estimating the toxicological impact. The discharge rate is dependent on multiple parameters, including (1) the hole area (2) the pressure within and outside the tank (3) the physical properties of the gas, and (4) the temperature of the gas, to name a few. The reality of the situation is that the maximum discharge rate of gas will occur when the leak first occurs, with the discharge rate decreasing as a function of time as the pressure within the tank decreases. The complete dynamic solution to this problem is difficult, requiring a mass discharge model cross-coupled to a material balance on the contents of the tank. An equation of state (nonideal) is required to determine the tank pressure given the total mass. Complicated temperature effects are also possible. A modelling effort of this detail is not necessarily required to estimate the consequence. A much simpler procedure is to calculate the mass discharge rate at the instant the leak occurs, assuming a fixed temperature and pressure within the tank equal to the initial temperature and pressure. The actual discharge rate at later times will always be less, and the downwind concentrations will always be less.

13.9 What a New Engineer Might Do

New engineers often get involved in conducting consequence analysis. Consequence modeling is often used in preparation for hazard identification studies to determine release rates for potential releases. Many tools are available to support consequence analysis as listed in Section 13.10. Although it may be tempting to access one of these tools and jump right into an analysis, it is best to spend some time up front getting ready. First, attend some training and practice using the software for performing the calculations. Then it is time to plan the analysis and gather the data. Refer to Figure 13.4 to understand the consequence mechanisms that should be modeled.

Plan the analysis including understanding what question the study is intended to answer. Understanding how the results are to be used can influence scope, tool selection, and resources. Determine if previous or similar analyses have been conducted that can support the current study.

The tools are applicable to specific situations and have limitations. This is due to the different bases and different data sets to which they have been validated. Understand these limitations as they will determine if the tool is appropriate for use in the scenario being modeled.

Gathering data is a very important, and sometimes tedious, step. Do not let the tedious nature prompt cutting corners. Remember the adage: garbage in, garbage out. Using the most relevant and accurate data will provide the best results for use in making decisions relating to risk reduction.

Finally, reach out to other consequence analysts for support and guidance. This is not a simple topic and talking with someone who has conducted similar analyses in the past can save time, confirm results, and increase understanding.

13.10 Tools

Tools that may be used in conducting consequence analysis include the following. Some tools are free and use self-study, others have a fee and require attendance at training courses. Often the cost of mitigation measures can be high, so the cost of the tool may be minor cost in the total effort.

CCPS Risk Assessment Screening Tool (RAST). RAST is designed to provide both qualitative and quantitative risk analysis for processes that handle hazardous chemicals, materials, and energies. RAST allows users to input the chemicals, reactivity data equipment type, operating conditions (e.g., pressures, temperatures), and facility layout. RAST then provides an initial list of risk scenarios based on generic Hazard and Operability (HAZOP)study information (refer to Chapter 12), estimates "worst" consequences, and creates a Layer of Protection Analysis (LOPA) worksheet (refer to Chapter 14) for the risk scenarios selected by the user.

RAST uses simplified models and empirical correlations. These relationships allow the RAST software to perform screening analyses on systems, processes, or scenarios in "real" time such that a study team has information available immediately to help make better informed decisions. The following information is entered and used for the risk screening evaluations in RAST:

- Chemical material safety data, including chemical reactivity data, either from a prepopulated list within RAST or by user-defined and entered information
- Type of equipment and specific equipment characteristics (used when evaluating failure mechanisms, failure likelihoods, and potential release rates)
- Processing conditions (such as temperatures, pressures, etc.), and
- Facility layout information (used to evaluate dispersion characteristics and distances).

A detailed user's manual is provided that fully explains how to use this tool at https://www.aiche.org/ccps/resources/tools/risk-analysis-screening-tool-rast-and-chemical-hazard-engineering-fundamentals-chef/rast-overview. (CCPS RAST)

CCPS Chemical Hazards Engineering Fundamentals (CHEF). The CHEF documentation provides the theoretical details of the methods, techniques, and assumptions which are used in RAST. CHEF also includes an Excel calculation aid to help users understand and utilize the – correlations used to model and evaluate the estimated airborne quantities, vapor dispersions, and explosion overpressures and their impact. These correlations provide the RAST users with a powerful screening tool to help an organization prioritize its overall process safety risks – a bridge between the qualitative approach and quantitative approach – before it invests in complex, detailed quantitative hazards and risk assessments (Discussed in Chapter 14). Detailed documentation that is an excellent explanation of RAST and also of consequence analysis in general is available at https://www.aiche.org/ccps/resources/tools/risk-analysis-screening-tool-rast-and-chemical-hazard-engineering-fundamentals-chef/chef-overview. (CCPS CHEF)

CAMEO®. The CAMEO® software suite is a system of software applications used widely to plan for and respond to chemical emergencies. It is one of the tools developed by the U.S. EPA and the National Oceanic and Atmospheric Administration (NOAA) to assist front-line chemical emergency planners and responders. They can use CAMEO® to access, store, and evaluate information critical for developing emergency plans. The CAMEO® system integrates four tools: a chemical database and a method to manage the data, an air dispersion model, and a mapping capability. (EPA CAMEO)

- CAMEO*fm*- Database and Information Management Tool. CAMEO*fm* is a database application that supports Emergency Planning and Community Right-to-Know Act (EPCRA) data management requirements.
- CAMEO Chemicals - Chemical Response Datasheets and Reactivity Prediction Tool. CAMEO Chemicals has an extensive chemical database with critical response information for thousands of chemicals and allows additional chemicals to be added.
- MARPLOT - Mapping Application for Response, Planning, and Local Operational Tasks. MARPLOT is a mapping application that comes with several global background basemap options, allows users to add objects to the map, and supports linking these objects with CAMEO*fm* records.
- ALOHA - Areal Locations of Hazardous Atmospheres. ALOHA is an atmospheric dispersion model that can estimate threat zones associated with toxic gas clouds, fires, and explosions. Threat zones can be displayed on MARPLOT maps.

Many additional software products support consequence modeling. Some of these are free as they provide a means to comply with regulations. Others are proprietary. A list of selected models is provided in Table 13.16. References for these models are provided in the reference section of this chapter. While all these models have been validated to some degree by their authors, only DEGADIS, PHAST and FLACS have been validated by a formal and extensive LNG Model Evaluation Protocol. (NFPA 2016)

Table 13.16. Selected consequence analysis models

Models	No cost	Discharge	Source	Neutrally Buoyant Dispersion	Dense Gas	VCE	Projectile	BLEVE & Fireball	Confined Explosions	Pool Fire	Jet Fire
ALOHA	X			X							
ARCHIE	X	X	X	X		X		X			
AutoReaGas (CFD)						X					
DEGADIS	X				X						
EFFECTS		X	X			X		X			X
FLACS (CFD)				X		X					
HGSYSTEM 3.0			X		X						
PHAST		X	X	X	X	X		X		X	X
SAFER One						X					
SAFER TRACE			X	X				X		X	X
SAFESITE		X	X	X	X	X	X				
SAFETI		X	X	X	X	X		X		X	X
Shell Shepherd		X	X	X	X	X		X	X	X	X
SLAB	X				X						
SUPERCHEMS		X				X	X	X		X	X

13.11 Summary

This chapter answers the question posed in Chapter 1, "how bad can it be?" It takes a hazardous material loss of containment scenario through the steps of consequence analysis to estimate the impact on people, property, and the environment. Models are available to estimate all of the steps: source, dispersion, consequence and impact. The models are based on chemical engineering and other scientific fundamentals but can be complex when addressing the specifics of real-world consequences.

Many software packages are available to facilitate consequence analysis. Some are available for free as they support modeling required by governmental regulations. Others are proprietary and typically offer greater capability, ease of use, and technical support. These models are sometimes deceptively easy to use. Care should be taken in understanding the limitations of the model and in gathering the best data for use in the model. Every model has uncertainties. Selecting data that will yield a conservative, and still realistic, result is the best approach.

13.12 Other Incidents

This chapter began with a description of the DPC Enterprises toxic release. Other incidents relevant to consequence analysis include all the incidents listed in Chapters 4, 5, and 6.

13.13 Exercises

1. List 3 RBPS elements evident in the DPC Enterprises L.P. chlorine release incident summarized at the beginning of this chapter. Describe their shortcomings as related to this accident.
2. Considering the DPC Enterprises L.P. chlorine release incident, what actions could have been taken to reduce the risk of this incident?
3. List the stages of a consequence analysis in the order they are conducted.
4. What hole size and leak duration should be used in a source model? Why?
5. An LNG storage tank leaks into the diked area around the tank. Describe the expected transport and consequence effects.
6. A refinery HF acid unit has a release of isobutane and HF acid. Describe the expected transport and consequence effects.
7. What 2 atmospheric conditions (wind direction, wind speed, and atmospheric stability) would you select to represent the conditions illustrated in the wind roses in Figure 13.15?

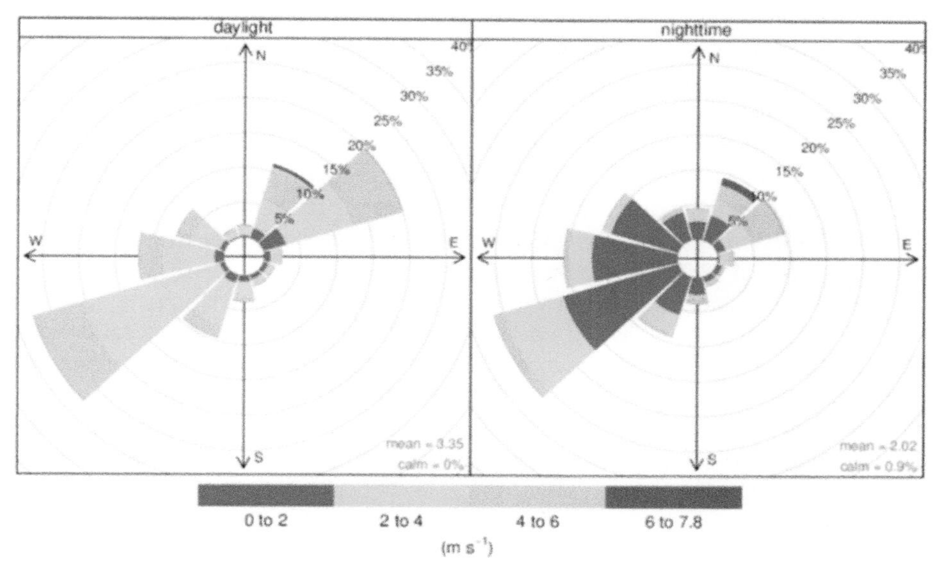

Figure 13.15 Wind roses
(Grange 2014)

8. Estimate the discharge rate of sulfur dioxide vapor (molecular weight of 64.1) at 25°C (77°F) and 200 kPa gauge (29 psig) pressure from a 25 mm (1 in) hole assuming a discharge coefficient of 0.61. Show your results.
9. A propane cloud is ignited in an area of the facility with few pieces of equipment and no surrounding structures. A second propane cloud ignition occurs in a process unit with rows of equipment so close together that the sunlight barely shows through. What differences are expected in the explosion strength?
10. Estimate the airborne rate for a 10 kg/s (22 lb/s) overflow release of acetone (molecular weight of 58.1) at 25°C (77°F) through a 51 mm (2 in) at 10 m above the ground into a 100 m^2 (1076 ft^2) diked area with wind speed of 3 m/sec (6.7 mi/hr). Show your results.
11. Estimate the concentration at the transition distance where jet mixing has diminished for a 2 kg/s (4.4 lb/s) release of ethylene (molecular weight 28.1) at 25°C (77°F) through a 100 mm (3.9 in) diameter pipe with wind speed of 3 m/s (9.8 ft/s). (Note you will need to estimate the density of ethylene at atmospheric pressure at 25°C (77°F).) Show your results.
12. Estimate the downwind distance to an ERPG-3 concentration of 25ppm for a release of 2.9 kg/s (6.4 lb/s) of sulfur dioxide vapor using ALOHA. Select a location of Ann Arbor, Michigan. Use: Wind Speed of 3 m/s (9.8 ft/s) at 10 m (33 ft) Measurement Height, Wind from W, Open Country, Cloud Cover of partly cloudy, Air Temperature of 25°C (77°F), No Inversion, Humidity of medium, Source height at ground level. Provide a screen shot of the Toxic Threat Zone.
13. Estimate the distance to 6.9 kPa (1 psi) overpressure from an explosion of 15 kg of acetylene with a heat of reaction of 190.92 kJ/mol. Use the TNT equivalency method. Show your results.

14. Estimate the maximum distance for a6.9 kPa (1 psi) blast overpressure isobar from leaking a medium fuel reactivity vapor into a medium-congestion 2930 m^3 (3832 yd^3) enclosed process structure. Assume the concentration within this confined space could exceed the lower flammable limit. Show your results.
15. Estimate the distance to 6.9 kPa (1 psi) blast overpressure from a 30 kg/s (66 lb/s) outdoor isopropyl amine (molecular weight 59.1, medium reactivity fuel) airborne rate into a medium congestion process area. The distance to the lower flammable limit of 2 volume % is estimated as 140 m (459 ft) at a wind speed of 3 m/sec (9.8 ft/s) and averaging time of 18.75 sec. Show your results.
16. Describe the effect on dispersion and downwind concentration by changing a) release height, b) wind speed), c) distance downwind, d) distance in the horizontal direction from centerline of plume, e) distance in the vertical direction from centerline of plume, and f) stability class.
17. For each parameter given in exercise 16, which direction of change leads to the most conservative estimate of downwind concentration from a release. Why?

13.14 References

API STD 521, "Pressure-Relieving and Depressuring Systems", American Petroleum Institute, Washington, D.C.

ARCHIE, "Automated Resource for Chemical Hazard Incident Evaluation", FEMAhttps://apps.usfa.fema.gov/thesaurus/main/termDetail.

ASTM E1529, "Standard Test Methods for Determining effects of Large Hydrocarbon Pool Fires on Structural Members and Assemblies", ASTM International, https://www.astm.org/Standards/E1529.htm.

Autoreagas, TNO, www.tno.nl.

Baker 1973, Baker, W. E., *Explosions in Air*, University of Texas Press, Austin.

Baker 2002, "Explosion Risk and Structural Damage Assessment Code (ERASDAC)", 30th DOD Explosive Safety Seminar, U.S. Department of Defense Explosive Safety Board, Arlington, VA.

BakerRisk 2005, A. Pierorazio et.al., "An Update to the Baker-Strehlow-Tang Vapor Cloud Explosion Prediction Methodology Flame Speed Table", Process Safety Progress, Volume 24, No. 1, American Institute of Chemical Engineers, New York, N.Y., March.

CCPS 1987, *Guidelines for Use of Vapor Cloud Dispersion Models*, Center for Chemical Process Safety, John Wiley & Sons, Hoboken, N.J.

CCPS, 1999, *Guidelines for Chemical Processes Quantitative Risk Analysis,* Center for Chemical Process Safety, John Wiley & Sons, Hoboken, N.J.

CCPS 1999, *Guidelines for Consequence Analysis of Chemical Releases*, Center for Chemical Process Safety, John Wiley & Sons, Hoboken, N.J.

CCPS 2010, *Guidelines for Vapor Cloud Explosion, Overpressure Vessel Burst*, BLEVE and Flash Fire Hazards, Center for Chemical Process Safety, John Wiley & Sons, Hoboken, N.J.

CCPS 2012, *Guidelines for Engineering. Design for Process Safety*, Center for Chemical Process Safety, John Wiley & Sons, Hoboken, N.J.

CCPS 2012, *Guidelines for Evaluating Process Plant Buildings for External Explosions, Fires, and Toxic Releases*, Center for Chemical Process Safety, John Wiley & Sons, Hoboken, N.J.

CCPS 2021, *Process Safety in Upstream Oil and Gas*, Center for Chemical Process Safety, John Wiley & Sons, Hoboken, N.J.

CCPS CHEF, https://www.aiche.org/ccps/resources/tools/risk-analysis-screening-tool-rast-and-chemical-hazard-engineering-fundamentals-chef/chef-overview.

CCPS RAST, https://www.aiche.org/ccps/resources/tools/risk-analysis-screening-tool-rast-and-chemical-hazard-engineering-fundamentals-chef/rast-overview.

Clancey 1972, "Diagnostic Features of Explosion Damage", 6th International Meeting on Forensic Sciences, Edinburgh, Scotland.

CSB 2003, "Investigation Report Chlorine Release DPC Enterprises, L.P.", Report No. 2002-04-I-MO, U.S. Chemical Safety Hazard and Investigation Board, Washington, D.C., May.

DEGADIS, https://www.breeze-software.com/Software/LFG-Fire-Risk/Product-Tour/DEGADIS-Model/.

DOT 1980, LNG Facilities, Federal Safety Standards, Federal Register, Vol. 45, No. 29. U.S. Department of Transportation, Washington, D.C.

EFFECTS, TNO, https://www.tno.nl/media/10741/effects-brochure.pdf.

EN 1363, "Fire Resistance Tests", Comite Europeen de Normalisation, https://standards.iteh.ai/catalog/standards/cen/

EPA 1970, *Workbook on Atmospheric Dispersion Estimates*, Environmental Protection Agency, https://nepis.epa.gov/Exe/ZyNET.exe

EPA AEGL, https://www.epa.gov/aegl.

EPA ALOHA, https://www.epa.gov/cameo/aloha-software.

EPA CAMEO, https://www.epa.gov/cameo/what-cameo-software-suite.

HGSYSTEM, https://19january2017snapshot.epa.gov/scram/air-quality-dispersion-modeling-alternative-models_.html#hgsystem.

Explosion Research Cooperative, https://www.bakerrisk.com/products/research-development/joint-industry-programs/explosion-research-cooperative-erc/

Finney 1971, *Probit Analysis. 3rd Edition*, Cambridge University Press, London, U.K., ISBN 0-521-080-41-X.

FLACS, FLame ACcelleration Simulator, GexCon, https://www.gexcon.com/products-services/flacs-software/.

Fryer and Kaiser 1979, Fryer, L. S., and G. D. Kaiser, Report SRD RI52: "DENZ - A Computer Program for the Calculation of Dispersion of Dense Toxic or Explosive Gases in the Atmosphere", Culcheth, U.K.: UKAEA Safety and Reliability Directorate.

Grange 2014, "Technical note: Averaging wind speeds and directions", 10.13140/RG.2.1.3349.2006.

IOGP 2019, "Risk Assessment Data Directory - Report 434-01 Process release frequencies", International Association of Oil & Gas Producers, https://www.iogp.org/bookstore/product/434-00-risk-assessment-data-directory-overview/.

ISO 22899, "Determination of the resistance to jet fire fires of passive fire protection materials", International Standards Organization, Geneva, Switzerland.

Lee, R. L., J. R. Albritton, D. L. Ermak, and J. Kim, 1995, "Computational Fluid Dynamics Modeling for Emergency Preparedness and Response", International Conference and Workshop on Modeling and Mitigating the Consequences of Accidental Releases of Hazardous Materials, September 26-29, New Orleans, LA, American Institute of Chemical Engineers.

NAS, National Academy of Sciences, http://www.nasonline.org/.

NIOSH, National Institute of Occupational Safety and Health, https://www.cdc.gov/niosh/idlh/default.html.

NFPA 68, "Standard on Explosion Protection by Deflagration Venting", National Fire Protection Association, https://www.nfpa.org/codes-and-standards/all-codes-and-standards/list-of-codes-and-standards/detail?code=68.

NFPA 2016, Ivings, et. al., "Evaluating vapor dispersion models for safety analysis of LNG facilities", National Fire Protection Association Research Foundation, Quincy, Massachusetts.

NIST 2005, "Final Report of the Collapse of the World Trade Center Towers", National Institute of Standards and Technology, September, https://nvlpubs.nist.gov/nistpubs/Legacy/NCSTAR/ncstar1.pdf.

PHAST, DNVGL, https://www.dnvgl.com/software/services/phast/index.html.

Rijnmond Public Authority 1982, "Risk Analysis of Six Potentially Hazardous Industrial Objects in the Rijnmond Area: A Pilot Study", Dordrecht, The Netherlands and Boston: D. Reidel, ISBN 90-277-1393-6.

SAFER One, Safer Systems, https://safersystem.com.

SAFER TRACE, Safer Systems, https://safersystem.com.

Safesite, BakerEngineering and Risk Consultants, https://www.bakerrisk.com/products/software-tools/safesite/.

SAFETI, DNVGL, https://www.dnvgl.com/services/qra-software-safeti-1715.

SLAB, EPA, https://19january2017snapshot.epa.gov/scram/air-quality-dispersion-modeling-alternative-models_.html#hgsystem.

Slade 1968, "TID-24l90: Methodology and Atomic Energy", U.S. Air Resources Laboratory and Division of Reactor Development and Technology, U.S. Atomic Energy Commission, Washington, D.C.

SuperChems, ioMosaic, https://www.iomosaic.com/services/enterprise-software/process-safety-office/superchems.

World Bank 1985, *Manual of Industrial Hazard Assessment Techniques*, ed. P. J. Kayes, Office of Environmental and Scientific Affairs, World Bank, Washington, D.C.

14
Risk Assessment

14.1 Learning Objectives

The learning objectives of this chapter are:

- Understand how to obtain frequency and probability data,
- Identify resources for frequency data,
- Understand the concept of a Risk Matrix and how it is used,
- Understand the concept of Layer of Protection Analysis (LOPA),
- Understand the basics of Quantitative Risk Assessment (QRA), and
- Understand the concept of risk tolerability.

14.2 Incident: Phillips 66 Explosion Pasadena, Texas, 1989

14.2.1 Incident Summary

At about 13:00 on October 23, 1989, a massive explosion followed by a major fire occurred at the Phillips 66 Chemical Company at Pasadena, Texas. Twenty-three workers were fatally injured, and more than 130 workers were injured. (CCPS 2008)

Key Points:

Safe Work Practices - Especially during maintenance work, it is important to "keep it in the pipes". Ensure that isolation from the process is secure and dependable.

Contractor Management - It is to everyone's benefit to make sure contractors are well managed. Their employees may be doing the work, but the operating company still owns the risk.

14.2.2 Detailed Description

The explosion occurred at a polyethylene reactor; a large diameter loop reactor fitted with a propeller-like driver to keep the fluid in motion. The explosion occurred while maintenance work was being done to clear a blockage in a settling leg at the bottom of a reactor loop. In this process, polyethylene particles are formed in a series of continuous reaction loops from ethylene gas dissolved in isobutane held as a liquid at elevated temperature and pressure. These particles fall to the bottom of the loops where they are collected in vertical settling legs. Each reactor loop has six settling legs, each fitted with a single DEMCO® ball valve fitted with an air operated actuator to isolate it from the reaction loop. The nature of the process is such that the settling legs become frequently blocked with plugs of solid polyethylene particles which have to be removed.

On the previous day, a maintenance crew was allocated the task of unplugging three of the six settling legs on Reactor Loop 6. Under Phillips's written procedures operations personnel were required to prepare the settling leg for unblocking by closing the DEMCO®

valve and removing the air hoses used to rotate the air actuator. The legs were then handed over to a contractor to be dismantled to remove the blockage.

The first leg was unblocked without incident. At approximately 08:00 on October 23, work started to unblock the second leg. After dismantling the leg, a polyethylene plug was removed from one section of the leg. However, a plug also remained lodged in the leg, 300 - 500 mm (12 - 18 in) below the DEMCO® valve. A contractor's employee went to the control room to ask for assistance. A short time later the release occurred, with 5 witnesses observing the vapor release from the disassembled leg. Because of the high operating pressure, the reactor loop dumped virtually all of its 40 tons of flammable contents within seconds. A massive vapor cloud was formed which was ignited about 90 seconds later from an unknown source as seen in Figure 14.1. Two other major explosions occurred, one 10 – 15 minutes after the initial release when two 76,000 l (20,000 gal) isobutane storage tanks failed, and the other 25 – 45 minutes into the event when another polyethylene reactor loop failed catastrophically.

Figure 14.1. Phillips 66 Company Houston Chemical Complex explosion (OSHA 1990)

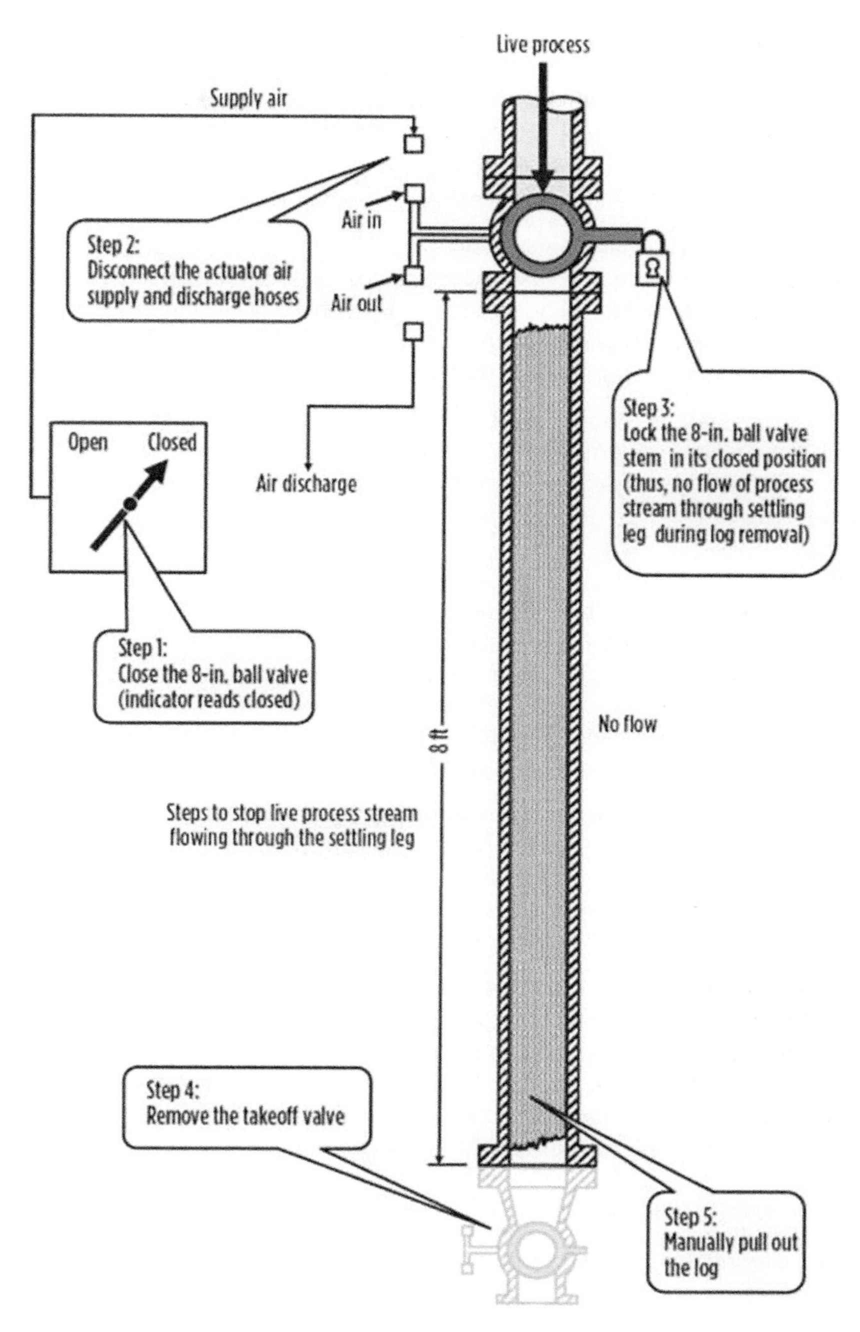

Figure 14.2. Illustration of a plugged settling leg prepared for maintenance (Bloch 2019)

After the incident, a physical examination of the DEMCO® valve revealed that it was open at the time of release, and that air hoses which supplied the air pressure to open or close the valve were improperly connected in the reverse position. The air hose connectors for the open and closed sides of the valve actuator were identical allowing the hoses to be cross connected with the result that the valve would open when they were reconnected in this manner. The DEMCO® valve was fitted with a "lock out" device to safeguard isolation during maintenance, but this was not in place and in any event could also be used to lock the valve into the open position and was thus considered by the investigators to be inadequate.

The company's corporate safety procedures and standard industry practice require back up protection in the form of a double valve, which can be locked in the closed position with the intervening space vented, or line blind inserted between flanges whenever a process line connected to operating process plan is opened. However, at Phillips 66 Pasadena, a local plant safety procedure for this work was in place that did not require the form of back up to be used.

The OSHA investigation report found that the company, in respect to their own and contractor employees, did not ensure that proper safety precautions were observed during maintenance operations, including the unblocking of settling legs, by enforcing an effective work permit system.

14.2.3 Lessons

Safe Work Practices. A formal task or Job Safety Analysis (JSA)carried out prior to work commencing, or even as part of a generic procedure for this type of frequent maintenance work would be expected to critically examine the isolation arrangements and stipulate the necessary precautions to be taken. A risk assessment would have revealed the inadequacies of the valve "lock out" device, when considering the hazardous consequences associated with opening the valve while maintenance work was being carried out.

Management of Change. The local operating management had put in place a procedure for isolating the settling leg from the live process system that was in direct violation of the company's corporate process safety rules and industry standards. Much of the process plant was built before the standards were developed. As part of an MOC, such a case would be reviewed to make recommendations to address the situation.

Contractor Management. Much concern was voiced over the use of contactors for essential and high hazard maintenance activities and the contractor company involved in this incident was fined by OSHA. While the use of contactors is seen as a commercially attractive way to resource skills and tools, companies employing contractors are accountable and responsible for the contractor's performance on site. Contracts should be arranged in such a way that safety performance is not compromised by commercial considerations.

14.3 Risk Analysis Overview

The CCPS RBPS element Hazard Identification and Risk Assessment spans from identification of a hazard to assessment of the risk. Hazard identification was discussed in Chapters 4, 5, 6, and 8. Consequence analysis which addresses "how bad can it be?" was addressed in Chapter 13. This chapter will answer the next question "how often can it happen" and then combine the "how often" and "how bad" to estimate the risk. Risk was defined in Chapter 1 as follows.

Risk = Frequency x Consequence

The typical flow of a risk analysis is illustrated in Figure 14.3. This simple approach is applicable to many types of risk although this book focuses on process safety risks.

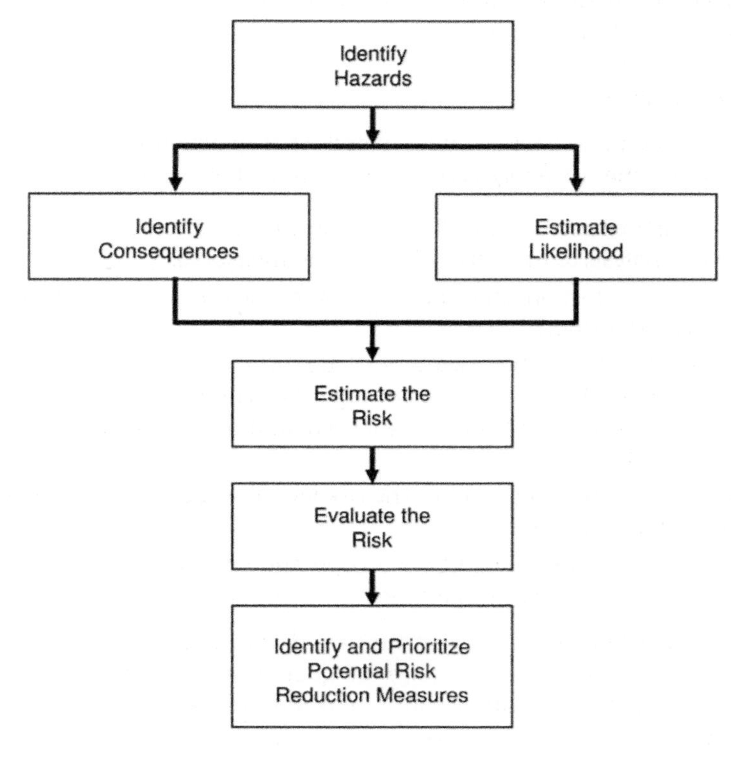

Figure 14.3. Risk analysis flowchart
(CCPS 1999)

Risk analysis is used to support risk management activities intended to reduce the risk to people, property, and the environment. It is important to understand the distinction between a few terms.

Risk Analysis - The estimation of scenario, process, facility and/or organizational risk by identifying potential incident scenarios, then evaluating and combining the expected frequency and impact of each scenario having a consequence of concern, then summing the scenario risks if necessary to obtain the total risk estimate for the level at which the risk analysis is being performed. (CCPS Glossary)

Risk Assessment - The process by which the results of a risk analysis (i.e., risk estimates) are used to make decisions, either through relative ranking of risk reduction strategies or through comparison with risk targets. (CCPS Glossary)

Risk Management - The systematic application of management policies, procedures, and practices to the tasks of analyzing, assessing, and controlling risk in order to protect employees, the general public, the environment, and company assets, while avoiding business interruptions. Includes decisions to use suitable engineering and administrative controls for reducing risk. (CCPS Glossary)

14.3.1 Risk Analysis Purpose

Before embarking on a risk analysis, understand what question the analysis is intended to answer. For example, the risk analysis could focus on the following questions.

- Identify major contributors to risk. This can support prioritization of activities as well as further analysis to evaluate risk reduction measures.
- Compare risk reduction alternatives to determine how best to spend resources for the best risk reduction.
- Define approval level. The risk level may be used to define the level of authority in a company required to make decisions regarding continued operation or risk reduction investment decisions. The level of authority would be commensurate with the level of risk.
- Evaluate the risk of an operation. The risk level may be used to compare the relative risk of company operations.
- Compare to risk criteria. The risk of an operation may be compared to company or regulatory risk tolerability targets. (Refer to Section 14.6)

The first four points in this list support business decisions of where to focus resources and who makes decisions. These do not require full quantification of the risk, but quantification may be useful for all of these. Examples include where severe consequences are possible or where the cost of risk reduction alternatives is high and qualitative judgments may be considered insufficient. The relative ranking of the risks or a semi-quantitative approach is sufficient to answer these questions.

The last point requires full quantification and the use of risk criteria. This distinction is made with respect to the understanding what question the risk analysis is intended to answer. If the risk analysis results are not intended to be compared to risk tolerability targets, then it is not necessary to generate the fully quantified data. This point is made because some regulatory regimes require comparison to risk tolerability criteria whereas other regimes do not. Additionally, some societies find quantified risk results to be sensitive and working with them challenging within their local legal environment.

14.3.2 Risk Analysis Quality

The importance of spending time to gather accurate and relevant data to support consequence analysis was discussed in Chapter 13.8. The same concepts apply to risk analysis. The level of complexity is increased beyond consequence analysis as the concept of frequency is included and all the scenario consequences and frequencies are combined to determine risk estimates. Many approaches are available for consequence and risk analysis, and expert risk analysts often have differing opinions on how best to approach a specific scenario. Many companies have defined guidelines for approaches, assumptions, and parameters with an aim to standardize risk analysis across the company. This supports comparing risk analysis results on an equal basis. Guidance can include software used, failure rate data, consequence modeling approaches, structural response to explosions, human impact criteria, and qualification requirements for risk analysts.

14.4 Frequency Analysis

In order to estimate the risk, the frequency of a scenario is needed along with the consequence of the scenario. A few definitions important to the topic of frequency analysis are listed.

Frequency - Number of occurrences of an event per unit time (e.g., 1 event in 1000 yr = 1 x 10^{-3} events/yr). (CCPS Glossary)

Probability - The expression for the likelihood of occurrence of an event or an event sequence during an interval of time, or the likelihood of success or failure of an event on test or on demand. Probability is expressed as a dimensionless number ranging from 0 to 1. (CCPS Glossary)

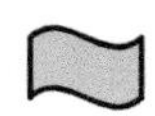

Likelihood - A measure of the expected probability or frequency of occurrence of an event. This may be expressed as an event frequency (e.g., events per year), a probability of occurrence during a time interval (e.g., annual probability) or a conditional probability (e.g., probability of occurrence, given that a precursor event has occurred). (CCPS Glossary)

Common frequency assessment techniques include the following:

- Review of historical records of similar events
- Fault tree analysis (Refer to Section 12.3.6)
- Event tree analysis (Refer to Section 12.3.7)
- Layer of Protection analysis (Refer to Section 14.7)

The first step of frequency analysis is clearly describing the incident scenario. This can be thought of as telling a story.

Scenario - A detailed description of an unplanned event or incident sequence that results in a loss event and its associated impacts, including the success or failure of safeguards involved in the incident sequence. (CCPS Glossary)

For example, the scenario story could be:

- the hazardous materials leaks from a pipe, and then,
- the materials form a vapor cloud that disperses downwind, and then,
- the vapor cloud may or may not be ignited, and if it is,
- it may flash back, or it may create a vapor cloud explosion, and finally,
- people may or may not be in the vicinity and be harmed.

Each of these steps has an associated frequency or probability. Typically, the initial loss of containment or initiating event is described by a frequency value and the subsequent steps in the scenario story are described by probabilities. Frequency analysis develops the specific frequencies associated with the specific potential scenario outcomes, as illustrated in Figure 14.4. Combining each frequency with the relevant consequence yields the risk.

14.4.1 Historical Records

This approach is typically used to quantify the initial step in the story, e.g. the loss of containment. However, it is not as simple as it sounds. Locating the data can be challenging. Companies may develop their own data sets; however, this requires significant effort and the availability of a large database in order to create a data set that is statistically valid. Publicly available data sets are described in Section 14.9.

Once data are located, they should be validated that they are applicable to the scenario being analyzed by considering points such as the following. Did the data come from analysis of the same type of equipment in the same environment? Is the data current or does it reflect equipment technology that is no longer predominately used?

14.4.2 Fault Tree Analysis

Fault tree analysis was described in Section12.3.6. A fault tree can be used to estimate incident frequencies. Fault tree analysis can calculate the hazardous incident (top event) frequency using the fault tree model of the system failure mechanisms. For example, the analyst may wish to calculate the frequency of a toxic release for a reactor overpressure, but this frequency is not available in the historical records. A fault tree could be constructed of the event using the frequency of loss of reactor cooling, the probability that safety interlocks fail, a runaway reaction occurs releasing toxic gas, and the probability that the emergency scrubber system fails. The fault tree describes the potential causes leading to the top event and that the logic (and/or gates) used. Once the fault tree structure is validated qualitatively, then the frequency of the top event may be calculated from the frequency values, probability values, and logic gates in the fault tree. FTA is well described in *Guidelines for Chemical Processes Quantitative Risk Assessment*. (CCPS 1999)

14.4.3 Event Tree Analysis

Event tree analysis was described in Section 12.3.7. An event tree can be used to quantitatively estimate the distribution of incident outcomes (e.g. explosion, flash fire, VCE, safe dispersion). In an event tree, the branches are typically a yes/no decision such as the following.

- The release ignites immediately, or it does not.
- The vapor cloud ignites sometime later, or it does not.
- People are in the impact area, or they are not.

These yes/no decisions can be described with probabilities as their sum must be equal to one. The branches may also be probabilities such as wind towards a vulnerable population. The initial incident is typically described by a frequency. The product of the initial frequency and the probabilities along each branch will result in frequencies for each of the incident outcomes.

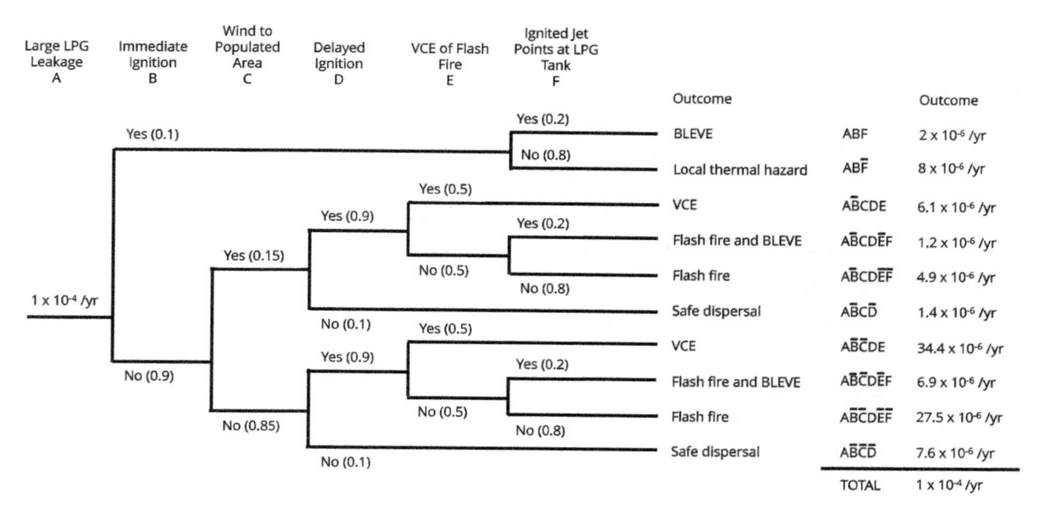

Figure 14.4. Event tree supporting frequency calculations (redrawn from CCPS 1999)

14.4.4 Human Reliability Analysis (HRA)

Human Reliability Analysis provides a means to quantify human performance issues for inclusion in fault tree analysis, event tree analysis, and risk assessment. Different HRA methods are reviewed in the HSE report "Review of human reliability assessment methods". Two of the more commonly used methods are the Technique for Human Error Rate Prediction (THERP) and the Human Error and Reduction Technique (HEART). THERP is focused on human performance issues in the nuclear power industry. (NUREG) HEART is a simple method to quantify human performance issues in any industry. (Williams 1985, 1986, 1988, and 1992)

14.5 Risk Analysis

Recalling Figure 14.3, at this point in the analysis both the consequence and the frequency of the scenario have been determined. It is now possible to estimate the risk. This can be done qualitatively, semi-quantitatively, or quantitatively.

A qualitative risk analysis may simply consider the frequency in qualitative terms such as the following.

- High, medium, or low
- Happened in this plant
- Never happened in this plant, but happened in this industry
- Never happened in this industry

A semi-quantitative risk analysis may consider the frequency in terms of broad bands such as order of magnitude bands, e.g. a fatality per 10^{-3} to 10^{-4} years, as opposed to a precise numerical value.

A quantitative risk assessment develops mathematical risk values, e.g. 3.5×10^{-5} fatalities per year. This approach is discussed in Section 14.5.2.

14.5.1 Risk Matrix

An early example of the risk matrix is in a military standard. (DOD 2012) The intent of the standard is to understand risks throughout the life cycle of an asset and use the risk matrix to evaluate when the risk level is tolerable. When the risk level is not tolerable, additional efforts can be made to eliminate or reduce the risks to a defined level and have the residual risk accepted by the appropriate level in the organization. The residual risk is the risk that remains after risk reduction measures have been implemented. For example, a refinery process unit handling flammable chemicals may be designed with high-pressure alarms, emergency shut-down systems, and fire suppression systems. This will reduce the risk of a fire, but there remains some level of residual risk.

The risk matrix combines the scenario consequence severity and scenario frequency in a simple matrix format as shown in Figure 14.5. This allows scenario risks to be compared and prioritized for allocation of resources intended to reduce the risks. In Figure 14.5, the matrix is color-coded, and the cells are given numerical designations to indicate the risk levels. Level 5 is the highest risk. Risk matrices vary from company to company. They may be three levels of consequence severity by three levels of frequency magnitude, they may be five by five, or some other dimension.

Companies often associate levels of authority within the company with the risk levels. The company authorities make decisions regarding the continued operation based on the residual risk and the approval of risk reduction measures. For example, the level of authority associated with risk level 'high' in Figure 14.5 may be a company executive whereas the level of authority associated with risk level 'low' may be a local facility manager.

The consequence severity levels are typically expressed in terms of harm to people, damage to property, and harm to the environment. Probability levels typically include both numerical frequency values as well as word descriptions that are easier for the non-risk analyst to understand. The levels suggested in the military standard are shown in Figures 14.6 and 14.7.

RISK ASSESSMENT MATRIX				
SEVERITY / PROBABILITY	Catastrophic (1)	Critical (2)	Marginal (3)	Negligible (4)
Frequent (A)	High	High	Serious	Medium
Probable (B)	High	High	Serious	Medium
Occasional (C)	High	Serious	Medium	Low
Remote (D)	Serious	Medium	Medium	Low
Improbable (E)	Medium	Medium	Medium	Low
Eliminated (F)	Eliminated			

Figure 14.5. Risk assessment matrix
(DOD 2012)

SEVERITY CATEGORIES		
Description	Severity Category	Mishap Result Criteria
Catastrophic	1	Could result in one or more of the following: death, permanent total disability, irreversible significant environmental impact, or monetary loss equal to or exceeding $10M.
Critical	2	Could result in one or more of the following: permanent partial disability,injuries or occupational illness that may result in hospitalization of at least three personnel, reversible significant environmental impact, or monetary loss equal to or exceeding $1M but less than $10M.
Marginal	3	Could result in one or more of the following: injury or occupational illness resulting in one or more lost work day(s), reversible moderate environmental impact, or monetary loss equal to or exceeding $100K but less than $1M.
Negligible	4	Could result in one or more of the following: injury or occupational illness not resulting in a lost work day, minimal environmental impact, or monetary loss less than $100K.

Figure 14.6. Severity categories
(DOD 2012)

PROBABILITY LEVELS			
Description	**Level**	**Specific Individual Item**	**Fleet or Inventory**
Frequent	**A**	Likely to occur often in the life of an item.	Continuously experienced.
Probable	**B**	Will occur several times in the life of an item.	Will occur frequently.
Occasional	**C**	Likely to occur sometime in the life of an item.	Will occur several times.
Remote	**D**	Unlikely, but possible to occur in the life of an item.	Unlikely, but can reasonably be expected to occur.
Improbable	**E**	So unlikely, it can be assumed occurrence may not be experienced in the life of an item.	Unlikely to occur, but possible.
Eliminated	**F**	Incapable of occurence. This level is used when potential hazards are identified and later eliminated.	Incapable of occurence. This level is used when potential hazards are identified and later eliminated.

Figure 14.7. Probability levels
(DOD 2012)

An attribute of the risk matrix is that it is graphical and easy for many people to understand and use. A challenge to this ease can be in the way the frequency is expressed. The frequency magnitude levels can be defined in a variety of ways, either qualitatively or semi-quantitatively. Additionally, quantitative QRA results may be plotted on a risk matrix to facilitate comparison with other risks. Quantitative risk values in terms of N fatalities per 10^{-x} per year can be challenging for people who are not risk analysts to understand. In order to retain the usefulness and understandability of the risk matrix, the frequency axis is often expressed in simpler terms relating to the life of a facility or the ability to imagine the scenario occurring. This is shown in the suggested probability levels in the military standard.

Another issue is which consequence to use. For example, in Figure 14.4 many potential outcomes are possible, each with different consequence severities and probabilities that will fall into different risk levels on the risk matrix. One approach is to select the worst-case scenario where no safeguards function as intended. Another is to consider the most optimistic outcome where all safeguards function as intended to reduce the severity and probability. The more common approach is to select a worse-credible case where safeguards fail however, the outcome is still plausible. Alternatively, two or three outcomes of increasing consequence and presumably declining frequency can be plotted. This is a simplistic approximation to an event tree highlighting multiple possible outcomes.

14.5.2 Quantitative Risk Analysis (QRA)

Quantitative risk analysis, in its fullest and most complex form, estimates the risk of all the potential scenarios. To illustrate the complexity, a QRA can include all of the following.

- All potential incident scenarios for all sources
- A variety of leak sizes for each scenario
- A variety of weather conditions for each leak size for each scenario

- All potential outcomes for each weather condition for each leak size for each scenario
- There will be a frequency associated with each outcome for each weather condition for each leak size for each scenario.

All these combinations lead to the complexity. A QRA can be simplified by selecting a smaller number of combinations. This is often done as a first step and followed with more detailed QRAs focusing on higher risks.

The basic steps of risk analysis as defined in Chemical Process Quantitative Risk Assessment (CPQRA) are as follows.

1. Define the potential event sequences and potential incidents.
2. Evaluate the incident outcomes (consequences) using tools such as vapor dispersion modeling and fire and explosion modeling.
3. Estimate the potential incident frequencies using databases, fault trees, or event trees.
4. Estimate the incident impacts on people, environment, and property.
5. Estimate the risk by combining the potential consequence for each event with the event frequency and summing over all events.

A QRA can be supported using spreadsheets. As discussed in Section 14.9, software tools are available to conduct QRAs. The risk estimate resulting from a QRA is presented in terms of individual risk or societal risk. A combination of the two provides a more complete picture of the risk.

Individual Risk - The risk to a person in the vicinity of a hazard. This includes the nature of the injury to the individual, the likelihood of the injury occurring, and the time period over which the injury might occur. (CCPS Glossary)

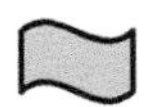

Societal Risk - A measure of risk to a group of people. It is most often expressed in terms of the frequency distribution of multiple casualty events. (CCPS Glossary)

Individual risk expresses the risk to a single person in a single location exposed to an incident or all the incidents. It is sometimes referred to as location specific individual risk (LSIR) to highlight this point. For example, the total individual risk to an individual working at a facility is the sum of the risks from all potentially harmful incidents considered separately, i.e., the sum of all risks due to fires, explosions, toxic chemical exposures, etc., to which the individual might be exposed. Individual risk is typically expressed as the frequency of fatal injuries per year. Individual risk is graphically displayed as risk contours of 10^{-6}, 10^{-7}, an 10^{-8} per year on a plot plan as shown in Figure 14.8. Individual risk appears simple, but it can be complex to interpret. Assumptions on ignition likelihood can greatly affect contour size and criteria suggested in regulations (i.e. toxicity concentrations) are suitable for emergency response but generally overstate fatality risk.

Societal risk measures the potential for effects to a group of people located in the effect zone of an incident or set of incidents. Thus, societal risk estimates include a measure of incident scale in terms of the number of people impacted.

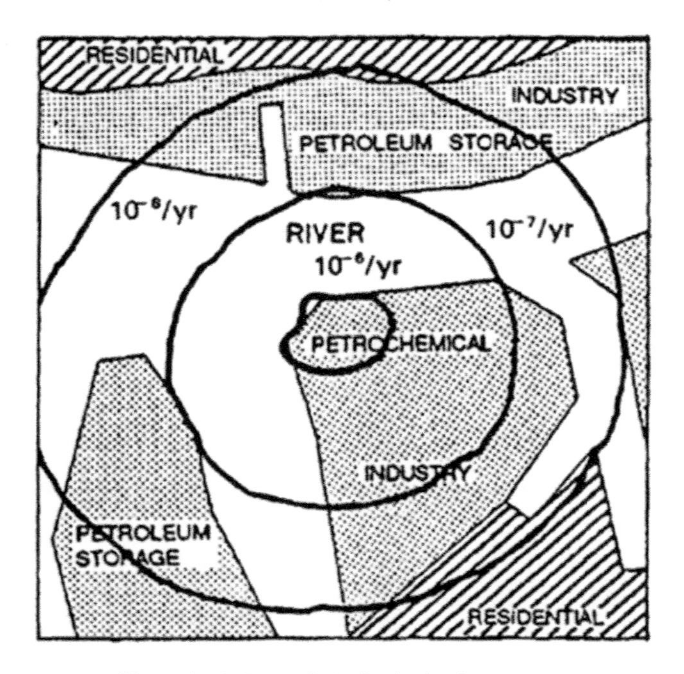

Figure 14.8. Example individual risk contours
(RPA 1982)

Societal risk is most commonly presented as F-N (Frequency - Number) curve. An F-N curve is a plot of the frequency distribution of multiple casualty events, where F is the cumulative frequency of all events leading to N or more casualties (typically expressed as the number of fatalities) as shown in Figure 14.9. F-N curves typically use log-log plots since the frequencies and number of fatalities often ranges over several orders of magnitude.

The calculation of societal risk requires the same frequency and consequence information as individual risk. Although they require the same data, they are determined in different ways such that it is not possible to calculate one from the other. Whereas individual risk requires details of an individual's occupancy within hazard zones, societal risk estimation requires a definition of the number of exposed populations within hazard zones. This definition can include factors such as the following.

- Number and geographical distribution of the population
- Population type (e.g., residential, school, industrial)
- Probability of people being present (i.e., the number of hours a day people are present)

Traditional emphasis has been on the calculation of societal risk for offsite populations; however, companies are increasingly recognizing the importance of the consideration of group risk for onsite personnel. As with individual risk, societal risk estimates are typically the summation of risk contributions from many incident outcomes.

F-N curves cover many orders of magnitude, no zero is plotted, and the N parameter must be "N or more". Because of this, they can be difficult to explain to the public and others who are not risk analysts. These plots are, however, good for decision making.

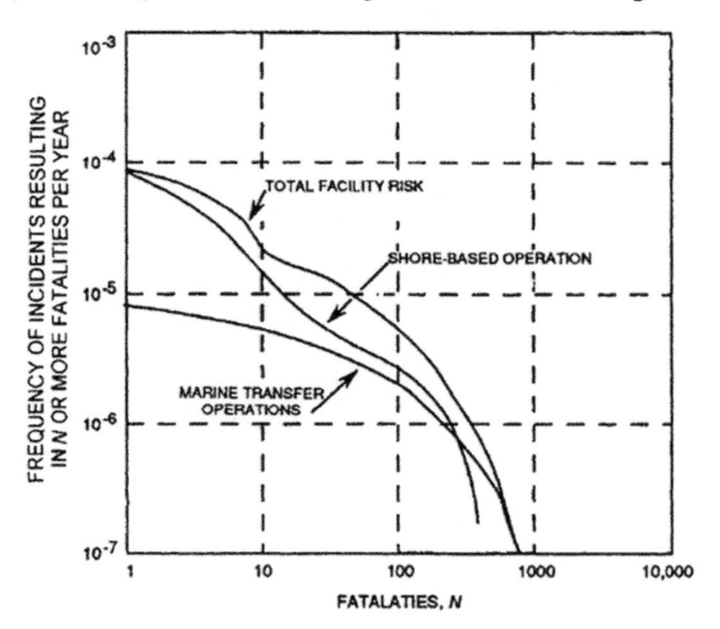

Figure 14.9. Example societal risk F-N curve

(Rasmussen 1975)

14.6 Risk Criteria

As identified in Section 14.3.1, risks can be compared to risk criteria as required by some regulations or by company practice. *Guidelines for Developing Quantitative Safety Risk Criteria* provides further detail on risk criteria. (CCPS 2009) By comparing to risk criteria, it can be decided if the risks are tolerable or if further risk reduction is warranted.

Risk Tolerance Criteria - A predetermined measure of risk used to aid decisions about whether further efforts to reduce the risk are warranted. (CCPS Glossary)

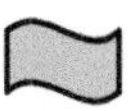

Risk criteria aids in making decisions about which risks to address first and if further risk reduction is warranted. In other words, risk criteria help inform when the risks are reduced "far enough".

An example of regulatory risk criteria is that used in Hong Kong as shown in Figure 14.10. This graph illustrates some interesting attributes of most risk criteria. First is the slope of the line. This reflects that society is more willing to tolerate higher frequency events impacting fewer people than lower frequency events impacting a greater number of people. An everyday example of this is the societal reaction to the number of traffic accident fatalities every day as compared to the reaction of a single plane crash that fatally injures hundreds of people in a single event. The Hong Kong criteria includes a cutoff at 1000 fatalities indicating that it does not permit industrial activities with the potential for accidents with greater than 1000 fatalities. This was based on its available hospital resources at the time the matrix was established.

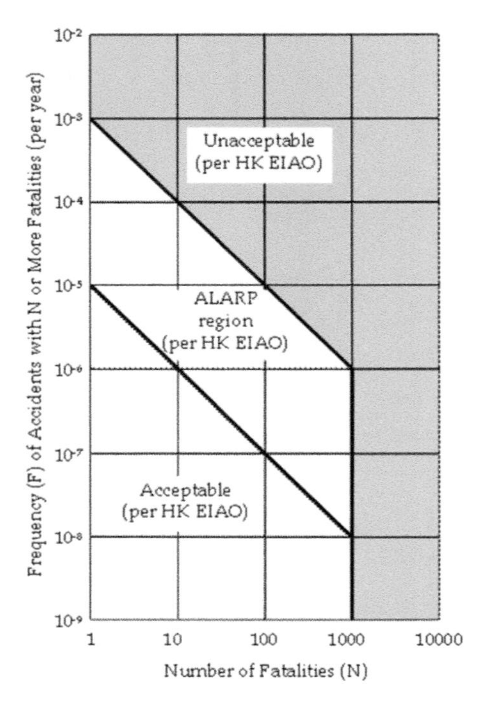

Figure 14.10. Hong Kong societal risk criteria (EPD)

Note: HK EIAO is the Hong Kong Environmental Impact Assessment Ordinance.

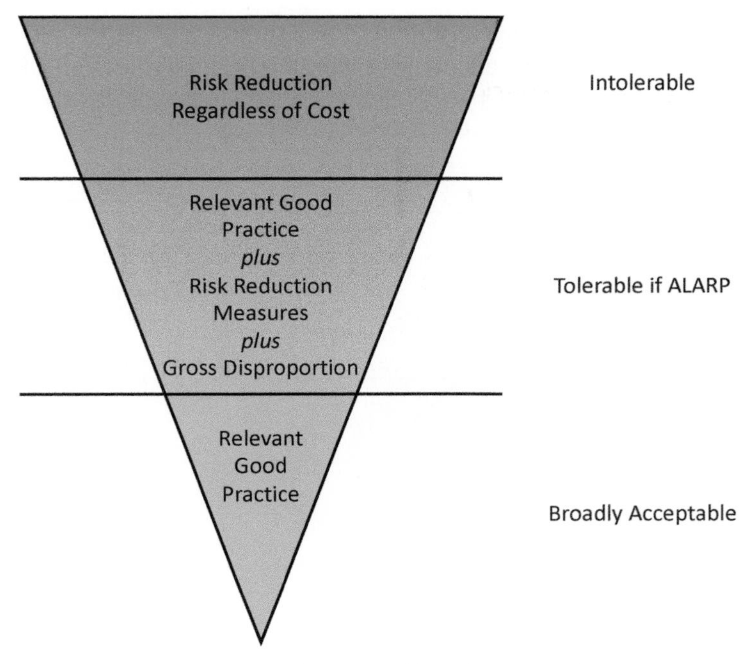

Figure 14.11. Types of ALARP demonstration
(HSE a)

The Hong Kong criteria shows that risks above a certain level (the gray area) are unacceptable. Below a certain level (bottom left), the risks are acceptable. The risks in the middle are in the ALARP region.

"ALARP" stands for "as low as reasonably practicable". The ALARP concept is illustrated in Figure 14.11. The intent is that risks in the ALARP region warrant further attention. They should be mitigated to a level, beyond which, it is not practicable to reduce the risk any further. The "practicable" aspect includes consideration of time, effort, and money. This requires a company to exercise judgment when making ALARP decisions. No simple method is available for determining if a risk is ALARP. Often cost-benefit analyses are used to aid in decision making. The Health and Safety Executive gives an example of ALARP as the following. (HSE b)

- To spend £1m to prevent five staff suffering bruised knees is grossly disproportionate; but
- To spend £1m to prevent a major explosion capable of causing 150 fatalities is proportionate.

14.7 Layer of Protection Analysis (LOPA)

LOPA is a simplified form of risk assessment. The purpose of LOPA is to determine if the scenario has sufficient layers of protection to make the scenario risk tolerable. The concept of layers of protection is illustrated in Figure 14.12.

LOPA evaluates single cause-consequence pairs, as compared to a QRA that calculates the cumulative risk. LOPA typically follows a hazard identification study which develops cause-consequence pairs for evaluation, such as a HAZOP. In a HAZOP, safeguards are identified. In LOPA, the independent protection layers (IPL) are identified from the lists of safeguards and only the IPLs are credited in reducing the risk. Safeguards that are not usually considered to be IPLs include: training, procedures, maintenance, and signage.

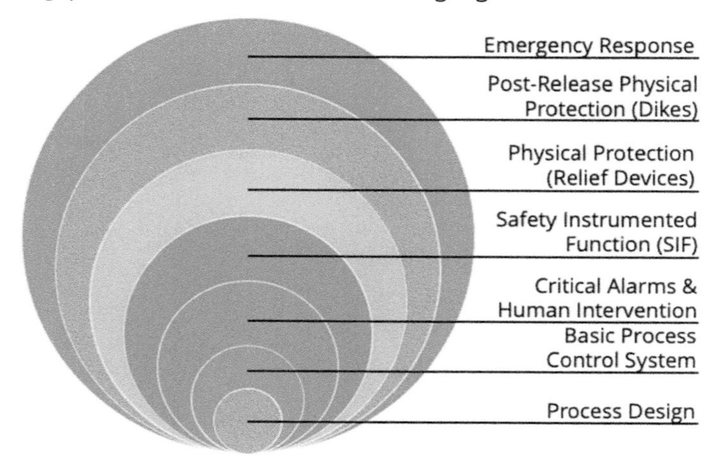

Figure 14.12. Typical layers of protection (redrawn from CCPS 2001)

Independent Protection Layer (IPL) - A device, system, or action that is capable of preventing a scenario from proceeding to the undesired consequence without being adversely affected by the initiating event or the action of any other protection layer associated with the scenario.

Note: Protection layers that are designated as "independent" must meet specific functional criteria. A protection layer meets the requirements of being an IPL when it is designed and managed to achieve the following seven core attributes: Independent; Functional; Integrity; Reliable; Validated, Maintained and Audited; Access Security; and Management of Change. (CCPS Glossary)

In order to be considered an IPL, a device, system, or action must be:

- effective in preventing the consequence when it functions as designed,
- independent of the initiating event and the components of any other IPL already claimed for the same scenario, and
- auditable; the assumed effectiveness in terms of consequence prevention and probability of failure on demand (PFD) must be capable of validation in some manner (by documentation, review, testing, etc.).

The following is an example of how the spectrum of tools may be used in supporting risk-based decision making.

- All scenarios should be analyzed using qualitative methods, such as a HAZOP or What-If analysis, where the consequence and frequency of the scenarios are estimated,
- The higher risk scenarios (10 – 20%) of the total scenarios should then be analyzed using LOPA (sometimes consequence instead of risk is used to select scenarios), and
- The highest risk scenarios (1%) should be subject to a QRA to more precisely estimate the risk and identify means to reduce it.

The steps in the LOPA method are described briefly below. Full details are provided in *Layer of Protection Analysis - Simplified Process Risk Assessment.* (CCPS 2001) Consistency in LOPA method and the values used is important in providing results that are repeatable and comparable. This is key because the results are compared to company risk criteria and used in making decisions regarding risk reduction. The numerical values in LOPA are expressed in orders of magnitude as opposed to precise values as seen in QRAs.

1. Identify how to screen the scenarios. Typically, the scenarios are selected from a HAZOP study. The scenarios selected could be based on consequence or on risk ranking.
2. Select a single incident scenario.
3. Identify the initiating event of the scenario and determine the initiating event frequency. Initiating event frequencies may be found in *Layer of Protection Analysis - Simplified Process Risk Assessment* (CCPS 2001), *Guidelines for Initiating Events and Independent Protection Layers in Layer of Protection Analysis* (CCPS 2015), or other failure frequency data sources.
4. Identify the independent protection layers and estimate the probability of failure on demand of each IPL. Probability of failure on demand values may be found in *Layer of Protection Analysis - Simplified Process Risk Assessment* (CCPS 2001) and other sources.
5. Estimate the risk of the scenario on the risk matrix by using the consequence, initiating event, and IPL data.
6. Determine if the risk should be further mitigated.

Consider the following hexane fire example.

1. A HAZOP is used as the data source and it is decided to select the scenario based on consequence.
2. A scenario involving the Hexane Surge Tank T-401 is selected. The consequence of a "high level in Hexane Surge Tank T-401" is a potential "release of hexane resulting in a fire affecting a large area. The consequence severity of this scenario was estimated to be the most severe in the company's severity categories.
3. The initiating event is "flow control valve fails open". An initiating event frequency is identified as: "Basic Process Control System Instrument Loop Failure" = 1×10^{-1} per year.

4. The HAZOP identified safeguards. However, none of the safeguards intended to prevent the tank overflow met the criteria of an IPL. A dike is provided that would contain the overflow and prevent a fire over a large area. This dike meets the IPL criteria and has a probability of failure on demand of 1×10^{-2}.
5. Combining the initiating event frequency of 1×10^{-1} per year and the probability of failure of the dike of 1×10^{-2}, the frequency is 1×10^{-3} per year.
6. Using the company's criteria, this would require risk reduction typically by the implementation of additional IPLs. The analyst could then consider the addition of a safety instrumented system to prevent the overflow. (Safety instrumented systems are discussed in Chapter 15.) With the addition of such a system with a probability of failure on demand of 1×10^{-2}, then the total probability of failure on demand would then be 1×10^{-4} and the frequency of the mitigated scenario, 1×10^{-5} per year. Per the company's risk criteria, this is a tolerable risk.

This is a very simple example. In reality, many factors can make a LOPA more complex. These include consideration of the following.

- The number of potential initiating causes in various modes of operation such that they are all accurately included and not double counted.
- Inclusion of IPLs that truly meet the requirements for an IPL.
- The use of enabling conditions or condition modifiers that are required to realize the consequence of concern, e.g. a process in recycle mode or the probability of maintenance personnel being present, respectively. A full description of these are given in *Guidelines for Enabling Conditions and Conditional Modifiers in Layers of Protection Analysis.* (CCPS 2013)

14.8 What a New Engineer Might Do

New engineers frequently participate in small unit projects or major capital projects which can include the use of hazard identification studies and risk assessments. This work can include plotting risks on a risk matrix in support of prioritization of resources all the way to gathering data for use in a QRA. As with consequence analysis, using the best data possible supports a quality QRA. Researching data sources to find current, relevant data is an important activity that is frequently supported by new engineers.

One thing that new engineers do not typically do is conduct detailed QRAs or analyze their results. The QRA results can be significantly influenced by the data, assumptions, and parameters used in the modeling. It can be easy to generate results, and it sometimes takes an experienced analyst to recognize that something is amiss in those results. Seeking the advice and review of an experienced analyst is a good approach in building risk analysis skills.

14.9 Tools

Risk analysis requires failure frequency data. Frequency analysis and risk analysis can be conducted manually or using simple spreadsheets; however, they are greatly facilitated using software packages. Resources include the following.

HSE Failure Rate and Event Data for use within Risk Assessments. The U.K. Health and Safety Executive Chemicals, Explosives and Microbiological Hazardous Division 5, has an established set of failure rates that have been in use for several years. This is available at https://www.hse.gov.uk/landuseplanning/failure-rates.pdf. (HSE Failure Rate)

IOGP Risk assessment data directory - Process release frequencies. The International Association of Oil and Gas Producers Report 434-01 provides frequencies of releases from process equipment. They are intended to be applied to equipment on the topsides of offshore installations and on onshore facilities handing hydrocarbons and can be applied to onshore facilities where hydrocarbons are processed. (IOGP 2019)

Offshore and Onshore REliability DAtabase. OREDA is a project organization with oil and gas companies as members. It is a comprehensive databank of reliability data covering process hardware, control systems, and electrical equipment that has been collected on offshore & onshore operations globally for 35 years. For non-members, access to selected data is available at https://www.oreda.com (OREDA)

Guidelines for Process Equipment Reliability Data. This book contains failure rate data for use in supporting QRAs. (CCPS 1989)

Fault Tree and Event Tree software. Many software packages are available on the internet to create fault trees and event trees, including one from Isograph.

LOPA software. LOPA may be conducted using worksheets to document scenarios and spreadsheets to calculate the values. Software tools are available such as LOPA Works from Primatech and AEShield from AE Solutions. (Primatech and AE Solutions)

SAFETI. This is a QRA software package available from DNV GL. It can be used to conduct a QRA of onshore process facilities as well as pipeline risks. SAFETI includes failure rate data similar to the IOGP data directory. (DNV GL)

Other options. Other software options include integrated PHA/ HAZOP and LOPA studies

14.10 Summary

This chapter discusses risk analysis, from qualitative to a fully quantitative. The methods include the risk matrix, LOPA, and QRA. Each of these methods has its own attributes that make it better suited for various needs. All of the methods share the purpose of assessing the risk with the intent to reduce it to a tolerable level.

The risk matrix is easy to use and easy to understand. It involves the least amount of time and resources in developing risk estimates but is a coarse approach.

LOPA can be used to analyze higher risk, or consequence, scenarios and is helpful in identifying if additional layers of protection are required or if a scenario is over-protected and resources could be saved. It is an order of magnitude approach which makes it simpler and quicker to use than a QRA but using a consistent method and values is imperative to having comparable results.

Quantitative risk assessment is typically reserved for the highest risks. It can be labor intensive, and its quality is dependent on the appropriateness of the data and parameters used. For those risk-based decisions regarding spending significant funds on project design or risk reduction measures, QRA can support prudent allocation of resources.

A challenging question for many process safety professionals, and their company colleagues, is when enough layers of protection have been implemented to yield a residual risk that is tolerable. Unless levels are defined in regulation, this can be a sensitive question. Resources are available to assist in creating criteria and precedents. Having criteria greatly supports the making of risk-based decisions and is required for LOPA.

All risk analysis is dependent on understanding both consequence and frequency. Several tools are available to support identification of frequency values including use of historical records, fault tree analysis, and event tree analysis.

14.11 Other Incidents

This chapter began with a description of the Phillips 66 Pasadena explosion. Other incidents relevant to consequence analysis include all of the incidents listed in Chapters 4, 5, and 6.

14.12 Exercises

1. List 3 RBPS elements evident in the Phillips 66 Pasadena explosion summarized at the beginning of this chapter. Describe their shortcomings as related to this accident.
2. Considering the Phillips 66 Pasadena explosion, what actions could have been taken to reduce the risk of this incident?
3. Use a simple fault tree to estimate the overall frequency of activation of a relief device caused by either failure of a pressure regulator or overheating of tank contents due to a failed temperature control. Use a failure frequency of 0.1/yr for the pressure regulator and 0.2/yr for the temperature control. Assume a high temperature interlock exists which shuts off heating of the temperature control with a Probability of Failure on Demand of 0.1. Show your results.
4. Use a simple event tree to estimate the frequency for an overfill event where Human Error of 0.1/year results in connection of a tank truck to an already full storage tank equipped with a high-level interlock to the feed pump of PFD=0.01 and flammable gas detection interlock to the feed valve of PFD=0.1. Show your results.
5. Estimate the frequency for a "full bore" leak (Full Leaks) of a 150 mm (5.9 in) diameter pipe with length of 1000 m (3280 ft) from Figure 14.13. Show your results.
6. For the following scenario statement, estimate the risk reduction factor or number of protective layers needed to meet a tolerable frequency. Use the risk matrix in Figure 14.14 to determine the tolerable frequency. Tank T-103, is involved in an overfill event caused by a level control failure with a subsequence airborne release of 1500 kg (331 lb) acrylonitrile. This incident may result in toxic Infiltration to a nearby occupied building which could result in up to 1 fatality. Show your results.

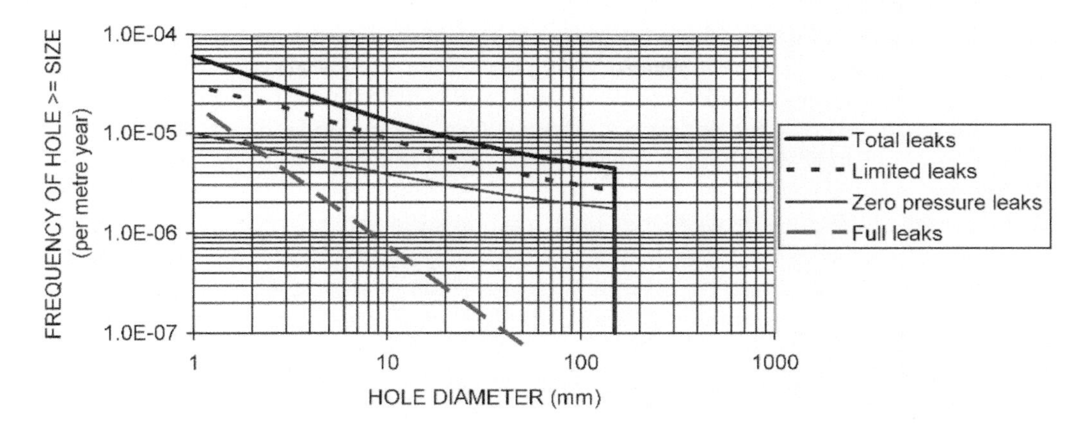

Figure 14.13. Frequency of hole sizes
(Spouge/Copyright IChemE 2006)

Risk Matrix - Consequence times Frequency

Consequence Severity Description				Frequency						
Description	Human Harm	Environmental Damage	Business Loss	10^-7/year	10^-6/year	10^-5/year	10^-4/year	10^-3/year	10^-2/year	10^-1/year
Minor	Minor Injury	Agency Reportable Incident Potential for Adverse Local Publicity	Property Damage and Business Loss < $50,000							
Moderate	Major Injury Public to Shelter or Evacuate	Contamination Confined to Site	Property Damage and Business Loss $50,000 to $500,000							
Severe	Fatality or Permanent Total Disability	Requires Significant Offsite Remediation	Property Damage and Business Loss $500,000 to $5 MM							
Extremely Severe	1 to 10 Fatalities	Major National Incident Not Recoverable for Years	Property Damage and Business Loss $5 MM to $50 MM							
Catastrophic	> 10 Fatalities	Major International Incident Not Recoverable for Decades	Property Damage and Business Loss > $50 MM							

Each company determines the Acceptable, Unacceptable and Tolerable frequences for the consequence severity adjusted to a single scenario.

Legend: Acceptable, Tolerable, Unacceptable

Figure 14.14. Risk Matrix
(First)

7. For the overfill scenario in question 4, your evaluation team has identified the following possible Protection Layers. Which one would you recommend as most effective? Explain the basis for your recommendation.
   a. SIL-1 interlock involving an analyzer to detect acrylonitrile in the air inlet to the building with automatic shutdown of the ventilation system
   b. SIL-2 interlock involving redundant high-level detection with automated shut-off of the fill pump.
   c. Operator sees material spilling into the diked area and manually closes a fill valve according to a simple well-documented procedure.
8. The scenario you are evaluating involves a large propane release resulting from a transfer hose failure into a low-congestion outdoor process area with "poor" ignition control. The cloud "footprint" covers approximately 4000 m^2 (1 acre). Estimate the probability of ignition for the following. Explain your results.

a. No known ignition sources are present
b. Fired heater is located within the potential area of the flammable cloud

9. Estimate the procedure reliability for a simple well-documented procedure assuming the operator is well-trained, under low stress, and sufficient time to diagnose the problem and execute the appropriate actions. Also assume no feedback such as a buzzer or bell indicates that the procedure has been executed properly. Explain your results.
10. Explain the difference between individual risk and societal risk.

14.13 References

AE Solutions, AEShield, https://www.aeshield.com/online-store.

Bloch 2019, "Looking Back at the Phillips 66 Explosion in Pasadena, Texas: 30 years later", Bloch, K. P., Vaughen, B. K., *Hydrocarbon Processing,* American Institute of Chemical Engineers, New York, N.Y., October.

CCPS Glossary, "CCPS Process Safety Glossary", Center for Chemical Process Safety, https://www.aiche.org/ccps/resources/glossary.

CCPS 1989, *Guidelines for Process Equipment Reliability Data*, Center for Chemical Process Safety, John Wiley & Sons, Hoboken, N.J.

CCPS, 1999, *Guidelines for Chemical Processes Quantitative Risk Assessment*, Center for Chemical Process Safety, John Wiley & Sons, Hoboken, N.J.

CCPS 2001, *Layer of Protection Analysis - Simplified Process Risk Assessment*, Center for Chemical Process Safety, John Wiley & Sons, Hoboken, N.J.

CCPS 2008, *Incidents That Define Process Safety*, Center for Chemical Process Safety, John Wiley & Sons, Hoboken, N.J.

CCPS 2009, *Guidelines for Developing Quantitative Safety Risk Criteria*, Center for Chemical Process Safety, John Wiley & Sons, Hoboken, N.J.

CCPS 2013, *Guidelines for Enabling Conditions and Conditional Modifiers in Layers of Protection Analysis*, Center for Chemical Process Safety, John Wiley & Sons, Hoboken, N.J.

CCPS 2015, *Guidelines for Initiating Events and Independent Protection Layers in Layer of Protection Analysis*, Center for Chemical Process Safety, John Wiley & Sons, Hoboken, N.J.

DNV GL, SAFETI, https://www.dnvgl.com/services/qra-software-safeti-1715.

DOD 2012, "Standard Practice, System Safety, Mil-STD-882E", U.S. Department of Defense.

EPD, https://www.epd.gov.hk/eia/register/report/eiareport/eia_1252006/html/eiareport/Part3/Section13/Sec3_13.htm

HSE a, "Guidance on ALARP Decision in COMAH", https://www.hse.gov.uk/foi/internalops/hid_circs/permissioning/spc_perm_37/.

HSE b, "ALARP at a glance, https://www.hse.gov.uk/managing/theory/alarpglance.htm

HSE c, "Review of human reliability assessment methods". U.K. Health and Safety Executive, Health and Safety Laboratory, 2009, https://www.hse.gov.uk/research/rrpdf/rr679.pdf

HSE Failure Rate, "Failure Rate and Event Data for use within Risk Assessments", U.K. Health and Safety Executive, Chemicals, Explosives and Microbiological Hazardous Division 5, https://www.hse.gov.uk/landuseplanning/failure-rates.pdf.

NUREG, U.S. Nuclear Regulatory Commission, "Handbook of Human Reliability Analysis with Emphasis on Nuclear Power Plant Applications - Final Report", CR-1278, https://www.nrc.gov/docs/ML0712/ML071210299.html.

IOGP 2019, "Risk assessment data directory - Process release frequencies", The International Association of Oil and Gas Producers Report 434-01.

First, Risk matrix, personal correspondence, 2021.

OREDA, Offshore and Onshore REliability DAtabase, https://www.oreda.com.

OSHA 1990, "Phillips 66 Company Houston Chemical Complex Explosion and Fire: A Report to the President", Occupational Health and Safety Administration, U.S. Department of Labor, Washington, D.C. 1990.

Primatech, LOPAWorks, https://www.primatech.com/software/lopaworks

Rasmussen 1975,"Reactor safety study. An assessment of accident risks in U. S. commercial nuclear power plants. Executive Summary", WASH-1400 (NUREG75/014), U.S. Nuclear Regulatory Commission.

RPA 1982, "Risk Analysis of Six Potentially Hazardous Industrial Objects in the Rijnmond Area, A Pilot Study", Rijnmond Public Authority, Springer.

Spouge/IChemE 2006, "Leak Frequencies from the Hydrocarbon Release Database", Symposium Series 151, Institute of Chemical Engineers, Rugby, England.

Williams 1985, "HEART – A Proposed Method for Achieving High Reliability in Process Operation by means of Human Factors Engineering Technology", Proceedings of a Symposium on the Achievement of Reliability in Operating Plant, Safety and Reliability Society, 16 September 1985, Southport, England.

Williams 1986, "A proposed Method for Assessing and Reducing Human Error", Proceedings of the 9th Advance in Reliability Technology Symposium, University of Bradford, England.

Williams 1988, "A Data-based method for assessing and reducing Human Error to improve operational experience", Proceedings of Institute of Electrical and Electronics Engineers 4th Conference on Human Factors in power Plants, Monterey, California.

Williams1992, "Toward an Improved Evaluation Analysis Tool for Users of HEART", Proceedings of the International Conference on Hazard identification and Risk Analysis, Human Factors and Human Reliability in Process Safety, Orlando, Florida.

15
Risk Mitigation

15.1 Learning Objectives

The learning objectives of this chapter are:

- Understand the basic methods used to mitigate process safety risks,
- Explain the difference between safeguards and IPLs,
- Identify where to find resources and tools describing risk mitigation measures, and
- Construct a Bow Tie diagram.

15.2 Incident: Celanese Explosion, Pampa, Texas, 1987

15.2.1 Incident Summary

An explosion occurred in a reactor at the Celanese Pampa, Texas plant on November 14, 1987 that led to a vapor release resulting in a vapor cloud explosion and several fires. Three fatalities and 39 injuries resulted. Extensive property damage occurred in the immediate area and severe damage occurred throughout the plant. The firehouse that contained the fire trucks was damaged so the trucks could not be driven out. Fixed firefighting equipment was damaged making it more difficult to control the fires. Figures 15.1 and 15.2 show the extent of the damage caused by the explosions (Forest, 2016).

Key Points:

Process Safety Competency – Humans are an important part of the system. Hence, it is important to understand human factors. Designing operations to help a human succeed, can also help to avoid process safety incidents.

Hazard Identification and Risk Analysis – Hazard identification methods should include scenarios involving human failures just as they do equipment failures. For cases in which a single human action or inaction may cause significant undesired consequences, the risk warrants management.

Figure 15.1. Oxidation reactor after explosion
(Celanese)

Figure 15.2. One of several units impacted by explosion
(Celanese)

15.2.2 *Detailed Description*

The Celanese plant was built in 1952 and produced acetic acid. The unit involved was a liquid phase oxidation (LPO) reactor in which butane was oxidized in the presence of air and a catalyst to make acetic acid and byproducts. This was an exothermic reaction. The reactor product was sent to several downstream units in the Pampa plant to make products that included acetic acid, acetic anhydride, and methyl ethyl ketone. The reactor operated at a relatively high temperature and pressure. Figure 15.3 is a schematic of the reactor.

On November 14, 1987, the reactor was prepared to start up following a shut down the previous day due to a problem in the steam system. Following the normal start-up process, the operators began heating the reactor contents. As the reactor approached start-up temperature, an explosion occurred in the air sparger inside the reactor. The explosion ruptured the 200 mm (8 in) diameter air piping at two places external to the reactor and one failure occurred internal to the reactor. The flammable reactor contents rapidly vaporized to the atmosphere. About 25 to 30 seconds after the initial explosion, a vapor cloud explosion occurred. The ignition source for the vapor cloud was thought to be the gas boilers that were immediately across the road from the reactor.

Extensive property damage occurred in the immediate area and severe damage occurred throughout the plant. Figure 15.4, shows the calculated extent of the flammable vapor cloud, extending to the boiler area.

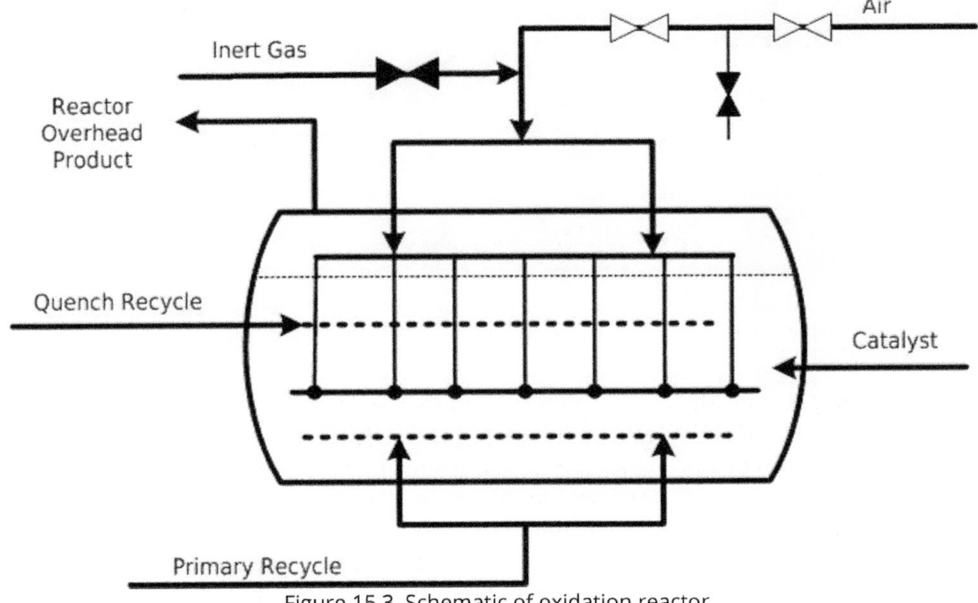

Figure 15.3. Schematic of oxidation reactor
(Celanese)

Figure 15.4. Predicted flammable vapor cloud from reactor explosion (Celanese)

In investigating this incident, it was determined that on November 13, a problem with the steam system occurred in the Pampa plant that led to the decision to shut down the reactor in question. The shutdown procedure was to close the air supply to the reactor with double block valves, open a bleed valve to further prevent air entry, and then purge the reactor with inert gas. Shutting off air and purging with inert gas were essential to make sure the reactor atmosphere was not flammable and to prevent backflow of the reaction mixture into the air line. Three ways to shut down the reactor included:

- a shutdown system designed to automatically shut down if safe limits were exceeded.
- a manual button that activated the shutdown system

- three manual buttons: one button to activate the double block, another to activate the bleed, and a third to activate the purge.

On the day of the incident, the operator chose to shut down the reactor using the three manual buttons on the control panel. The activation of these three buttons was equivalent to the activation of the manual shutdown button or the automatic shutdown. The first step was to close the process air valves to the reactor. The second step was to open the air bleed after the air to the reactor was blocked in. The third step was to activate the timed nitrogen purge.

The operator pushed the first two buttons, but mistakenly did not push the inert gas purge button. The standard operating procedure, or SOP for this critical step was not followed by the operator. Failure to initiate the inert gas purge allowed the flammable contents of the reactor, including the catalyst, to enter the air sparger system. The air sparger pipe contained an inventory of reactor contents for about a day.

As the reactor was started up on November 14 and approached start-up temperature, an explosion occurred in the air sparger inside the reactor. Oxygen was available because the reactor had not been purged, fuel was available from the reactor contents, and the ignition source was probably the catalyst that was plated on the inside of the air sparger.

1987 was a time when industry was converting from pneumatic control to computer control. The use of push buttons was common. As a result of the Pampa incident, Celanese implemented a detailed risk assessment and risk management methodology to identify and mitigate risks including the ones described at Pampa. Another aspect of the process safety management system includes rigorous controls around safety instrumented systems designed to mitigate similar hazards.

15.2.3 Lessons

Process Safety Competency. The shutdown system activated an indicator light when the shutdown started and another light when the shutdown and purge were complete when either the automatic system or the one button manual system was activated. When the three-button manual shutdown was used, the system gave no status feedback. In order to detect the lack of inert gas purge, the next shifts would have had to detect the absence of the purge from the computer activity log printed in another room. Current knowledge of human performance recognizes that one should not design a system in which a single error can lead to potential catastrophic consequences. Also, there should be obvious feedback systems for operations for critical actions.

Hazard Identification and Risk Analysis. The independent, manual shutdown buttons were identified in a Process Safety Review prior to the event as a potential source of human performance issues which could adversely impact shutdown, but changes were not recommended. The consequences of not doing the purge were well understood, but because the scenario was well known, the review team underestimated the likelihood of the error. The initiating event was that the operator neglected to start the inert gas purge cycle during shutdown.

15.3 Safeguards, Barriers, IPLs, and Other Layers of Protection

The concept of layers of protection was discussed in Section 14.7, Layer of Protection Analysis, and illustrated in Figure 14.10. Over the years, there have been many terms used to describe these layers including safeguards, barriers, and independent protection layers. Although these terms sound similar, they do have different attributes as defined below and in Figure 15.5.

Layers of protection - Independent devices, systems, or actions which reduce the likelihood and severity of an undesired event. (CCPS Glossary)

Safeguard - Any device, system, or action that interrupts the chain of events following an initiating event or that mitigates the consequences. (CCPS Glossary)

Barrier - A control measure or grouping of control elements that on its own can prevent a threat developing into a top event (prevention barrier) or can mitigate the consequences of a top event once it has occurred (mitigation barrier). A barrier must be effective, independent, and auditable. (CCPS Glossary)

Independent Protection Layer (IPL) - A device, system, or action that is capable of preventing a scenario from proceeding to the undesired consequence without being adversely affected by the initiating event or the action of any other protection layer associated with the scenario.

Note: There are specific functional criteria for protection layers that are designated as "independent." A protection layer meets the requirements of being an IPL when it is designed and managed to achieve the following seven core attributes: Independent; Functional; Integrity; Reliable; Validated, Maintained and Audited; Access Security; and Management of Change. (CCPS Glossary)

The term safeguard is a generic, all-encompassing term that is frequently used in HAZOPs to capture all the potential layers of protection. The term barrier is more specific in that a barrier prevents the incident or mitigates the consequences on its own. The term is used in Bow Tie Analysis and this specificity supports the integrity of Bow Ties. The term IPL was introduced in Section 14.7. It is even more precise which is required to support the integrity of a Layer of Protection Analysis.

Many devices, systems, and actions are used to reduce process safety risk and can be described by these terms. These are discussed in Section 15.4.

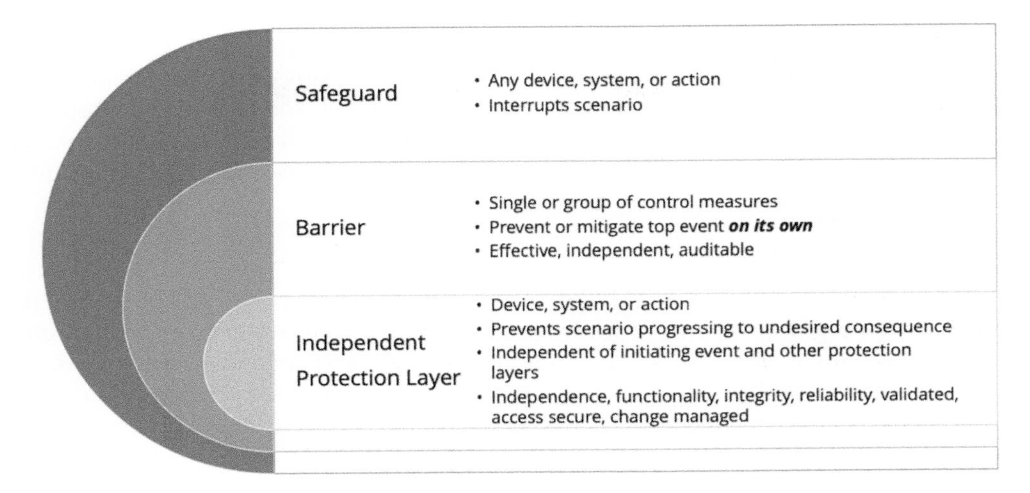

Figure 15.5. Terminology describing layers of protection

15.3.1 Swiss Cheese Model

James Reason (1990 and 1997) developed the Swiss cheese model which uses layers of Swiss cheese to represent layers of protection. The layers of protection can protect the hazard being realized and the consequence from occurring. The holes in the Swiss cheese indicate that these layers may have weaknesses and degrade over time and thus are not 100% effective. When the holes in the Swiss cheese align, representing each layer being compromised, then the consequence can occur. A Swiss cheese model is shown in Figure 15.6.

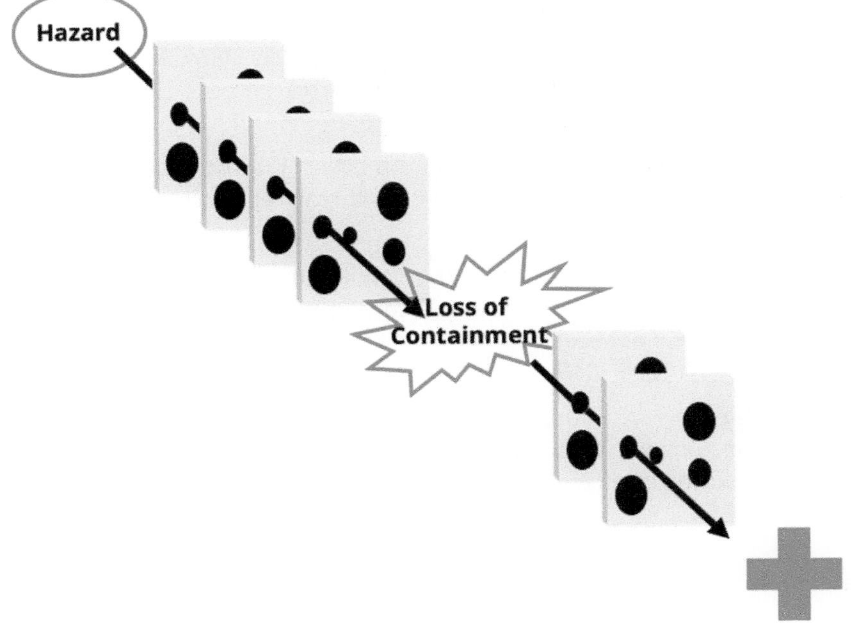

Figure 15.6. Swiss cheese model

The Swiss cheese model can help managers and workers in process industries understand the events, failures, and decisions that can cause an incident or near miss to occur by identifying key layers of protections and noting how they can fail (and providing appropriate defenses). The model illustrates that incidents are typically the result of multiple failures and highlights the importance of maintaining the integrity of all the layers through process safety management system. (Refer to Chapters 2 and 3.)

15.3.2 Bow Tie Analysis

The Bow Tie Model supports risk management and is a very good communication tool. The bow tie is graphical and qualitative which lends it to be easily understood by employees and external stakeholders. Bow Ties facilitate operational and maintenance teams in understanding how their day-to-day activities relate to risk management and barrier integrity. Bow Ties also illustrate a full picture of a risk as opposed to a single cause-consequence as illustrated in a Swiss Cheese Model.

A Bow Tie model is shown in Figure 15.7. The undesired top event is located in the middle. The threat legs are to the left of the top event and include the prevention barriers that can prevent the top event from occurring. The threat legs leading to the top event can be thought of as being similar to a Fault Tree Analysis as discussed in Section 14.4.2. The consequence legs are located to the right of the top event and include the mitigation barriers that can mitigate the various consequences resulting from the top event. The top event and the consequence legs can be thought of as being similar to Event Tree Analysis as discussed in Section 14.4.3. All of the prevention and mitigation barriers are subject to degradation (recall the holes in the Swiss Cheese Model).

Bow Tie Model - A risk diagram showing how various threats can lead to a loss of control of a hazard and allow this unsafe condition to develop into a number of undesired consequences. The diagram can show all the barriers and degradation controls deployed. (CCPS 2018)

Prevention Barrier - A barrier) located on the left-hand side of bow tie diagram and lies between a threat and the top event. It must have the capability on its own to completely terminate a threat sequence. (CCPS 2018)

Mitigation Barrier - A barrier located on the right-hand side of a bow tie diagram lying between the top event and a consequence. It might only reduce a consequence, not necessarily terminate the sequence before the consequence occurs. (CCPS 2018)

Degradation Factor - A situation, condition, defect, or error that compromises the function of a main pathway barrier, through either defeating it or reducing its effectiveness. (CCPS 2018)

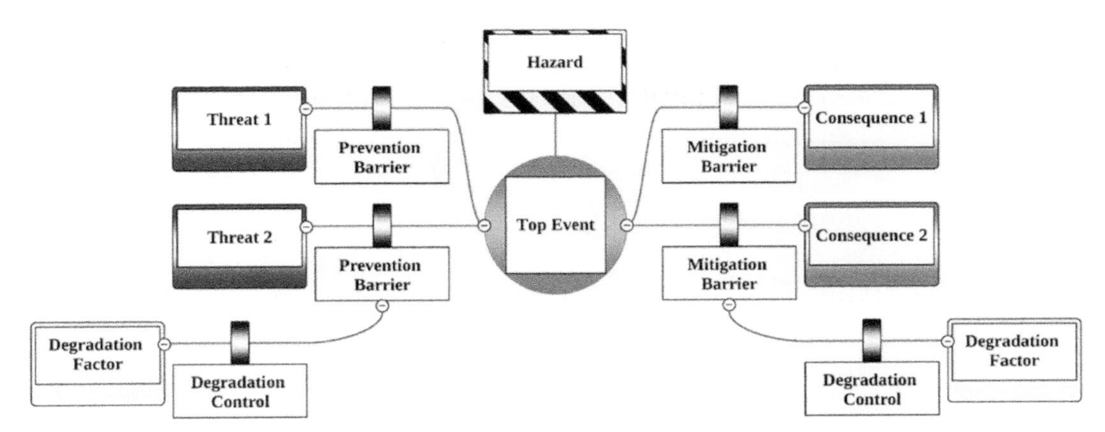

Figure 15.7. Example bow tie model
(CCPS 2018)

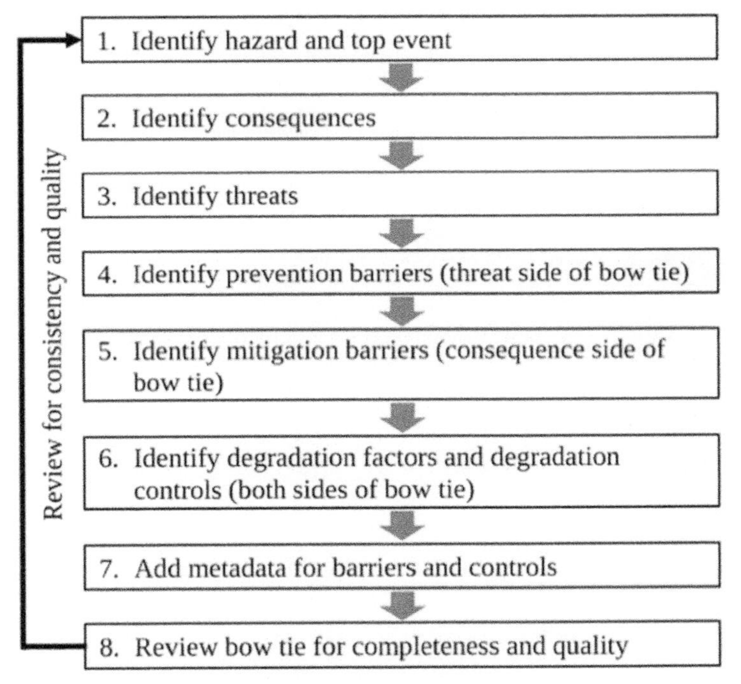

Figure 15.8. Steps in constructing a bow tie model
(CCPS 2018)

The steps in creating a Bow Tie Model are listed in Figure 15.8. A Bow Tie Model can get complex as all the threat legs and consequence legs are included. An example is shown in Figure 15.9 a and b. When constructing a Bow Tie Model, recall the definition of barrier – that it functions on its own. It is tempting to include a long list of safeguards and that would be incorrect and result in an unmanageable and confusing Bow Tie Model. For example, most human and organizational factors are controls for degradation factors and do not qualify as a full barrier. Thus, procedures and training are not barriers on their own, but they are degradation controls intended in support of an operator action which may be a barrier.

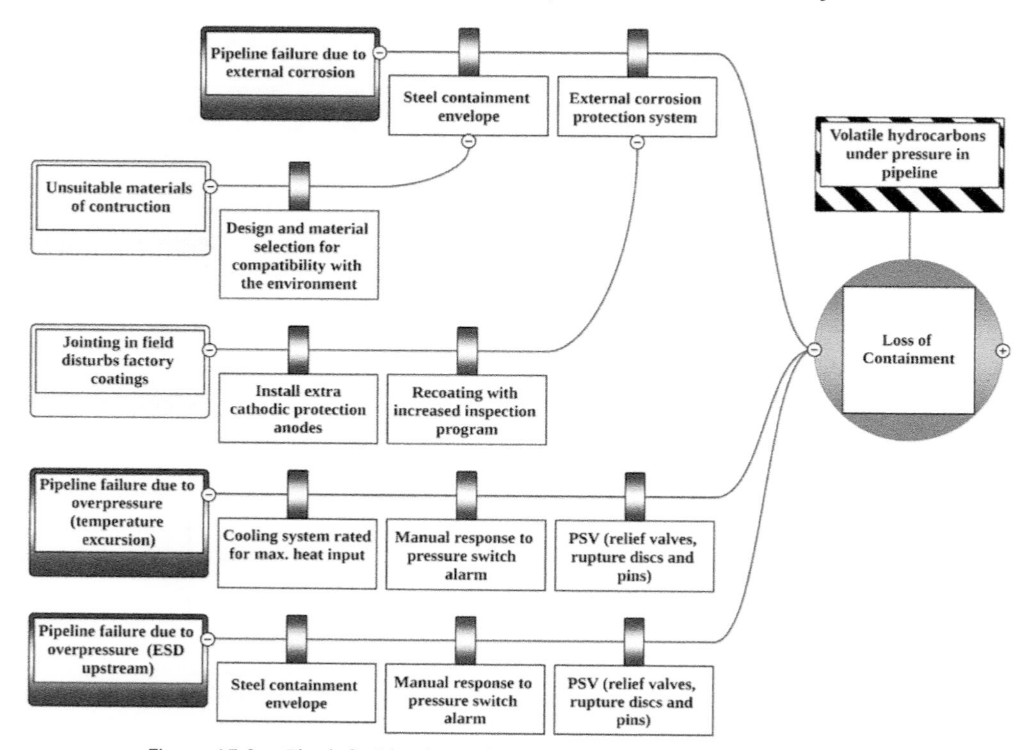

Figure 15.9 a. The left side (threat legs) of a bow tie for loss of containment (CCPS 2018)

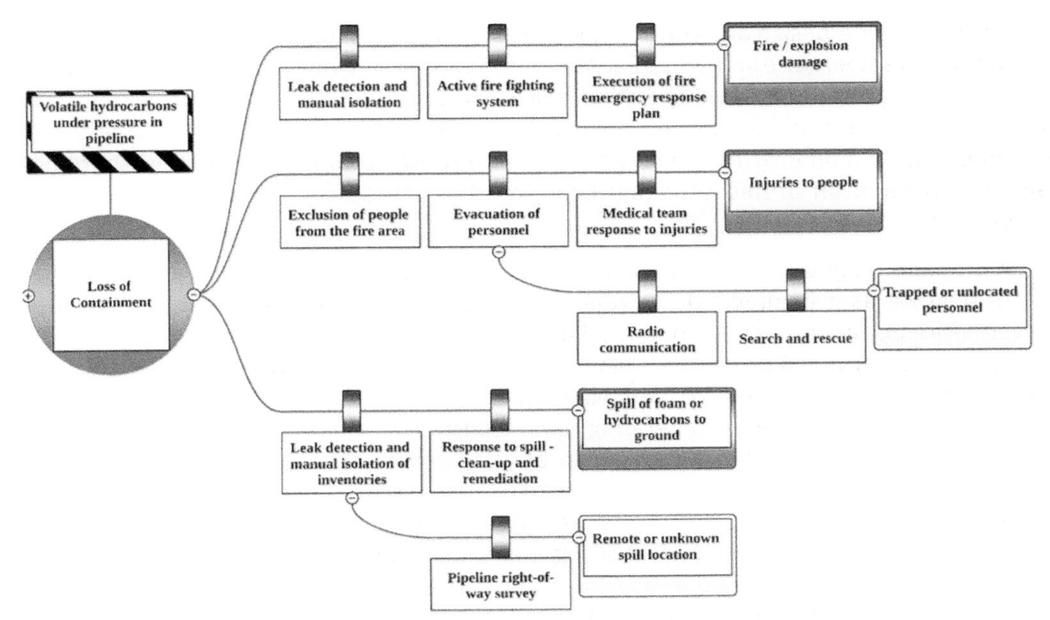

Figure 15.9 b. The right side (consequence legs) of a bow tie for loss of containment (CCPS 2018)

15.4 Risk Reduction Measures

Processes that pose risk are provided with risk reduction measures. These measures may prevent an incident from occurring or mitigate the consequences as illustrated in Bow Tie Analysis. They may be safeguards, barriers, or IPLs as discussed in section 15.3. Additionally, they may be passive or active.

Passive Hardware - A barrier system that is continuously present and provides its function without any required action. (CCPS 2018)

Active Hardware - A barrier system that requires some action to occur to achieve its function. All aspects of the barrier detect-decide-act functions are achieved by hardware or software. (CCPS 2018)

Active barriers may be pieces of equipment, human action, or a combination of the two, for example, an operator closing a valve in response to an alarm. Table 15.1 provides a list of potential risk reduction measures and classifies them in these categories. Many of these measures are fundamental to process safety. Select the appropriate risk reduction measure for the application. It is also important to recall the concepts of inherently safer design (Section 10.7.2) and the hierarchy of controls (Section 10.7.3) when making decisions on the implementation of risk reduction measures.

Not all measures meet the requirements to be considered an IPL; however, that does not make them less important in the overall management of process safety. Items such as training, procedures, and signage are key aspects of preventing process safety incidents.

The mechanical design and control of process equipment should maintain the process at the operating temperature and pressure. Alarms are typically provided to alert operators when a temperature, pressure, or level operating limit is being reached. The operators can then take action to bring the process back to normal operating conditions. If the level continues to exceed limits, then an instrumented system may be provided to take action (e.g. shut/open valves) to prevent an unsafe condition being reached. These concepts were discussed in Section 10.5.

A pressure relief valve is an active mitigation device that is often referred to as the last line of defense as it is intended to protect equipment should all other systems fail. If an overpressure occurs in a vessel and associated piping, pressure relief valves are designed to open at a predetermined pressure to allow the pressure to be relieved before the equipment fails. The three common types of relief valves are conventional, balanced bellows, and pilot operated. They have different operating characteristics that make them appropriate for different operating conditions. Common to all is that they are designed to operate at a specified set pressure that is related to the MAWP and operating limits (refer to section 10.6). In some processes, pressure relief valves discharge to a flare system where the diverted gas and fluid are safely burnt. API STD 520 Sizing, Selection and Installation of Pressure-relieving Devices and API STD 521 Pressure-relieving and Depressuring Systems provide guidance on these topics. (API STD 520 and API STD 521)

If a loss of containment occurs such as a tank overfilling, a containment dike (passive) can be provided to contain the fluid and avoid its flowing in areas containing ignition sources or environmentally sensitive areas. If a loss of containment and fire occurs in a process unit, an automatic suppression system (active) may be activated to control the fire and sloped drainage provided to limit the extent of the fire.

Occupied buildings may be protected by blast or fire walls (passive) to mitigate the potential impacts on building occupants.

Table 15.1. Typical risk reduction measures

Device, System, or Action	Prevention	Mitigation	Passive	Active	Safeguard	May be IPL if meets definition	IPL
Equipment mechanical design*	X		X				
Training	X	X		X	X		
Procedures	X	X		X	X		
Inspection, testing, and maintenance	X	X		X	X		
Communications	X	X		X	X		
Signage	X	X	X		X		
Basic Process Control System	X			X	X	X	
Critical Alarms and Human Intervention	X			X	X	X	
Containment Dike		X	X		X	X	
Blast or fire wall		X	X		X	X	
Automatic fire suppression system		X		X	X	X	
Safety Instrumented Functions	X			X	X		X
Pressure relief valves	X			X	X		X

*Equipment mechanical design is fundamental to the inherently safer design of a process although it may not fit the definition of a layer of protection.

One specifically defined type of risk reduction measure is the Safety Instrumented Function (SIF). Historically generic terms such as safety interlock and emergency shutdown system were used. With the creation of "Functional Safety of Electrical/Electronic/Programmable Electronic Safety-related Systems" (IEC 61508), terminology was defined, and requirements specified for these systems and other instrumented functions.

Safety Instrumented Function (SIF) - A system composed of sensors, logic solvers, and final control elements for the purpose of taking the process to a safe state when predetermined conditions are violated. (CCPS Glossary)

Safety Instrumented System (SIS) - A separate and independent combination of sensors, logic solvers, final elements, and support systems that are designed and managed to achieve a specified safety integrity level. A SIS may implement one or more Safety Instrumented Functions (SIFs). (CCPS Glossary)

Safety Integrity Level (SIL) Discrete level (one out of four) allocated to the SIF for specifying the safety integrity requirements to be achieved by the SIS. (CCPS Glossary)

IEC 61508 is applicable to all industries and describes methods to apply, design, deploy and maintain instrumented safety-related systems. IEC 61511 supports IEC 61508 with a focus on the process industry. ANSI/ISA 84.00.01 is essentially the same as IEC 61511.

IEC 61508 defines an engineering process called the safety life cycle intended to maintain the integrity of the safety instrumented system over its life cycle. This includes direction on assessment, design, and verification of SIS, amongst other topics.

Safety integrity levels are defined as shown in Table 15.2. SIL can be thought of as a performance level for the safety instrumented function. A typically refinery process unit likely has several SIL 2 SIFs protecting the highest risk process units. SIL 4 systems are seen in the protection of nuclear power plants.

Table 15.2. Safety integrity levels (IEC 61508)

SIL	Low-Demand Mode Probability of Failure on Demand (PFD_{avg})	High-Demand (Continuous) Mode Probability of Failure on Demand (PFD_{avg}) per hour
4	$>=10^{-5}$ to $<10^{-4}$	$>=10^{-9}$ to $<10^{-8}$
3	$>=10^{-4}$ to $<10^{-3}$	$>=10^{-8}$ to $<10^{-7}$
2	$>=10^{-3}$ to $<10^{-2}$	$>=10^{-7}$ to $<10^{-6}$
1	$>=10^{-2}$ to $<10^{-1}$	$>=10^{-6}$ to $<10^{-5}$

15.5 What a New Engineer Might Do

A new engineer will likely be engaged in hazard identification and risk analysis studies, MOCs, or design activities where safeguards, barriers, and IPLs are identified and evaluated. Understanding the difference between these terms and making sure that only the appropriate measures are credited is important to the integrity of LOPA studies and SIF specification.

New engineers may be involved in explaining process safety concepts or initiatives to operators, maintenance technicians and others. Models such as the Swiss Cheese Model and Bow Tie Model are simple to create and easy to understand and can be very helpful in process safety communication.

New engineers may be tasked with engineering calculations related to the specification of common risk reduction measures such as the sizing of pressure relief valves and the specification of instrument settings and actions. Remember that these are important steps in the reduction of process safety incidents.

15.6 Tools

Risk analysis requires failure frequency data. Frequency analysis and risk analysis can be conducted manually or using simple spreadsheets; however, they are greatly facilitated using software packages. Resources include the following.

IEC 61511, *Functional safety - Safety instrumented systems for the process industry sector.* This standard provides complete definition and instruction related to Safety Instrumented Systems in the process industries. (IEC 61511)

Bow Tie Software. It is possible to use tools such as PowerPoint to draw Bow Tie models, but this can be challenging. Software packages specifically designed to support the drawing and use of Bow Tie Models include:

- Bow Tie Software. BowTie Pro from BowTie Pro Ltd. (BowTie Pro), and
- BowTieXP & IncidentXP CGE Risk Management Solutions, Netherlands www.cgerisk.com. (BowTieXP)

15.7 Summary

The progression of a risk scenario can be interrupted by the provision of prevention measures to prevent the potential incident and mitigation measures to lessen potential consequences.

This risk scenario progress is illustrated by the use of layers of Swiss cheese in the Swiss cheese model where the holes in the cheese represent that the layers are subject to failure. A more complex, but still simple and easy to understand model, is the Bow Tie Model where all the potential threat pathways leading to a potential incident and all the potential consequences resulting from a potential incident are included. Bow Tie analysis uses specific terminology where the preventive and mitigative barriers are sufficient to function *on their own*.

Many prevention and mitigation measures can be implemented to manage risk. All are important in the overall aim to reduce process safety incidents. Each has its own attributes that may make it more appropriate in a given risk scenario. These include whether it is active or passive and whether it is a simple safeguard, a barrier, or a safety instrumented function with a defined safety integrity level. Different safeguards result in different levels of risk reduction. In some cases, additional safeguards may result in diminishing reduction of overall risk as well as add unnecessary cost and complexity.

Engineers are often involved in projects where risk is considered, and risk reduction measures are specified. The concepts of inherently safer design and the hierarchy of controls (Sections 10.6.2 and 10.6.3) should be kept in mind when risk reduction measures are specified.

15.8 Other Incidents

This chapter began with a description of the Celanese Pampa, Texas explosion. Other incidents relevant to risk mitigation include the incidents listed in Chapters 4, 5, and 6.

15.9 Exercises

1. List 3 RBPS elements evident in the Celanese Pampa explosion summarized at the beginning of this chapter. Describe their shortcomings as related to this accident.
2. Considering the Celanese Pampa explosion, what actions could have been taken to reduce the risk of this incident?
3. Redundant instrumentation is an important design concept for improving the safety of a system (e.g. airplanes, nuclear). Give two examples where redundant instrumentation would have prevented a major chemical plant incident.
4. In a HAZOP, safeguards are identified that prevent or mitigate the consequences. In a LOPA, IPLs are credited. How are safeguards and IPLs related?
5. How are SIS, SIL and IPL related?
6. Draw a swiss cheese model for the Arkema incident in 2017 as described in the CSB investigation report (https://www.csb.gov/arkema-inc-chemical-plant-fire-/).
7. Draw a bow tie diagram for filling of a tank with a flammable chemical. The tank is filled from a pipeline. The tank is fitted with a manual level gauge and a pressure relief valve. The tank is located on the edge of facility near the main road.
8. Estimate the Probability of Failure on Demand (PFD) for a firewater sprinkler system with failure rate of 0.1 failures/year and inspection interval of 2 years. Show your results.
9. Estimate the Unavailability or Probability of Failure on Demand (PFD) for a flow controller that is continually monitored with failure rate of 0.5 failures/year and mean time to repair (MTTR) of 5 days. Show your results.
10. Table 15.1 lists some risk reduction measures. These all cost money. How would you determine which ones should be implemented in your design of a facility that handles flammable materials?

15.10 References

API STD 520, "Sizing, Selection and Installation of Pressure-relieving Devices", American Petroleum Institute, Washington, D.C., www.api.org.

API 521 "Pressure-relieving and Depressuring Systems", American Petroleum Institute, Washington, D.C., www.api.org.

BowTie Pro, BowTie Pro Ltd., www.bowtiepro.com.

BowTieXP, CGE Risk Management Solutions, www.cgerisk.com

CCPS 2018, *Bow Ties in Risk Management: A Concept Book for Process Safety*, Center for Chemical Process Safety, John Wiley & Sons, Hoboken, N.J.

CCPS Glossary, "CCPS Process Safety Glossary", Center for Chemical Process Safety, https://www.aiche.org/ccps/resources/glossary.

Celanese 2020, private correspondence

Forest 2016, private correspondence

IEC 61508, "Functional Safety of Electrical/Electronic/Programmable Electronic Safety-related Systems", International Electrotechnical Commission, Geneva, Switzerland.

IEC 61511, "Functional safety - Safety instrumented systems for the process industry sector", International Electrotechnical Commission, Geneva, Switzerland.

Reason 1990, *Human Error*, Cambridge: Cambridge University Press, U.K.

Reason 1997, *Managing the Risks of Organizational Accidents*, Ashgate Publishing, U.K.

16
Human Factors

16.1 Learning Objectives

The learning objectives of this chapter are:

- Understand how facilities & equipment, people, and management systems interact and can combine to deliver excellent process safety, and
- Explain the importance of human factors in the conduct of operations.

16.2 Incident: Formosa Plastics VCM Explosion, Illiopolis, Illinois, 2004

16.2.1 Incident Summary

An explosion and fire at the Formosa Plastics Corporation, IL (Formosa-IL) PVC manufacturing facility in Illiopolis, Illinois caused five fatalities and three injuries. "The explosion occurred after a large quantity of highly flammable vinyl chloride monomer (VCM) was inadvertently released from a reactor and ignited. The explosion and fire that followed destroyed much of the facility and burned for two days (Figure 16.1). Local authorities ordered residents within one mile of the facility to evacuate." (CSB 2007). The damage led to the facility being permanently closed.

Key Points:

Conduct of Operations – Conducting work as planned is key to keeping that system running safely and smoothly. A worker, whether Operator or Engineer, is merely one part of an integrated system.

Measurement and Metrics – Leading indicators can warn of an impending incident. In this case, prior instances of opening the wrong valve could have been recognized and prompted actions to support operators in better understanding the valving arrangement and the potential consequences of incorrect action.

16.2.2 Detailed Description

Polyvinyl chloride (PVC) was made by heating vinyl chloride monomer (VCM), water, suspending agents, and polymerization reaction initiators under pressure in a batch reactor. VCM is a highly flammable material. The building contained 24 reactors in groups of 4 with a control station for every two reactors (Figure 16.2). When a reaction was complete, the PVC solution was transferred through the bottom valve to a vessel for the next step in the process.

After the transfer, the reactor was purged of hazardous gases and cleaned by power washing it through an open manway. The wash water was emptied to a drain through the reactor bottom valve and a drain valve. All of these steps were done manually.

Figure 16.1. Smoke plumes from Formosa plant
(CSB 2007)

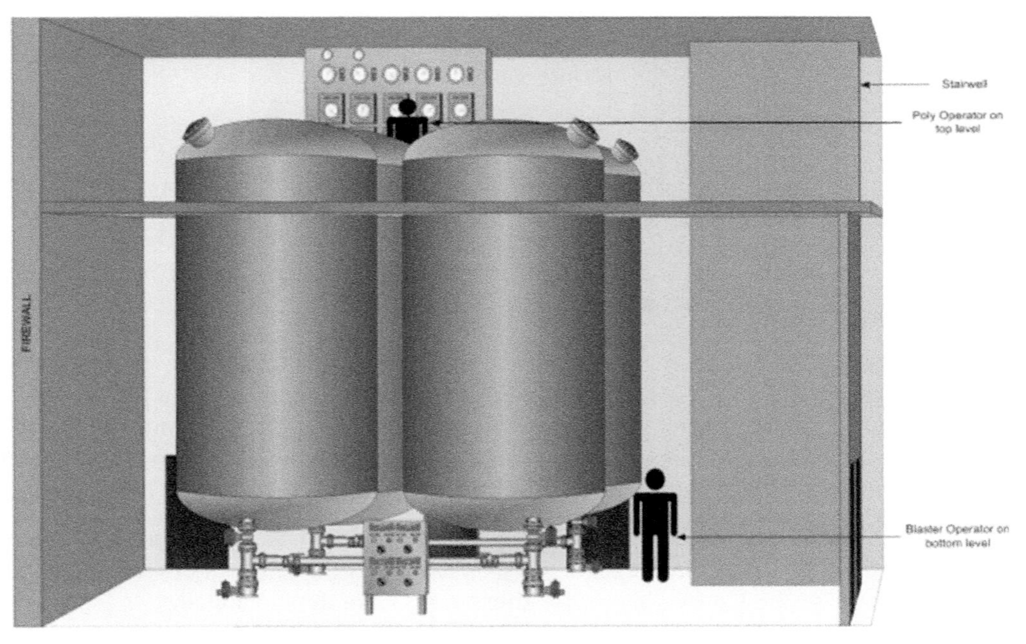

Figure 16.2. Formosa reactor building elevation view
(CSB 2007)

On the day of the incident, the transfer of the reaction mass and the power washing had been completed in one reactor, 306. The blaster operator went downstairs to drain the reactor. At the bottom of the stairway, he turned in the wrong direction towards an identical set of four reactors that were in the reaction phase of the process. See Figure 16.3. The operator mistakenly attempted to empty reactor 310 by opening the bottom and drain valves. The bottom valve; however, was interlocked to remain closed when the reactor pressure was above 68.9 kPa (10 psi), that is, when it was processing a batch of PVC. Consequently, it did not open. The air supply to the bottom valve had been equipped with quick disconnect fittings and a separate supply of emergency air was provided to allow operators to transfer a batch from one reactor to another in an emergency.

When the bottom valve did not open, the blaster operator switched to the backup air supply thereby overriding the interlock. This was done without contacting the upstairs reactor operator or shift foreman to check on the status of the reactor.

As the bottom valve was opened, VCM poured out of the reactor and the building rapidly filled with flammable liquid and vapor. A deluge system in the building alarmed but failed to activate (the deluge system might not have prevented the explosion even if it did activate). A shift supervisor came to the area to investigate. VCM detectors in the building were reading above their maximum measurable levels. The shift foreman and reactor operators took measures to slow the release rather than evacuate. The VCM vapors found an ignition source and several explosions occurred.

The operator overrode an interlock without consulting with shift supervision, leading to a release of hot, pressurized VCM. Several factors that made this error more likely to occur include:

- The similar layout of the reactor groupings (see Figure 16.3) making it error prone.
- The operators on the lower levels were not given radios to make communication with the reactor control operators on the upper level easier. (Similar Formosa plants had radios or an intercom system.)
- Formosa eliminated an operator group leader position and shifted their responsibility to the shift supervisors, who were not always as available as the group leaders had been. This reduced the amount of support available to operators.

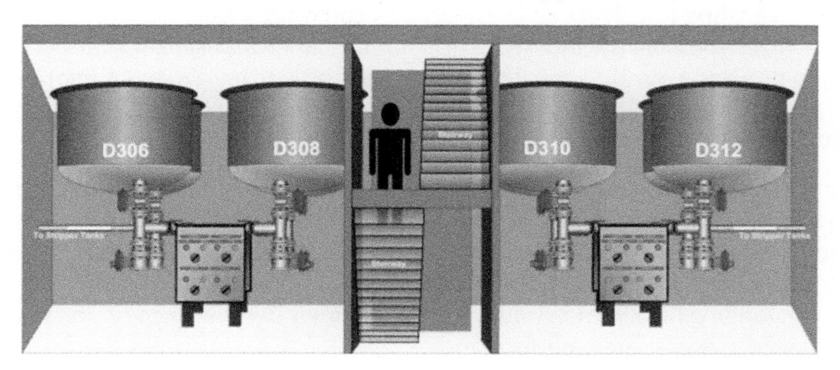

Figure 16.3. Cutaway of the Formosa reactor building (CSB 2007)

16.2.3 Lessons

Conduct of Operations. Conduct of Operations means doing tasks deliberately and correctly. That means following operating procedures and protocols, and, when the process moves outside the operating envelope, to stop work, think about the response, get experienced advice as needed, and shut down as appropriate. In the Formosa event, the blaster operator was supposed to get supervisory approval to override the interlock.

Conduct of Operations does not apply to just operators on the process floor; it also applies to engineers, supervisors, and managers. Conduct of Operations includes the design of controls and systems needed to support operators and others in doing their jobs safety and correctly. In the case of the Formosa VCM explosion, the blaster operators had to cope with an error prone design. The reactor layout made it easier for a mix-up to occur. An emergency transfer procedure required bypassing the bottom valve interlock, and an easy means was provided to do this. Engineers who design and run plants should try to provide engineering controls and monitor shift notes and logs for instances of bypassing interlocks. In this case, a reactor status indication on the operating floor could have been provided along with more rigorous enforcement of procedures and interlock management. Operators were not provided tools (radio communication) to make it easier for them to follow their procedures. It is the responsibility of management to provide the tools and controls necessary for operators to do their jobs safely.

Management of Change. When Formosa Plastics took over the plant, a reduction in staff and changes in responsibilities were made. The new management made no formal management of organizational change review of the staffing changes to analyze the impact of these changes.

Emergency Response. The Formosa Plastics VCM explosion also illustrates the importance of emergency response planning and training. When the VCM release occurred, operators responded by trying to mitigate the release. Operators should clearly understand if the planned response is to mitigate the release or safely evacuate.

Measurement and Metrics. Formosa Plastics plants had two previous incidents that involved operators opening the wrong valves. A metrics system that tracked leading and lagging indicators, could have alerted Formosa Plastics to a systemic problem and enabled the company to take steps to correct it. These actions could have been to train operators at all plants about why bypassing the interlocks was not safe and/or placing a guard over the hose connection to prevent overriding the interlock without proper review and authorization.

16.3 Introduction to Human Factors

One factor is central to process safety, but as yet in this text, has only been briefly mentioned. That is the human. Humans design, operate, and maintain the equipment in process facilities. They create the management systems in which the work takes place. Engineers spend quite a bit of time understanding how equipment works and improving its performance. Spending time on understanding how humans work to improve their performance can help support the overall workplace process safety performance.

Human Factors - A discipline concerned with designing machines, operations, and work environments so that they match human capabilities, limitations, and needs. This includes any technical work (engineering, procedure writing, worker training, worker selection, etc.) related to the human factor in operator-machine systems. (CCPS Glossary)

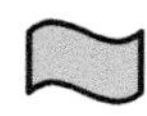

Many terms are used in the area of human factors. They are related but are not synonymous.

The term "ergonomics" is typically used to describe the relationship between the human and the physical work environment such as locating valve handles within easy reach.

"Human factors" includes ergonomics and is broader than the human's relationship with the physical environment. It also includes the way we perceive, process information, and respond – the way our brains work.

"Human performance" is another term often used. Human performance is the result. Humans can perform a task successfully, or poorly. By understanding human factors, we can strive to support the human to perform their task correctly. This support of successful human performance then supports good process safety performance.

Human factors considers the human as part of a system. *Human Factors Methods for Improving Performance in the Process Industries* features the 3-part system shown in Figure 16.4. (CCPS 2007) The three parts are 1) people, 2) the facilities & equipment that the human works with, and 3) the management systems that the human works within.

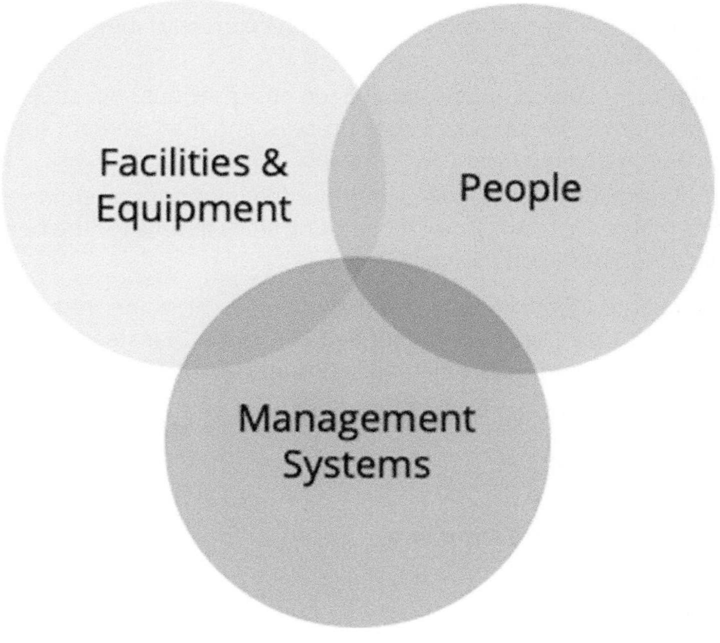

Figure 16.4. Model for human factors
(CCPS 2007)

Each of these areas relates to human performance in specific ways.

People:

- Human anthropometry (size and shape) impacts a person's ability to do different types of work
- Humans have physical needs of food, water, air, and rest
- The 5 human senses provide information, but the senses can be degraded
- The human brain processes information, but it too has limitations
- Humans have environmental limits, e.g. temperature, noise, light as well as boring or stressful situations

Facilities & Equipment: This is the human machine interface (HMI), i.e. matching the human and the hardware in the workplace.

Management Systems:

- Operating Procedures affect performance; they should be accessible and clear
- Work handover communication impacts the workers understanding of the process state
- Training should focus on correct operations and process safety risks
- Work motivations should be consistent with process safety goals

Putting all of this together, a workplace can be designed to support successful human performance. For example,

- An operator can follow a procedure, identify the correctly labeled valve, and reach the handle to easily turn it.
- An operator can safely complete a job begun on a previous shift supported by a clear shift handover, documented work process, and management support in providing adequate time to complete the job.
- The control panel design is such that it draws the operator's attention to a process parameter that is exceeding normal operating limits in time for the operator to take corrective action based on training.

Human Factors is not identified as a stand-alone element in the management systems identified in Chapter 2 or the regulations identified in Chapter 3. Instead, it can be seen as a part of these Risk Based Process Safety elements. (Bridges 2010)

- Process Safety Culture
- Workforce Involvement
- Operating Procedures
- Training and Performance Assurance
- Operating Readiness
- Conduct of Operations

Although the term 'human factors' may not be commonly used in guidance, it is frequently cited as a cause of process safety incidents. Companies and organizations such as CCPS have focused efforts specifically on increasing the understanding of human factors as it relates to process safety in efforts to reduce process safety incidents.

Historically it was common to cite the cause of an incident as "operator error". The objective of many of those investigations was to assign blame to an individual or a group. We recognize now that blaming an operator does not address the root cause of an incident. If an operator made an error, why did the operator make the error? Was the procedure incorrect? Was the training insufficient? Was the alarm overload such that the operator could not understand what was happening? Was there a motivation for speed over safety?

The Institute of Nuclear Power Operations (INPO) identified five principles that moved beyond the old approach of assigning blame while continuing to focus on the worker. (INPO 2006) Their approach was to integrate these five principles into management, leadership and worker practices.

1. People are fallible, and even the best people make mistakes.
2. Error-likely situations are predictable, manageable, and preventable.
3. Individual behavior is influenced by organizational processes and values.
4. People achieve high levels of performance largely because of the encouragement and reinforcement received from leaders, peers, and subordinates.
5. Events can be avoided through an understanding of the reasons mistakes occur and application of the lessons learned from past events (or errors).

16.4 The Human Individual

This section will address the following human factors topics related to the individual human as listed in Figure 16.5. The intent is to inform the reader that these topics influence human performance such that they can then consider these topics as they design systems in which humans work. Much of this material is based on *Fundamental Human Factors Concepts* (CAA).

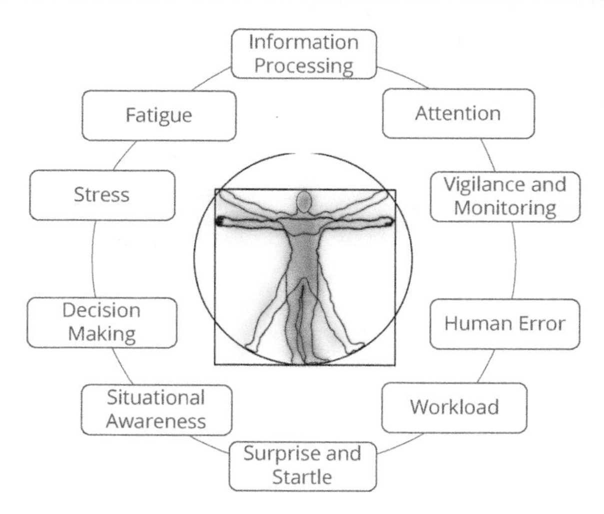

Figure 16.5. Human factors topics related to the human individual

Information Processing. Understanding information processing can help us understand why people do what they do. It explains how the brain takes incoming information, processes it, and responds.

Information comes in through the senses and is held in memory. The information is held for just long enough for a person to unconsciously decide (through perception) if the stimulus warrants paying attention. For incoming information to make it to processing, it has to be sensed, perceived, and noticed.

Processing the information involves comparing it to long-term memories as a means to decide what to do. This takes mental effort and can include assessing, calculating, and decision-making. Some well-learned perceptions can skip processing by triggering an automatic response. For example, have you ever driven to work or school and, once you arrived, noted that you could not remember the details of the drive?

Response can be saying or doing something. Well-learned responses can be almost automatic.

Attention. Paying attention to something is concentrating on it. Attention is limited. Sometimes things that would appear to be perfectly audible or visible do not catch our attention and do not receive any further processing. The brain only has the capacity to process a small amount of what is sensed, and it can be surprising what it can miss.

Vigilance and monitoring. Vigilance refers to sustaining attention on something to be able to notice when something changes. For example, this could be monitoring a control panel to notice when a pressure is exceeded. The human brain is not set up to monitor unchanging information well – it gets bored. Control interfaces should be designed with the human in mind including clear signals that make immediate attention easy to sense (e.g. seen, heard) above the background information.

Human error. Over the years, the term "human error" has evolved to the term "human performance." This recognizes that humans are involved in every aspect of the process throughout its entire life time. Humans design, fabricate, install, commission, operate, maintain, change, and decommission each part of a process.

It is clear that people make mistakes. This simple statement is being included in many company human factors educational material and policy statements. If you admit this truth, then you must address it by making systems more robust.

For further information on the academic study on human error, refer to the many publications by Jens Rasmussen and James Reason who are recognized experts on this topic. (Rasmussen 1979 and 1986, Reason 1991) One approach taken is to ask the question, "Could a well-motivated, equally competent, and comparably qualified individual have made the same kind of error in similar circumstances?" If yes, then this human performance issues should be addressed in risk management activities.

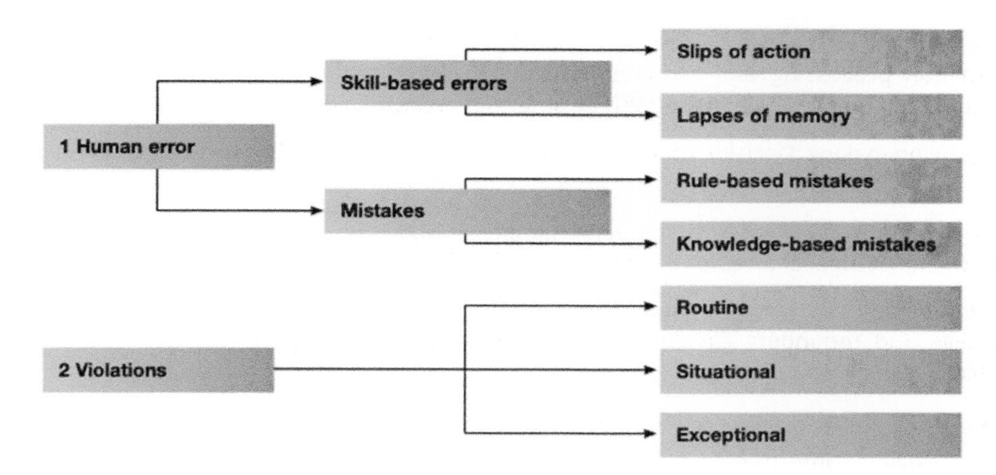

Figure 16.6. Human failure types
(HSE a)

Figure 16.6 identifies the types of human error and violations.

- Skill-based errors include slips such as conducting an action too soon in a sequence or lapses such as forgetting to perform a task.
- Mistakes occur when someone does the wrong thing, believing it to be appropriate. This includes rule-based mistakes such as conducting a procedure differently than written or knowledge-based mistakes such as responding based on knowledge which is incomplete in this instance.
- Violations are intentional. These failures can be routine in that it is the way a task is conducted (despite what the procedure says); situational or exceptional in that the operator, in that situation, thought it was appropriate to perform the task that way.

Workload challenges. Workload is the amount of mental effort used to process information. Problem solving, decision making, and thinking cause workload. The more attention required, the higher the workload. Sustained high workload is associated with increased errors, fatigue, and stress.

Multi-tasking causes competition between tasks for mental attention. Communicating and conducting a manual operation at the same time is an example. Although this sounds easy, the two tasks compete for the same brain space and this can result in poor performance of one or both tasks.

Reducing the difficulty of the task, the number of tasks running in parallel, the number of tasks in a series, and time constraints all serve to minimize workload challenge.

Surprise and startle. This is related to the human fight or flight response. In this situation all mental capacity becomes focused on the threat and/or the escape from it. In a fight or flight state, time is key to survival. Today, that is seen in an urge to be engaged in the active solution. To act fast, the brain requires a quick understanding of the problem that doesn't require problem solving. Providing very clear instruction on what to do and when to do it is helpful in this situation.

Situational awareness. Simply put, this is "knowing what is going on". It is related to information processing and requires attention. Good situational awareness depends on sufficient data and time for the data to be sensed, perceived, and interpreted. (Endsley)

Decision making. Decision making can be thought of along a continuum as shown in Figure 16.7.

Rational decision making is when a person applies reasoning and logic, which takes time, to make the most ideal choice, for example when deciding which car to buy.

Quicker decision-making uses biases and shortcuts including the following. Using checklists and reminders can lessen the potential negative impacts of these biases and shortcuts.

- Recency - more recent information is given priority
- Neglect – information is overlooked
- Availability - information that is easier to recall, is more influential
- Small samples – hypotheses are created based on only one or two experiences
- Confirmation bias - once a hypothesis or response is decided on, more weight is given to evidence that confirms the hypothesis, and less given to evidence that conflicts with the hypothesis

Figure 16.7. Decision making continuum

Intuitive decisions are made without mental processing or using shortcuts. Consider a decision in which someone has a feeling that it should be safe enough to 'bend the rules' in this case. For example, if they have motivation to get the task done quickly, they may be more inclined to bend the rules. The decision will appear reckless in hindsight but feel acceptable at the time. Setting limits can protect against inappropriate intuitive decisions.

Stress. Stress is the response to unfavorable environmental conditions. If excessive demands are placed on a human, it is possible to exceed the individual's capacity to meet them. Sources of stress include:

- Environmental sources of stress such as temperature, vibration, noise, humidity, glare
- Life stressors such as social pressures, financial pressures, family arguments, death of a close relative, smoking or drinking to excess, as well as physiological factors such as hunger, thirst, pain, lack of sleep and fatigue
- Organizational stressors such as poor communication, role conflict, workload, lack of career development, pay inequality, bureaucratic processes

Stress can cause a person to miss a step in a procedure or conduct steps in the wrong order. As stress increases, attention is focused on a single thing and other information may not be sensed or perceived. Managing and designing the workload within the constraints of the time and environment helps to manage stress levels.

Fatigue. Excessive workload, shift work or time zone changes can lead to cumulative sleep disruption of the body cycles. When fatigued, an individual's sleep quality is degraded, they have slower reactions and more likely to make mistakes. Fatigue can be managed by controlling the length of shifts, shift patterns, and the amount of overtime permitted.

16.5 The Work Team

Factors impacting a work team's performance are shown in Figure 16.8 and discussed in this section.

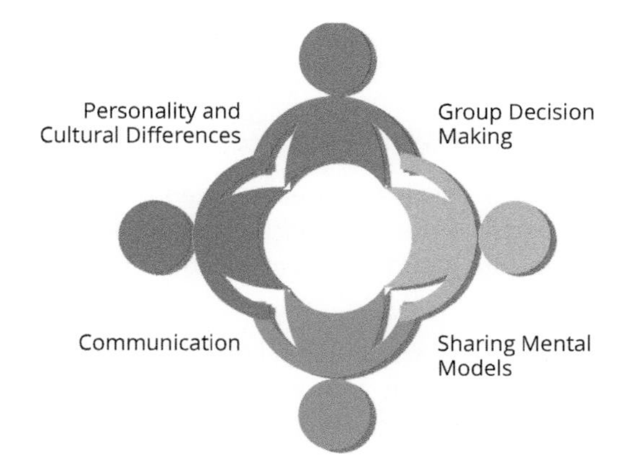

Figure 16.8. Factors impacting work team performance

Personality and cultural differences. People are different. Cultures have different norms when it comes to respect of authority, importance of the individual versus the group, avoidance of uncertainty, focus on long versus short term, and indulgence. These differences can influence team behavior. The benefits of working as a team are increased resources and synergy. Teams can also have challenges working together. Clearly defining roles in the team and encouraging communication can minimize these challenges.

Group decision making. Groups can suffer the same biases as individuals when making decisions, as well as additional effects due to the group dynamics. Groupthink is a common issue where members of a group are reluctant to challenge the decision of the group and instead agree, even when in doubt. People do this to maintain good relationships with team members. The other extreme is risk-polarization where groups err towards very high or very low risk strategies, rather than moderate risk strategies which individuals are more likely to make alone. Using standard practices to support decision-making can lessen the impact of these biases.

Communication. Communication is key. Teams who communicate more often commit fewer errors during critical operations. Communication can fail because the message is long and complex, not clearly transmitted, or because it is misinterpreted. It can be easy for primary

English speakers to forget that English is not their colleague's first language. Colloquialisms, slang phrases, poor pronunciation, and speaking quickly can cause communication problems. Air Traffic Control communications are a good example of excellent communications. Each communication is acknowledged. They are clear and brief and use a standard glossary of words and phrases.

Mental models. A mental model is a human's understanding or mental image of a situation, task, or environment. The same situation can look different to two people, depending on what each senses and understands. In process safety, making sure people have the same mental model with respect to critical safety elements is important. Communicating well and having access to the same data are important in sharing a common mental model. A sobering example of this is the NASA Challenger incident where managers perceived the engineers as raising unwarranted concerns and the engineers perceived the managers as disinterested. The investigation concluded that NASAs safety system had been "silent" including misrepresentations of criticality (of known safety problems) and lack of involvement in critical discussions. (CCPS 2012)

16.6 Human Factors in the Process Workplace

The goal is to improve human performance as a means to reduce process safety incidents. As humans are involved in all aspects of the workplace, many opportunities arise to use the human factors that influence individuals and work teams in the workplace. A few examples are described here.

HAZOP Study. The HAZOP study was discussed in Section 12.3.4. Human factors appear in two ways in a HAZOP. First is that they can be a potential cause of process deviations analyzed in a HAZOP such as a valve being inadvertently left open or a safeguard where human response to an alarm is required. Checklists are often used to address human factors in a HAZOP. Other approaches include having a walk-through of the unit being reviewed in advance of the study to help the HAZOP team develop a common mental model of the process. If it is a study of a unit yet to be built, a 3D model review can be used to provide this common understanding. During the study itself, particular focus on safeguards that are entirely human dependent when considering if recommendations are warranted is desirable.

The second way is that human factors impact the conduct of the HAZOP study. The HAZOP is conducted by a team, thus the factors influencing both the individual and the work team should be kept in mind. Understanding human factors can lead to a more effective study.

- HAZOPs can be a significant time commitment. Consider the participants' workloads and set the schedule to minimize stress and fatigue. Manage the time and stick to the agenda, providing appropriate meeting breaks
- Select a location to minimize distractions and allow people to focus attention.
- Clearly state the scope and objective. With a shared mental model of the, the study will be more effective and efficient.

- Facilitate the communications. Keeping the team on topic will keep everyone's attention focused and mind engaged. Ensure that everyone contributes. In some cultures, it might be necessary to encourage people to speak up. Control dominant personalities so that the entire team can contribute to the brainstorming. Avoid groupthink.

Human Factors Engineering. Figure 16.4 showed a model of the three-part system of people, facilities and equipment, and management systems. The design of the human interfaces between these three areas is human factors engineering. Human factors engineering aims to support the human in completing a task successfully. Human factors engineering can include designing facilities, equipment, and systems so that the human can access, operate, understand, and use them effectively. It can be seen in valves that are accessible, display screens that are easy to interpret, and procedures that are clear and concise. Two examples of human factors engineering are as follows.

- A car is a classic example of a man-machine interface. Car design has been optimized over the years to improve the mechanical design, and also to improve the operability, comfort, and safety of the driver and passengers. For example, the car radio may be in the dashboard, but now the controls for volume and station selection are also often on the steering wheel. This makes it easier to reach, and it also allows the driver to keep their eyes on the road – which improves safety.
- In a process plant emergency, it may be important to quickly isolate the process flow. Emergency isolation valves (EIV) are installed for this purpose. The valve itself may not be sufficient during the busy and critical time of an emergency. The type of valve and its location may also be important. For example, the EIVs can be grouped in a single location outside of the hazardous area, labeled clearly, located at grade or provided stair access (not a ladder), and, if they are large, automated to make them easier to close and minimize the time at risk for the operator.

Critical Task Analysis. Simply put, critical task analysis is a human factors tool that dissects a task into individual steps, analyzes how the task is completed, what could go wrong, and what are the opportunities for improvement. Critical task analysis is typically conducted on those tasks with the potential for a higher risk outcome if not performed correctly.

A critical task analysis will typically involve a walk-through of the part of the process plant where the task would be carried out. This allows the analysts to see the lighting, signage, and accessibility. It also allows the team to envision, first-hand, what the task entails much more directly than working with a paper procedure in an office.

Critical task analysis can identify specific ways to improve the likelihood of human success in completing a task. This is much more helpful than continually trying to write better procedures, but with little guidance on how to make them better. Critical task analysis can identify improvements in many of the areas influencing human performance discussed in this chapter such as labeling to help operators quickly identify equipment, managing lighting and noise levels to enable better sensory signals, the appropriateness for tools such as a checklist for critical steps, and potentially improved human machine interfaces.

Several critical task analysis methods are available. (NOPSEMA 2020, HSE 2000) A simplified approach called Task Improvement Process, TIP, follows the steps shown in Figure 16.9 which are similar to many critical task analysis approaches. (Miller 2019)

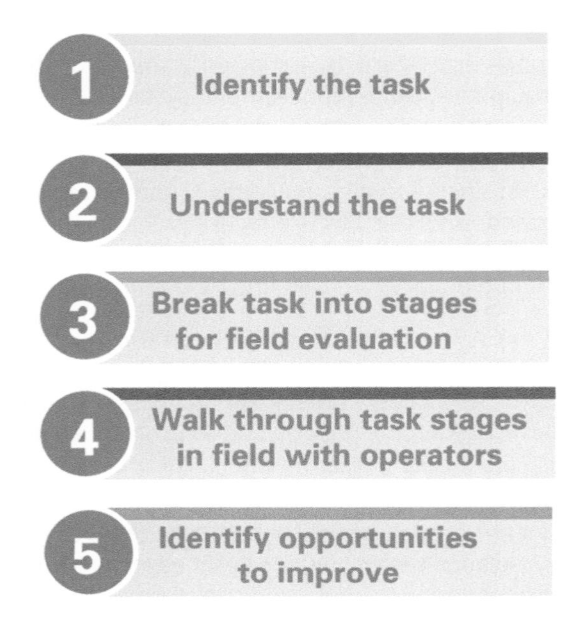

Figure 16.9. Task Improvement Plan steps
(Miller 2019)

Human reliability analysis. Several analysis methods are used; two of the more commonly used methods are THERP and HEART.

Human Reliability Analysis - A method used to evaluate whether system-required human-actions, tasks, or jobs will be completed successfully within a required time period. Also used to determine the probability that no extraneous human actions detrimental to the system will be performed. (CCPS Glossary)

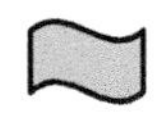

The Technique for Human Error Rate Prediction (THERP). THERP addresses task analyses, error identification, and quantified human error probabilities (HEP). The THERP includes tables of HEPs and uses event trees to determine overall failure probabilities. THERP was developed in the nuclear industry and has been applied to the process industries. (HSE 2009) The THERP Handbook is publicly available via the U.S. Nuclear Regulatory Commission website. (NUREG)

Human Error Assessment and Reduction Technique (HEART). HEART is a simpler error prediction method that can be applied by non-specialists. Nine Generic Task Types (GTTs) are described in HEART, each with an associated nominal human error potential (HEP), and 38 Error Producing Conditions (EPCs) that may affect task reliability, each with a maximum amount by which the nominal HEP can be multiplied. HEART classifies a task into one of the 9 Generic Task Types (GTT) and assigns the nominal human error potential (HEP) to the task. Error Producing Conditions (EPC) that may affect task reliability are identified. The task HEP is calculated. HEART can help identify areas for improvement and includes strategies to reduce errors._HEART is used in nuclear, process, medical, and transportation industries and is described in several papers by J. C. Williams. (Williams 1992 and HSE 2009)

16.7 What a New Engineer Might Do

A new engineer should look beyond equipment design and consider the role of the human in the system of people, facilities and equipment, and management systems that defines the workplace. Even in small projects and simple systems, engineer the human machine interface to support human success.

New engineers will likely be involved in HAZOP studies and various projects that involve teamwork. Helping to facilitate meetings and supporting good team communications can lead to more efficient and effective teams. Simple critical task analysis methods can be led by new engineers which would not only improve tasks but also build a relationship with operators and maintenance technicians.

16.8 Tools

Human factors resources include the following.

CCPS Human Factors for Process Plant Operations: A Handbook. This book describes human factors concepts and principles in an easy to understand manner. It describes how to support human capabilities including identification and design of job aids to do so. (CCPS expected 2022)

Flight-crew human factors handbook, CAP 737. The aviation industry has built significant knowledge in human factors. (CAA) This handbook addresses both individual and work team human factors. "The knowledge in the handbook was intentionally simplified to make the document more easily accessible, readable and more usable in the practical domain." (CAA)

Critical Task Analysis. Several approaches for critical task analysis are mentioned in Section 16.6. (NOPSEMA 2020, HSE 2000, Miller 2019)

Human reliability assessment tools. Refer to the discussion on THERP and HEART in Section 16.6.

HSE. HSE provides guidance and many references on human factors topics and human reliability on the Health and Safety Executive webpage at hse.gov.uk.

16.9 Summary

It is helpful to think of the workplace as a three-part system of people, facilities and equipment, and management systems. Humans are an important part of this engineered system. Focusing on how the human works (both physically and mentally), and how the system can be designed to support their success, can also support successful process safety performance.

An ethical aspect of engineering is to work within your area of expertise and to call for the assistance of others when outside of that area of expertise. Human factors may not be an area of expertise for the typical engineer. Reach out to human factors experts, outside of engineering, that can assist and build competency in the area.

16.10 Other Incidents

This chapter began with a description of the Formosa Plastics explosion. Other incidents relevant to human factors include the following.

- NASA Challenger Disaster, Florida, U.S., 1986
- Celanese Pampa Explosion, Texas, U.S., 1987
- Piper Alpha Platform, North Sea, U.K., 1988
- Texaco Oil Refinery Explosion and Fire, U.K., 1994
- Mars Climate Orbiter, U.S., 1999
- NASA Columbia Loss, Texas, U.S., 2003
- Buncefield Storage Tank Overflow and Explosion, U.K., 2005
- BP Isomerization Unit Explosion, Texas City, Texas, U.S., 2005
- Air France Flight 447, Brazil, 2009
- U.S. Airways 1549, U.S., 2009
- Deepwater Horizon Well Blowout, Gulf of Mexico, U.S., 2010
- Chevron Richmond Refinery Fire, California, U.S., 2012
- MGPI Processing Plant, Kansas, U.S., 2016

16.11 Exercises

1. List 3 RBPS elements evident in the Formosa Plastics VCM explosion summarized at the beginning of this chapter. Describe their shortcomings as related to this accident.
2. Considering the Formosa Plastics VCM, what actions could have been taken to reduce the risk of this incident?
3. Training, incentives, and supervision can prevent humans making mistakes. True or false? Explain your answer.
4. A refinery explosion investigation found that in the 11 minutes before the explosion, 275 alarms requiring operator action activated. This illustrates which human factors topic? Explain your answer.
5. The BP Refinery explosion in Texas City is described in Chapter 2. What human factors topics were illustrated in this incident? Explain your answer.
6. The Deepwater Horizon well blowout is described in Chapter 21. What human factors topics were illustrated in this incident? Explain your answer.
7. A complicated catalyst regeneration procedure takes place over three days. This process is done only once every few years. What human factors topics could impact this procedure? Explain your answer.

8. Explain the three parts of situational awareness.
9. In times of financial challenges, companies may try to "do more with less". Which human factors topics should be considered? Explain your answers.
10. Consider an engineering design project. What human factors topics can influence the human performance on the project?
11. Describe at least 3 ways human factors could contribute to process safety incidents during a) a unit startup, b) normal operation, and c) unit shutdown.

16.12 References

Bridges 2010, "Human Factors Elements Missing from Process Safety Management (PSM)", American Institute of Chemical Engineers 2010 Spring Meeting, San Antonio, Texas, March 22-24.

CAA, *Fundamental Human Factors Concepts,* CAP 719, Civil Aviation Authority Safety Regulation Group, http://publicapps.caa.co.uk/docs/33/CAP719.PDF.

CSB 2007, "Vinyl Chloride Monomer Explosion, Investigation", Report No. 2004-10-I-IL, U.S. Chemical Safety and Hazard Investigation Board, March.

CCPS Glossary, "CCPS Process Safety Glossary", Center for Chemical Process Safety, https://www.aiche.org/ccps/resources/glossary.

CCPS 2007, *Human Factors Methods for Improving Performance in the Process Industries*, Center for Chemical Process Safety, John Wiley & Sons, Hoboken, N.J.

CCPS 2012, "Building Process Safety Culture Tool Kit, Incident Summary: Challenger Case", https://www.aiche.org/ccps/topics/elements-process-safety/commitment-process-safety/process-safety-culture/challenger-case-history.

CCPS expected 2022, *Human Factors for Process Plant Operations: A Handbook*, Center for Chemical Process Safety, John Wiley & Sons, Hoboken, N.J.

Endsley, https://upload.wikimedia.org/wikipedia/en/6/61/Endsley-SA-model.jpg.

HSE a, "Leadership and worker involvement toolkit, "Understanding human failure", https://www.hse.gov.uk/construction/lwit/assets/downloads/human-failure.pdf.

HSE 2000, "Human Factors Assessment of Safety Critical Tasks", Offshore Technology Report – OTO 1999 092, U.K. Health and Safety Executive, 2000, https://www.hse.gov.uk/research/otopdf/1999/oto99092.pdf

HSE 2009, "Review of human reliability assessment methods", RR679, U.K. Health and Safety Executive, https://www.hse.gov.uk/research/rrpdf/rr679.pdf.

INPO, *Human Performance Reference Manual*, INPO 06-003, Institute of Nuclear Power Operations, October 2006.

Miller 2019, "Helping humans get it right", Miller, L. and Grounds, C., *Process Safety Progress,* 38: e12003. Viewed December 9, 2020, https://doi.org/10.1002/prs.12003

NOPSEMA 2020, "Critical Task Analysis", N-06300-IP1704 A500978, Australia National Offshore Petroleum Safety and Environmental Management Authority.

NUREG, "Handbook of Human Reliability Analysis with Emphasis on Nuclear Power Plant Applications - Final Report", NUREG/CR-1278, U.S. Nuclear Regulatory Commission, 2011, https://www.nrc.gov/docs/ML0712/ML071210299.html.

Rasmussen, 1979, "On the structure of knowledge – a morphology of mental models in a man-machine system context", Riso National Laboratory, Riso-M, No. 2192, Denmark.

Rasmussen 1986, "Information Processing and Human-machine Interaction: An Approach to Cognitive Engineering", North-Holland, New York, 1986.

Reason 1991, *Human Error*, Cambridge University Press, U.K.

Williams 1992, "Toward an Improved Evaluation Analysis Tool for Users of HEART". Proceedings of the International Conference on Hazard identification and Risk Analysis, Human Factors and Human Reliability in Process Safety, 15-17 January, Orlando, Florida.

17
Operational Readiness

17.1 Learning Objectives

The learning objective of this chapter is:

- Understand the importance of operational readiness.

17.2 Incident: Piper Alpha Explosion and Fire, Scotland, 1988

17.2.1 Incident Summary

An explosion occurred July 6, 1988 on the Piper Alpha offshore platform, owned by Occidental Petroleum and located off the Scottish coast in the North Sea. The initial explosion set off a chain of fires and explosions resulting in the loss of 167 lives and near-total destruction of the platform. The explosion began as a release of flammable gas through a poorly installed flange on a line that was improperly put into service. This set off a chain of escalating events which included more explosions and fires that eventually destroyed the platform. Sixty-one members of the crew survived the event by jumping into the water and being rescued by boat.

The platform layout consisted of a drilling derrick at one end, a processing area in the center, and living accommodation for its crew on the opposite end (Figure 17.3). Piper Alpha acted as a gas gathering facility for two other platforms in the area: Tartan and Claymore "A", receiving high-pressure gas through risers leading from undersea pipelines. Piper Alpha processed the gas from the risers together with its own gas and oil and piped the final products to shore in two separate pipelines (CCPS 2008).

Key Points:

Operational Readiness - Whether starting up for the first time, after a change, or following maintenance work, it is imperative to make sure the equipment and workers are ready for safe operation.

Emergency Management - Plan for and practice emergency response, with all those potentially involved, both internal and external to the company. Emergencies can escalate quickly. There may not be means or time to communicate. Knowing what to do is key to minimizing losses.

Figure 17.1. Piper Alpha platform
(CCPS 2005)

Figure 17.2. Piper Alpha platform after the incident
(CCPS 2005)

17.2.2 Detailed Description

The chain of events started when a standby pump was taken off-line for maintenance. The relief valve had also been removed for maintenance with blind flanges isolating the pipework connections, but with far fewer than the number of bolts required to hold full operating pressure. Later, a condensate pump, reinjecting hydrocarbon liquids from the gas/liquid separation process back into the oil export line, stopped in the late evening. Attempts to restart it were unsuccessful, and a decision was made to start up the standby pump as liquid levels were rising rapidly in the process vessels. If not reversed, this would have resulted in total shutdown of the platform. The night shift crew was aware that the standby pump had been taken out of service for maintenance earlier the same day but believed that the maintenance work had yet to commence. They re-energized the pump motor, which had not been locked out, and started the pump. Within seconds a large quantity of condensate and gas began to escape from the pump discharge pressure relief valve location, in the module above and out of sight of the pump.

The condensate pumps were located at the 21 meter (68 ft) deck support frame level, below the modules. The condensate pump relief valves were located inside the Gas Compression Module "C", with the connecting pipework entering and exiting Module "C" through the floor. Module "C" was separated from Module "D" containing the control room and emergency facilities with a non-structural firewall consisting of 3 sheets of a composite plating with mineral wool laid between steel sheets designed to be fire and blast resistant. The fire walls between modules "C" and "B", and "B" and "A" were built from a single plate coated with a fireproofing insulation material. The firewalls installed between the modules were not designed to withstand blast from within any of these modules. An explosion blew down the firewall containing the processing facility and separating it from the control room. As a result, the control room was destroyed, and important emergency control was lost. Large quantities of stored oil were quickly burning out of control.

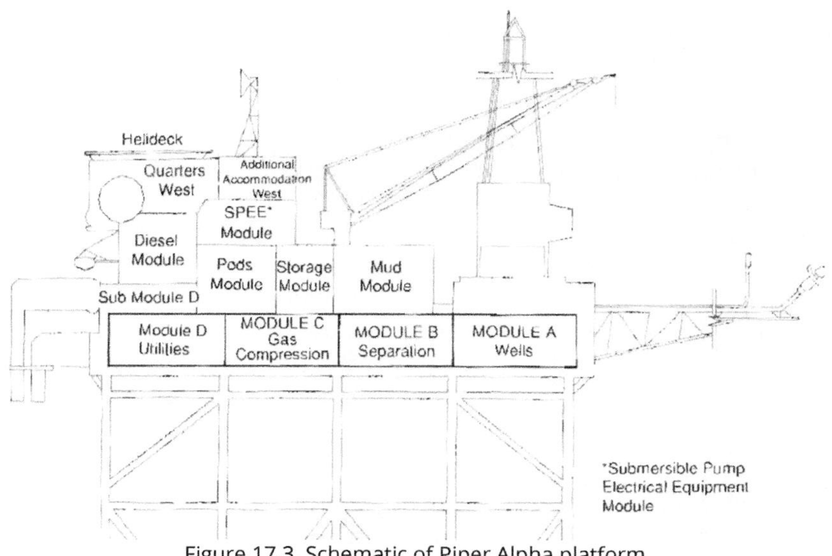

Figure 17.3. Schematic of Piper Alpha platform (Cullen 1990)

The automatic seawater deluge system, which was designed to extinguish such a fire, was unable to be activated as it had been previously isolated to protect divers carrying out inspection and maintenance on the platform supporting structure near the fire pumps submerged inlets.

About twenty minutes after the initial explosion, the fire had spread to the gas risers generating sufficient heat to cause them to fail catastrophically. The risers, which carried very large quantities of gas from the seabed to the platform, were constructed of 610 and 915 mm (24 and 36 in) diameter steel pipe containing flammable gas at 138 bar-g (2000 psig). When these risers failed, the resulting release of fuel dramatically increased the size of the fire to a towering inferno. At the fire's peak, the flames reached a height of 90 to 120 meters (300 – 400 ft) three to four hundred feet. The heat was felt from over 1.6 km (1 mi) away, and reflections in the clouds could be seen from 137 km (85 miles).

The crew began to congregate in the platform's living accommodation area, which was the farthest from the blaze and seemed the least dangerous, awaiting helicopters to rescue them. However, the fire prevented helicopters from landing. The accommodation was not smoke-proof and due to lack of training, people repeatedly opened and shut doors allowing smoke to enter. Some crew members decided that the only way to survive would be to leave the accommodation immediately. However, they found that all routes to lifeboats were blocked by smoke and flames and, lacking any other instructions, they jumped into the sea hoping to be rescued by boat. Sixty-one men survived by jumping. Most of the 167 fatalities occurred due to carbon monoxide and smoke exposure in the accommodation area. Two fatalities occurred from a rescue vessel as well.

The gas risers that were fueling the fire were finally shut off about an hour after they had burst, but the fire continued as the oil on the platform and the gas that was already in the pipes burned off. Three hours later the majority of the platform had burned down to sea level with the derricks and modules, including the accommodation, sliding off and sinking to the sea floor below. Only the drilling part of the platform remained standing above sea level. Oil continued to burn on the sea due to leakage from Piper Alpha's oil production risers (CCPS 2008).

The investigation found that the immediate cause of this incident was failure of the Work Permit system to control maintenance and inspection work on the platform. At the beginning of July 6, a work permit had been issued for the maintenance of the standby condensate pump. The pump's process connections had been isolated only by valve. A second work permit was issued for removal of the pump's discharge pressure relief valve for maintenance. (LBP 2018) When the pressure relief valve was removed, only four bolts instead of the full set required for operation were used to fasten the blind flanges fitted over the open ends of the connecting pipework, most likely just to keep the system clean. This pressure relief valve was located in the module above and out of sight of the pump. After removal, the pressure relief valve was taken to the platform workshop for inspection but had not been replaced by the end of the working day.

When the Maintenance Supervisor returned the Work Permit to the Control Room after he and his crew finished their shift, the Process Supervisors and Operators were in deep conversation. Consequently, he left the Work Permit lying on the desk without making any verbal or written hand over. The investigation found that vital communications systems on Piper Alpha had become too relaxed, with the result that the Work Permit was left on the manager's desk instead of it being personally given to him to enable proper communication at the subsequent shift change. If the system had been implemented properly, the initial gas release would not have occurred. However, once this had occurred, many other factors conspired together to cause the fatalities and loss of platform (CCPS 2008).

17.2.3 Lessons

Safe Work Practices. Good safe work practices are needed to control hazards due to maintenance work and make sure equipment is ready before starting up. These work practices need to include communication between the people doing the work and production personnel. In Piper Alpha, the night shift crew was not informed that the relief valve had been removed and the pump was not ready to be returned to operation. Additionally, the blind flange put in the line was not properly installed, so it could not hold the pressure.

Emergency Management. The Offshore Installation Manager (OIM) did not order an evacuation immediately resulting in his fatality shortly after. Fire boats responding to the event waited for orders from the OIM, which delayed response. Many of the evacuation routes were blocked. Other platforms in the area were feeding material to Piper Alpha and did not turn off their feeds, providing a continuing source of fuel to the fire. Even though they could see the fire on the horizon, they believed they needed permission from onshore management to turn off their feeds. The workers on the platform were not adequately trained in emergency procedures, and management was not trained to provide good leadership during a crisis situation. Evacuation drills were performed, but not every week as required by regulations. A full drill had not taken place in over three years. The place where the crew gathered was not safe. Smoke could enter and this caused the fatalities. After Piper Alpha, the U.K. Government required that there be a Temporary Safe Refuge (TSR) protecting staff sheltering there from explosions, fire, and toxic smoke until safe evacuation can be organized.

The Piper Alpha fire and explosion led to development of stronger offshore safety requirements in the U.K. Offshore Installations (Safety Case) Regulations. The Safety Case regulations are goal-based and replaced the previous prescriptive regulations. A Safety Case is the documentation that a production organization must submit in the U.K. to demonstrate that their operation is safe. Another change made was having responsibility for enforcing safety case moved from the U.K.'s Department of Energy to the Health and Safety Executive (HSE) to avoid potential conflicts between production and safety.

17.3 Introduction to Operational Readiness

Chapters 10 through 15 addressed project design, methods to identify hazards and analyze risk, and risk prevention and mitigation measures. Whether these concepts are applied to a new project, following a change, or during maintenance, the facility will be started. This chapter addresses the topic of verifying the facility is ready for a safe start up and safe operation.

Process safety incidents occur five times more often during startup than during normal operations. (CCPS 1995)

The RBPS Operational Readiness element corresponds to the Pre-Startup Safety Review in the U.S. OSHA PSM and U.S. EPA RMP regulations. In this case, however, operational readiness is defined more broadly than the OSHA process safety management pre-startup safety review element in that it specifically addresses startup from all shutdown conditions as illustrated in Figure 17.4.

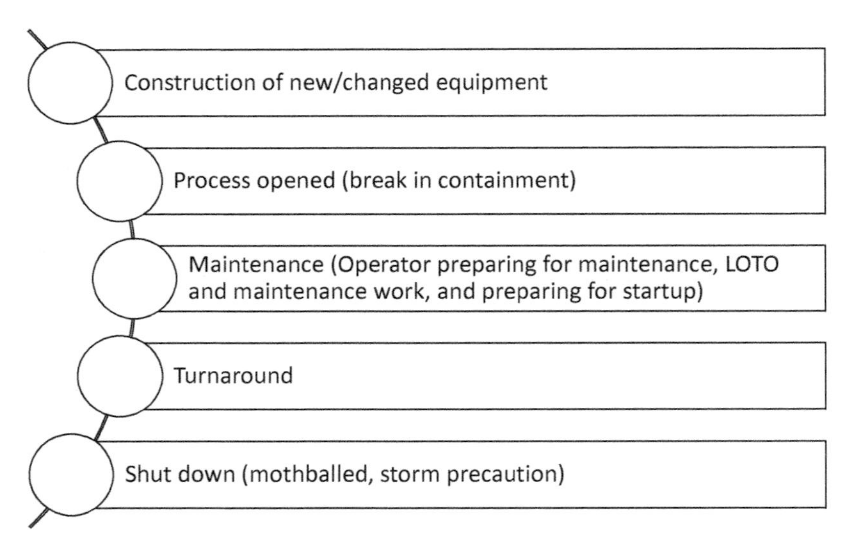

Figure 17.4. An operational readiness review should follow these activities

Pre-Startup Safety Review (PSSR) - A systematic and thorough check of a process prior to the introduction of a highly hazardous chemical to a process. The PSSR must confirm the following: Construction and equipment are in accordance with design specifications; Safety, operating, maintenance, and emergency procedures are in place and are adequate; A process hazard analysis has been performed for new facilities and recommendations have been resolved or implemented before startup, and modified facilities meet the management of change requirements; and training of each employee involved in operating a process has been completed. (CCPS Glossary)

Operational Readiness - A PSM program element associated with efforts to ensure that a process is ready for start-up/restart. This element applies to a variety of restart situations, ranging from restart after a brief maintenance outage to restart of a process that has been mothballed for several years. (CCPS Glossary)

Operational readiness reviews vary depending on the amount and complexity of equipment involved, the length of the shutdown, and the amount of change that occurred. In the case of a new process, operational readiness may take the form of a comprehensive commissioning plan. For a startup following a change managed through a Management of Change (Chapter 18), it may take the form of a PSSR.

An Operational Readiness Review includes the following topics and activities that are relevant to the startup.

- Confirming that the construction and equipment of a process are in accordance with design specifications.
- Ensuring that adequate safety, operating, maintenance, and emergency procedures are in place.
- Ensuring that any safeguards that may have been bypassed during the outage are verified to be in service and operational.
- Confirming that all sensors, instruments, blinds, and valves are properly reset to the proper state or condition.
- Ensuring that training has been completed for all workers who may affect the process.
- Verifying that management of change actions are completed.
- Verifying that hazard identification and risk analysis study actions are completed.

A similar approach is appropriate for conducting maintenance work. This would involve readying the equipment for maintenance, delivering it in a safe state to the maintenance technicians, developing work plans and conducting training specific to the work. After the work is completed, the operational readiness activities would be conducted in readiness for startup.

Operational readiness reviews of simple startups may involve only one person walking through the process to verify that equipment is in its intended operational state, and the equipment is ready to resume operation. They should ensure that the process is safe to be operated by examining issues such as the equipment lineup, leak tightness, blind positions, and proper isolation from other systems not yet ready for startup. In other cases, multi-disciplinary teams may be involved to ensure that equipment has been installed correctly and is ready for start-up.

Ensuring the operational readiness for the initial startup of a newly constructed process unit typically involves a commissioning review. These complex reviews may extend over many weeks or months as licensors, contractors, construction specialists, equipment vendors, regulators, and the owner's engineers, operators, and maintenance technicians verify equipment conformance to design intent, construction quality, procedure completion, training competency, and other requirements. Typically, extensive checklists, multi-stage verification, and multiple functional signoffs are required for startup authorization. Table 17.1 lists typical construction, pre-commissioning, and commissioning tasks.

Typical questions that may be incorporated in an operational readiness review include the following.

- Have all PHA recommendations been addressed?
- Has safety-critical equipment been properly installed, functional and documented?
- Has outstanding maintenance work been completed?
- Has equipment involved been physically verified to be safe to start?
- Has all computer control logic been tested?

- Have personnel received orientation of process and health hazards?
- Have operators been sufficiently trained in operating procedures?
- Have emergency response plans been prepared, responsibilities assigned, practice drills run?
- Have operating manuals, procedures been completed, updated and approved?
- Are field change authorization procedures in place?

Regardless of the type of operational readiness review, they should ensure that shutdown processes and equipment are verified to be in a safe condition for re-start and include a physical inspection of the process as part of the verification.

Table 17.1. Typical construction, pre-commissioning, and commissioning tasks (Mannan 2012)

Phase	Task
Construction and Pre-Commissioning	
A	Prepare plant/equipment for pre-commissioning/mechanical testing
B	Prepare services: clean and pressure test systems
C	Check and prepare major mechanical equipment, instrumentation, and protection systems
D	Final preparations for start-up
Commissioning	
E	Charge with feedstock, etc. Start-up plant and operate
F	Performance test and plant acceptance
G	Remainder of maintenance period

17.4 What a New Engineer Might Do

New engineers are often involved in operational readiness reviews for projects or in support of maintenance work. They may be part of a commissioning team and be asked to help complete some of the many checklists. Being part of a commissioning team is a good opportunity to learn about a facility and how it is, literally, put together.

New engineers are often given smaller projects to manage on their own. The startups after these smaller projects are just as important as those after the complex projects.

A new engineer should be diligent in conducting operational readiness activities. These activities come when the site has pressure to startup the facility (a turnaround has overrun the schedule) or when personnel are excited to start it up after the long project design and construction. Diligently completing the verifications and checklists is important in making sure the equipment, the people, and the management systems are ready to support safe operation.

17.5 Tools

Resources to support Operational Readiness include the following.

CCPS *Performing Effective Pre-Startup Safety Reviews*. This book provides guidance to those with responsibility for scheduling and executing a Pre-Startup Safety Review (PSSR). It outlines a protocol and tool for use by project or turnaround teams, to effectively and efficiently schedule and execute a PSSR. It features how-to checklists and provides guidance on hazard assessment, batch and continuous processes, validation, and documentation. (CCPS 2007)

17.6 Summary

Operational readiness is about verifying that equipment and people are ready for the process to start up safely. This can apply after a short shutdown for minor maintenance work or a managed change in which case a PSSR might be used. For more complex startups, such as for the initial startup of a new process unit, a comprehensive commissioning plan may be used. In all cases, operational readiness involves reviewing the state of the equipment as well as the readiness of the personnel related topics including training and operating procedures. Operational readiness activities are the last steps before startup and should be conducted diligently to support a safe startup and safe operations.

17.7 Other Incidents

This chapter began with a description of the Piper Alpha Explosion. Other incidents relevant to operational readiness include the following.

- NASA Challenger Disaster, Florida, U.S., 1986
- BP Amoco Thermal Decomposition, Georgia, U.S., 2001
- First Chemical Corporation Reactive Explosion and Fire, Mississippi, U.S., 2002
- Bayer CropSciences Pesticide Chemical Runaway Reaction and Pressure Vessel Explosion, West Virginia, U.S., 2008
- Pike River Coal Mine Explosion, South Island, New Zealand, 2010

17.8 Exercises

1. List 3 RBPS elements evident in the Piper Alpha explosion and fire summarized at the beginning of this chapter. Describe their shortcomings as related to this accident.
2. Considering the Piper Alpha explosion and fire, what actions could have been taken to reduce the risk of this incident?
3. Why is operational readiness important?
4. Give three examples of when an operational readiness review should be conducted.
5. Give three examples of topics that should be included in an operational readiness review.

17.9 References

CCPS Glossary, "CCPS Process Safety Glossary", Center for Chemical Process Safety, https://www.aiche.org/ccps/resources/glossary.

CCPS 1995, *Guidelines for Safe Process Operations and Maintenance*, Center for Chemical Process Safety, John Wiley & Sons, Hoboken, N.J.

CCPS 2005, "Building Process Safety Culture: Tools to Enhance Process Safety Performance, Piper Alpha", American Institute of Chemical Engineers, Center for Chemical Process Safety, New York, NY.

CCPS 2007, *Performing Effective Pre-Startup Safety Reviews,* Center for Chemical Process Safety, John Wiley & Sons, Hoboken, N.J.

CCPS 2008, *Incidents That Define Process Safety*, Center for Chemical Process Safety, John Wiley & Sons, Hoboken, N.J.

Cullen 1990, The Hon. Lord W.D. Cullen, "The Public Inquiry into the Piper Alpha Disaster", HMSO, London, U.K.

LPB 2018, Piper Alpha – What have we learned?, Loss Prevention Bulletin 261, June, icheme.org/media/1237/lpb261_pg03.pdf.

Mannan 2012, *Lees' Loss Prevention in the Process Industries, Volumes 1-3 - Hazard Identification, Assessment and Control (4th Edition)*, Elsevier, Netherlands.

18
Management of Change

18.1 Learning Objectives

The learning objectives of this chapter are:

- Explain Management of Change (MOC) and what it involves,
- Describe what changes trigger an MOC, and
- Understand how managing change impacts process safety.

18.2 Incident: Nypro Explosion, Flixborough, England, 1974

18.2.1 Incident Summary

On the evening of June 1, 1974, the Nypro site in Flixborough, U.K. was severely damaged by a large explosion that resulted in 28 employee fatalities and 36 injuries. The site office building was demolished, but fortunately it was empty because the incident occurred on the weekend. Outside of the plant, injuries and damage were widespread, with 53 people being reported injured. No offsite fatalities occurred. Varying degrees of damage was caused to 1,821 houses and 167 business premises (CCPS 2008).

Key Points:

Management of Change - Think before making changes. Verify that hazards introduced by a change have been identified and managed.

Compliance with Standards - Standards codify knowledge and experience. Take advantage of the opportunity to learn from others and implement good practice through using codes and standards.

18.2.2 Description

The part of the plant on which the explosion occurred involved the oxidation of cyclohexane to cyclohexanone using air injection at 896 kPag (130 psig) and 155°C (311°F). The oxidation reaction was carried out in the liquid phase in six reactor vessels arranged in series, each set 360 mm (14 in) below its predecessor to allow the flow to progress through the reaction train by gravity as shown in Figure 18.1.

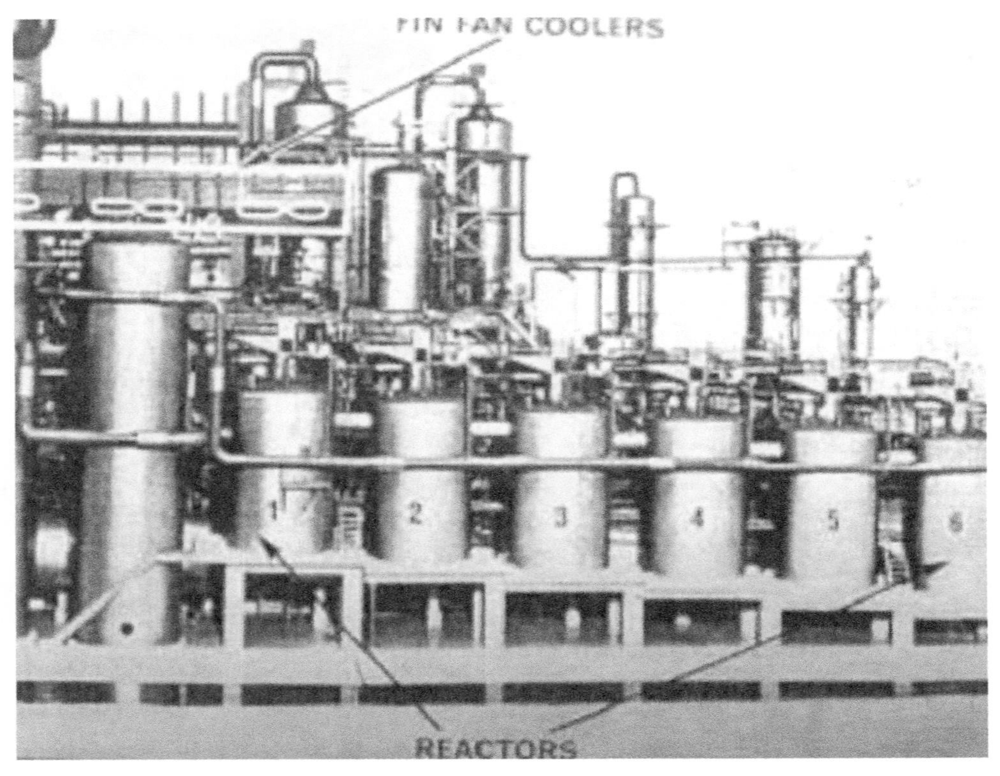

Figure 18.1. Flixborough Reactors
(CCPS 2005)

On March 27, 1974, a crack was detected on Reactor No.5. A Maintenance Engineer recommended complete closure for 3 weeks. The Maintenance Manager, whose job had been filled for several months by the head of the laboratory while awaiting a reorganization of the company, proposed dismantling Reactor No. 5 and connecting numbers 4 and 6 together by a 500 mm (20 inch) diameter temporary connection. To support the piping, the proposal was to use a structure made from conventional construction industry scaffolding, Figure 18.2. This resulted in a piece of bent piping with two bellows, one at either end. During start-up activities this piping was subjected to asymmetrical loads and failed as the double bellows design allowed the pipe connection between the bellows to be unsupported and unrestrained. A single bellows would have been effective to absorb thermal expansions with structural rigidity coming from fixed connections at either end and only a short length of piping.

The temporary connection was not adequate for the forces and temperatures involved, and failed, releasing 30 metric tons of cyclohexane in 30 seconds. Of the 28 fatalities, 18 were in the control room. The fire lasted over three days with 40,000 m^2 (10 acres) affected. See Figure 18.3. (CCPS 2008).

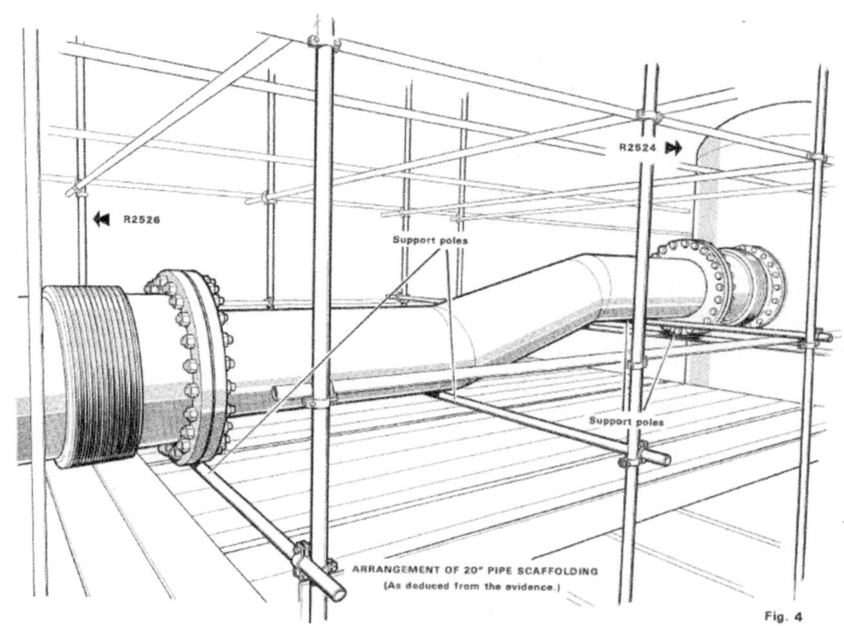

Figure 18.2. Schematic of Flixborough piping replacement
(HMSO 1975)

Figure 18.3. Flixborough site after explosion
(CCPS 2005)

The design was based on chalk drawings on the workshop floor. No engineering review by a qualified mechanical engineer was undertaken to review the mechanical adequacy of the connection. The investigation found that "No consideration was given to the bending moments or hydraulic thrusts that would be imposed on the assembly due to its dogleg design. There was no reference made to vendor manuals for the expansion bellows, nor to relevant British Standards. No drawing was made for the design." (HMSO 1975).

18.2.3 Lessons

Management of Change. Changes to a process or equipment must be reviewed and implemented by people with knowledge appropriate to the situation. This incident is important in the history of process safety as the prime example of the importance of an MOC program. The site made no engineering review of this change. As seen in the cause section, important mechanical design features were not considered during the change.

Flixborough highlights the importance of Management of Organizational Change (MOOC) as well as physical change. At Flixborough, "the works engineer had left early in the year and had not yet been replaced. At the time the bypass line was being planned and installed, no engineer was on site with the qualifications to perform a proper mechanical design, or to provide critical technical review on related issues. There were chemical and electrical engineers on staff, but no other mechanical engineers." A statement often used in relation to the modifications at Flixborough is that "they didn't know what they didn't know". Although the presence of a mechanical engineer may not have changed the outcome if no MOC review was held, it is more likely that the significance of the change could have been recognized by someone at the plant. MOOC covers modification of work schedules, personnel turnover, task allocation changes, organizational hierarchy changes, and organizational policy changes. *Guidelines for Managing Process Safety Risks During Organizational Change* (CCPS 2013) covers this topic in more detail.

Compliance with Standards. As stated in the summary, the site office building was destroyed. At the time, 1974, no facility siting and layout standards existed. This event is an example of why such a standard, API RP 752, "Management of Hazards Associated with Location of Process Plant Buildings" was developed.

18.3 Introduction to Management of Change

Much thought goes into the design and engineering of a facility to support it operating safely. This includes a design review with PHA, a pre-startup safety review, sound operating procedures, and ongoing mechanical integrity processes. Management of change addresses any changes in design, equipment, and operation. Examples of types of changes are given in Table 18.1.

One of the key things to understand is what constitutes a change. Many things may be changed over the life cycle of a facility for example, to improve operations, expand production, add new products, and replace worn equipment. The one example that is not a change is replacement-in-kind. This deserves a close review as changes that may seem like a replacement-in-kind may have minor variations that warrant consideration as a change.

Making a change can unknowingly negate prevention and mitigation measures and introduce new hazards. MOC reviews all safety barriers and confirms these are not degraded by the change and identifies where new hazards warrant additional safety barriers.

MOC is specifically covered by the U.S. OSHA PSM and U.S. EPA RMP regulations, and by regulations in other countries. These regulations typically include updating the Process Safety Information (PSI) and operating procedures affected by the change and informing the workforce of the change.

Management of Change - A management system to identify, review, and approve all modifications to equipment, procedures, raw materials, and processing conditions, other than replacement-in-kind, prior to implementation to help ensure that changes to processes are properly analyzed (for example, for potential adverse impacts), documented, and communicated to employees affected. (CCPS Glossary)

Change - Any addition, process modification, or substitute item (e.g. person or thing) that is not a replacement-in-kind. (CCPS 2008)

Replacement-in-kind - An item (equipment, chemicals, procedures, organizational structures, people, etc.) that meets the design specifications, if one exists, of the item it is replacing. (CCPS 2008)

The purpose of management of change is to assess the risk associated with change and mitigate that risk to tolerable levels. This is usually accomplished by a team using a defined methodology and approval process as shown in the flowchart in Figure 18.4.

The process starts with the identification of the need for a change. A change request can come from an individual on the facility floor, a project team, or a research or development team. The initial review considers whether a change is necessary and whether it is a change based on the definition of a change in that MOC system.

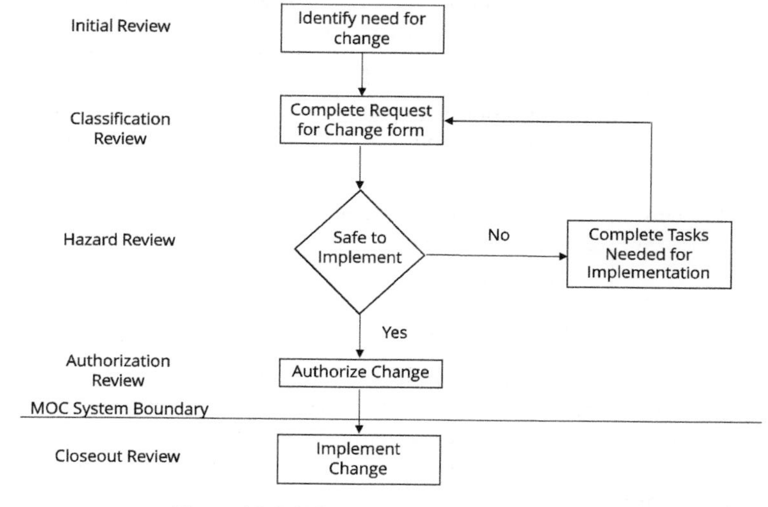

Figure 18.4. MOC system flowchart
(redrawn from CCPS 2008)

Table 18.1. Types of changes and examples
(adapted from CCPS 2005)

Change Type	Examples
Process equipment	materials of construction design parameters, and equipment configuration such as changing from a carbon steel pipe to stainless steel, adding a thermowell to a reactor, or changing the size of an impeller in a pump
Process control	instrumentation, controls, interlocks, and computerized systems including logic solvers and software. such as changing an instrument set point or changing the software logic
Safety systems	alarm and detection systems, fire suppression systems such as allowing process operation while certain safety systems are out of service
Site infrastructure	permanent and temporary buildings, roads, and service systems such a changing the number of people working in a building or closing an access road
Operating and technology	process conditions, process flow paths, raw materials and product specifications, introduction of new chemicals on site, and changes in packaging such as adding a bypass or changing the feedstock composition
Asset Integrity	Inspection, testing, and preventive maintenance, or repair requirements such as lengthening an inspection interval or changing the lubricant type used in a compressor
Procedures	standard operating procedures, safe work practices, emergency procedures, administrative procedures, and maintenance and inspection procedures such as changing the order of steps in a procedure or changing who is responsible for steps in a procedure
Organization	organizational and staffing changes such as reducing the number of operators on a shift, changing the maintenance contractor for the site, changes from a 5-day operation to a 7-day operation, or changing roles and responsibilities
Management Systems and Policies	Documentation describing the management system and the supporting policies, procedures, and practices such as changes in content, revision frequency, or responsibilities
Other changes	anything that is not a replacement-in-kind

A management of change review can be performed by a single person; however, it is more effective when performed by a team of two or more. The people involved may be based on the risk level of the equipment or procedure being changed or impacted. (Refer to Chapter 14.) The MOC process involves several people and stages thus the documentation completion and hand-off are key aspects of MOC.

The hazard review involves reviewing the request, typically by someone independent from the requester, to determine if any potentially adverse risk impacts could result from the change and may suggest additional measures to manage risk. The nature of the review depends on the extent of the change. Simple changes may only need a few knowledgeable people using a checklist. A significant change may require a more detailed HIRA study such as a HAZOP.

The hazard review may identify actions to prevent or mitigate the risk such as addition of protective systems, revision of procedures, addition of equipment to maintenance management systems, and training. These actions should be completed before the change is authorized. A wide variety of personnel and technical skills are normally involved in making the change, notifying or training potentially affected employees, and updating drawings and documents affected by the change.

When all the actions are completed, final authorization can be granted. This final approval for implementing the change comes from a designated individual, independent of the hazard review team.

Lastly the management of change is closed out, an operational readiness review such as a Pre-startup safety review, is conducted, and the equipment can be started up.

Changes can be permanent or temporary. Temporary changes should have a defined end date. If an extension is requested, or if the temporary change is requested to become permanent, the change should be reviewed again to ensure the change is still warranted and any risk reduction measures are still valid.

18.3.1 Management of Organizational Change

Although the importance of management of change is understood by many organizations today, the importance of Management of Organizational Change (MOOC) is less well understood. MOOC covers personnel changes, task allocation changes, organizational structure changes and policy changes. OSHA issued a letter of interpretation making clear that organizational changes that "may affect PSM at the plant level ... would therefore trigger a PSM MOC procedure" (OSHA 2009) The HSE lists the following potential organizational changes that should trigger an MOOC review as follows.

- business process re-engineering
- delayering
- introduction of 'self-managed' teams
- multi-skilling
- outsourcing
- mergers, de-mergers and acquisitions
- downsizing
- changes to personnel with process safety responsibilities
- centralization or dispersion of functions
- changes to communication systems or reporting relationships (HSE)

The risk assessment in an MOOC should focus on the competence of the people involved and the system in which they conduct their work including the time available. The risk assessment should consider not only the risks of the change but also the risks during the change. An organizational change puts demands on top of uncertainties resulting in increased stress and the potential for more human performance issues (Chapter 16). The risk assessment should include consideration of the change on all operating conditions including emergency response, e.g. impacts on emergency response crew numbers.

18.4 What a New Engineer Might Do

New engineers will likely be involved in the Management of Change process. They may be requesting a change, participating in the hazard review of a change, closing action items created by the change, or tracking changes as they progress through the MOC system. In all cases, a new engineer should understand the MOC system at their facility. A key part of this is understanding what is, and what is not, a change. It can be easy to believe that a change is minor enough that it won't impact process safety; however, many significant process safety events have resulted from what was considered to be a minor change at the time.

18.5 Tools

Resources to support Management of Change include the following.

CCPS *Guidelines for the Management of Change for Process Safety*. This book provides guidance on the implementation of effective and efficient Management of Change (MOC) procedures, which can be applied to improve process safety. In addition to introducing MOC systems, the book describes how to design an MOC system, including the scope of the system and the applications over a plant life cycle and the boundaries and overlaps with other process safety management systems. (CCPS 2008)

CCPS *Guidelines for Managing Process Safety Risks During Organizational Change*. This book provides an understanding of the management of organizational change which is essential for successful corporate decision making with little adverse effect on the health and safety of employees or the surrounding community. Addressing the myriad of issues involved, this book helps companies bring their MOOC systems to the same degree of maturity as other process safety management systems. Topics include corporate standard for organizational change management, modification of working conditions, personnel turnover, task allocation changes, organizational hierarchy changes, and organizational policy changes. (CCPS 2013)

18.6 Summary

It is inevitable that changes, permanent or temporary, will be made to a facility over its life cycle. The intent of Management of Change is to ensure those changes don't inadvertently introduce new hazards or remove any existing risk prevention or mitigation measures. The MOC process starts with identifying a change. This is a key step as changes that are not identified will not be managed. The MOC involves a hazard review which can be simple or can be as complex as a HAZOP. Action items identified in the MOC process may be required either pre-start up or post-start up. It should be verified that the pre-start up action items have been implemented before the equipment is started up.

18.7 Other Incidents

This chapter began with a description of the Flixborough Explosion. Other incidents relevant to management of change include the following.

- Union Carbide MIC Release, Bhopal, India, 1984
- Hickson Welsh Jet Fire, Yorkshire, U.K., 1992
- Texaco Oil Refinery Explosion and Fire, U.K., 1994
- Esso Longford Gas Plant Explosion, Australia, 1998
- Georgia Pacific Hydrogen Sulfide, Pennington, Alabama, U.S., 2002
- Hayes Lammerz Dust Explosion, Indiana, U.S., 2003
- Formosa Plastics VCM Explosion, Illinois, U.S., 2004
- BP Isomerization Unit Explosion, Texas City, Texas, U.S., 2005
- Buncefield Storage Tank Overflow and Explosion, U.K., 2005
- T-2 Laboratories Reactive Chemicals Explosion, Florida, U.S., 2007
- Valero-McKee LPG Refinery Fire, Texas, U.S., 2007
- Imperial Sugar Dust Explosion, Georgia, U.S., 2008
- Deepwater Horizon Well Blowout, Gulf of Mexico, U.S., 2010
- Williams Olefins Heat Exchanger Rupture, Louisiana, U.S., 2013
- DuPont MMA Release, LaPorte, Texas, U.S., 2014

18.8 Exercises

1. List 3 RBPS elements evident in the Flixborough explosion and fire summarized at the beginning of this chapter. Describe their shortcomings as related to this accident.
2. Considering the Flixborough explosion and fire, what actions could have been taken to reduce the risk of this incident?
3. What is a "replacement-in-kind"?
4. What changes were made in the Esso Longford gas plant explosion incident?
5. What change was made in the Imperial Sugar dust explosion incident?

18.9 References

API RP 752, "Management of Hazards Associated with Location of Process Plant Buildings", American Petroleum Institute, Washington, D.C., 2009.

CCPS Glossary, "CCPS Process Safety Glossary", Center for Chemical Process Safety, https://www.aiche.org/ccps/resources/glossary.

CCPS 2005, "Building Process Safety Culture: Tools to Enhance Process Safety Performance, Flixborough", American Institute of Chemical Engineers, Center for Chemical Process Safety, New York, NY.

CCPS 2008, *Incidents That Define Process Safety*, Center for Chemical Process Safety, John Wiley & Sons, Hoboken, N.J.

CCPS 2008, *Management of Change for Process Safety*, Center for Chemical Process Safety, John Wiley & Sons, Hoboken, N.J.

CCPS 2013, *Guidelines for Managing Process Safety Risks During Organizational Change*, Center for Chemical Process Safety, John Wiley & Sons, Hoboken, N.J.

HMSO 1975, *The Flixborough Disaster – Report of the Court of Inquiry*, Her Majesty's Stationery Office.

HSE, "Chemical Information Sheet No CHIS7, Organisational change and major accident hazards", https://www.hse.gov.uk/pubns/chis7.pdf.

OSHA 2009, https://www.osha.gov/laws-regs/standardinterpretations/2009-03-31-0.

19

Operating Procedures, Safe Work Practices, Conduct of Operations, and Operational Discipline

19.1 Learning Objectives

The learning objectives of this chapter are:

- Understand conduct of operations and how it is supported through operating procedures and safe work practices,
- Understand the importance of operational discipline, and
- Assist in the development and revision of operating practices and safe work practices.

19.2 Incident: Exxon Valdez Oil Spill, Alaska, 1989

19.2.1 Incident Summary

On March 24, 1989, the Exxon Valdez ran aground on a reef in the Prince William Sound, off the coast of Alaska, at 12:04 AM. See Figure 19.1. Approximately 40,860 cubic meters (257,000 barrels) of oil was spilled. Approximately 2,092 km (1,300 mi) of shoreline were impacted, 321 km (200 miles) of it heavily or moderately. Exxon spent $2.1 billion on cleanup costs. (EVOSTC)

An outcome of the Exxon Valdez oil spill was Exxon's development of its Operational Integrity Management System (OIMS). Exxon's OIMS precedes the OSHA PSM regulation in the U.S. and approximately matches the CCPS Risk Based Process Safety elements. (Refer to Chapters 2 and 3)

Key Points:

Conduct of Operations – Conduct of operations depends on human performance with respect to both the performance itself and the systems supporting their performance. In this case, the equipment and the workload management aspects of the system resulted in fatigued and inexperienced personnel who did not perform well.

Management of Change – Both equipment and organizational changes were made but not managed and these changes resulted in degradation of safety performance.

Figure 19.1. Exxon Valdez tanker leaking oil
(EVOSTC)

19.2.2 Description

At 11:25 PM, the state pilot (who guides the ship out of the harbor) left the ship and the captain informed the Vessel Traffic Center that he was increasing to sea speed. He also reported that the Exxon Valdez would divert from the outbound lane and end up in the inbound lane if it had no conflicting traffic due to icebergs. The traffic center indicated concurrence, stating the inbound lane had no reported traffic. The ship actually went beyond the inbound lane.

At 11:52 PM, the command was given to place the ship's engine on "load program up" a computer program that, over a span of 43 minutes, would increase engine speed from 55 RPM to sea speed full ahead at 78.7 RPM. It is normal practice for the Captain to be present on the bridge when maneuvering in restricted waters. However, after conferring with the helmsman about where and how to return the ship to its designated traffic lane, the Captain of the Valdez left the bridge. At about midnight, the ship struck the reef. The grounding was described by the helmsman as "a bumpy ride" and by the third mate as six "very sharp jolts". Eight of 11 cargo tanks were punctured. Computations aboard the Exxon Valdez showed that 21,955 cubic meters (138,095 barrels) had gushed out of the tanker in the first few hours. (EVOSTC). Figures 19.2 is a photograph of the cleanup operations.

The National Transportation Safety Board investigated the incident and determined that the probable causes of the grounding were as follows.

- The failure of Exxon Shipping Company to supervise the master and provide a rested and sufficient crew for the *Exxon Valdez*. Notes: The average size of an oil tanker crew was 40 in 1977, the Exxon Valdez had a crew of 19. The Exxon Valdez crews routinely worked 12-14 hour shifts and had rushed to get the tanker loaded and out of port.

- The failure of the U.S. Coast Guard to provide an effective vessel traffic system. Note: The radar station in Valdez had replaced its radar with a less powerful one, the location of tankers near Bligh reef could not be monitored with this equipment.
- The lack of effective pilot and escort services. Note: The practice of tracking ships out to the Bligh reef had been discontinued; tanker crews were never informed of this. (Notes from Leveson 2005.)

Figure 19.2. Shoreline cleanup operations in Northwest Bay, west arm, June 1989 (Shigenaka 2014)

19.2.3 Lessons

Conduct of Operations. In the Exxon Valdez grounding, several operational requirements for operating the vessel were not followed. Another conduct of operation issue was the failure of the Exxon Shipping Company to supervise the master. Plant management should spend some time in the field observing the conditions of the plant and behavior of personnel and communicating with them. It is through such activities that a manager can observe whether or not the most up to date set of instructions are being followed. In this case, whether or not the crew was sufficiently rested, and the officers were following procedures were questions of Conduct of Operations.

Management of Change. Two major changes were made in the way ships were guided out of Prince William Sound; the onshore radar was downgraded and the practice of tracking ships out of the sound was discontinued. It was noted that ship crews were not informed of the end of that practice. A management of change review, which included a representative of the shipping company, could have either prevented the change or allowed ships to adjust their operations accordingly.

19.3 Operating Procedures

19.3.1 Introduction

Operating procedures are practical instructions that describe how to safely operate a facility.

> **Operating Procedures** - Written, step by step instructions and information necessary to operate equipment, compiled in one document including operating instructions, process descriptions, operating limits, chemical hazards, and safety equipment requirements. (CCPS Glossary)

Operating procedures are an element of the U.S. OSHA PSM and U. S. EPA RMP regulations. These rules require that there be procedures for the various phases of operation, and that they be kept up to date. Local authorities and other countries are likely to have similar requirements. Operating procedures should be written for all phases of operation including the following.

- Routine operating procedures – intended to support normal operation within safe operating limits, (refer to Section 10.5)
- Non-routine operating procedures – addressing operations that happen infrequently such as catalyst regeneration
- Batch operations – where operator interaction may be required when making additions and taking samples
- Emergency procedures – explaining what actions to take when the operations exceed safe operating limits or when utilities are lost
- Start-up and shut-down procedures – addressing these phases of the operation where operating parameters (temperature, pressure, level) may be changing significantly, more operator intervention is required, and many engineered safety systems may be unavailable
- Transient operations – such as extended holds for maintenance or where specific equipment is out of service

In addition to the step-by-step instructions, good operating procedures describe the process parameters, associated hazards, tools, protective equipment, and controls. The operating procedures should describe what is expected, so the operators can confirm that the process is responding as expected, and what actions to take if the process does not respond as expected. Operating procedures also provide instructions for troubleshooting when the process exceeds normal operating limits. Operating procedures should specify when an emergency shutdown is appropriate and how it should be executed.

Operating procedures should be available to the operators and others that use them. They can be in manuals at the worksite, available electronically, and printed on demand. It is common for critical tasks to use 'in hand' procedures where the operator has a copy in hand to check off each step as each task is performed. Where the workforce is not fluent in a single language, the operating procedures should be provided in multiple languages.

19.3.2 Writing Operating Procedures

Operating procedures translate the engineering design intent into instructions for the operator to achieve the process results, safely. Operating procedures are intended to support human performance by describing the intended methods to be used by each operator in an effort to achieve consistently executed tasks.

Operating procedures are often jointly written by operators (who use the procedures) and engineers who have a high degree of involvement and knowledge of process operations. Operators, supervisors, engineers, and managers are often involved in the review and approval of new procedures or changes to existing procedures. Other work groups, such as maintenance, should also be involved if the operating procedures could potentially affect them.

Risk associated with the various operating phases should be identified in the HIRA. (Refer to Chapter 12.) HIRAs should include startup and shutdown, loss of utilities, and should define the responses to the identified process upsets. This information can be written into the operating procedures.

Once procedures have been developed and approved, the procedures should be followed. Where potential changes are identified to improve the procedures, these should be subject to the Management of Change process. A common problem occurs when a procedural change is taken with no review, and the change seems to work without incident. This deviation then becomes accepted and may present an unknown risk. This is known as normalization of deviance.

Operating procedures should be routinely reviewed, annually per OSHA, to verify that they reflect current operations. If operating procedures require a change, they should be changed through a formal management of change process as described in Chapter 18.

19.4 Safe Work Practices

Safe work practices are formalized processes to help control hazards and manage risk associated with work that is not directly involved with process operations. Maintenance and inspection activities within process areas are examples of work that would be managed with a safe work practice. Making and breaking connections to unload a railcar would likely be covered by an operating procedure, whereas breaking a connection to a pressure transmitter would be considered maintenance and inspection activity and included in the scope of the safe work practice.

Safe Work Practices - An integrated set of policies, procedures, permits, and other systems that are designed to manage risks associated with non-routine activities such as performing hot work, opening process vessels or lines, or entering a confined space. (CCPS Glossary)

Safe work practices are not specific to a single activity or work instruction. They are generic and often written for use across a facility or the entire organization. Many countries may have specific regulations for safe work practices which industries are required to follow. Hot work practices (for work involving welding, flames, or generating sparks) are required by U.S. OSHA PSM and U.S. EPA RMP while other safe work practices may be required by other regulations. Examples of safe work practices include those listed in Table 19.1.

Table 19.1. Example Safe Work Practices

Control of general hazards or protection of personnel from a hazard	• Lockout/tagout and/or control of energy hazards • Line breaking/opening of process equipment • Confined space entry • Work authorization (hot work, cold work, safe work) • Access to process areas by unauthorized personnel • Hot tapping lines and equipment (i.e. drilling into process piping)
Protection against mishaps that could have catastrophic secondary effects	• Excavation in or around process areas • Operation of vehicles in process areas • Lifting over process equipment • Use of other heavy construction equipment in or around process areas
Control of special hazard	• Use of explosives/blasting operations • Use of ionizing radiation (e.g., to produce x-ray images of process equipment)
Prevention of unauthorized impairment of safety systems	• Fire protection system impairment • Temporary isolation of relief devices • Temporary bypassing of interlocks

Although an organization's occupational safety and health departments often oversee and manage safe work procedures, these procedures are essential to process safety. The layers of protection provided in the design can be intentionally negated by the opening of vessels and use of hot work. These safe work procedures support the maintenance, inspection, and repair of process equipment, vessels, controls, and piping in a consistent and safe manner. Safe work procedures are often supplemented with permits (i.e., a checklist that includes an authorization step).

The development of safe work practices can involve multiple engineering disciplines. Safety engineers have a lead role in developing and implementing work permit procedures.

Chemical engineers advise on the process adjustment, isolation, and preparation needed for safe work practices to be implemented in a process area. They may develop lockout plans, confined space entry plans, rescue plans, isolation schemes, and equipment and process preparation activities associated with maintenance, repairs, and inspection.

Mechanical engineers may focus on avoiding confined spaces in a design, effective means for isolation of equipment, and providing additional support for piping and equipment disconnected as part of the isolation. Mechanical engineering may also support the location of equipment to allow maintenance access, tie off points, and ease of entry.

Electrical engineers often have a role in identifying lockout devices, energy isolation and coordination, and lockout design for power sources such as lighting, control panels, and power panels.

Instrument and Control engineers may ensure that control devices can be isolated and secured, that personnel have limited access to PLCs and other logic solvers, adequate control of bypassing is in place, and proper labelling and drawings are completed to ensure safe maintenance repair activities. They may also focus on developing and following validation protocols for new or modified software or programming changes for control systems. They may also be called upon to ensure the proper electrical classification has been identified and that instrumentation is appropriate for the classification and fit for use in the process.

19.5 Conduct of Operations and Operational Discipline

19.5.1 Introduction

Conduct of operations involves the relationship between what operators do, engineers check, and managers plan as illustrated in Figure 19.3. "Conduct of Operations is a term that describes human factors and tools that are necessary to produce repeatability in performance and consistency in results and working within the defined operational boundaries. These skills and tools help the manufacturing unit reduce incidents and take the guess work and human error out of operations." (Forest 2018)

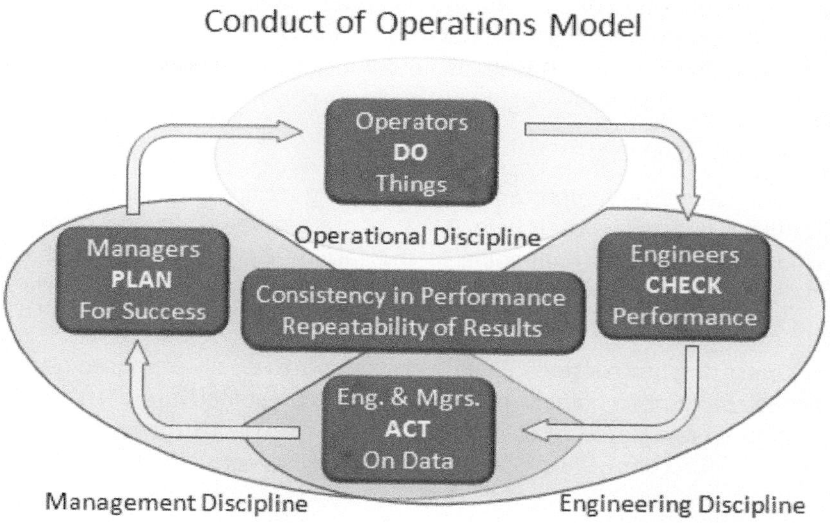

Figure 19.3. Conduct of operations model
(Forest 2018)

Conduct of Operations - The embodiment of an organization's values and principles in management systems that are developed, implemented, and maintained to (1) structure operational tasks in a manner consistent with the organization's risk tolerance, (2) ensure that every task is performed deliberately and correctly, and (3) minimize variations in performance. (CCPS Glossary)

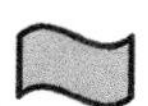

Operational Discipline - The performance of all tasks correctly every time. (CCPS Glossary)

Conduct of operations institutionalizes the pursuit of excellence in the performance of every task and minimizes variations in performance. Personnel at every level are expected to perform their duties with alertness, due thought, full knowledge, sound judgment, and a sense of pride and accountability. Conduct of operations is the execution of operational and management tasks in a deliberate and structured manner. The conduct of operations includes operating procedures and safe work practices, and much more.

Operational discipline is closely tied to an organization's culture. It refers to the operations being conducted correctly, every time, by everyone in the organization.

Operational discipline should not be confused with blindly following a procedure. Operators should think about what they are doing. Some procedures, such as for critical operations, are intended to be used in-hand (often in a checklist format) and followed exactly as written. Other procedures provide general guidance and should be thoughtfully followed, but not literally step by step. For procedures that are expected to be followed step by step, operators should be trained on how to obtain approval for a deviation from the procedure.

Many topics important to the conduct of operations go beyond operating procedures and safe work practices discussed in Sections 19.3 and 19.4. A list of these topics is presented along with a short discussion on each. This list is not all inclusive. It is organized in categories of operational continuity, process oversight, and business performance.

19.5.2 Operational Continuity

Operational continuity is very important in an operating facility, especially across shifts. This includes the things (examples listed below) that operators do during and between shifts so that they understand what is happening in the facility. By monitoring of shift operating instructions, logs, and handover, the engineer can gain a clear understanding of what is happening in a facility.

Shift or Operating Instructions. Shift instructions are usually produced once per day by an operations specialist or frontline supervisor. They are presented in written format either by checklist/form, free-hand notes, or electronic instructions. The instructions give guidance on the activities that will occur on shift. Good practice is for the instructions to have the following attributes.

- A standard format with pre-defined sections and a means for operators and engineers to review and acknowledge the instructions.
- Specific operating conditions, operating constraints, and production targets.
- Key performance indicator targets and operating parameter targets.
- Special instructions for transient operations such as start-up or shutdown.
- Special instructions relating to MOC's, unit tests, special bypass of equipment or safety systems, and any other notable events.
- Any special process safety, safety, environmental, or reliability considerations.

Operating shift log or shift notes. A good practice is for each operator to prepare, and sign, notes or a log each shift. The log contains a record of all significant events that occurred during the shift, including actions taken and observations made. As with shift instructions, engineers should review and acknowledge operator's shift notes. Examples of significant events which should be documented are as follows.

- Any process safety concerns or hazardous conditions with latest status and actions taken
- Unscheduled equipment shutdowns, interlock actuation, SIS or PRD activation, and significant alarm conditions
- Any environmental problems or concerns with latest status and actions taken
- Details of non-standard operating modes or line-ups
- Status of ongoing maintenance work, as well as any non-routine contractor activities. (example, open work permits)
- Any problems requiring follow-up or unusual operating occurrences that may require additional investigation (example, developing reliability issues)
- Product quality problems

- All contacts (direct and by phone) with outside parties (neighbors, regulatory agencies, etc.) concerning company business or complaints
- All incidences of safeguard bypass or disabling. Any exceedance of a safe operating limit,
- Any security issues
- MOC's/ PSSR's completed or begun on shift
- Unit specific operating parameters or KPI's
- Important transfers or production information

Good shift notes detail the problems encountered during the shift and the steps taken to correct. Often unit engineers are called upon to solve problems that arise in the unit. If written thoroughly, the shift notes may give valuable clues that aid the engineer in finding the solutions. Similarly, shift notes provide a platform for proactive problem solving when engineers see subtle changes in comments and notes from operators.

In addition to an engineer's monitoring of operator shift notes, engineers often keep their own shift notes. Research engineers will most certainly keep lab notes and pilot plant observations. The most common application for the unit production engineer is to keep shift notes that detail engineering activities during outages, turnarounds, and during trial or test runs.

Shift Handover. Shift handover (turnover) is closely related to the shift notes in that the content can be used as a template for discussion. A practice should be in place to ensure the shift handover is thorough and promotes safe operations. Good practice includes the following.

- Shift Handover takes place on the job, after the oncoming operator is outfitted for work, and in an area protected from the environment and excessive unit noise.
- Adequate time is given to provide a thorough handover report to the oncoming shift.
- The content elements of the operating shift log are the outline for the handover report.
- Process Leaders discuss all elements of the operating shift log.
- Process safety and plant EHS and/or security concerns are emphasized.
- The status of key operations, maintenance activities, and outstanding work permits (including non-routine contractor activities) at shift handover such as equipment start-ups, batch charging, chemical transfers to tanks, railcars and trucks, etc. is identified.
- The status of MOC's is communicated.
- The bypass log is updated for any safety devices temporarily bypassed or deactivated and what compensating measures are in place.

Alarm Disabling & Management. Good practice is to have a written procedure for disabling alarms with the appropriate level of approval. The procedure should also include: a means to make operators aware of disabled alarms, alternate process indicators to be used while alarms are disabled, and requirements to return the disabled alarm to service.

Car Seals. Car seals, are devices for physically locking valves in position where the position is critical to safe operation. (see Figure 19.4) They play an important role in the effectiveness of process safety systems. P&IDs should indicate whether the valve is car sealed open or car sealed closed (CSO or CSC). By itself, a car seal does not necessarily prevent a position critical device from being moved to an unsafe position. It is what the seal represents and how it is managed that keeps the device in the appropriate position. The seal can be broken in an emergency to change the position of a valve. Chains and locks are sometimes used instead of car seals.

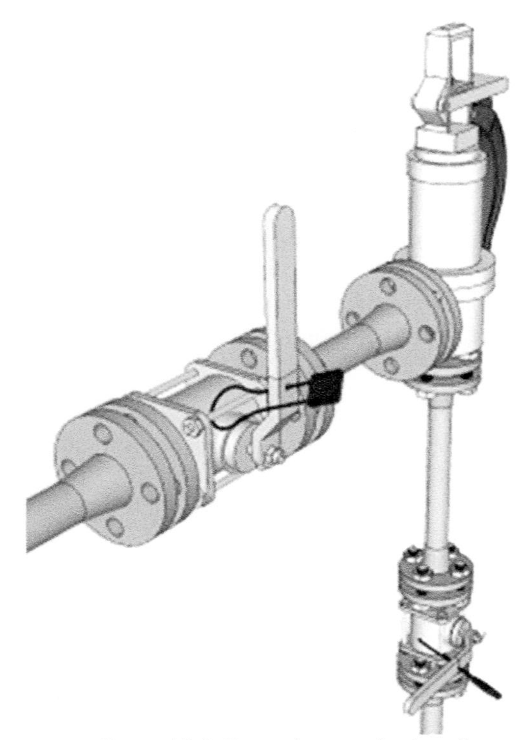

Figure 19.4. Car seal on a valve handle

(Wermac)

Good practices for position critical devices include having a written procedure, frequently verifying correct position, using MOCs for changing the position, and verifying position as part of a PSSR.

Equipment Labeling. The purpose of proper equipment and piping labeling is to support plant operations and maintenance activities. Good practices include labeling of all equipment (including spares), safety instrumented system components, piping, utilities, and safety critical double block and bleed.

19.5.3 Process Oversight

Process Readings and Evaluation. Operators collect information on the process status and evaluate that information to determine if processes are running efficiently and meeting performance targets. The process parameter readings and evaluation portion of operating discipline and engineering discipline are closely related. Engineers should define what operators need to evaluate and define how the evaluation is made. Operators or engineers should be thinking about what the process parameter readings mean, not just recording them by rote. When the criteria are not met, the operator makes decisions (or informs supervision to receive instructions) to take actions to return the process to the predetermined values. For example, if a pump should have a certain discharge pressure to maintain a safe operating limit, that range is defined by the engineer and included on the evaluation sheet. If the range is exceeded, the operator notes the exceedance and notes on the evaluation sheet what is done to return to normal. The engineer (or unit supervision) then closes the loop by shift/daily review of the evaluation sheets to ensure these concerns are adequately addressed.

One of the most common failures in conduct of operations is to have ineffective process evaluation sheets that have operators simply "taking readings" without evaluating the equipment. When this happens, the value of understanding how the process is performing and the opportunity to identify potential upsets before they happen, can be missed.

Good practice for operator evaluation is to include both outside operator evaluations and board operator evaluations. Since data from the DCS can be printed or stored, it is often assumed that operators do not need to collect or write down that information. However, the purpose of writing the information down isn't to collect it for someone else. The purpose is for the board operator to evaluate that variable or parameter and take pre-determined action if the expected value isn't observed. Suggestions on how to create an effective process readings and evaluation sheet are included in Appendix F.

Process Indicators. Engineers should identify process indicators used to track process safety, environmental, reliability, and economic optimization and have a defined schedule of review and action taken. Documentation on these indicators should include the operating range, steps to correct the situation if out of range, who is responsible to monitor the indicator and at what frequency. Safe operating limits are an example of a process indicator. (Refer to Section 10.5.)

Sample Collection. The process sample schedule should be defined by the engineers, documented, and be part of standard operating procedure. A good practice is to develop a process map of defined sample points to optimize the round. PPE and other safety precautions are identified for each sample point. The schedule should include: the sample point, frequency, technical analysis and target ranges, and actions to take if the sample is out of range.

Operator Line-Up. The most fundamental responsibility of an operator is to understand and know the position of every valve in their area of responsibility and to control the energy among all points of material transfer. This responsibility is commonly referred to as "line-up." While engineers usually don't operate or "line-up" equipment, they should clearly understand this aspect of an operator's job and take it into account when designing equipment, developing operating procedures, and conducting operator training.

Practical activities to support consistency in operations and minimize line-up errors, collectively called "walk the line" are discussed in two papers by J. Forest. (Forest 2014 and Forest 2018)

Operators should have a defined responsibility to evaluate all valve positions periodically. They should be trained on this expectation. During steady state operations, the valve positions will not change much, if at all, which makes learning their position a relatively simple task.

However, during non-steady state operation or transient conditions, valve positions will be changed, leading to a higher probability of a process safety incident. Whenever any material is transferred from any point to another the operator must understand and control the transfer of energy. Examples include opening or closing a valve, starting a pump or compressor, putting any equipment into or out of service, bypassing equipment, and others.

An example valve line-up is shown in Figure 19.5. It illustrates that even in a simple transfer from one tank to another, all valve positions should be confirmed to verify that the fluid will be transferred as intended.

Restrict Access to Console Operators/ Minimize Distraction. A board operator can have literally thousands of control points to monitor in a process, each with various attributes. The mode, set point, process variable and alarm status should be known at all times. Unnecessary distractions divert the board operator's attention from their primary role; safe operation of the manufacturing unit.

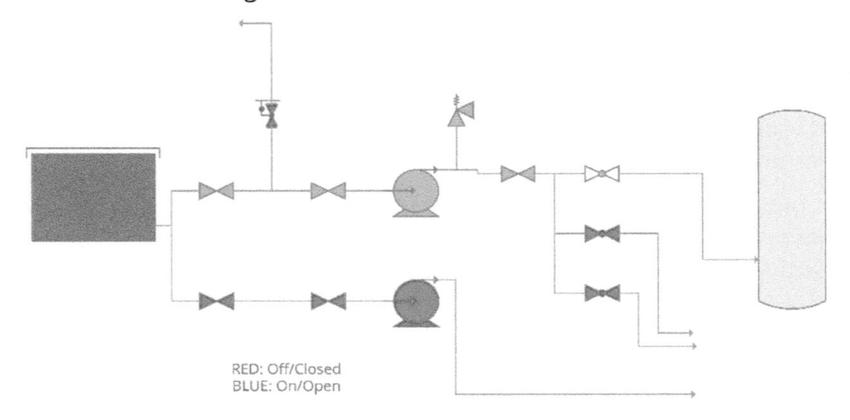

Figure 19.5. Example valve line-up

19.5.4 Business Performance

Fatigue Management. While not explicitly an established topic in operating discipline, working overtime is a reality for many operators and unit production engineers. Many process facilities operate continuous processes around the clock, every day of the year. Sometimes maintenance is needed on the entire process and the operation is shut down for a turnaround to make repairs and maintain equipment. While production is shut down, no product, nor money, is being made. Thus, time is of the essence and work is typically conducted around the clock.

The fact is that tired and overworked employees make more errors. This is especially true for major project work such as turnarounds during the most critical phase of operation; startup following a long working period. Shift work may include, 12-hour shifts, straight night shifts, or some other work schedule that is outside of a normal 40-hour straight day job. With non-standard work schedules also comes fatigue management. API RP 755 "Fatigue Risk Management Systems for Personnel in the Refining and Petrochemical Industries" provides guidance on this topic.

Incident, Event, and Near Miss Notification. Significant events should be reported to shift supervision. Each company has specific definitions of what type of event should be reported in what time frame and to which level of management. Understand and follow these expectations as they are important to process safety and also to regulatory compliance.

Near Miss - An event in which an accident (that is, property damage, environmental impact, or human loss) or an operational interruption could have plausibly resulted if circumstances had been slightly different. (CCPS Glossary)

Daily Review of Operations. Many companies have daily production reviews with staff to define the current manufacturing issues and plans to address them. A more important aspect of the daily meeting is a review of all incidents to ensure proper classification, investigation, and resources devoted to prevent a repeat incident. Correction of near miss causes prevents process safety incidents. Daily review provides a good tool for the site director to communicate directly to section leaders on what is important to work on & evaluate effectiveness of closure.

Safety & Process Safety Evaluation. A good practice is to have separate occupational safety and process safety evaluations in addition to the operator evaluations.

Occupational safety evaluations might include inspection of routes of ingress and egress, safety signs, lighting, ladders, hoses, and evaluation of unit safety equipment. Process safety evaluations might include safety critical equipment, such as ensuring key process safety alarms are correctly set and SIS field bypasses and position critical devices are secured in the safe position.

Defining the appropriate safety and process safety evaluation to be performed is another example of how engineering discipline is closely related to operating discipline. Engineers perform specific reviews with the appropriate process safety information to determine what should be evaluated, and operators complete the inspection. Management discipline then periodically inspects the evaluation sheet to ensure that action is taken on deficiencies noted.

Process Safety Officer. A good practice for conduct of operations is the use of a Process Safety Officer (PSO). A PSO is used as another layer in preventing process safety incidents during startups, shutdowns and other critical operating modes by providing an objective independent safety-oriented perspective of those activities. Each unit should have a written PSO policy describing qualifications, roles, responsibilities, and a PSO reference manual. The unit operations requiring a PSO should be documented and the PSO qualifications and responsibilities defined.

Emergency Accountability. An aspect of the process safety operational discipline that deserves attention is the emergency accountability system. Engineers are often involved in development and participation of emergency drills response. This topic is discussed in Chapter 20.

Housekeeping. Poor housekeeping is a warning sign of an ineffective safety program. Providing an orderly, clean worksite improves performance and morale, and it reduces injuries and process safety incidents. Provide clear expectations for all personnel regarding housekeeping and inspect performance daily or on a shift basis. Performance will follow. Housekeeping applies to process units as well as keeping documentation such as procedures and practices in order.

19.6 What a New Engineer Might Do

A new engineer, like everyone else in a facility, should follow the procedures and practices that relate to the equipment or areas where they work. The number of procedures and practices can be daunting at first. A new engineer may be supporting the operations of a specific unit and should understand the operating procedures and process evaluation reporting for that unit. The other unit operating procedures can wait until more time is available or the engineer's responsibilities change.

Of most importance are the safe work practices as they may prevent immediate harm to the engineer as well as prevent a process safety incident. For example, an engineer may inspect a vessel's internal trays/packing during maintenance work. This is a confined space and understanding confined space safe work practices is imperative before starting this work.

A new engineer will have many procedures and practices to learn in order to help support safe operations at a facility. Education does not end with a diploma. A new engineer should strive to continue to learn. This includes formal learning offers as well as on-the-job learning. This means being observant and paying attention to detail during unit walk-arounds, recognizing hazards in the facility, listening to operators, and reviewing operator logs and process safety information such as P&IDs. One of the best ways to learn is from operators and engineers who have years of experience. To confirm your understanding, ask them about practices and procedures before engaging in the work.

Formal learning offers may include company provided training courses and seminars. External opportunities from AIChE SAChE courses and the Global Congress on Process Safety provide a wealth of information on process safety practices in many industries worldwide.

A new engineer can benefit from reviewing the CSB investigations and videos relevant to this chapter as listed in Appendix G.

19.7 Tools

Resources to support operating procedures, safe work practices, and conduct of operation and operational discipline include the following.

CCPS *Conduct of Operations and Operational Discipline: For Improving Process Safety in Industry.* This book details management practices which help ensure rigor in executing process safety programs in order to prevent major accidents. (CCPS 2011)

BP and IChemE Process Safety Series. Produced jointly by BP and IChemE, this award-winning series includes 16 paperback books, which guide the user in the application of safe operating practices and procedures throughout a range of industries. A few of the topics include the following. (IChemE)

- Hazards of Steam, Fourth edition, 2004
- Safe Furnace and Boiler Firing, Fifth edition, 2012
- Hazards of Trapped Pressure and Vacuum, Third edition, 2009
- Confined Space Entry, First edition, 2005
- Control of Work, Second edition, 2007
- Safe Tank Farms and (Un)Loading Operations, Fourth edition, 2008
- Safe Ups and Downs for Process Units, Seventh edition, 2009

Occupational Safety and Health Administration publications. The OSHA mission is safe working conditions. In addition to standards, OSHA publishes booklets, Fact Sheets, and QuickCards on safe work practices. Figure 19.6 is the OSHA QuickCard on permit-required confined space. (OSHA)

19.8 Summary

Process safety depends on the day-to-day ability of the organization to rigorously conduct operations correctly every time. The failure of one person in completing a job task correctly one time can, unfortunately, lead to serious injuries and process safety incidents. The conduct of operations includes operators taking actions in the facility, engineers checking data, and managers planning work. They should all have the operational discipline to conduct work correctly, every time.

Work can be performed correctly and consistently by following operating practices, safe work practices, and other practices and procedures supporting communication, process oversight, and business performance. These practices and procedures should reflect the current facility design and operations. If they do not, a management of change process should be used to update them.

Operating procedures provide step-by-step guidance on operating the process including what should be done to keep it performing inside of operating limits and what to do if it exceeds those limits. Operating procedures should address all phases of the operation.

Safe work practices address maintenance, inspection, and other work that can defeat safeguards (such as breaking containment) and introduce hazards (such as hot work).

Many other practices and procedures support safe operations including those that support communications across shifts such as shift handover and shift operating notes. As many tasks take place over multiple shifts, having a clear record and understanding of what has happened and what should happen next is important to support safe operations.

Permit-Required Confined Spaces

A confined space has limited openings for entry or exit, is large enough for entering and working, and is not designed for continuous worker occupancy. Confined spaces include underground vaults, tanks, storage bins, manholes, pits, silos, underground utility vaults and pipelines. See 29 CFR 1910.146.

Permit-required confined spaces are confined spaces that:

- May contain a hazardous or potentially hazardous atmosphere.
- May contain a material which can engulf an entrant.
- May contain walls that converge inward or floors that slope downward and taper into a smaller area which could trap or asphyxiate an entrant.
- May contain other serious physical hazards such as unguarded machines or exposed live wires.
- Must be identified by the employer who must inform exposed employees of the existence and location of such spaces and their hazards.

What to Do

- Do not enter permit-required confined spaces without being trained and without having a permit to enter.
- Review, understand and follow employer's procedures before entering permit-required confined spaces and know how and when to exit.
- Before entry, identify any physical hazards.
- Before and during entry, test and monitor for oxygen content, flammability, toxicity or explosive hazards as necessary.
- Use employer's fall protection, rescue, air-monitoring, ventilation, lighting and communication equipment according to entry procedures.
- Maintain contact at all times with a trained attendant either visually, via phone, or by two-way radio. This monitoring system enables the attendant and entry supervisor to order you to evacuate and to alert appropriately trained rescue personnel to rescue entrants when needed.

You have a right to a safe workplace.
If you have questions about workplace safety and health, call OSHA.
It's confidential. We can help!

For more information:

OSHA® Occupational Safety and Health Administration

www.osha.gov (800) 321-OSHA (6742)

OSHA 3214-07R 2013

Figure 19.6. OSHA QuickCard on permit-required confined spaces (OSHA)

19.9 Other Incidents

This chapter began with a description of the Exxon Valdez Oil Spill. Other incidents relevant to operating procedures, safe work, conduct of operations, and operational discipline include the following.

- Nypro Explosion, Flixborough, U.K., 1974
- ICMESA Dioxin Release, Seveso, Italy, 1976
- Piper Alpha Platform, North Sea, U.K., 1988
- Port Neal AN Explosion, Sioux City, Iowa, U.S., 1994
- Elf Refinery BLEVE, Feyzin, France, 1996
- Shell Hydrocracker Explosion, Martinez, California, U.S., 1996
- Motiva Enterprises Sulfuric Acid Tank Failure, Delaware City, Delaware, U.S., 2001

19.10 Exercises

1. List 3 RBPS elements evident in the Exxon Valdez oil spill summarized at the beginning of this chapter. Describe their shortcomings as related to this accident.
2. Considering the Exxon Valdez oil spill, what actions could have been taken to reduce the risk of this incident?
3. Why use a double block and bleed system to isolate a vessel vs. just one isolation valve?
4. Why does OSHA require that standard operating procedures be written?
5. Name three kinds of work permits.
6. An engineer will be entering a chemical plant vessel to do an internal inspection. Name two safe work practices that should be used in this work task.
7. Maintenance work is to be conducted on a chemical pump. What safe work practices should be used?
8. Welding or cutting pipe in a process unit would require use of what safe work practice?
9. The U.S. Nuclear Navy has never had a reactor incident. They must be very good at what?
10. Name two sources of information that you could use as a means to understand what happened on the night shift.

19.11 References

API RP 755, "Fatigue Risk Management Systems for Personnel in the Refining and Petrochemical Industries", American Petroleum Institute, Washington, D.C., 2019.

CCPS Glossary, "CCPS Process Safety Glossary", Center for Chemical Process Safety, https://www.aiche.org/ccps/resources/glossary.

CCPS 2011, *Conduct of Operations and Operational Discipline*, Center for Chemical Process Safety, John Wiley & Sons, Hoboken, N.J.

EVOSTC, Exxon Valdez Oil Spill Trustee Council website, http://www.evostc.state.ak.us/.

Forest 2014, "Walk the Line", *Process Safety Progress*, Wiley Online Library, DOI 10.1002/prs.11724.

Forest 2018, "Don't Walk the Line – Dance it!", *Process Safety Progress*, Volume 37, No. 4, December.

IChemE, https://icheme.myshopify.com/collections/bp-process-safety-series/products/bp-process-safety-series-set-of-16-books-paperback.

Leveson 2005, Leveson, Nancy G, "Software System Safety", http://web.archive.org/web/20101108055426/http://ocw.mit.edu/courses/aeronautics-and-astronautics/16-358j-system-safety-spring-2005/lecture-notes/class_notes.pdf.

OSHA, https://www.osha.gov/OshDoc/data_Hurricane_Facts/confined_space_permit.pdf.

Shigenaka 2014, "Twenty-Five Years After the Exxon Valdez Oil Spill: NOAA's Scientific Support, Monitoring, and Research", NOAA Office of Response and Restoration, Seattle, WA.

Wermac, http://www.wermac.org/valves/valves_car_seal.html

20
Emergency Management

20.1 Learning Objectives

The learning objective of this chapter is:

- Understand the importance of planning for and managing emergencies.

20.2 Incident: West Fertilizer Explosion, West, Texas, 2013

20.2.1 Incident Summary

On April 17, 2013, a fire occurred at the West Fertilizer Company (WFC) in West, Texas that triggered an explosion of about 27 metric tonne (30 ton) fertilizer grade ammonium nitrate (FGAN) at 7:51 PM. The explosion registered as a 2.1 on the Richter scale. (See Figure 20.1.) Fifteen people were fatally injured, 12 of them were emergency responders, 3 members of the public. One of the public fatalities was in a nursing home (from a stress induced heart attack) and the other two were in an apartment complex. An additional 260 people were injured. The overpressure from the blast damaged 150 buildings offsite, including 4 schools, a nursing home (later demolished), an apartment complex, and 350 private residences (142 beyond repair) (CSB 2013).

This was a significant incident in the U.S., due to the extensive public impact, and the prevalence of FGAN storage and handling facilities in the U.S. The CSB identified over 1,300 facilities handling ammonium nitrate (AN) within close proximity to a community, so the U.S. President issued Executive Order EO-13650. This established a working group consisting of the U.S. Department of Homeland Security, the U.S. Environmental Protection Agency, and the U.S. Department of Labor (under which OSHA is located), Justice, Agriculture and Transportation. The purpose of the working group was to improve the identification and response to the risks of chemical facilities (EO 2013).

Figure 20.1. Video stills of WFC fire and explosion
(CSB 2013)

Key Points:

Process Safety Culture - A poor safety culture will have consequences. It can lead to incidents, bad press coverage, and failure to receive new operating permits. It could be much worse. Process safety should be valued and seen as important by all.

Stakeholder Outreach - Work together to prevent incidents. It is important that local planners understand the hazards of neighboring facilities and that enforcement agencies identify shortfalls in compliance. Stakeholders communicating with each other can create a mutual understanding on managing risks.

Emergency Management - Do the responders understand the hazards? Inform your local emergency responders of the risks on your site so that when they respond to help you, you do not put them in harm's way.

20.2.2 Description

West Fertilizer Company (WFC) stored and handled ammonium nitrate (AN) in a fertilizer building along with several other fertilizers including diammonium phosphate, ammonium sulfate, and potash. The fertilizer building was a wood frame building. AN was stored in two plywood bins. Figure 20.2 shows an overview of the building layout.

In addition to receiving and storing the various fertilizers, West Fertilizer also made fertilizer blends, delivered, and sometimes applied the fertilizers. West Fertilizer also stored and handled anhydrous ammonia in two pressurized storage tanks.

When the facility was first built, 1962, it was surrounded by open land. Over the years the town grew and WFC came to be surrounded by residences and schools (Figure 20.3). This contributed to the high impact of this incident. Most of the fatalities and injuries occurred within 457 to 610 m (1,500 to 2,000 ft) of the explosion (CSB 2013).

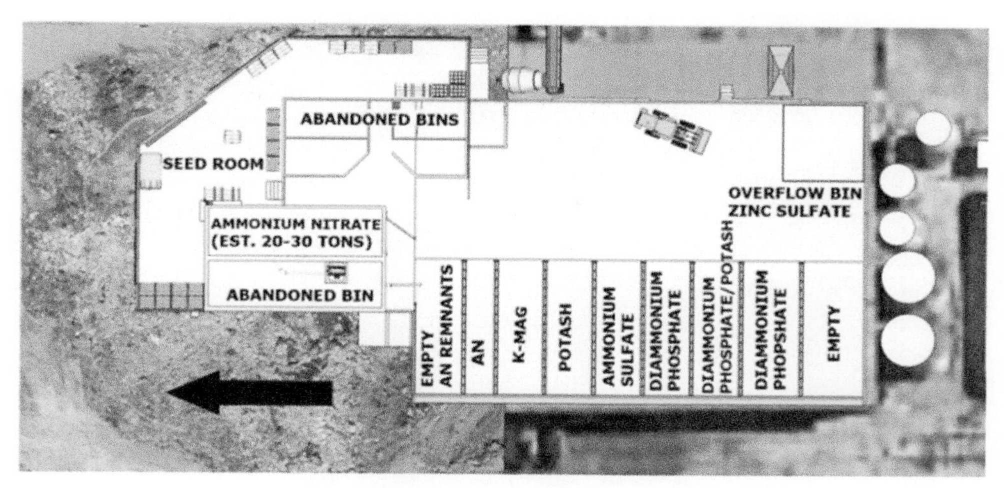

Figure 20.2. Fertilizer building overview
(CSB 2013)

Figure 20.3. WFC and community growth (left - 1970; right - 2010)
(CSB 2013)

Figure 20.4. Overview of damaged WFC
(CSB 2013)

Figure 20.5. Apartment complex damage
(CSB 2013)

WFC itself was destroyed (see Figure 20.4). An FGAN railcar was overturned. Fortunately, the two anhydrous ammonia tanks on-site were not damaged. A large amount of off-site property was damaged. The following were severely damaged.

- an apartment complex, 122 m (450 ft) from WFC (see Figure 20.5)
- an intermediate school, 168 m (552 ft) from WFC
- a nursing home, 183 m (600 ft) from WFC
- a high school, 385 m (1,263 ft) from WFC

The cause of the initial fire itself is unknown. The ATF concluded that the cause was arson (Ellis 2016). The CSB developed three theories as to why the AN exploded that did not involve arson (CSB 2013).

The first scenario is that during the early part of fire, soot and other organics contaminated the FGAN and served to keep heat in. This could have caused formation of hot liquid FGAN at the top of the pile. The liquid layer could have produced oxidizing gases, which would have created a cloud of oxidizers, NO_2, O_2 and HNO_3. All are the decomposition products of AN. This gas cloud may then have detonated.

The second scenario is that the detonation was caused by heat from the exterior walls of the bin. Photos show that just prior to the detonation, the exterior walls of the bin were penetrated, which allowed more air in and caused the fire to become even hotter. There could have been some melting of the FGAN along the exterior wall.

The third scenario focuses on an elevator pit; a bucket elevator was used to unload FGAN and other materials. There could have been FGAN remnants in the pit. FGAN could have spilled into the pit if the wall of the AN bin collapsed. The remnant of FGAN could have been contaminated by burning rubber and the falling FGAN, plus the confinement by concrete elevator walls might have caused the detonation. This is considered the least likely scenario.

20.2.3 Lessons

The RBPS management systems are interlinked, and the West Fertilizer explosion shows how important this linkage is.

Process Safety Culture. Prior to 2009, WFC had insurance through Triangle Insurance Company. In 2009 Triangle stopped insuring WFC because of losses and a lack of compliance with Triangle's recommendations from their loss control surveys. Several of the recommendations involved electrical problems, such as corroded wires and grounds. In one of its evaluations, a Triangle consultant noted that WFC had no safety program and "had no positive safety culture". (CSB 2013).

Compliance with Standards. AN is covered by OSHA's "Blasting and Explosive Agents" standard (OSHA 1998); however, this is not widely known throughout the fertilizer industry. AN is also covered by NFPA 495, "Code for the Manufacture, Transportation, Storage, and Use of Explosives and Blasting Agents" (NFPA 495) and NFPA 400, "Hazardous Material Code" (NFPA 400). Prior to 2002, AN was covered NFPA 490 "Code for the Storage of Ammonium Nitrate" (NFPA 490).

A weakness in the OSHA standard is that it allows the use of wood "protected against impregnation by ammonium nitrate" for the walls of the bin (the floor must be non-combustible) (OSHA 1998). NFPA 400 was updated in 2016 and now requires buildings be of non-combustible construction, automatic sprinklers, and fire detection systems, the last two being *retroactive* requirements.

The fact that the OSHA standard covers AN is not well known in the fertilizer industry as reported by the industry itself (CSB 2013). OSHA did not have a history of citing fertilizer facilities under the *Blasting and Explosive Agents* standard, contributing to this lack of knowledge. This contributed to a lack of process safety knowledge in the industry, which in turn led to inadequate hazard identification and emergency response planning.

AN is not covered by OSHA PSM, or EPA RMP. This means that facilities handling AN do not need a process safety management program. The lack of a PSM program led to several safety management gaps.

Stakeholder Outreach. WFC shared little information with emergency responders and the community. The lack of process safety knowledge on WFC's part contributed to this. Without an understanding of the potential hazards of ammonium nitrate at the WFC facility, they had no motivation to prevent the community from building up near the facility.

Process Knowledge Management. Since AN was not on the PSM or RMP highly hazardous chemical list, and because the fertilizer industry was not familiar with the OSHA *Blasting and Explosives Agents* standard, neither the WFC management and employees, nor the emergency responders, were familiar with the AN hazard. The Emergency Responders did not know that AN could detonate.

Process safety knowledge includes collecting and disseminating information and learnings from incidents with similar technologies and chemicals from throughout the industry. AN producers and handlers should learn from the long history of AN related incidents. In Texas in 2009, a fire occurred at another facility that stored and handled AN. The firefighters decided not to fight the fire but to evacuate the area. About 80,000 people were evacuated. A review of that emergency response was conducted, and an after-action report was issued that emphasized the need for emergency responders to "reflect on protection, response and recovery activities" that occurred in the 2009 fire (CSB 2013). This report apparently was not known by the West Fire Department.

Emergency Management. The absence of AN from the PSM and RMP rules led to no emergency planning, which also would have been required by these regulations. When responding, the fire department initially tried to fight the fire, but only the fire engines internal tanks could be used until a hose could be connected to the hydrant, which was 490 m (1,600 ft) away. They did not have enough hose to reach the fire. Developing an emergency response plan should have exposed these problems and allowed them to be addressed before an incident occurred.

20.3 Introduction to Emergency Management

Emergency management is a necessary element of process safety. Despite all the effort put into preventing and mitigating potential process safety incidents, they still occur. Be prepared.

An emergency can cause harm to people or the environment or damage to property. It can also prompt a facility's license to operate within the community to be questioned. Effective emergency management can save lives, protect property and the environment, and reassure stakeholders that a facility is well managed. Emergency management is part of the U.S. OSHA PSM and U.S. EPA RMP regulations and are covered by regulations in other countries.

Managing human performance is challenging during an emergency and having a well-practiced plan can improve this performance. Emergency management strives to protect workers, neighbors, and emergency responders. It also focuses on communication so that both those involved in the response and external to it are aware of what is happening.

Many types of emergencies can impact process safety either directly, such as a fire, or indirectly by lessening the ability of those supporting the safe operation of a facility to do so.

- Process safety incident (refer to Chapter 9)
- Natural disasters such as floods, hurricanes, and tornedos (as seen in Figure 20.6)
- Incident at a neighboring property
- Pandemic
- Intentional attack or sabotage

Emergency management encompasses activities that occur at the facility and in the community before, during, and after an emergency including the following.

1. Emergency response planning for potential emergencies
2. Emergency response resources to execute the plan
3. Emergency response exercises for practicing and continuously improving the plan
4. Emergency response training or informing employees, contractors, neighbors, and local authorities on what to do, how they will be notified, and how to report an emergency
5. Emergency response communications with stakeholders in the event an incident does occur

Figure 20.6. Coffeyville Refinery 2007 flood
(KDA)

20.3.1 Emergency Response Planning

Emergency response plans should be developed collaboratively with experts aware of potential hazards, operations personnel that could be involved in an emergency response, and emergency responders (internal and external). The Local Emergency Planning Committees (LEPC), more than 3000 across the US, develop emergency response plans and interact with stakeholders. The following are the steps in developing an emergency response plan.

Identify accident scenarios based on hazards. Emergency response plans can address a few scenarios involving each type of hazard to cover the range of potential scenarios. Process safety emergency scenarios can be selected from hazard identification studies (refer to Chapter 12) and from industry incident history. Other emergencies, such as those noted in section 20.3, may be identified through focusing on the specific hazard. The CCPS Monographs on *Assessment of and Planning for Natural Hazards* and *Risk Based Process Safety During Disruptive Times*, and CCPS *Guidelines for Analyzing and Managing the Security Vulnerabilities of Fixed Chemical Sites* provide helpful guidance. (CCPS 2020 and CCPS 2003)

Plan response actions. Response actions should be identified, reviewed, and optimized in advance of a potential incident as opposed to trying to decide what to do in the heat of the moment. Response actions include, but are not limited to, the following.

- Emergency recognition and reporting – Identifying what is considered an emergency and when and to whom it is to be reported.

- Authority and methods for raising a standby or evacuation alarm – Stating who has the authority to call an emergency or various emergency response actions and how they are to do so.
- Personal protective equipment to assist in response and evacuation.
- Safe havens or shelter-in-place locations – Identification of who has the authority to call for this strategy that can be applied onsite, offsite or both and when this should be used.
- Emergency egress – Identification of evacuation routes, assembly points, and actions that should be taken at the assembly points, such as verifying headcounts.
- Establishment of an emergency operations center (EOC) – Identification of when, where, and by whom an EOC is to be established. This includes the incident command structure (see Figure 20.7) along with names, responsibilities, and contact information.
- Operational responses - Operations personnel are typically responsible for immediate emergency response activities, such as shutting down the process and isolating hazardous material inventories. They may also be members of an on-site emergency response team.
- Firefighting preplans – The provision of firewater, firefighting foam and other fire suppressants appropriate to fight a fire should be identified in advance so that the firefighting personnel, equipment, and supplies are available. This includes both onsite personnel and equipment and mutual aid response teams from nearby facilities. In addition, facility reviews should be conducted to predetermine where run-off firewater, foam, and contaminants will be captured so as to avoid creating cascading problems from a potential incident.
- Exclusion zones - Defining the boundaries for the "hot" and "warm" zones including control of access into and out of these zones.
- Response team communications – The equipment and protocols to be used should be defined with particular emphasis on communications between responders, operations, and supporting groups.
- Decontamination procedures – This includes identification of equipment and methods to decontaminate any chemical hazards to protect emergency responders or those involved in clean-up activities.
- Crisis management and business continuity – These may be included in an emergency response plan or be the subjects of separate plans. They address continuing business during and after the emergency and are normally led by a senior manager.

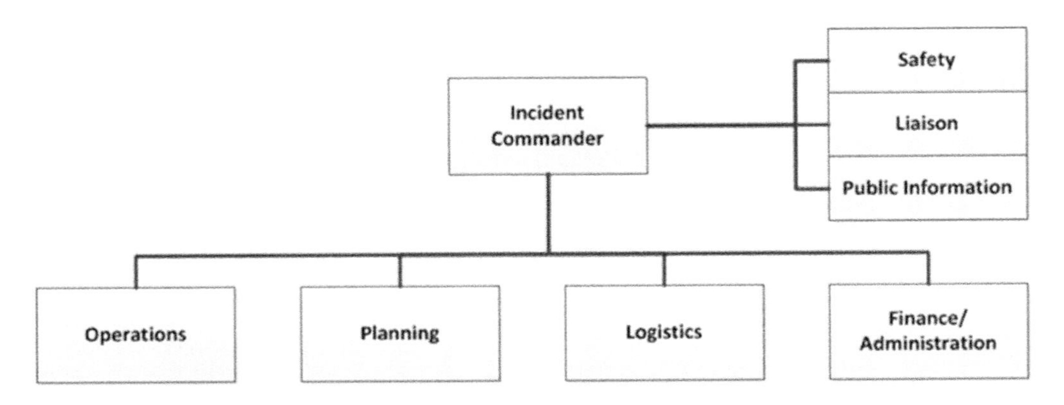

Figure 20.7. Incident command structure
(listo)

Develop written emergency response plan. The response actions should be written in an emergency response plan. This not only formally documents the plan but also forms the basis for determining what facilities, equipment, staffing, training, communication, coordination, and other resources or activities are required. The emergency response plan should clearly address the response actions along with who is responsible to perform them.

Operational discipline is important to ensure that personnel adhere to emergency procedures in the same manner as for other procedures. (Refer to Section 19.5.) During an emergency event, things happen quickly, many people are involved, and communications can be difficult. Performing the emergency response actions as planned can be critical as it is the combination of all the actions that can bring the emergency under control.

Emergency response plans, as with other procedures, should be reviewed on a routine basis as well as after drills when improvement opportunities may have been identified.

20.3.2 Emergency Response Resources

The resources needed for emergency response should be described in the response plan. Resources can include firefighting equipment, chemical spill containment supplies, personal protective equipment and more. Facilities should also give considerable thought to the location of equipment: siting it too far from likely response areas will increase response time, while siting it too close to locations where hazardous materials may be released can make it difficult impossible to use safely during an incident.

The response equipment must be maintained, periodically inventoried, and tested to ensure that it will function when it is needed.

20.3.3 Emergency Response Exercises

Effective emergency response requires practicing emergency response roles, actions, and communications. This not only provides practice but can also identify areas for improvement in the emergency response plans. For example, a site may have appropriate fire equipment, but the time and staff required to deploy this may be lacking (e.g. layout of firehoses, make all necessary connections, set-up firewater monitors).

Emergency response exercises can be based on scenarios identified in process hazard analysis or on past incidents. Exercises can include tabletop exercises, tests of communication systems, and field drills. Drills can be simple involving a small portion of a response team or can be quite large involving mutual aid supporters and external agencies. Each drill should be followed by a critique of the response, the communications, and the plan. Findings should be used to improve the emergency response plan.

20.3.4 Emergency Response Training

Emergency response training should be conducted for those with roles defined in the emergency response plan and others affected by the potential emergency. This may include employees, contractors, neighbors, and local authorities. The training should include how they will be notified and how they should respond in an emergency. Training should be provided initially and refreshed periodically.

20.3.5 Emergency Response Communications

Communications are critical to an emergency and having effective communications requires planning. This includes what communications equipment will be used, where this equipment will be located, who will be communicating, and maintaining a list of current names and contact information. Although this may be relatively simple inside a facility, emergency communications will include contractors, local authorities, neighbors, and other stakeholders which can increase the complexity. A strategy should be developed to ensure that facility responders can quickly and easily communicate with other responders (community, mutual aid, etc.). It should also be kept in mind that the emergency itself can challenge communications. For example, natural disasters can interrupt power supplies and communication towers leaving cell phones inoperable.

20.4 Recovery and Recommissioning

After the emergency has passed, it is time to manage the aftermath and resume operations. There will likely be damaged equipment to repair, contamination to address, hidden or silent failures, new hazards associated with old equipment, and typical startup challenges. The first step in this phase is to stabilize and secure equipment to make it safe for investigators and those working to clean up and repair it. It also preserves potential evidence, physical and electronic, that could be helpful in understanding what happened and preventing its recurrence. The next step is to repair the facility.

The recommissioning plan for a facility following an emergency must be at least as comprehensive as that for the initial start-up of a new facility. Before starting operations, those responsible for the startup should be properly trained, it should be verified that the equipment is ready to receive the chemicals and utilities, and all operational and safety systems should be functional. Things that may have worked properly before the emergency may not work after it. Do not assume that equipment will perform as expected. Confirm it. Refer to Chapter 17 on operational readiness.

20.5 What a new Engineer Might Do

Every person in a facility, including new engineers, should know what to do during an emergency. At the simplest level, this can be recognizing alarm signals and evacuating using escape routes or sheltering in place as per instructions. Beyond that, a new engineer may be trained and participate on an emergency response team as a firefighter or as a technical expert advising the incident command. Whatever the role of a new engineer in emergency management, they know their role and perform it per the response plan.

A new engineer can benefit from reviewing the CSB investigations and videos relevant to this chapter as listed in Appendix G.

20.6 Tools

Resources to support emergency management include the following.

CCPS Monograph: *Assessment of and planning for natural hazards.* This monograph provides basic information, an approach for assessing natural hazards, means to address the hazards, and emergency planning guidance. It applies to both new and existing facilities. It provides several checklists for preparation ahead of meteorological hazards such as hurricanes Katrina and Harvey in the U.S. and many such events across the globe. It also addresses recovery in the aftermath. (CCPS 2019)

CCPS Monograph: *Risk Based Process Safety During Disruptive Times.* This CCPS Monograph provides insights for managing Process Safety during the COVID-19 pandemic and other similar crises. It incorporates input from company experience and learnings during these events. It is organized by the RBPS elements and human factors impact is addressed in multiple areas. (CCPS 2020)

Ready.gov. This U.S. Government webpage provides guidance for development of an emergency response plan and includes many helpful links.

National Incident Management System. "The National Incident Management System (NIMS) provides a systematic, proactive approach to guide departments and agencies at all levels of government, non-governmental organizations, and the private sector to work seamlessly to prevent, protect against, respond to, recover from, and mitigate the effects of incidents, regardless of cause, size, location, or complexity, in order to reduce the loss of life and property and harm to the environment." (FEMA) NIMS includes a section on incident command systems.

20.7 Summary

Even with much process safety effort, emergencies do occur. Be prepared for them to minimize potential harm and damage. Potential emergencies include process safety incidents as well as natural disasters, incidents at a neighboring property, pandemics, and intentional attacks or sabotage. Emergency response actions should be planned collaboratively with those that would be involved in the emergency response and written in an emergency response plans. Personnel should be trained on the plans and exercises should be conducted to both test the plans and provide practice for the responders.

Communication is critical to an effective emergency response. It involves the responders and many external stakeholders during a time when events may be confusing and moving quickly. Communication equipment warrants consideration as it should be available when needed and not be made ineffective by the emergency itself.

20.8 Other Incidents

This chapter began with a description of the West Fertilizer Explosion. Other incidents relevant to emergency response include the following.

- Grandcamp Freighter Explosion, Texas City, Texas, U.S., 1947
- Sandoz Storehouse Fire, Basel, Switzerland, 1986
- Gulf Oil Refinery Fire, Philadelphia, Pennsylvania, U.S., 1975
- Deepwater Horizon Well Blowout, Gulf of Mexico, U.S., 2010

20.9 Exercises

1. List 3 RBPS elements evident in the West Fertilizer explosion summarized at the beginning of this chapter. Describe their shortcomings as related to this accident.
2. Considering the West Fertilizer explosion, what actions could have been taken to reduce the risk of this incident?
3. Name three types of emergencies that a chemical facility should prepare for?
4. What are key steps a company should take to prepare for an emergency?
5. Why is it important to have an Emergency Response Plan?
6. Name two leading indicator metrics that would help you judge the 'health' of the emergency response preparedness in a given facility?
7. When planning a drill what external stakeholders might you invite to participate?
8. Considering process safety emergencies, how might you decide what emergencies to include in the emergency response plan?

20.10 References

CCPS Glossary, "CCPS Process Safety Glossary", Center for Chemical Process Safety, https://www.aiche.org/ccps/resources/glossary.

CCPS 2003, *Guidelines for Analyzing and Managing the Security Vulnerabilities of Fixed Chemical Sites*, Center for Chemical Process Safety, John Wiley & Sons, Hoboken, N.J.

CCPS 2019, "Monograph Assessment of and planning for natural hazards", https://www.aiche.org/sites/default/files/html/536181/NaturalDisaster-CCPSmonograph.html.

CCPS 2020, Monograph Risk Based Process Safety During Disruptive Times, https://www.aiche.org/sites/default/files/html/544906/RBPS-during-COVID-19-and-Similar-Disruptive-times.html.

CSB 2013, "West Fertilizer Company Fire and Explosion, Chemical Safety and Hazard Investigation Board", Investigation Report, Report 2013-02-I-TX, U.S. Chemical Safety and Hazard Investigation Board, Washington, D.C.

EO 2013, Executive Order 13650 Improving Chemical Facility Safety and Security, White House, Washington, D.C., August 1.

Ellis, Ralph, 2016, "Fire that led to Texas fertilizer blast set on purpose, officials say", cnn.com. CNN. Retrieved May 11.

FEMA 2008, National Incident Management System, https://www.fema.gov/pdf/emergency/nims/NIMS_core.pdf.

HSE 1999, "Emergency planning for major accidents: Control of Major Accident Hazards Regulations", Health and Safety Executive, U.K.

KDA, https://agriculture.ks.gov/divisions-programs/dwr/floodplain/resources/historical-flood-signs/lists/historical-flooding/coffeyville.

Listo, https://www.listo.gov/es/node/344.

NFPA 400, "Hazardous Material Code", National Fire Protection Association, Quincy, MA.

NFPA 495, "Code for the Manufacture, Transportation, Storage, and Use of Explosives and Blasting Agents", National Fire Protection Association, Quincy, MA.

OSHA 1998, 29 CFR 1910.109, Blasting and explosive agents, Occupational Safety and Health Administration, Federal Register 33450, June 18.

Ready.gov, https://www.ready.gov/business/implementation/emergency.

21
People Management Aspects of Process Safety Management

21.1 Learning Objectives

The learning objectives of this chapter are:

- Describe how process safety competency requirements may be similar/different across a company,
- Understand the importance of managing process safety at the owner/contractor interface, and
- Identify process safety activities that involve the workforce and other stakeholders.

21.2 Incident: Deepwater Horizon Well Blowout, Gulf of Mexico, 2010

21.2.1 Incident Summary

Most of the information in this section comes from a report by the Bureau of Ocean Energy Management Regulation and Enforcement (BOEMRE 2011), the BP report (BP 2010), the CSB reports (CSB 2014 a, b, c, and d), and the Transocean report (TO 2011).

At approximately 9:50 PM on the evening of April 20, 2010, an undetected influx of hydrocarbons escalated to a blowout on the Deepwater Horizon rig at the Macondo well. A cement barrier was set in the process of temporarily abandoning the well for future production. Tests of the cement barrier integrity were misinterpreted, and the failure of the cement barrier allowed hydrocarbons to flow up the wellbore, through the riser and onto the rig, resulting in the blowout. Shortly after the blowout, hydrocarbons that had flowed onto the rig floor through a mud-gas vent line ignited. Flowing hydrocarbons fueled a fire on the rig that continued to burn until the rig sank on April 22. (See Figure 21.1.)

The event resulted in 11 fatalities and 17 injuries. Over the next 87 days, an estimated five million barrels of oil were discharged from the Macondo well into the Gulf of Mexico. (BOEMRE 2011) This was one of the worst environmental incidents in U.S. history. The incident aftermath had a devastating impact on the gulf coast region economy, and studies of the environmental impact continue. BP, Transocean and MOEX Offshore LLC (10% owner of the well) agreed to pay the following fines. (DOJ 2015)

- $5.5 billion as a Clean Water Act penalty, 80% of which goes to restoration efforts (BP)
- $8.1 billion for natural resource damages (BP)
- $600 million for other claims (BP)
- $4 billion in criminal fines (BP)
- $90 million (MOEX)
- $400 million as a Clean Water Act penalty (Transocean Deepwater Inc.)

At the time, both the revenue management and safety and environmental protection were managed by the U.S. Department of Interior - Mineral Management Service (MMS). The incident caused a reorganization for offshore drilling regulation, and the creation of the U.S. Department of Interior - Office of Natural Resources Revenues (ONRR) which is responsible for the revenue function, the U.S. Department of Interior - Bureau of Ocean Energy Management (BOEM) who is responsible for resource planning and leasing and the U.S. Department of Interior - Bureau of Safety and U.S. Environmental Enforcement (BSEE) who is responsible for safety and environmental protection. (CSB 2014 a)

Key Points:

Process Safety Culture – The way we do things around here. Understanding that company values cannot be ignored. Knowing what the culture actually is 'on the shop floor' and if it is consistent across a company may identify opportunities for improvement.

Asset Integrity and Reliability – Is that last line of defense truly a defense? The integrity of barriers that are critical to safety and safe shutdown should be assured through systematic analysis and maintenance.

Contractor Management – Have a clear interface. Many workplaces involve multiple contractors and numerous interfaces. Is there complete clarity on who is in handling what? Are communication paths defined and used so that all are informed

Figure 21.1. Fire on Deepwater Horizon
(CSB 2014d)

21.2.2 Description

The Macondo well was owned by BP (leaseholder and operator). Transocean was the owner and operator of Deepwater Horizon, the drilling rig. Halliburton was responsible for the well monitoring and cementing operations. Cameron, contracted by Transocean, was responsible for providing testing and repairs for the Blowout Preventer (BOP), a key safety and environmental protection device. Other subcontractors involved as well, but these were the main companies involved.

At the time of the incident the well was being shut down temporarily, with the intention of being reopened for production later, a process known as temporary abandonment. The production casing, a high strength steel pipe set in a well to ensure well integrity and allow future production, was installed on April 18 - 19. The bottom of the well was located in a laminated sand-shale zone, which has an increased likelihood of cement channeling, which can prevent a strong bond. (BOEMRE 2011)

On April 19, cementing began using novel nitrogen foamed cement. The purpose of the cement is to seal the well and prevent hydrocarbons from flowing out of the well. The cement operation was monitored by comparing the amount of material flowing into the well with what comes out. The crew believed they had seen a full return of everything that went in, indicating a successful cementing job.

After the cementing was completed, personnel ran a positive well integrity test run to see check for outflow from the well to its surroundings. This well passed the positive test. The positive well test cannot determine if the cement is sealing the well at the very bottom. A negative pressure test was conducted. The test was repeated several times with poor results, but eventually the pressure did stop increasing. A final test, a cement bond log, was cancelled on the belief the cement barrier injection was successful.

After deciding the cementing was successful, the crew began to complete the temporary abandonment procedures. During this time, the well was supposed to be monitored for abnormalities, specifically, a "kick" (an influx of hydrocarbon into the well that forces drilling mud back up into the well). Kicks can be detected by imbalances in the inflow and outflow of the well. During this time, volume in some of the tanks and pits was increasing. Eventually the blowout occurred.

Gas alarms began going off on the rig. The general alarm system was not activated automatically which required the control room to manually sound the general alarm. Personnel were told to abandon the rig 12 minutes after the first gas alarm went off.

The Blowout Preventer (BOP) is a large 17 m (57 ft) tall, and 363 metric tons (400 tons) apparatus at the ocean floor that is designed to seal a well in an emergency. The BOP had Variable Bore Rams (VBR) designed to seal around the drill pipe and annulars designed to close around the drill pipe (Figure 21.2). The annulars and VBR were activated by the crew. It also had a Blind Shear Ram (BSR), designed to cut the drill pipe and seal the well. A subsequent investigation showed that the BOP had been actuated too late. Multiple sections of drill pipe had become wedged in the BOP and had pushed the pipe out of the blind shear ram cutting zone. The BOP failed to seal the well.

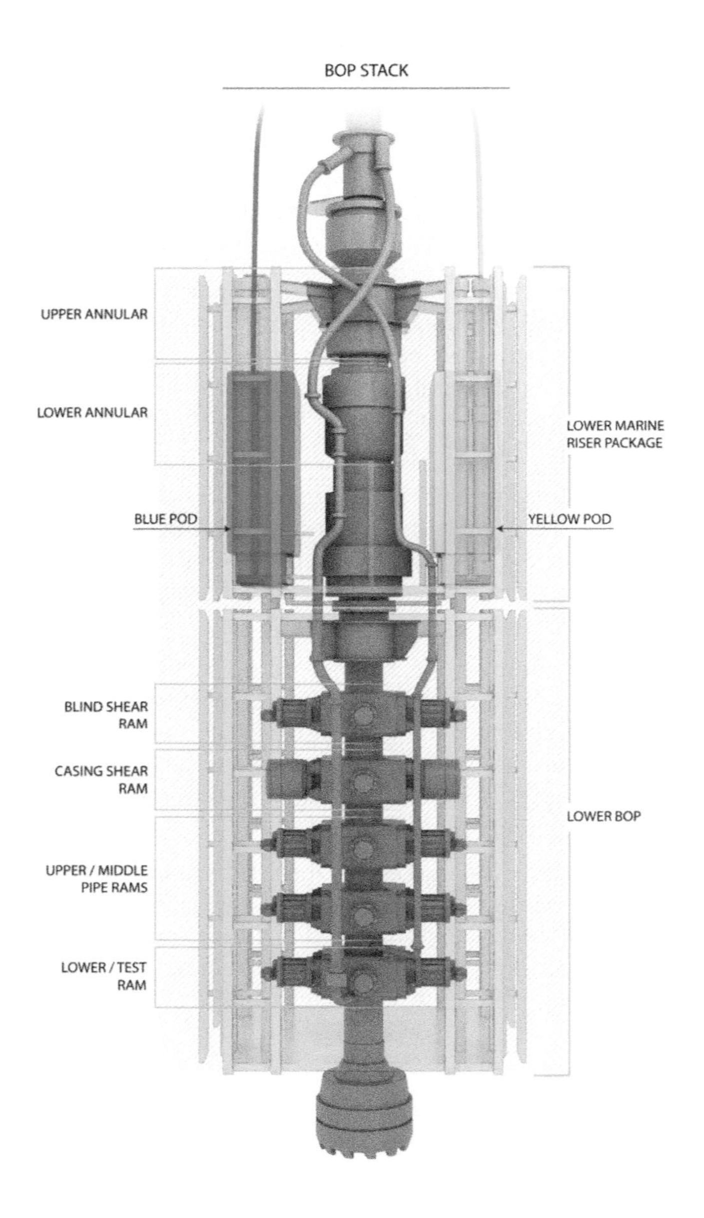

Figure 21.2. Macondo Well blowout preventer
(CSB 2014 c)

BP evaluated several options for plugging the well, however, no risk assessment was done for the plan chosen. (CSB 2014 a) During the drilling of the well, there had been significant losses of drilling mud into the formation. BP engineers and Halliburton studied a way to do the cementing in a way that would minimize additional losses. To do this they used a different cement mixture than originally planned, a foamed cement slurry that is injected with nitrogen bubbles. An MOC review was not done on the change. After the blowout, investigations showed the cement mixture was not stable. The conclusion that the cement job was successful was based on an assumption of the volume of a pump stroke. This value was found to be incorrect during the investigation.

After the cementing was completed, a positive well integrity test was run to check for outflow from the well to its surroundings. This well passed the positive test. The positive well test cannot test if the cement is sealing the well at the very bottom. A negative pressure test was conducted although it was not called for in the abandonment plans and was not required by regulations. The results of the negative pressure test showed that drill pipe pressure was increasing; this was an indication the cement barrier had failed, and material was flowing into it. The test was repeated several times and eventually the pressure did stop increasing. Not believing the initial test results, a member of the crew of the rig put forward a theory (which became known as the bladder effect) to explain the differences and the well leaders accepted it. A final test, a cement bond log, was cancelled on the belief the cement barrier injection was successful. The BOEMRE investigation states that the "central cause of the blowout was failure of a cement barrier in the production casing string". (BOEMRE 2011)

A simplified explanation of this behavior is that, based on the original monitoring of material in and out, and the successful positive test, the crew believed the cement job was successful, and any evidence to the contrary was rationalized away. This is called "confirmation bias" and is discussed in Section 16.4.

During the temporary abandonment operations, the well was supposed to be monitored for kicks by measuring the outflow of the well into a pit. At this time, the crew began directing the mud to two pits instead of one, and from them to other pits and from the rig to another ship, reducing the ability to rapidly detect a kick. A mudlogger, an employee from a different contractor, was supposed to do this monitoring. He questioned directing the mud to two pits, but was told this was how it was done, and let the matter go at that. The result was that a kick was hard to detect. When the kick did occur, it was not detected. All of this was a violation of the rig owner's policies regarding well monitoring.

During this time, the pit level rose by 15.8 m^3 (4,190 gal) in 15 minutes. The crew's response was to try to bleed off pressure by opening the well, an indication they still did not know that the well was actually flowing. The capacity of the mud gas separator was overwhelmed, and the hydrocarbon flowed onto the rig. The blowout could have been sent to diverter lines which would have directed it off of the rig (Figure 21.3), which would have reduced the likelihood of ignition of the release and reduced the consequences if ignition did occur. The diverter system can be routed to direct well fluids containing flammable gas to the mud gas separator (green on the figure) so that gas can be vented away from rig floor or drilling fluids can be directed routed overboard (red on the figure). Procedures on when to use the diverter instead of the mud gas separator were not clear and were overly complicated (the normal procedure to switch to overboard flow took 10 steps). (BOEMRE 2011, CSB 2014 d)

As the alarms were sounding, the engine room operators called for instructions, but were never told to shut down the engines. The engines were determined to be the likely ignition source.

Operators had three ways to operate the BOP in an emergency mode. The explosions likely disabled the first method. Later investigation showed that the second method which should have worked automatically without operator action, likely did not function due to critical control pods on the BOP that were faulty. One had a fault in the solenoid valve, and one had insufficient battery charge. Lastly a remote operated vehicle was used to again try to close the blind shear rams, but by this time (33 hours later), the drill pipe had buckled in the BOP and was forced outside of the zone of the blades of the blind shear ram. The failure to stop flow through the BOP resulted in the prolonged oil spill.

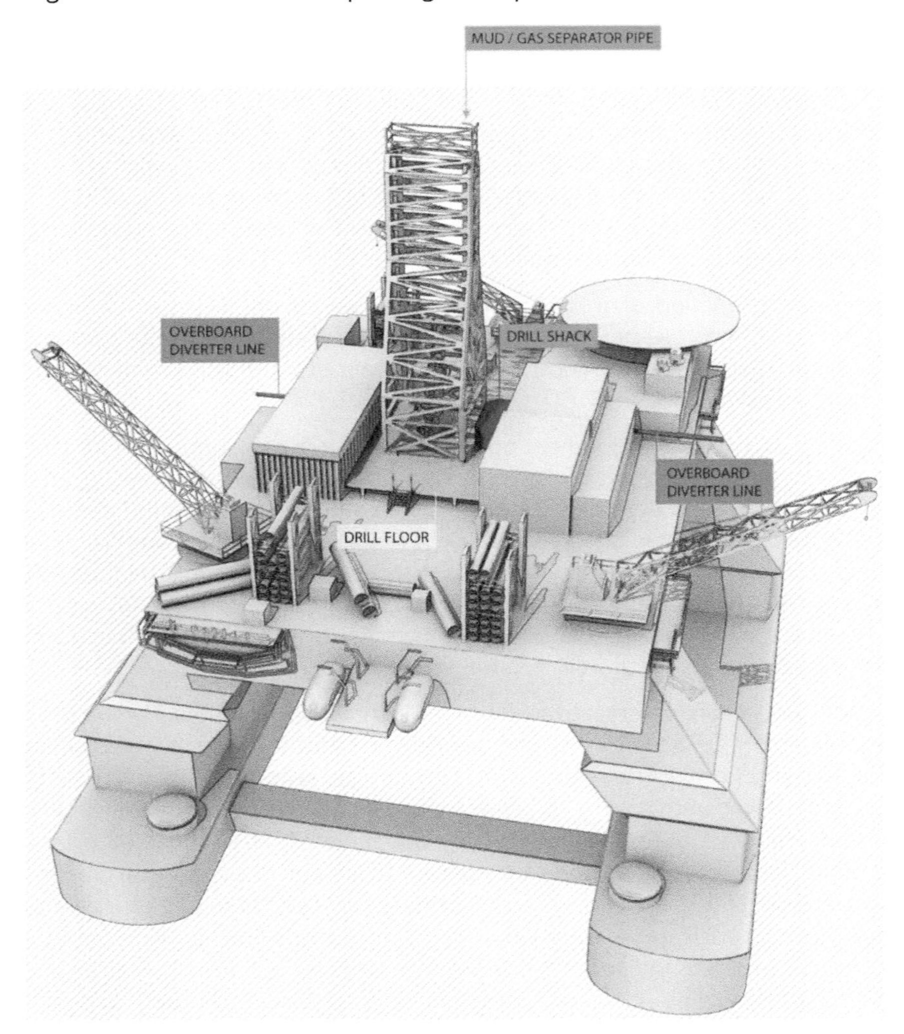

Figure 21.3. The diverter system on a rig
(CSB 2014d)

21.2.3 Lessons

Process Safety Culture. The characteristics of a good process safety culture include maintaining a sense of vulnerability and establishing a learning/questioning environment. The Baker Panel report written after the BP Texas City Refinery explosion focused on process safety culture. (Baker 2007) BP was in the process of implementing the recommendations of the Baker Panel in 2010. In 2008 BP overhauled its management system and developed a new system called the Operating Management System Framework (OMS), and by 2009, OMS was about 80% implemented. BP intended to have OMS applicable to drilling rigs. However, BPs requirements were just being rolled out when the Macondo Well was drilled and were not applied to the Macondo well.

The confirmation bias, which prevented the crew from recognizing the failure of negative pressure test as valid, is another symptom of a lack of a learning/questioning environment and a lack of a sense of vulnerability.

Further illustrating this point was the BOEMRE report statement that "in the weeks leading up to the blowout on April 20, the BP Macondo team made a series of operational decisions that reduced costs and increased risk" and that the investigation team "found no evidence that the cost-cutting and time-saving decisions were subjected to the various formal risk assessment processes that BP had in place".

Compliance with Standards. Neither BP nor Transocean implemented their own process safety management policies. Both had MOC guidelines that were not followed during the abandonment procedure.

The CSB noted Transocean's "minimal guidance and unclear expectations of the risk management tools its personnel should use". The crew at Macondo well did not apply the techniques identified as Transocean's risk management tools; HAZID/HAZOP, Major Hazard Risk Assessment, Safety Case, and Operation Integrity Case. These tools were supposed to demonstrate the risk was As Low As Reasonably Practicable (ALARP), but Transocean did not provide guidance on what tools to use.

Hazardous processes should be designed with multiple safeguards. BOPs are designed with multiple rams that close in various ways and are intended to shut off the flow from the well either through the annular space or the central drill pipe. At the time of the incident, Transocean had BOPs with two BSRs on 11 out of 14 of its rigs, and BP had two BSRs on all the other rigs it was leasing. The Deepwater Horizon rig only had one BSR. Although it was normal industry operating practice at the time, relying on such a vulnerable layer of protection as the final layer is an example of poor COO. One could argue that reliance on the BOP may have reduced the crew's "sense of vulnerability" as they believed it was the ultimate layer of protection, when, in fact, it was a flawed safeguard.

Operating Procedures. The Deepwater Horizon crew were not supplied with a procedure for testing the cement barrier. The crew did not, therefore, have a criterion for deciding if the test was positive or negative, or actions to take following a negative test. (CSB 2014 a) The abandonment procedure was written 24 hours in advance. This was partly due to the nature of the strata at the bottom of the well, which could not be known until the well was drilled. No

MOC or process hazard review was done for the procedure, with exception of an occupational safety review.

Asset Integrity and Reliability. The BOP was not managed as Safety Critical Equipment, though it was the only equipment on the rig designed to be able to stop a blowout. One of each pair of redundant solenoid systems were inoperable at the time of the blowout. The BOP was overdue for vendor-recommended preventive maintenance, and no effective testing or monitoring process was in place to confirm the availability of the redundant systems in the emergency Automatic Mode Function (AMF)/Deadman system if called upon to function. (CSB 2014 b)

Contractor Management. An offshore drilling rig employs many contractors, hence communications and management of the relationship between owner and the various contractors is very important. The CSB report (2014 d) states, referring to BP and Transocean, that "while both companies had more rigorous corporate policies for risk management, neither assumed effective responsibility for ensuring their implementation at Macondo."

One safeguard against a blowout was supposed to be the monitoring of well conditions by the mudlogger. The mudlogger was from a subcontractor. He was not included in the discussions that occurred during the well testing, so was unaware there had been issues with the negative pressure test, diminishing the reliability of this safeguard. When he raised concerns about how the outflow was being directed to multiple locations, he was disregarded.

Management of Change. The temporary abandonment procedure was changed several times, but no MOC review was done on any changes. Changes included using foamed cement (which is known to be less stable than non-foamed cement), and the cement left-over from a previous well.

Conduct of Operations (COO). Transocean relied upon operator response to sound alarms rather than automated shutdowns for its most critical safeguards against catastrophic reservoir blow-out and gas in the riser, yet when the blow-out actually occurred the operating staff hesitated to engage them. The delay in activating the general alarm and the failure to shut down the two operating diesel generators, which seem to be the likely ignition source, shows a failure of Conduct of Operations (COO).

In addition to these two, the valves to divert flow from the inboard mud separators to the outboard, emergency discharges were remote operated but required operator action. A robust design would have automated this. This is another example of inadequate COO with respect to engineering design. (The CSB's conclusion was that Transocean was concerned about preventing environmental releases from inadvertent discharges of drilling mud to the ocean.) Finally, it seems unclear even now when and by whom the final safeguard, the Blind Shear Ram, was actuated, only that it failed to seal off the well pipe.

Neither BP nor Transocean ensured that sufficient, robust safeguards were in place.

Incident Investigation. The Deepwater Horizon well blowout was an informative illustration of the need for learning from experience. The simplest example of not learning from experience concerns an earlier kick at the Macondo Well. The kick had occurred on March 8, 2010 and was also not detected for 30 minutes. Detection and response to a kick is a key safety barrier in well operations. The failure to detect the kick of March 8 should have been investigated. This was required by BP's own internal requirements. The failure to do an investigation was cited as a contributing cause to the incident by the BOEMRE report (BOEMRE 2011).

The next level of failure to learn was not learning the lessons from similar incidents at other rigs. In 2008, a blowout occurred on a BP rig in the Caspian Sea. It was reported to be due to a poor cement job and resulted in 211 people being evacuated from the rig and the field being shut down for 4 months. The Montrara blowout was a similar incident in Australian waters. (AU 2010). However, in the risk matrix for the Macondo Well, an uncontrolled well incident was considered a medium risk event (cost of $ 1-3 million).

In December 2009, an event similar to the Deepwater Horizon's occurred on an offshore rig operated by Transocean in the United Kingdom. The crew had finished displacing mud and conducted a pressure test. They stopped monitoring and were surprised when mud began flowing onto the rig. In this event they were able to shut down the well. Transocean, the owner and operator of the drilling rig, prepared a presentation on this event, and issued an operations advisory to its North Sea fleet. However, the lessons from these events were not learned by the crew and engineers running the Deepwater Horizon.

21.3 Overview

The previous chapters have explained what process safety is, why it is important and provided details on various process safety topics. Clearly process safety involves many people conducting their work diligently and in concert. The challenge is how to manage and sustain this focus on process safety performance over time.

The intent of process safety management is to not only pull all of the aspects of process safety together, but also to sustain process safety performance over time.

Section 1.5 discussed process safety management systems and Chapter 2 detailed CCPS *Risk Based Process Safety Management*. These process safety management systems include topics on the practice of process safety as addressed in Chapters 3 through 19.

Management System- A formally established set of activities designed to produce specific results in a consistent manner on a sustainable basis. (CCPS Glossary)

Some additional topics relate more to the management of, instead of the practice of, process safety. Those topics that focus on the people management aspects of process safety management are listed as follows and addressed in this Chapter.

- Process safety competency
- Training and performance assurance
- Process safety knowledge management
- Contractor management
- Workforce involvement
- Stakeholder outreach

Those topics that focus on the management aspects of process safety management are addressed in Chapter 22. As this book is intended for undergraduates and new engineers and these topics are typically the responsibility of management, they are treated more briefly than the process safety topics in the previous seventeen chapters.

21.4 Process Safety Competency

People should be competent in the field in which they work. This does not mean that they need to be expert in all topics but instead that they should possess the appropriate level of understanding of a topic to perform the work that they are conducting. Consider a facility that uses complex chemistry to produce products and some of the chemicals used are explosive. All people working with those chemicals should be competent to understand the chemical hazards. In addition, some experts would be competent in complex chemistry – but likely not all of the personnel.

This competency applies to all personnel at all levels of an organization. For example, Operators should be expert in operating practices and safe work practices. Engineers should be expert in codes and standards. Those supporting process safety studies such as HAZOP and risk analysis should be expert in those areas. Managers, although they may not be expert in all of these areas, should have a sufficient competency to be able to consider all these topics when making business decisions.

Managing process safety competency requires defining the process safety competency requirements for each role, providing opportunities to help individuals increase their process safety competency, and assuring that they can perform as desired (refer to Section 21.5). The CCPS *Guidelines for Defining Process Safety Competency Requirements* contains useful information on the topic. Typically, companies develop matrices that list job roles on one axis and topics on the other. Each cell then indicates the level of competence required such as awareness, practitioner, expert.

21.5 Training and Performance Assurance

Building on the topic of competency, personnel should be provided learning opportunities to support development of the required competencies. A learning and development model suggests that 70% of learning comes from work experience, 20% from mentoring or coaching, and 10% from coursework. (Lombardo 1996) Thus, while this RBPS element uses the word "training", it is more appropriate to consider the broader term "learning".

Companies may identify learning opportunities as part of a competency matrix or may include them in job description requirements. The learning opportunities could be working in certain roles or attending courses. Progressing through the roles or courses may enable workers to meet minimum initial performance standards, to maintain their proficiency, or to qualify them for promotion to a more demanding position. A challenge can be in ensuring that employees are able to access the learning. Time should be allowed in work schedules to attend training or access online opportunities. Caution should be taken in cutting training during periods of financial constraint as this will have longer term impacts on retaining a competent workforce.

Table 21.1 is an example of a listing of process safety training course for new employees. This is an abbreviated example; a full training matrix will likely include information such as prerequisite course and whether the course is computer based or classroom training.

Table 21.1. Example process safety training course list

Course	Target Audience	Triggers
Understanding and Managing Flammable Atmospheres	Required for all Engineers, Chemists, involved in design, maintenance and operations	First Two Years
PHA Methodology & Team Leader Training	Recommended for technical people involved in design, operations, and safety reviews, including MOCs and PHAs Required for PHA Team Leaders	First Two Years as well as PHA Team Leader Requirement
MOC Safety Review Team Leader Training	Recommended for MOC Core Team Members. Required for MOC Safety Review Team Leaders that have not taken the PHA Team Leader Training Class	First Two Years
Consequence Assessment	Recommended for people involved in modeling releases of chemicals and energy	Prior to use of consequence modeling tools
Pressure Relief Device (PRD) Application	Required for engineers and recommended for designers involved in PRD design, application, sizing and selection	Prior to involvement in design, application, sizing and selection of PRDs.
Design and Application of SCAI and Safety Instrumented Systems	Required for I&E, Control, and Process Engineers and recommended for designers involved in shutdown system design, review, and specification	Required prior to involvement in Shutdown System review, design or operation OR recommended within the first two years
Fire Protection and Fire Suppression	Required for engineers and recommended for designers involved in fire protection systems	Prior to involvement in design of fire suppression systems
Incident Investigation	Recommended for incident investigators and participants	Prior to leading or participating in incident investigations

All employees, contractors and visitors are typically required to attend training on the occupational safety and process safety basics at a facility. This is intended to prevent harm from workplace accidents by providing information on safe work practices and instruction on emergency response procedures.

Performance assurance is about confirming that workers can conduct the tasks expected of them. This can take the form of a written test following a training course or practical demonstration of a task. Performance assurance can also identify where additional training is required. In some cases, a competency matrix may list the means through which performance of that competency is assured. For example, engineers may be asked to conduct a HAZOP with an expert HAZOP leader as a member of the HAZOP team before being allowed to lead HAZOP studies on their own. Other forms of performance assurance include testing, field observation, and work product quality reviews.

21.6 Process Knowledge Management

Understanding risk depends on accurate process knowledge. Thus, this element underpins many of the other RBPS elements such as hazard identification and risk analysis, operating procedures, and management of change. Process knowledge primarily focuses on process safety information that can be recorded in the following.

- Technical documents and specifications.
- Engineering drawings and calculations.
- Specifications for design, fabrication, and installation of process equipment.
- Selection of safe operating limits for pressure, temperature, level, concentration, etc.

It is preferred that this information be electronically available in data systems or electronic documents as this supports revision control.

The term process knowledge management includes efforts to create, organize, maintain, and provide information. The use of the term "knowledge" implies understanding, not simply compiling, data. In that respect, the competency and knowledge elements are closely related. The documents recording the process knowledge are often used in training to develop competency.

Development and documentation of process knowledge starts early and continues throughout the life cycle of the process. For example, early laboratory efforts to develop new materials, characterize these materials, and evaluate the synthesis route (including the potential for runaway reaction or other inherent hazards) normally become part of the process knowledge. Efforts continue through the design, hazard review, construction, commissioning, and operational phases of the life cycle. Many facilities place special emphasis on reviewing process knowledge for accuracy and thoroughness immediately prior to conducting a risk analysis or management of change review.

Process knowledge is also closely related to process safety information which includes information on the hazards, the technology, and the equipment in a process. (Refer to Section 10.3.) Process Safety Information (PSI) can be considered a subset of process knowledge that is specific to process safety.

21.7 Contractor Management

Industry often relies upon contractors for everyday operations, certain specialized skills, and during periods of intense activity, such as maintenance turnarounds. These considerations, coupled with the potential lack of familiarity that contractor personnel may have with facility hazards and operations, pose unique challenges for the safe use of contract services. Contractor management. is a system of controls to ensure that contracted services support both safe facility operations and the company's process safety and occupational safety performance goals. This element addresses the selection, acquisition, use, and monitoring of such contracted services.

Companies are increasingly leveraging resources by contracting a diverse range of services, including design and construction, maintenance, inspection and testing, and staff augmentation. In doing so, a company can achieve goals such as: 1) accessing specialized expertise that is not continuously or routinely required, 2) supplementing limited company resources during periods of unusual demand, and 3) providing staffing increases without the overhead costs of direct-hire employees.

Using contractors brings an outside organization into the realm of the company's risk control activities. The use of contractors can place personnel who are unfamiliar with the facility's hazards and protective systems into locations where they could be affected by process hazards. Conversely, as a result of their work activities, the contractors may introduce new hazards to a facility, such as new chemicals or different procedures. Also, their activities onsite may unintentionally defeat or bypass facility safety controls.

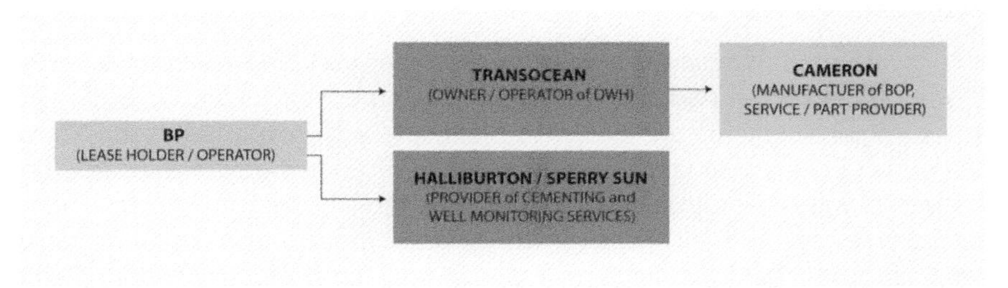

Figure 21.4. Key company relationships associated with the Deepwater Horizon accident (CSB 2014 b)

Companies should recognize new challenges associated with using contractors and take the following actions.

- Before engaging the contractor,
- Review the contractor's process safety management system, and
- Review the contractor's occupational safety and process safety incident safety records.
- Establish, document, and communicate occupational and process safety expectations, roles, and responsibilities for the contractor.
- Ensure that contractor personnel are properly trained.

- Supply appropriate information to the contractor to ensure that the contractor can safely provide the contracted services.

Contractual relationships can extend many layers as sub-contractors engage in their own contracts with other sub-contractors. It can be difficult, and very important, to ensure that process safety (and other topics such as quality) expectations are communicated through the layers of contractors. As an example, Figure 21.4 depicts the primary contractual relationships involved in the Deepwater Horizon incident.

21.8 Workforce Involvement

Personnel, at all levels and in all positions in an organization, have a role in the safety of the organization's operations. Workforce involvement provides a system for enabling the active participation of company and contract workers in the design, development, implementation, and continuous improvement of process safety.

Those personnel directly involved in operating and maintaining the process are those most exposed to the hazards of the process. These workers are also frequently the most knowledgeable people with respect to the day-to-day details of operating the process and maintaining the equipment and facilities. They may be the sole source for some types of knowledge gained through their unique experiences. Workforce involvement provides management a formalized mechanism for tapping into this valuable expertise. This proactive engagement would illustrate at least two positive things, the right people are involved in the review, and the workforce, down to the operating staff, is able to provide candid views without fear of adverse consequences.

Workforce involvement is specifically mentioned in the U.S. OSHA PSM and U.S. EPA RMP regulations (termed "employee participation") as well as other regulations regarding process safety. This is a requirement for the presence of people with "experience and knowledge specific to the process being evaluated" at a hazard identification study of a covered process. Proactive companies will expand that to include workers directly involved in maintenance and operations at these reviews and will encourage their honest input. Subject matter experts such as process engineers, mechanical engineers, material engineers, etc. should be relied on for technical information and other process safety information. Operators and mechanics should be relied on for evaluating the understanding of the process, the clarity and efficiency of procedures, and an understanding of what is being done in the field vs. what engineering and management think is being done.

Some opportunities to involve the workforce include as members of a hazard identification study team, in writing operating procedures and safe work procedures, during management of change reviews, and on incident investigation teams.

Many companies implement a stop-work authority which allows anyone to stop an activity, without fear of repercussion, if they believe the activity is potentially dangerous. This can be seen as an ultimate workforce involvement activity.

21.9 Stakeholder outreach

Stakeholder outreach . seeks to include external stakeholders in process safety efforts in the same way that workforce involvement seeks to involve the workforce. It is a process for:

1) seeking individuals or organizations that can be or believe they can be affected by company operations and engaging them in a dialogue about process safety including understanding their needs,
2) establishing a relationship with community organizations, other companies and professional groups, and local, state, and federal authorities,
3) providing accurate information about the company and facility's products, processes, plans, hazards, and risks, and
4) communicating emergency response plans and preparing emergency responders and the community for their roles in a potential emergency.

This process ensures that management makes relevant process safety information available to a variety of organizations. This element also encourages the sharing of relevant information and lessons learned with similar facilities within the company and with other companies in the industry group. Finally, stakeholder outreach promotes involvement of the facility in the local community and facilitates communication of information and facility activities that could affect the community.

A good example of stakeholder outreach is from the late 1990s, when many facilities handling hazardous materials engaged in communications with their neighboring communities as part of the U.S. EPA's Risk Management Plan (RMP) requirements (EPA 1996). In highly populated areas having a large industrial base, many companies collaborated by having regional RMP communication events. One of the largest such activities was conducted in the Houston, Texas, area where more than 120 facilities coordinated their RMP outreach activities over a 4-year period. Those activities helped nurture relationships with communities, regulators, local emergency response agencies, and nongovernmental community groups.

21.10 What a New Engineer Might Do

New engineers often have defined programs to gain process safety competency. Take advantage of these opportunities to continue learning.

Building competency can be done in many ways. In those few quiet moments with no immediate deadline, it is good to pick up some of documents mentioned in the section on process knowledge management and learn about practices, processes, and other details of the operating facility. These documents are often updated by new engineers. Recognize how important they are to process safety and take on the task diligently.

It is also important to have a mentor or coach. Having a relationship with a more senior expert that you can ask questions of or seek advice from can be a great way to learn more efficiently.

It has been said that any given plant actually has three plants – the plant that the engineers and managers think is there, the plant that the operators think is there, and the real plant.

Operators know and work with the real plant process every day, but they may not understand the complex chemistry or thermodynamics involved. New engineers should work to make the first plant concept lines up with the third plant reality. Operators not only understand how the plant works, but how it can fail. Sometimes things that engineers think are true about the way the process is operated may not actually be true.

The same is true for maintenance personnel, electricians, and other craftsmen. Such personnel work directly with the process equipment and instrumentation. Asset integrity depends on them. They can provide input to a reliability and/or process engineer on what equipment might be fit or unfit for certain kinds of service.

An anecdote about one engineer's experience with an operator in a PHA illustrates an operator's value.

> A HAZOP was being done on an existing process for the first time. The HAZOP team was quite large, including plant engineers, a senior process design engineer, the plant maintenance and reliability engineer, and a maintenance technician. There was extensive discussion about the potential consequences and safeguards during several deviations, with engineers debating the effects of the deviations. Finally, the operator said, "I don't know about all your logic, but I know that when these events occur, the outlet vent valve opens fully, and we get an alarm." During a break the team went to the control room, where the operator deliberately caused the deviations, and in each case, the outlet vent valve opened fully, and an alarm went off. It turned out the vent line to a treatment system was a pinch point where the first effect of many deviations announced themselves.

The anecdote illustrates at least two things; first, operators, maintenance personnel and other craftsmen may very well understand the actual cause-effect relationships of process deviations better than the engineers. Second, all people working in the plant, and their opinions should be treated with respect. Their knowledge is important. Avoid a know-it-all attitude and actively listen. If operators and other craftsmen are treated with respect, they will be more willing to tell engineers how a plant really works during a PHA.

In addition to technical subjects, non-technical skills in which a new engineer should be competent include as the following.

- Writing skills – to support writing concise project status summaries and reports, documenting process safety information
- Public speaking skills – to support delivering training, presenting project results
- Leadership skills – to support leading project teams, facilitating meetings
- Interpersonal skills – to support all work through good listening, assertiveness, and respect for other people's opinion

A new engineer can benefit from reviewing the CSB investigations and videos relevant to this chapter as listed in Appendix G.

21.11 Tools

Resources to support process safety competency, training and performance assurance, process knowledge management, contractor management, workforce involvement, and stakeholder outreach include the following.

CCPS *Guidelines for Defining Process Safety Competency Requirements*. This Guideline presents the framework of process safety knowledge and expertise versus the desired competency level in a "super-matrix" format, vertically and diagonally. The matrix references for potential remedies/required training may be tailored to a company's internally developed training, reference externally available training, or some combination of the two. (CCPS 2015)

CCPS *Contractor and Client Relations to Assure Process Safety*. Written and edited by engineering contractors and industry project/maintenance managers as an easy-to-use guide for other industry professionals, this book identifies important process safety issues in the contractor-client relationship, which are not addressed by other groups and publications. While the issues may arise at any point in the life cycle of a plant, they should be resolved early in the relationship to permit a clearer focus on process safety issues. Topics covered are a general discussion of contractor safety programs; EPC (engineering, procurement, construction) contractual bases and work division as they address regulatory PSM issues; subcontractor relationships; and managing contractor-client risks. (CCPS 1996)

National Contract Management Association "The Contract Management Standard™" (CMS™). This ANSI-accredited standard describes contract management skills and competencies across the phases of a contract life cycle. (NCMA)

21.12 Summary

The topics addressed in this chapter relate to the personnel management aspects of, as opposed to the practice of, process safety. They include process safety competency, training and performance assurance, process knowledge management, contractor management, workforce involvement, and stakeholder outreach.

Good process safety performance depends not only on good process safety tools and practices being performed well, it also depends on competent people developing, performing, and improving them. The course work knowledge gained in school is only the beginning. There is much to learn in the field of process safety. Many companies will create a process safety competency matrix to define the competency expectations for company personnel in the various process safety topics. Not everyone is expected to be an expert in everything, but people should be competent in the specific areas required to perform their job.

This competency is gained primarily through on the job experience and also through training and coaching. In support of the competency expectations, many companies will define job experience and training requirements for roles and promotions into future roles. Performance assurance verifies that workers are competent to perform their tasks appropriately.

These competent people are then dependent on accurate process information to perform their work. Process knowledge management includes efforts to create, organize, maintain, and provide information – specifically the information required to safely operate the facility. This process knowledge underpins many process safety activities including hazard identification, management of change, and training.

The workforce in many facilities includes contractors. These contractors may be supporting major construction projects, maintenance activities, or specialty engineering work. Before the contract is agreed, the occupational safety and process safety of the contractor be reviewed. Once the contractor is onboard, then communicate safe work practices and other expectations to them. This is important not only for a primary contractor, but also for their subcontractors and their subcontractors and so on.

Including the workforce in process safety is likely not only a requirement but also the best way to get "real world" experience into process safety activities. Beyond the workforce, it is also important to engage with other stakeholders including facility neighbors, regulators and others that may impact or may be impacted by process safety in a facility.

21.13 Other Incidents

This chapter began with a description of the Deepwater Horizon Well Blowout. Other incidents relevant to people management aspects of process safety management include the following.

- Exxon Valdez Oil Spill, Alaska, U.S., 1989
- BLSR Deflagration and Fire, Texas, U.S., 2007
- Pike River Coal Mine Explosion, South Island, New Zealand, 2010

21.14 Exercises

1. List 3 RBPS elements evident in the Deepwater Horizon Well Blowout incident summarized at the beginning of this chapter. Describe their shortcomings as related to this accident.
2. Considering the Deepwater Horizon Well Blowout incident, what actions could have been taken to reduce the risk of this incident?
3. For what roles should process safety competencies be defined?
4. Name two sources where process knowledge may be stored.
5. Why is it important to manage contractors?
6. Why is it important to engage the workforce in process safety activities?
7. Name two external stakeholders that it might be important to engage with on the topic of process safety.

21.15 References

AU 2010, "Report of the Montara Commission of Inquiry", Commonwealth of Australia, June.

BOEMRE 2011, "Report Regarding the Causes of the April 20, 2010 Macondo Well Blowout", The Bureau of Ocean Energy Management, Regulation and Enforcement, U.S. Department of the Interior, September 14, https://www.bsee.gov/sites/bsee.gov/files/reports/blowout-prevention/dwhfinaldoi-volumeii.pdf.

Baker 2007, Baker James et al., "The Report of the BP U.S. Refineries Independent Safety Review Panel", www.csb.gov.

BP 2010, "Deepwater Horizon Accident investigation report", BP, London, U.K., September 8.

CCPS Glossary, "CCPS Process Safety Glossary", Center for Chemical Process Safety, https://www.aiche.org/ccps/resources/glossary.

CCPS 1996, *Contractor and Client Relations to Assure Process Safety*, Center for Chemical Process Safety, John Wiley & Sons, Hoboken, N.J.

CCPS 2015 *Guidelines for Defining Process Safety Competency Requirements*, Center for Chemical Process Safety, John Wiley & Sons, Hoboken, N.J.

CSB 2014a, "Explosion and fire at the Macondo Well; Overview", U.S. Chemical Safety and Hazard Investigation Board, Case Study, Report No. 2010-10-I-OS, June 5. http://www.csb.gov/investigations.

CSB 2014b, "Explosion and fire at the Macondo Well; Vol. 1, Macondo-specific incident events", U.S. Chemical Safety and Hazard Investigation Board, Case Study, Report No. 2010-10-I-OS, June 5. http://www.csb.gov/investigations.

CSB 2014c, "Explosion and fire at the Macondo Well; Vol. 2, Technical findings on the Deepwater Horizon blowout preventer (BOP)", U.S. Chemical Safety and Hazard Investigation Board, Case Study, Report No. 2010-10-I-OS, June 5. http://www.csb.gov/investigations.

CSB 2014d, "Explosion and fire at the Macondo Well; Vol 3, Human, organizational and safety system factors of the Macondo blowout", U.S. Chemical Safety and Hazard Investigation Board, Case Study, Report No. 2010-10-I-OS, June 5. http://www.csb.gov/investigations.

EPA 1996, Accidental Release Prevention Requirements: Risk Management Programs Under Clean Air Act Section 112(r)(7), 20 CFR 68, U.S. Environmental Protection Agency, June 20, Fed. Reg. Vol. 61[31667-31730]. www.epa.gov

Lombardo 1996, Lombardo, Michael M; Eichinger, Robert W, *The Career Architect Development Planner* (1st ed.), Minneapolis: Lominger.

NCMS 2019, *The Contract Management Standard™ (CMS™)*, National Contract Management Association, https://www.ncmahq.org/docs/default-source/standards-certification-files/the-contract-management-standard.pdf.

TO 2011, "Macondo Well incident", Transocean Investigation Report, Vol. 1, June.

22
Sustaining Process Safety Performance

22.1 Learning Objectives

The learning objectives of this chapter are:

- Understand the importance of investigating and learning from incidents, and
- Understand methods to sustain and continuously improve process safety performance.

22.2 Incident: Space Shuttle Columbia, 2003

22.2.1 Incident Summary

The NASA Space Shuttle, Columbia, was destroyed during its re-entry into the Earth's atmosphere at the end of a 16-day voyage, just 16 minutes before scheduled touchdown. During the launch, a large piece of insulation foam became detached from the area where the shuttle had been attached to the external fuel tank and hit the leading edge of the left wing. After the incident, it was discovered that a fragment of the thermal protective panel drifted away from the wing while in space. At the critical part of re-entry when friction with the Earth's atmosphere is at its greatest, superheated air entered the left wing, destroying the structure and causing the spacecraft to lose aerodynamic control, and break up (Figure 22.1). All seven of the crew were fatally injured. Within two hours of loss of signal from Columbia, the independent Columbia Accident Investigation Board (CAIB) was established following procedures that had been put in place after the Challenger disaster 17 years earlier. (CCPS 2008)

Key Points:

Process Safety Culture – It can take time to build a strong process safety culture but not long for it to degrade. Having a strong process safety culture requires constant vigilance and leadership.

Measurement and Metrics – "You get what you measure" is a quote attributed to Peter Drucker, a widely recognized management consultant. However, if you are not measuring the right things, you likely won't get the best results. Choose what you measure carefully.

22.2.2 Description

Columbia was launched on January 16, 2003 for the 28th time. At 81.7 seconds into the flight, a large piece of insulation foam became detached. The detached piece of foam hit the leading edge of the left wing 0.2 seconds later (Figure 22.2). The event was not observed in real time.

This event was not detected by the crew or ground support functions until detailed examination of the launch photographs and videos took place the following day. A Debris Assessment Team was created to determine whether the event had caused critical damage to the shuttle. No adverse effects were noticed by the crew or support staff as the mission continued. What they did not know was that on the second day of the flight, an object drifted away from the shuttle. The radar signature of this object, discovered after the incident, was consistent with it being a 900 cm^2 (140 in^2) fragment of the protective panel from the left wing of the shuttle. At the critical part of re-entry when friction with the Earth's atmosphere is at its greatest, superheated air entered the left wing, destroying the structure and causing the spacecraft to lose aerodynamic control leading to break up.

The question was asked as to how a piece of lightweight foam material could catastrophically damage something as apparently strong as a spacecraft designed for one of the most aggressive operating environments. Calculations showed that, at the time of separation, the foam was traveling at the same speed as Columbia - about 700 m/s (1,568 mph) and the rapid deceleration of the foam combined with continued acceleration of the shuttle explained the severity of the impact. It was also found that insulation foam loss had occurred on previous Space Shuttle flights, from small "popcorn" sized pieces, to briefcase sized chunks, and that, on 10% of flights, foam loss had occurred at the bipod attachment area. The original design team had strong concern that foam loss would result in significant damage to the shuttle. Since the specification for the large external fuel tank contained a requirement that "no debris shall emanate from the critical zone of the external tank on the launch pad or during ascent", no protection had been provided to the leading edges of the shuttle's wings. Despite this, there had been a lot of damage to Columbia's protective tiles during its first mission – more than 300 had to be replaced. One engineer stated that if they had known in advance the extent of the foam debris shower that occurred, they would have had difficulty in getting the Space Shuttle cleared for flight.

Figure 22.1. Columbia breaking up
(CAIB 2003)

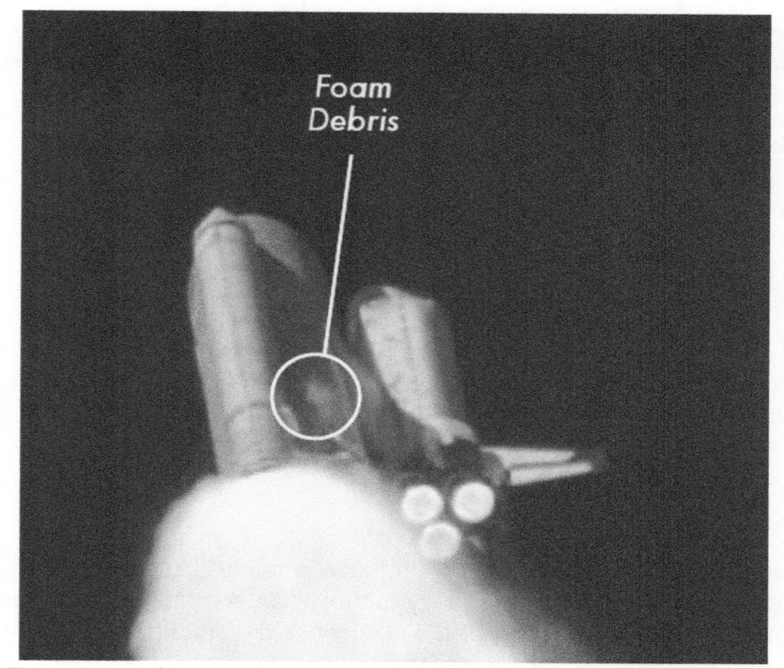

Figure 22.2. A shower of foam debris after the impact on Columbia's left wing. (CAIB 2003)

Over the previous decade, NASA was placed under severe pressure to reduce costs. The focus on measuring costs resulted in losing about 40% of its budget and workforce. Part of the response was for NASA to hand over much of its operational responsibilities to a single contractor, replacing its direct involvement in safety issues with a more indirect performance monitoring role. NASA managers continued to talk about the importance of safety, but their actions sent the opposite signal.

Despite the cutbacks, personnel felt pressure to keep the Space Shuttle program on schedule, particularly to complete the International Space Station (ISS). The uncertainty over the long-term future of the program resulted in reduced investment, with safety upgrades delayed or deferred. The CAIB found that the infrastructure had been allowed to deteriorate, and the program was operating too close to too many margins.

Technically, the cause of the incident was the failure of the foam insulation at the bipod attachment. No non-destructive testing (NDT) of hand-applied foam was carried out other than visual inspection at the vehicle assembly building and at the space center, even though NDT techniques for foam adherence had been successfully used elsewhere. The CAIB concluded that too little effort had gone into the understanding of foam fabrication, adhesion, and failure modes.

Culture also played a key role in the incident. In spite of cutbacks and deadline pressures, the organization continued to pride itself on its "can do" attitude, which had contributed to former successes. This enabled the phenomenon known as "Normalization of deviance". The failure of the foam without significant consequences was observed so many times that it became an accepted part of every flight and with each successful landing the original concerns seem to have faded away. They loss their sense of vulnerability to a major incident related to foam failure.

Normalization of deviance - A gradual erosion of standards of performance as a result of increased tolerance of nonconformance. (CCPS Glossary)

In the words of the CAIB report, "Cultural traits and organizational practices detrimental to safety were allowed to develop, including: reliance on past success as a substitute for sound engineering practices (such as testing to understand why systems were not performing in accordance with requirements); organizational barriers that prevented effective communication of critical safety information and stifled professional differences of opinion; lack of integrated management across program elements; and the evolution of an informal chain of command and decision-making processes that operated outside the organization's rules." (CAIB 2003)

22.2.3 Lessons

Process Safety Culture. An important aspect of a good safety culture is maintaining a sense of vulnerability. An example of the poor safety culture at NASA is the denial of requests by the Debris Assessment Team for imaging of the wing while the shuttle was in orbit. The team concluded, based on modeling that "some localized heating damage would most likely occur during re-entry, but they could not definitively state that structural damage would result." The Mission Management Team eventually concluded the debris strike was a "turnaround" [time between launches] issue. As stated in the CAIB report "Organizations that deal with high-risk operations must always have a healthy fear of failure – operations must be proved safe, rather than the other way around. NASA inverted this burden of proof."

The CAIB found "NASA's safety culture has become reactive, complacent, and dominated by unjustified optimism. Over time, slowly and unintentionally, independent checks and balances intended to increase safety had been eroded in favor of detailed processes that produce massive amounts of data and unwarranted consensus, but little effective communication. Organizations that successfully deal with high-risk technologies create and sustain a disciplined safety system capable of identifying, analyzing, and controlling hazards throughout a technology's life cycle."

22.3 Overview

Chapter 21 addresses topics that focus on the personnel management aspects of process safety management. This chapter addresses topics that focus on the business management activities used to sustain process safety management including the following.

- Incident investigation
- Measurement and metrics
- Auditing
- Management review and continuous improvement

22.4 Incident investigation

Process safety incidents can lead to loss of lives, money, and a company's reputation. Incident investigation is a process for reporting, tracking, and investigating incidents and near misses. (Refer to Chapter 9 for information on near misses and classification of process safety incidents.) This includes a formal process for conducting incident investigations including staffing, performing, documenting, and tracking investigations of process safety incidents. It also includes the trending of incident and incident investigation data to identify recurring incidents. The purpose of incident investigation is to identify and eliminate the causes of incidents to prevent their recurrence and sustain or improve process safety performance. The incident investigation process also manages the resolution and documentation of recommendations and action generated by the investigations.

Incident investigation provides a way of learning from incidents that occur over the life of a facility or business and communicating the lessons learned to both employees and other stakeholders. Depending upon the depth of the analysis, this feedback can apply to the specific incident under investigation or a group of incidents sharing similar root causes at one or more facilities.

Incident investigation should not be used to assign blame. Assigning blame serves to stop the investigation short of identifying the root causes. The failure to identify root causes results in ineffective recommendations being implemented. A more effective approach is to pursue the investigation to the root causes and develop recommendations that address the management causes of the incidents.

Root Cause - A fundamental, underlying, system-related reason why an incident occurred that identifies a correctable failure(s) in management systems. Typically, more than one root cause can be found for every process safety incident. (CCPS Glossary)

Incident investigation begins after the emergency is contained and the site is stabilized. At this point, preserve evidence in the field and electronic data or images that could be overwritten. Investigation personnel entering the scene should be qualified to enter a potentially hazardous area as there may be additional hazards present following an incident. The incident investigation team should also interview witnesses and those present during the incident as practical after the incident as memories fade quickly.

Several types of incident investigation methods are available. The method used will usually depend on the perceived severity of the incident or near miss. These methods can range from simple brainstorming to creating logic trees. One simple technique is the "5 Whys" method as shown in Figure 22.3. (Serrat 2009)

Using the 5 Whys method to illustrate the point of not pursing an investigation until the root causes are understood, a recommendation might have been to replace the pipe or adjust the inspection frequency. But this would not have addressed the root cause that the MOC procedure should be improved to clarify its applicability to changes resulting from budget cuts. By doing this, a broader range of potential incidents, those impacted by budget cuts, can be addressed.

Assembling an incident timeline, a time-sequenced listing of all relevant data that occurred before, during, and immediately after an incident, is also a good method to understand what happened and identify potential gaps in the time sequence that should be further investigated.

Why? did the fire happen? Because the pipe leaked.

Why? did the the pipe leak? Because it corroded.

Why? was the corrosion not found in time? Because inspection frequency was reduced.

Why? was inspection frequency reduced? Because budgets were cut across the board without review.

Why? were budgets cut without review? Because it was not clear that the Management of Change procedure applied to budgets.

Figure 22.3. Example of 5 Whys technique

The investigation team should consist of people with expertise in the investigation method being used and with technical expertise appropriate to the event. For example, the investigation into the pipe rupture should have a metallurgist and someone familiar with the company's asset integrity program. Consequence Analysis (see chapter 13) might be used to better understand the details of what might have happened during the incident, for example, how large a hole would be required to release sufficient flammable material to cause an explosion with sufficient energy to match the observed damage.

Recommendations should be tracked until they are implemented. No risks are actually reduced until recommendation actions are completed. Lessons learned should be shared with similar facilities within the organization and externally where findings may be beneficial to broader industry. The lessons learned from these events drive growth and understanding of process safety management concepts.

Finally, companies should also track incidents and near misses in a database to enable them to analyze the events for trends that can be causing repeat incidents.

Refer to Appendix H for a table listing causes for selected major process safety incidents.

22.5 Measurement and metrics

Metrics provide a means to monitor the performance of a management system. This topic includes identifying which metrics to consider, how often to collect data, and how to use the information to sustain and improve process performance.

It is too easy to create a long list of metrics that might be helpful. By having too many metrics, the importance of all are minimized and the work involved in compiling the data can result in not enough time or energy to make improvements from metric learnings.

The key process safety metrics are the Process Safety Incident (PSI) Tier 1 through 4 which are discussed in detail in Chapter 9. These PSI are both lagging, meaning the incident has occurred, or leading, meaning it is in indication that an incident could occur.

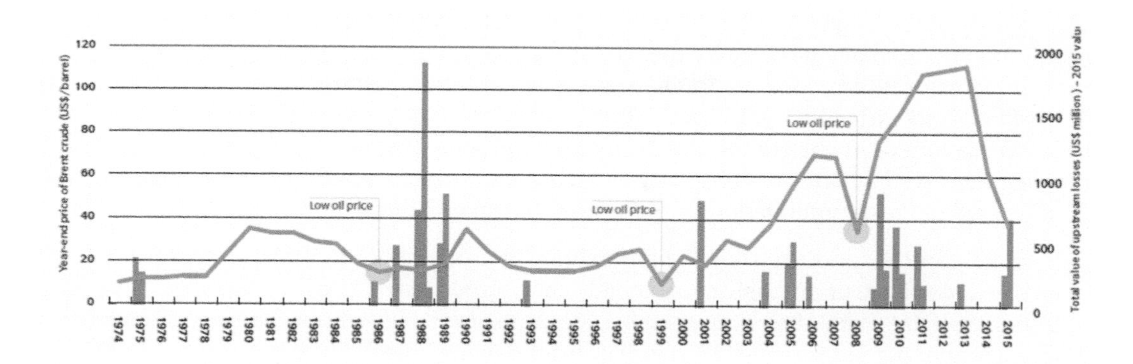

Figure 22.4. Crude oil price versus upstream losses by year (Marsh 2016)

Metrics will be required from a corporate level; however, they may not be focused on the problems at an individual facility. At a facility level, consider what problem warrants solving. This may be indicated through, for example, incident trends or production data. Then consider what leading metrics could be created relative to this problem. For example, production data could indicate that production levels are being reduced because pressure relief valves are relieving frequently which diverts product to the flare. Leading metrics could be created to track relief valve lifts and high operating pressure limit excursions. Lagging metrics could track the number of relief valve lifts. Through the attention that these metrics focus, it might be identified that the alarms and safety instrumented systems are set too close to the relief valve set pressure giving insufficient time for operators or the instrumented systems to respond. This could lead to an action to reset set pressures on the alarms and instrumented systems.

Figure 22.4 shows the relationship between oil price and the value of losses in the upstream hydrocarbon industry. Historically, as the oil price declines, the resources allocated to maintenance and training are reduced. The figure shows the correlation between these reductions and increased process safety incidents.

The *Guidelines for Risk Based Process Safety* chapter 20 provides many examples of metrics related to sustaining process safety performance. (CCPS 2007)

22.6 Auditing

The purpose of process safety auditing is to identify management system and performance gaps in the process safety management system and allow correction of those gaps before an incident occurs.

Audit - A systematic, independent review to verify conformance with prescribed standards of care using a well-defined review process to ensure consistency and to allow the auditor to reach defensible conclusions. (CCPS Glossary)

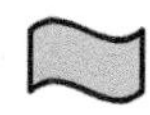

Auditing employs a well-defined review process to ensure consistency and to allow the auditor to reach defensible conclusions. An audit involves a methodical, typically team-based, assessment of the implementation status of one or more RBPS elements against established requirements, normally directed by the use of a written protocol. Data are gathered through the review of program documentation and implementation records, direct observations of conditions and activities, and interviews with individuals having responsibilities for implementation or oversight of the element(s) or who might be affected by the RBPS management system. The data are analyzed to assess compliance with requirements, and the conclusions and recommendations are documented in a written report.

Audits can be conducted by a team of qualified personnel selected from a variety of sources, depending upon the scope, needs, and other aspects of the specific situation. Auditors may be from the facility itself (1st party), from elsewhere in the company but independent of the facility itself (2nd party) or from outside organization completely (3rd party). While internal company personnel may be familiar with the organization and the process, they may lack independence from the topic or facility being auditing. On the other hand, an external consultant may be independent and bring external views, but may not be as familiar with the process, documentation systems, or organization.

Audits require a system for scheduling, staffing, and effectively evaluating all RBPS elements. A system should be in place for implementing any resulting findings or recommendations and verifying their effectiveness.

Auditing is one of the elements in the U.S. OSHA PSM and U.S. EPA RMP regulations, and audits are required at least every three years for covered processes. Other regulations also include auditing and may specify details including scope, frequency, and auditors.

22.7 Management review and continuous improvement

Management reviews have many of the characteristics of an audit described in the previous section. The difference between management reviews and audits is the reviews are more broadly focused and more frequent than audits, and they are typically conducted in a less formal manner. This is because the objective of a management review is to spot current or incipient deficiencies and address them promptly. The management review assesses process safety performance and determines if it is meeting planned expectations. If not, then additional resources can be brought to bear to meet the expectations and also achieve continuous improvement.

Management Review and Continuous Improvement - A PSM program element that provides for the routine evaluation of other PSM program management systems/elements with the objective of determining if the element under review is performing as intended and producing the desired results as efficiently as possible. It is an ongoing due diligence review by management that fills the gap between day-to-day work activities and periodic formal audits. (CCPS Glossary)

Effective performance is a critical aspect of any process safety program; however, a breakdown or inefficiency in a safety management system may not be immediately obvious. For example, if a facility's training coordinator unexpectedly departed, required training activities might be disrupted. The existing trained workers would continue to operate the process, so there would be no outward appearance of a deficiency. An audit or incident might eventually reveal any incomplete or overdue training, but by then it could be too late. The management review process provides regular checkups on the health of process safety management systems in order to identify and correct any current or incipient deficiencies before they might be revealed by an audit or incident.

Management reviews typically occur periodically, i.e. quarterly or monthly, and involve the facility leadership. They involve review of process safety metrics, incident investigations, and improvement initiatives. A good practice is to have the process safety management review as a separate meeting from financial and other business topics to avoid these other topics overrunning and leaving too little time for an adequate management review of process safety.

22.8 What a New Engineer Might Do

All engineers in a plant or process must learn the definition of process safety incidents and near misses in their organization. New engineers from many disciplines will likely get a chance to participate in process safety near miss and incident reviews. Chemical engineers contribute the knowledge of the process technology, and process chemistry, chemical interactions and kinetics that may have contributed to the incident. Mechanical engineers contribute with the knowledge of what equipment failed, what the failure modes and causes are. Instrument and Control engineers will contribute their knowledge of how control and SCAI systems work and what new controls may be needed. Electrical engineers may identify electrical component failures, substation/MCC electrical coordination issues, contributions of electrical noise and harmonics, as well as solutions and action items to address electrical reliability.

As a new engineer, you may be involved in collecting and analyzing the data for leading and lagging metrics. This data may be used in management reviews.

As a new engineer it is unlikely that you will actually be conducting an audit, but you may be involved in interviews and responsible for providing information and documentation to an auditor.

22.9 Tools

Resources to support incident investigation (and learning from incidents), measurement and metrics, auditing, and management review and continuous improvement include the following.

CCPS. Acknowledging that performance metrics continue to evolve, CCPS has created an evergreen webpage resource for process safety metrics. The CCPS webpage contains links to resources, reports, and research in multiple languages. It also includes a link to a Process Safety Incident Evaluation Tool which assists the user in determining how to classify an incident. Consult the CCPS Metrics webpage at https://www.aiche.org/ccps/process-safety-metrics.

CCPS *Guidelines for Auditing Process Safety Management Systems, 2nd edition*. This book discusses the fundamental skills, techniques, and tools of auditing, and the characteristics of a good process safety management system. A variety of approaches are given so the reader can select the best methodology for a given audit. This book is accompanied by an online download featuring checklists for both the audit program and the audit itself. This package offers a vital resource for process safety and process development personnel, as well as related professionals such as insurers. (CCPS 2011)

CCPS *Guidelines for Investigating Process Safety Incidents, 3rd Edition*. This book provides a comprehensive treatment of investigating process incidents. It presents explanations, techniques, and examples that support successful investigations. Issues related to identification and classification of incidents (including near misses), notifications and initial response, assignment of an investigation team, preservation and control of an incident scene, collecting and documenting evidence, interviewing witnesses, determining what happened, identifying root causes, developing recommendations, effectively implementing recommendations, communicating investigation findings, and improving the investigation process are addressed in the third edition. While the focus of the book is investigating process safety incidents the methodologies, tools, and techniques described can also be applied when investigating other types of events such as reliability, quality, occupational health, and occupational safety incidents. (CCPS 2019)

CCPS *Incidents That Define Process Safety* and *More Incidents That Define Process Safety*. These books describe nearly one hundred events. While the majority of events are from across the process industries, these books also describes event from other industries such as mining and transportation, illustrating the point that many process safety elements are universal in their relevance to safe operations. (CCPS 2008 and CCPS 2019).

CCPS Process Safety Beacon. The Process Safety Beacon is aimed at delivering process safety messages to plant operators and other manufacturing personnel. It is also a source of incident descriptions and learnings. Each issue presents a real-life incident and describes the lessons learned and practical means to prevent similar incidents in your plant. The monthly one-page Process Safety Beacon covers the breadth of process safety issues. Each issue presents a real-life accident, and describes the lessons learned and practical means to prevent a similar accident. The Process Safety Beacon webpage is www.aiche.org/ccps/process-safety-beacon.

Chemical Safety and Hazard Investigation Board. The CSB is a U.S. government agency charged with investigating significant process safety incidents in the U.S. The reports of their investigations are available for download from their website (www.csb.gov). The website also includes a series of videos that describe many of the process safety incidents they have investigated.

API RP 754 *Process Safety Performance Indicators for the Refining and Petrochemical Industries* addresses both leading and lagging metrics and includes detailed definitions and classifications. (API RP 754)

IOGP Report 456 – *Process safety – recommended practice on key performance indicators* addresses both leading and lagging metrics and includes detailed definitions and classifications. It focuses on the upstream oil and gas industry and is closely aligned with API RP 754. (IOGP)

22.10 Summary

Process safety, just as any other aspect of business, should be managed. This involves creating a management system such as that described in Section 1.5. The management system performance should then be evaluated, sustained, and improved.

Incident investigations can provide learnings on management system failures. Measurement and metrics such as tracking Process Safety Incidents and related Tier 3 and Tier 4 leading indicators can focus attention on management system aspects that may be weakening. Audits, and the more frequent management reviews, can also indicate opportunities for improvement in process safety element performance and thus, process safety performance.

22.11 Other Incidents

This chapter began with a description of the Space Shuttle Columbia disaster. Other incidents relevant to sustaining process safety performance include the following.

- Union Carbide MIC release, Bhopal, India, 1984
- NASA Challenger Disaster, Florida, U.S., 1986
- Esso Longford Gas Plant Explosion, Australia, 1998
- Mars Climate Orbiter, U.S., 1999
- Motiva Enterprises LLC, Delaware, U.S., 2001

22.12 Exercises

1. List 3 RBPS elements evident in the Space Shuttle Columbia incident summarized at the beginning of this chapter. Describe their shortcomings as related to this accident.
2. Considering the Space Shuttle Columbia incident, what actions could have been taken to reduce the risk of this incident?
3. Why is it important to identify the root causes of a process safety incident?
4. How does one ensure that learnings from incident investigations are retained at the facility, as well as other company operations, to prevent a reoccurrence?
5. Name a non-analytic method for conducting RCAs, and note how you would decide which to choose?
6. Identify a difference between and audit and a management review.
7. Incident investigations, audits, and management reviews can all generate actions. How should these actions be handled?

22.13 References

API RP 754, "Process Safety Performance Indicators for the Refining and Petrochemical Industries", American Petroleum Institute, Washington, D.C., U.S., 2016.

CAIB 2003, "Columbia Accident Investigation Board Report Volume 1", http://www.nasa.gov/columbia/caib/html/start.html.

CCPS Glossary, "CCPS Process Safety Glossary", Center for Chemical Process Safety, https://www.aiche.org/ccps/resources/glossary.

CCPS 2007, *Guidelines for Risk Based Process Safety*, Center for Chemical Process Safety, John Wiley & Sons, Hoboken, N.J.

CCPS 2008, *Incidents that Define Process Safety*, Center for Chemical Process Safety, John Wiley & Sons, Hoboken, N.J.

CCPS 2011, *Guidelines for Auditing Process Safety Management Systems, 2nd edition*, Center for Chemical Process Safety, John Wiley & Sons, Hoboken, N.J.

CCPS 2019, *Guidelines for Investigating Process Safety Incidents, 3rd Edition*, Center for Chemical Process Safety, John Wiley & Sons, Hoboken, N.J.

CCPS 2019, *More Incidents that Define Process Safety*, Center for Chemical Process Safety, John Wiley & Sons, Hoboken, N.J.

CCPS Beacon, Process Safety Beacon, https://www.aiche.org/ccps/process-safety-beacon.

CSB, U.S. Chemical Safety and Hazard Investigation Board, www.csb.gov.

IOGP 2018 "Report 456 – Process safety – recommended practice on key performance indicators", International Association of Oil and Gas Producers, London, U.K.

Marsh 2016, "The 100 Largest Losses 1974-2015", Marsh.

Serrat 2009, "The Five Whys Technique", Asian Development Bank.

23 Process Safety Culture

23.1 Learning Objectives

The learning objective of this chapter is:

- Understand the concept of Process Safety Culture.

23.2 Overview

Process safety culture, put succinctly, is "How we do things around here" or "How we behave when no one is watching."

Process Safety Culture - the common set of values, behaviors, and norms at all levels in a facility or in the wider organization that affect process safety. (CCPS Glossary)

Process safety culture weaknesses have been identified through investigations such as in the Space Shuttle Challenger and Columbia disasters and the BP Texas City Refinery Explosion. As seen in these incidents, many of the individual elements of process safety were weak. Process safety culture impacts and is impacted by the process safety management system elements as well as other business management systems, e.g. financial.

It is common for organizations to perform culture surveys as a method to determine the current level of culture and then conduct subsequent surveys to monitor improvement based on action taken. This approach was taken following the BP Isomerization unit explosion in Texas City and the survey approach is documented in the Baker Panel report. (Baker 2007)

It is not possible to write a policy requiring a good process safety culture or a procedure that tells someone how to achieve it. What would be the requirements? It is, however, possible to see a good process safety culture in action. The following are examples of what a good process safety culture looks like.

- **Leadership sets the tone ('tone at the top")** - Management demonstrates process safety is a priority. They do this through their own actions. They are personally involved in process safety. In other words, they walk the talk. A literal example of this is when management walks through the plant, discusses process safety concerns with operators and follows up on those concerns.
- **Metrics, Organization and Incentives support strong safety culture -** Process safety is at the same level as other business functions. Just as with finance, employee relations, and other functions, process safety is included in top level business management. This means that persons with process safety responsibility are included in the meetings and that process safety metrics are included in the discussions. Safety metrics promote strong safety priorities/behaviors, and discourage excess risk taking.
- **Conduct of operations is valued -** The organization clearly defines safety-related responsibilities. Accordingly, employees are provided the resources needed and are

expected to fulfill their individual process safety responsibilities. For example, operators and mechanics fully follow and properly complete procedural checklists, engineers follow engineering practices, and managers diligently consider process safety decisions and resources.

- **People are pre-occupied to identify the next failure or deviation -** Everyone is vigilant. The organization maintains a high awareness of process hazards and their potential consequences, maintains a sense of vulnerability, and is constantly vigilant for indications of system weaknesses that might foreshadow more significant safety events. Deviations are not tolerated, instead they are investigated, and actions taken to address them.
- **The organization values learning vs. blaming -** The organization learns from smaller problems and views failures as opportunities to improve, not blame. Everyone wants to learn. All involved want to improve their own and the overall facility performance. Expertise is sought and valued. Personnel attend training and have coaches to support on the job learning. Investigations and audits are viewed as opportunities to learn and improve.
- **Employees feel comfortable to 'speak up', point out problems, allow for dissenting views -** Open communication is encouraged. Healthy communication channels exist both vertically and horizontally within the organization. Vertical communications are two way – managers listen as well as speak. Horizontal communications ensure that all workers have the information. The organization emphasizes promptly observing and reporting non-standard conditions to permit the timely detection of weak signals that might foretell safety issues.

These examples of what a good process safety culture looks like are similar to the characteristics of High Reliability Organizations (HRO). HRO's are organizations with strong safety performance in high- risk environments. Examples include U.S. Navy aircraft carriers, U.S. forest firefighters, Federal Aviation Administration (FAA) traffic controllers, and also some private companies. They all possess the following characteristics. (Weick 2001)

- **Preoccupation with failure –** Personnel are always alert to early warning signs and envision where the next potential failure will occur.
- **Reluctance to simplify-** Personnel are not quick to accept simple explanations to anomalies, and probe for deeper understanding.
- **Sensitivity to operations –** The organization recognizes and understands how different elements of an organization interact/impact the front line and others in safety critical roles.
- **Commitment to resilience –** The organization not only is able to prevent failures but also recover quickly from them. This requires strong and fast learning capabilities.
- **Deference to expertise –** senior decision makers recognize that lower level employees have relevant knowledge and expertise to address problems and make them feel comfortable for speaking up.

23.3 Beyond the Management of Process Safety

Process safety culture extends beyond the regulations and process safety management systems. Figure 23.1 illustrates this concept with the CCPS Vision 20/20 which describes the characteristics of companies with great process safety performance. Vision 20/20 recognizes not only responsibilities of industry (industrial tenets) involved, but also responsibilities of external stakeholders (societal themes) necessary to achieve this great process safety performance. These stakeholders . include regulatory and investigative authorities, labor organizations, communities, research institutions, and academia. Industry working collaboratively with these stakeholders enables great process safety performance.

Table 23.1. CCPS Vision 20/20 industry tenets and societal themes

Vision 20/20 Industry Tenets
In a **Committed Culture**, executives involve themselves personally, managers and supervisors drive excellent execution every day, and all employees maintain a sense of vigilance and vulnerability.
Vibrant Management Systems are engrained throughout the organization. Vibrant systems readily adapt to the organization's varying operations and risks.
Disciplined Adherence to Standards means using recognized design, operations, and maintenance standards. These standards are followed every time, all the time, and are continually improved.
Intentional Competency Development ensures that all employees who impact process safety are fully capable of meeting the technical and behavioral requirements for their jobs.
Enhanced Application & Sharing of Lessons Learned communicates critical knowledge in a focused manner that satisfies the thirst for learning.
Vision 20/20 Societal Themes
Enhanced Stakeholder Knowledge promotes understanding of risk among all stakeholders, including the public, government, and industry leaders.
Responsible Collaboration is a cooperative relationship among regulatory and investigative authorities, labor organizations, communities, research institutions, universities, and industries.
Harmonization of Standards for the safe design, operation, and maintenance of equipment streamlines practices, eliminates redundancy, and cooperatively addresses emerging issues.
Meticulous Verification by knowledgeable independent parties helps companies evaluate their process safety programs from an independent perspective.

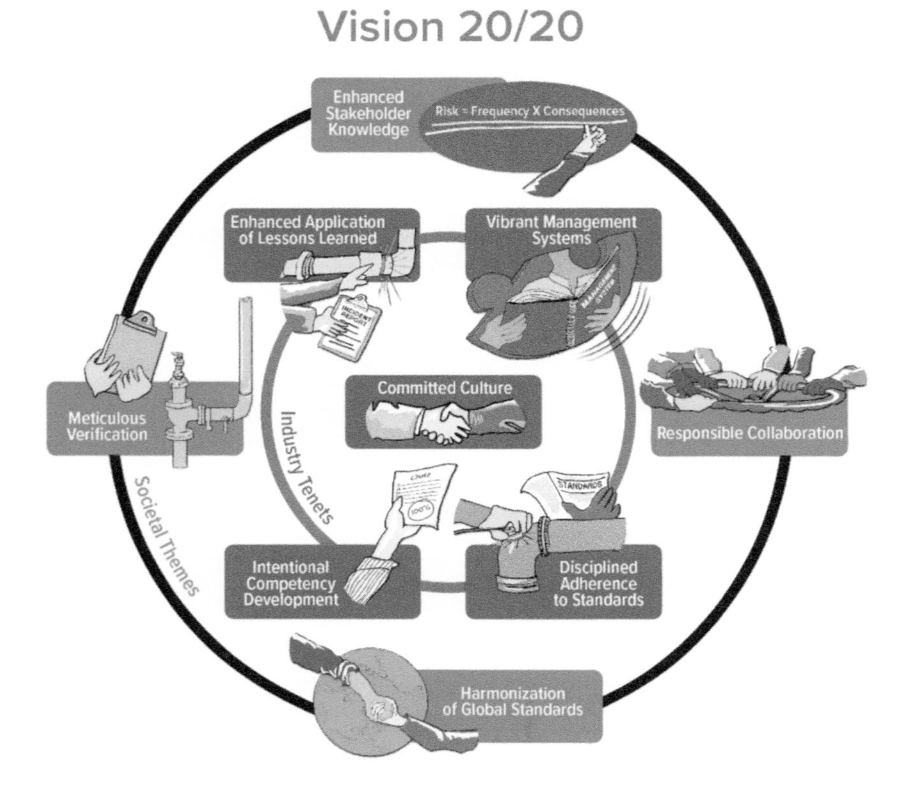

Figure 23.1. CCPS Vision 20/20

23.4 What a New Engineer Might Do

A new engineer can benefit from reviewing the CSB investigations and videos relevant to this chapter as listed in Appendix G.

23.5 Tools

Resources to support process safety culture include the following.

CCPS Vision 20/20 Assessment Tool. This tool is intended to help a company assess its process safety implementation as compared to the Vision 20/20 elements. It can be used in various operating locations or parts of a business to compare implementation across the company. The tool is available at https://www.aiche.org/ccps/vision-2020.

23.6 Exercises

1. In simple terms, define process safety culture.
2. List three things you might look for when evaluating the process safety culture in a facility.
3. Describe how a weak process safety culture was seen in the BP Texas City Refinery explosion.
4. Describe how a weak process safety culture was seen in the Space Shuttle disasters.
5. A new operator is learning the job from the experienced operator. The experienced operator uses his experience as a reference, instead of the operating procedures for the unit, and describes the short-cuts he uses. When the new operator asks about procedures, the experienced operator explains that they don't describe the way the facility works today. They came straight from engineering without any input from operations. What observations can you make about the process safety culture at this facility? Explain your thoughts.

23.7 References

Baker 2007, Baker James et al., "The Report of the BP U.S. Refineries Independent Safety Review Panel", www.csb.gov.

CCPS Vision 20/20, https://www.aiche.org/ccps/vision-2020.

Weick 2001, Weick, Karl E., Kathleen M. Sutcliffe, "Managing the Unexpected – Assuring High Performance in an Age of Complexity", Jossey-Bass, San Francisco, California.

Appendix A – Concluding Exercises

These concluding exercises bring together topics from multiple chapters of this book illustrating how the various process safety elements are relevant to a single facility.

Exercise 1: LNG Value Chain

The LNG value chain involves offshore production of gas, pipeline transportation of the gas to shore, treatment and liquefaction of the gas, storage and loading of the gas onto ships, shipping the LNG, and finally a receiving terminal where the gas will be used.

You are involved in the project to design and construct the LNG receiving terminal. The chemical process is relatively simple.

- The LNG is offloaded from ships into large LNG storage tanks.
- The LNG storage tanks consists of a stainless steel inner tank that contains the LNG, about 1 meter thickness of insulation, and an outer reinforced concrete tank to provide secondary containment and store the LNG at atmospheric pressure and -260 F.
- From the tanks, the LNG is vaporized, also referred to as regassified.
- Before the natural gas is delivered by pipeline to the customer, it is odorized with an unpleasant smelling odor to aid in leak detection.

References that may be helpful for this exercise include those from The International Group of Liquefied Natural Gas Importers available at these links.

https://giignl.org/sites/default/files/PUBLIC_AREA/About_LNG/3_LNG_Safety/giignl2019_infopapers4.pdf
https://giignl.org/sites/default/files/PUBLIC_AREA/About_LNG/4_LNG_Basics/giignl2019_infopapers2.pdf

1. Name 2 codes or standards that might apply to this project. One code should be specific to the design of the facility. The other should address management of the risks.
2. Describe the physical properties of LNG. Is it hazardous? Cite your sources.
3. Beyond chemical hazards, what other hazards might warrant consideration?
4. Have there been any LNG accidents in industry that you can learn from?
5. Make a plan for what process safety studies and activities you will do, or have done, at what stage of the project.
6. As this is early in the project and details are not available, a Preliminary Hazards Analysis is being conducted. List 10 specific questions that should be considered. For each, identify potential consequences.
7. For the scenarios identified in the Preliminary Hazards Analysis, list potential methods to prevent or mitigate the consequences.
8. What inherently safer design options might be considered for this project?

9. The main pieces of equipment on the site will be LNG tanks, unloading arms, heat exchangers, pumps, and piping. Name three failures that might occur with this equipment.
10. A consequence analysis is to be performed. List 3 potential scenarios including source, transport, consequence effects, and potential outcomes.
11. Draw a swiss cheese diagram for one of the scenarios identified in the Preliminary Hazards Analysis.
12. Suggest three aspects of human factors that should be considered in the project team and their design of this facility.
13. List 5 things you expect to be on the operational readiness plan for this project.
14. As the project is 50% through the detailed engineering, a proposal is made to add an additional LNG tank. How should this be handled?
15. List three operating practices and three safe work practices that would be appropriate for this facility when it is operational.
16. List 3 emergencies that should be addressed in the Emergency Response Plan for this facility.
17. List 2 means to engage the workforce in the project. List 2 stakeholder groups that should be involved in the project.
18. List 3 leading and 3 lagging process safety metrics that might be appropriate for this facility when it is operational.
19. What action might you take to foster a good process safety culture on the project?

Exercise 2: Polymerization Reactor

You have been assigned to a HIRA study team to evaluate hazards associated with a continuous solution polymerization reactor at your manufacturing facility which is located near the Houston, Texas ship channel. This reactor is located within a 2000 m^3 (71000 ft^3) enclosed process structure and roughly 250 m (820 ft) from a 50-person housing complex.

Styrene monomer and ethylbenzene solvent are added to the 2000-gallon reactor by flow control from their respective storage tanks via pump (not shown). The monomer-solvent mixture is heated in a shell and tube heat exchanger with 10 barg steam to the normal reactor operating temperature of 90 C. The exothermic reaction is maintained at 90 °C within the reactor by temperature control of the vessel jacket with cooling water. The reactor is well mixed with a 15-horsepower agitator. The reactor is also maintained under an inert atmosphere at 0.25 bar gauge using nitrogen by a series of back pressure regulators.

A catalyst solution is added to the reactor from drums by a small metering pump. Drum weight is monitored by the Catalyst Scale and a low weight alarm. Once a drum is empty, the operator manually stops the reactor feeds, replaces the empty catalyst drum with a full one, and restarts the system.

The flow from the reactor to the evaporator is controlled to maintain a constant level in the reactor. The Reactor Pump is also used during start-up to circulate the reactor contents through the Feed Heater (with monomer feed, solvent feed and catalyst feed manually shut off) until the reactor reaches the normal operating temperature. Once the reactor reaches the desired 90 °C operating temperature, all feeds are resumed, the recirculation valve is closed, and level control of the reactor begins. During start-up from an empty reactor, the reactor is charged with solvent and a small fraction of monomer prior to heat up and initiation of feeds.

This simple Piping and Instrument Diagram which shows the equipment, interconnected piping (including utility piping such as nitrogen, cooling water and steam) and control instrumentation for the continuous solution polymerization reactor.

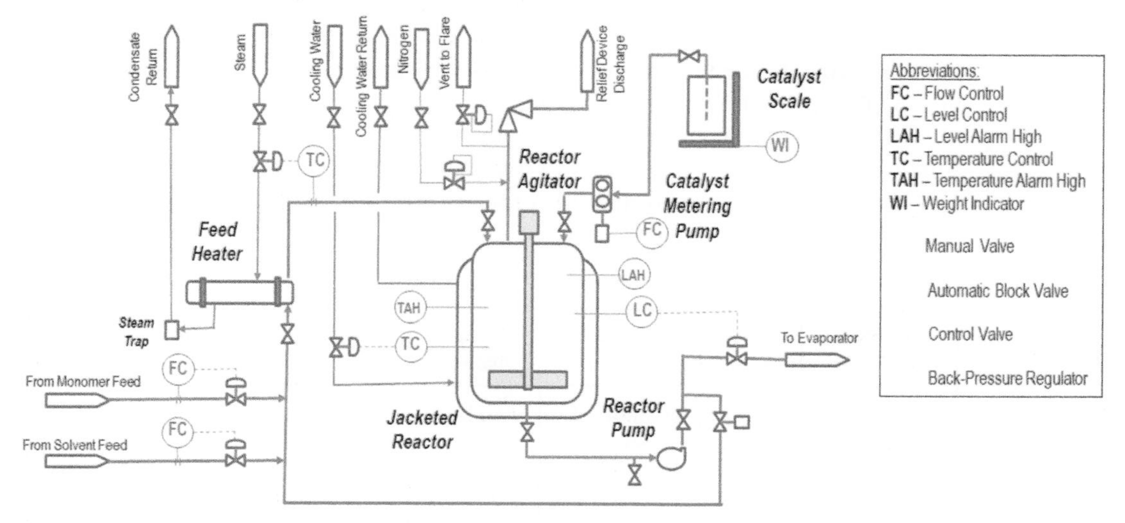

Figure A.1. Polymerization Reactor P&ID

1. Name 2 codes or standards that might apply to this project. One code should be specific to the design of the facility. The other should address management of the risks.
2. Describe the physical properties of the chemicals being handled. Are they hazardous? Cite your sources.
3. Estimate the maximum reaction temperature and pressure starting from the initial conditions of the normal operating temperature and pressure. (Note that a "worst case" concentration would need to be assumed based on addition of monomer feed during an upset where no reaction is occurring.) Use a relationship for vapor pressure of the reaction mixture as:

   Ln (P) = 9.42 - 3365.5 / (T - 55.3) for T in degree K and P in bar

   Use a liquid heat capacity for the styrene-ethyl benzene of 2.0 Joules per gram.

   Use a heat of polymerization of -1120 Joule per gram styrene.

Note that at the 90 °C normal operating temperature, the vapor pressure of the styrene-ethyl benzene mixture is 0.22 bar absolute with the balance of the normal operating pressure from the nitrogen pad.

4. Beyond chemical hazards, what other hazards might warrant consideration?
5. Have there been any accidents in industry that you can learn from?
6. Make a plan for what process safety studies and activities you will do, or have done, at what stage of the project.
7. A HAZOP is to be conducted. Using the jacketed reactor as a node,
   - Develop a design intent statement. Include as appropriate intended volumes, flow rates, composition, temperature, pressure, and other process information.
   - Add to the HAZOP Log Sheet other parameter/deviations and causes as you can think of. Complete the remaining columns for each scenario added. Refer to the P&ID for existing safeguards. Document any recommendations you think would help eliminate the scenario or significantly reduce the severity of the consequence.
   - Using a scale of high/medium/low, rank the severity and likelihood for each scenario.

Table A.1. HAZOP Log Sheet

Scenario	**Parameter/ Deviation**	**Cause**	**Consequences**	**Existing Safeguards**	**Severity**	**Likelihood**	**Recommend-ations**
1	Level-High	Level Control Loop Failure	Overflow of styrene-ethyl benzene mixture onto floor of the enclosed process area. Evaporation of spill with ignition leading to a Building Explosion.	High Level Alarm with a procedure to stop all feeds.	High Level Alarm with a procedure to stop all feeds.		Route the pressure relief device to a "safe" location outdoors.

8. What inherently safer design options might be considered for this project?
9. Name three failures that might occur with this equipment.
10. A consequence analysis is to be performed. List 3 potential scenarios including source, transport, consequence effects, and potential outcomes.
11. Draw a swiss cheese diagram for one of the scenarios identified in the HAZOP.
12. Suggest how human factors could be considered in conducting the HAZOP.
13. List 5 things you expect to be on the operational readiness plan for this project.
14. As the project is 50% through the detailed engineering, a proposal is made to increase the reactor size. How should this be handled?
15. List three operating practices and three safe work practices that would be appropriate for this facility when it is operational.
16. List 3 emergencies should be addressed in the Emergency Response Plan for this facility.
17. List 2 means to engage the workforce in the project. List 2 stakeholder groups that should be involved in the project.
18. List 3 leading and 3 lagging process safety metrics that might be appropriate for this facility when it is operational.
19. What action might you take to foster a good process safety culture on the project?

Exercise 3: Ethylene Buffer Tank

An operator is preparing an outdoor ethylene buffer tank for maintenance by evacuating the vessel of ethylene to an acceptable level. While lining up the vessel vent line to a flare header, an ethylene release to atmosphere occurred due to a ¾″ bleed valve being inadvertently left open.

When you arrive at work, your boss has several questions regarding the incident, and has given you 80 minutes to give him the answers. The first thing you do is gather the process safety information related to the incident from the Cameo database, MSDS, and CRW. This is given in Table A.2.

Table A.2. Ethylene Chemical Properties

Property	Ethylene
Chemical Formula	C_2H_4
MW	0.028 kg/gmole
$\gamma = C_p/C_v$	1.22
Flash Point	-213 ºF
Boiling Point	-154.7 ºF
Autoignition Temperature	842 ºF
Lower Explosive Limit	2.75 %vol
Upper Explosive Limit	28.6 %vol
UN dangerous goods hazard class	Division 2.1 – Flammable gas
Specific Gravity Vapor	0.569 @ -154.7 ºF
Pressure of ethylene inside the buffer tank	300 psig
Temperature of ethylene inside the buffer tank	78 ºF
MIE	0.08 mJ
Heat of Combustion	1322.6 kJ/mol

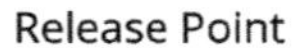

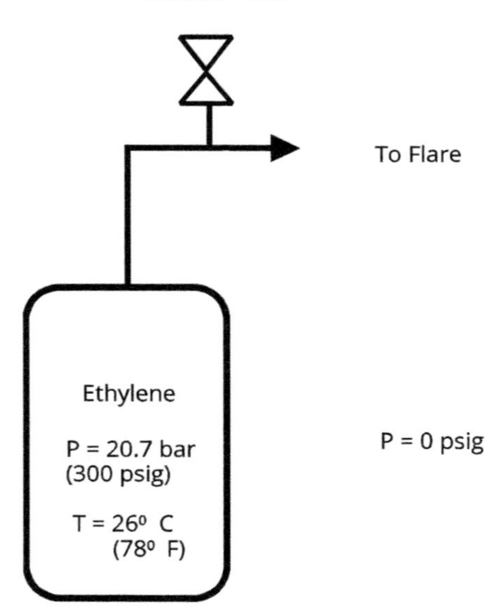

Figure A.2. Ethylene Buffer Tank

Use the information in Table A-2 and Figure A-2 to answer the following questions and build your report for your boss.

1. Ethylene is best described as: corrosive, flammable, combustible, or toxic?
2. The information related to hazards associated with the equipment, technology and chemicals in the process is called: Process Safety Information (PSI), Process Hazard Analysis (PHA), Management of Change (MOC), or Hazard Identification and Risk Analysis (HIRA).

3. What is the appropriate electric classification (Class and Division) using article 500 of the NFPA National Electric Code if a low probability of producing an explosive or ignitable mixture exists during abnormal conditions?

4. Name a source that can be used to get more information about the chemical hazards of ethylene.

5. What is the applicable API RP 754 outdoor Tier 1 TQ for this loss of primary containment incident?

6. If the amount released to atmosphere is greater than the outdoor Tier 1 TQ in a 60-minute period, what is the API RP 754 process safety incident classification?

7. If the amount released is less than the outdoor Tier 2 TQ, but the incident causes an OSHA recordable injury, what is the API RP 754 process safety incident classification?

8. If the ethylene ignites and causes direct damage to surrounding equipment of $25,000 U.S. dollars, what is the proper process safety incident classification?

9. If the ethylene ignites and causes direct damage to surrounding equipment of $90,000 U.S. dollars, what is the proper process safety incident classification?

10. An LEL monitor located 50m from the release point measures 5%. Will the ethylene/air vapor support combustion at this location?

11. The cause of this incident is most like which of the following historical process safety incidents? Phillips 66, Pasadena Texas, 1989; Flixborough Explosion, U.K. 1974; Bhopal India, 1984; BP Refinery Explosion, or Texas City, 2005.

12. If operator response to an alarm is used to reduce the consequence of release, what type of safeguard is this? Administrative, Passive Engineered, Active Engineered, or Safety Instrumented System (SIS)

13. At some point during the release, a water monitor was positioned to knock down the vapor cloud. Your boss asked for an evaluation of the chemical reactivity of ethylene and water. You run a chemical reactivity worksheet with the following result. Answer the following question to help your boss interpret the chemical reactivity.

Mixture Manager | Mixture Report | Compatibility Chart | Reactive Groups

Print Chart

Export Chart Data

NFPA | Chemical Pairs

Health	Flammability	Instability	Special	Ethylene water Compatibility Chart	ETHYLENE	WATER
2	4	2		ETHYLENE	SR	
				WATER	C	

Figure A.3. Chemical Reactivity Worksheet for Ethylene and water compatibility

14. Which statement best describes the reactivity of water with ethylene? Compatible, caution, chemical reactive, or incompatible?
15. Which statement best describes the reactivity of ethylene with itself? Selective reduction, selective reactivity, saturated radicals, or self-reactive?
16. During the release, your boss is concerned with potential ignition sources. Considering the MIE, what do you tell him about the likelihood of ignition?
17. You look up the maximum intended inventory of the process for ethylene and see that it is 20,000 lbm. If the process is in the United States, would it be covered under OSHA 1910.119, the "PSM" or Process Safety Management regulation? There is no OSHA PSM TQ for ethylene.
18. What is the damage estimate for common structures if the side on pressure is 2.0 psig?
19. If the wind speed during the release is 2 m/s on a still night, what is the atmospheric stability class?
20. The ethylene release scenario is analyzed. It is determined that the consequence could be an explosion causing $5 million in damages. What severity category is this using the risk matrix provided in section 14.5?
21. This ethylene release scenario involves an equipment failure that is estimated to occur sometime in the life of the piece of equipment. What probability category is this using the risk matrix provided in section 14.5?
22. Considering the severity and probability categories from the previous two questions, what is the risk using the risk matrix provided in section 14.5?
23. It is decided that this ethylene release risk should be reduced. What strategy should be considered first to reduce the risk?
24. It is decided to install an independent protection layer (IPL) as a means to reduce the risk. What functional criteria must this protection layer have to be considered an IPL?

Exercise 4: Wastewater Equalization Tank

You are assigned to conduct a HAZOP of a process. Your colleagues help you to break the process into nodes for the study. Node 1 is shown in Figure A.4 and the Node 1 intention and parameters are described in Table A.3.

Table A.4 is the HAZOP Worksheet for Node 1. The "high level" deviation has been completed. Complete the other deviations listed on the worksheet. Include the risk ranking for both the unmitigated and mitigated scenario. Use the risk matrix provided in Figure A.5 and Table A.5 and Table A.6.

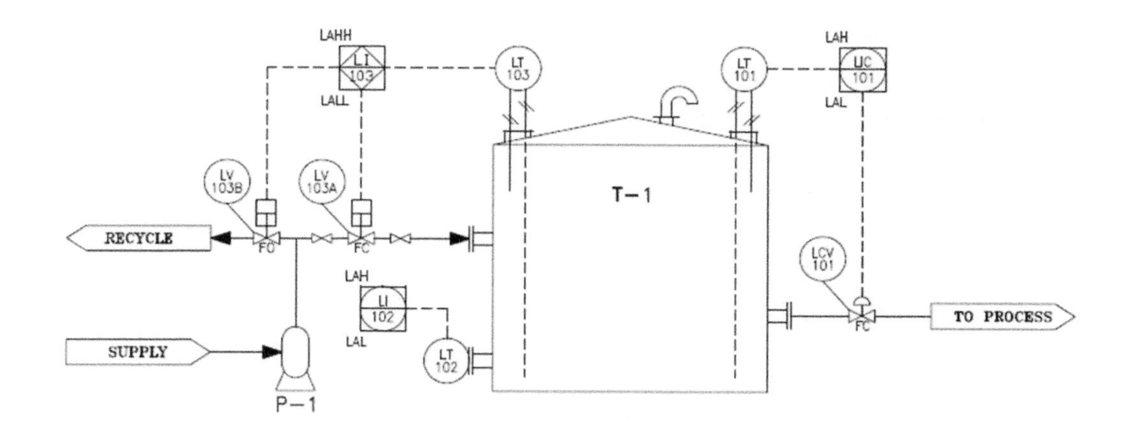

Figure A.4. Node 1 – T-1 WWT Equalization Tank

Table A.3. Node 1 – T-1 Intention, Boundary, Design Conditions and Parameters

Node #	Node	Node Intention	Node Boundary	Design Conditions/ Parameters	Operating Conditions/ Parameters	Drawings	
						Drawing	Rev
1	T-1 WWT Equalization Tank	Supplying variable low pH wastewater with P-1 to T-1, which serves to equalize the feed pH being fed forward to the process on level control (LIC-101).	P-1 Wastewater Feed Pump to T-1 WWT Equalization Tank including the level control valve (LCV-101)	P-1, Centrifugal Pump, 200 GPM, 80 ft TDH, Ductile Iron. T-1, Atmospheric Storage Tank, CS, 33,050 gallons, 25' Dia. x 15' Height	Normal Operating Range - 50% level - Ambient T 65 to 85F - pH 4.5 to 6	P&ID Example	0

Table A.4. HAZOP Worksheet Node 1 – T-1 WWT Equalization Tank

Node: 1. T-1 WWT Equalization Tank

Drawing Number: P&ID Example

Deviation	Causes	Consequences	Unmitigated Severities		Unmitigated RR		Safeguards Non-IPL	Mitigated RR		Recommendations
			S	E	L	RR		L	RR	
1. Low Pressure		1.								
2. High Temperature		1.								
3. Low Temperature		1.								
4. High Level	1. Malfunction of LIC-101 low closing LCV-101 or LCV-101 fails closed.	1. Potential for overfill of T-1 with liquid flow of low pH wastewater out through the atmospheric vent with splash and rundown into diked area. Potential for personnel injury due to chemical exposure to low pH wastewater. With continued feed from P-1 potential to overfill diked area into stormwater basin with exceedance of stormwater discharge permit.	4	4	2	M	2. LAH-102 High Level Alarm in the Basic Process Control System with operator response to check line up and shutdown the process if necessary. 3. LIT-103 in the Safety Instrumented System closes (LV-103A) the feed to T-1 and opens (LV-103B) the recycle back to the Process Tank 1. Plant PPE	3	L	
5. Low Level		1.								
6. High Flow		1.								
7. Low/No Flow		1.								
8. Reverse/ Misdirected Flow		1.								
9. Composition		1.								
10. Reaction		1.								
11. Relief		1.								
12. Abnormal Operations (Startup / Shutdown / Emergency)		1.								
13. Corrosion/ Erosion		1.								
14. Maintenance		1.								
15. Loss of Utilities		1.								
16. Loss of Containment		1.								
17. Other		1.								

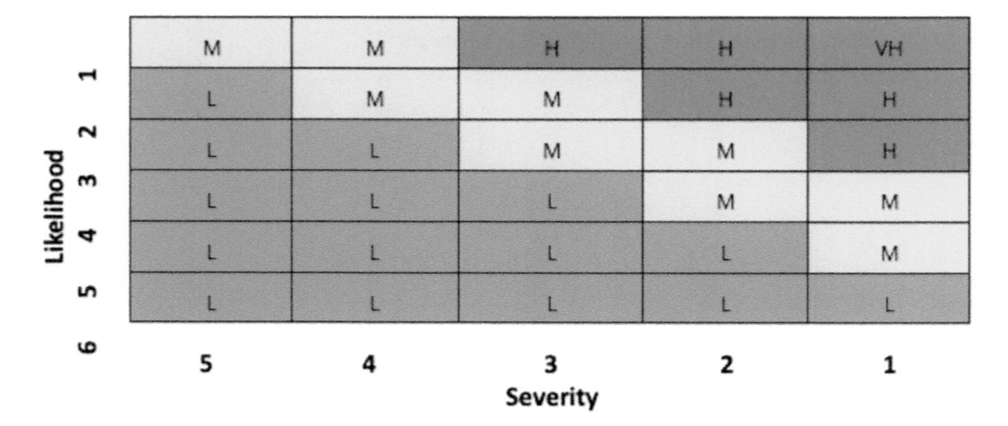

Figure A.5. Risk Matrix

Table A.5 Risk Matrix Severity

Code	Description	Safety	Environmental	Property	TMEL
1	Catastrophic	Multiple Fatalities Single Offsite Permanent Disabling Injury (irreversible)	Long-term offsite impact Extended clean-up/containment required Significant media interest	> $100 million	1.00E-05
2	Significant	Single Fatality or Multiple Permanent Disabling Injuries Multiple Offsite First Aid Single Offsite Hospitalization (reversible)	Offsite impact resulting in permanent damage to ecosystem Event requiring substantial clean up Substantial media interest	Up to $100 million	1.00E-04
3	Moderate	Single Permanent disabling injury Multiple lost time injuries/illness Single offsite First Aid	On-site release requiring extended clean-up Plant Evac Off-site release resulting in Offsite Shelter in Place	Up to $10 million	1.00E-03
4	Minor	Single Lost Time or Multiple First Aid	Short-term onsite impact Permit exceedance Reportable quantity Plant Shelter in Place	Up to $1 million	1.00E-02
5	Negligible	Single First Aid, Injury, or Illness	Minor onsite impact, fully contained, readily mitigated, no reportable quantity	< $ 100,000	1.00E-01

Table A.6 Risk Matrix Likelihood

Code	Description	Description
1	Frequent	Could be expected to occur several times within the facility lifetime
2	Probable	Could be expected to occur once during the facility lifetime
3	Unlikely	Might occur once in ten facility lifetimes
4	Rare	Known to have occurred, but unlikely within 10 facility lifetimes
5	Improbable	Conceivable, but not known to have occurred
6	Unforeseen	Highly Improbable event that is unforeseen

Exercise 4 Content, tables and figures provided by AESolutions (https://www.aesolns.com/)

Appendix B – Relationship Between Book Content and Typical Engineering Courses

This book is intended to support both the teaching of a process safety course and as a materials resource for the inclusion of process safety topics in typical engineering courses. To support the later, this matrix relates the chapters in this book with typical engineering courses.

Table B.1. Typical engineering course relationship with book contents

		Introduction to Chemical Eng. Introduction to Eng.	Organic Chemistry Inorganic Chemistry	Engineering Materials (Materials Science)	Corrosion Engineering	Thermodynamics	Heat Transfer	Material and Energy Balance	Reaction Engineering	Fluid Mechanics	Numerical Methods	Particle Technology	Process Design (Engineering Design or Plant Design)	Unit Operations Laboratory	Mass Transfer	Process Control	Statistical Design of Experiments	Chemical Eng. Process Modelling/Simulation	Separation Processes (e.g. Distillation)	Biochemical Engineering	Pollution Control Technology (Green Engineering)	Leadership in Chemical Engineering	Special Topics in Process Safety
	Section I: What is Process Safety and Why should I care?																						
1	Introduction	X				X																	
2	Risk Based Process Safety - Four Pillars and 20 Elements																						
3	Process Safety Codes, Standards, and Regulations																						
	Section II: What are the Hazards?			X		X		X	X			X		X					X	X	X		
4	Fire and Explosion Hazards	X				X			X					X					X				
5	Reactive Chemical Hazards	X	X						X					X					X				
6	Toxic Hazards	X	X											X					X				
7	Chemical Hazards Data																						
8	Other Hazards																						
9	Process Safety *Metrics*																						
	Section III: How do I address the hazards in my design?																						
10	Project Design basics	X					X		X	X			X	X	X	X			X		X		
11	Equipment Failure			X	X								X	X					X				
12	Hazard Identification			X		X	X	X	X	X		X	X	X	X		X		X	X	X		
13	Consequence Analysis			X		X	X	X	X	X	X	X	X	X	X		X	X	X	X	X		X
14	Risk Assessment	X							X	X			X	X	X	X	X	X	X		X		
15	Risk Prevention and Mitigation						X			X			X	X	X	X			X				
	Section V: How do I manage risk in operations?																						
16	Human Factors												X	X		X	X		X		X	X	X
17	Operational Readiness																						
18	Management of Change																						
19	Operating Procedures, Conduct of Operations, and Safe Work Practices																						
20	Emergency management																						
	Section VI: How do I sustain the focus on Process Safety?																						
21	Management Side of Process Safety Management																						
22	Managing Process Safety Performance																						

Based on CCPS resources and:

J Forest, Risk-Based Process Safety in Chemical Engineering Curriculum, Chemical Engineering Education, Vol 52, No 4, 2018

Klein, James A.; Davis, Richard A., Conservation of Life as a Unifying Theme for Process Safety in Chemical Engineering Education, Chemical Engineering Education, Vol 45, No. 2, 2011 Gainseville, Florida

Appendix C – Example RAGAGEP List

Recognized and Generally Accepted Good Engineering Practice (RAGAGEP) is a term used by OSHA, stemming from the selection and application of appropriate engineering, operating, and maintenance knowledge when designing, operating and maintaining chemical facilities with the purpose of ensuring safety and preventing process safety incidents. OSHA does not provide a specific list of RAGAGEP practices. Instead, these are inferred from OSHA letters of interpretation and OSHA audit findings. Once a company specifies a RAGAGEP standard, then it is committed to implementing it. An example RAGAGEP list is provided in Table C.1.

Table C.1. Example RAGAGEP list

Topic	Code or Standard
Atmospheric Tanks	API 620: Design and Construction of Large, Welded, Low-pressure Storage Tanks
Fired Equipment	NFPA 85: Boiler and Combustion Systems Hazards Code NFPA 86: Standard for Ovens and Furnaces FM 6-0: Industrial Heating Equipment, General FM 6-9: Industrial Ovens and Dryers FM 6-10: Process Furnaces FM 7-99: Hot Oil Heaters API 521: Pressure-Relieving and Depressuring Systems API 537: Flare Details for General Refinery and Petrochemical Service
Flammable Liquids	NFPA 30: Flammable and Combustible Liquids Code NFPA 77: Recommended Practice on Static Electricity
Heat Exchangers	TEMA: Standards of the Tubular Exchanger Manufacturers Association API 510: Pressure Vessel Inspection Code: In-Service Inspection, Rating, Repair, and Alteration
Instrumentation and Controls	ISA-18.2 Management of Alarm Systems for the Process Industries ISA-84.91.01 Identification and Mechanical Integrity of Safety Controls, Alarms, and Interlocks in the Process Industry ISA-84.00 Functional Safety: Safety Instrumented Systems for the Process Industry Sector ISA-101 (Draft) Human Machine Interfaces for Process Automation Systems
Plant Buildings	API 752: Management of Hazards Associated with Location of Process Plant Permanent Buildings API 753: Management of Hazards Associated with Location of Process Plant Portable Buildings
Pressure Vessels	ASME Section VIII – Pressure Vessels API 510: Pressure Vessel Inspection Code: In-Service Inspection, Rating, Repair, and Alteration

Table C.1 *continued*

Topic	Code or Standard
Solids Handling Equipment	NFPA 654: Standard for the Prevention of Fires and Dust Explosions from the Manufacturing, Processing, and Handling of Combustible Particulate Solids NFPA 68: Standard on Explosion Protection by Deflagration Venting NFPA 69: Standard on Explosion Prevention Systems FM 7-76: Prevention and Mitigation of Combustible Dust Hazards
Chemical Specific Codes	
Chlorine	Chlorine Institute Pamphlet 5) Bulk Storage of Liquid Chlorine Chlorine Institute Pamphlet 6) Piping Systems for Dry Chlorine Chlorine Institute Pamphlet 9) Chlorine Vaporing Systems
Peroxides	NFPA 430: Code for the Storage of Liquid and Solid Oxidizers
Compressed Gases	Compressed Gas Association P-22: The Responsible Management and Disposition of Compressed Gases and their Cylinders

Appendix D – Reactive Chemicals Checklist

This checklist is adapted from a CCPS Safety Alert; *A Checklist for Inherently Safer Chemical Reaction Process Design and Operation*, March 1, 2004. For additional information on chemical reactivity tools, see section 5.8.

D.1 Chemical Reaction Hazard Identification

1. Know the heat of reaction for the intended and other potential chemical reactions.

Several techniques are available for measuring or estimating heat of reaction, including various calorimeters, plant heat and energy balances for processes already in operation, analogy with similar chemistry (confirmed by a chemist who is familiar with the chemistry), literature resources, supplier contacts, and thermodynamic estimation techniques. You should identify all potential reactions that could occur in the reaction mixture and understand the heat of reaction of these reactions.

2. Calculate the maximum adiabatic temperature for the reaction mixture.

Use the measured or estimated heat of reaction, assume no heat removal, and that 100% of the reactants actually react. Compare this temperature to the boiling point of the reaction mixture. If the maximum adiabatic reaction temperature exceeds the reaction mixture boiling point, the reaction is capable of generating pressure in a closed vessel and you will have to evaluate safeguards to prevent uncontrolled reaction and consider the need for emergency pressure relief systems.

3. Determine the stability of all individual components of the reaction mixture at the maximum adiabatic reaction temperature.

This might be done through literature searching, supplier contacts, or experimentation. Note that this does not ensure the stability of the reaction mixture because it does not account for any reaction among components, or decomposition promoted by combinations of components. It will tell you if any of the individual components of the reaction mixture can decompose at temperatures which are theoretically attainable. If any components can decompose at the maximum adiabatic reaction temperature, you will have to understand the nature of this decomposition and evaluate the need for safeguards including emergency pressure relief systems.

4. Understand the stability of the reaction mixture at the maximum adiabatic reaction temperature.

Are there any chemical reactions, other than the intended reaction, which can occur at the maximum adiabatic reaction temperature? Consider possible decomposition reactions, particularly those which generate gaseous products. These are a particular concern because a small mass of reacting condensed liquid can generate a very large volume of gas from the reaction products, resulting in rapid pressure generation in a closed vessel. Again, if this is possible, you will have to understand how these reactions will impact the need for safeguards, including emergency pressure relief systems. Understanding the stability of a mixture of components may require laboratory testing.

5. Determine the heat addition and heat removal capabilities of the pilot plant or production reactor.

Don't forget to consider the reactor agitator as a source of energy – about 2550 Btu/hour/horsepower. Understand the impact of variation in conditions on heat transfer capability. Consider factors such as reactor fill level, agitation, fouling of internal and external heat transfer surfaces, variation in the temperature of heating and cooling media, variation in flow rate of heating and cooling fluids.

6. Identify potential reaction contaminants.

In particular, consider possible contaminants which are ubiquitous in a plant environment, such as air, water, rust, oil and grease. Think about possible catalytic effects of trace metal ions such as sodium, calcium, and others commonly present in process water. These may also be left behind from cleaning operations such as cleaning equipment with aqueous sodium hydroxide. Determine if these materials will catalyze any decomposition or other reactions, either at normal conditions or at the maximum adiabatic reaction temperature.

7. Consider the impact of possible deviations from intended reactant charges and operating conditions.

For example, is a double charge of one of the reactants a possible deviation, and, if so, what is the impact? This kind of deviation might affect the chemistry which occurs in the reactor – for example, the excess material charged may react with the product of the intended reaction or with a reaction solvent. The resulting unanticipated chemical reactions could be energetic, generate gases, or produce unstable products. Consider the impact of loss of cooling, agitation, and temperature control, insufficient solvent or fluidizing media, and reverse flow into feed piping or storage tanks.

8. Identify all heat sources connected to the reaction vessel and determine their maximum temperature.

Assume all control systems on the reactor heating systems fail to the maximum temperature. If this temperature is higher than the maximum adiabatic reaction temperature, review the stability and reactivity information with respect to the maximum temperature to which the reactor contents could be heated by the vessel heat sources.

9. Determine the minimum temperature to which the reactor cooling sources could cool the reaction mixture.

Consider potential hazards resulting from too much cooling, such as freezing of reaction mixture components, fouling of heat transfer surfaces, increases in reaction mixture viscosity reducing mixing and heat transfer, precipitation of dissolved solids from the reaction mixture, and a reduced rate of reaction resulting in a hazardous accumulation of unreacted material.

10. Consider the impact of higher temperature gradients in plant scale equipment compared to a laboratory or pilot plant reactor.

Agitation is almost certain to be less effective in a plant reactor, and the temperature of the reaction mixture near heat transfer surfaces may be higher (for systems being heated) or lower (for systems being cooled) than the bulk mixture temperature. For exothermic reactions, the temperature may also be higher near the point of introduction of reactants because of poor mixing and localized reaction at the point of reactant contact. The location of the reactor temperature sensor relative to the agitator, and to heating and cooling surfaces may impact its ability to provide good information about the actual average reactor temperature. These problems will be more severe for very viscous systems, or if the reaction mixture includes solids which can foul temperature measurement devices or heat transfer surfaces. Either a local high temperature or a local low temperature could cause a problem. A high temperature, for example, near a heating surface, could result in a different chemical reaction or decomposition at the higher temperature. A low temperature near a cooling coil could result in slower reaction and a buildup of unreacted material, increasing the potential chemical energy of reaction available in the reactor. If this material is subsequently reacted because of an increase in temperature or other change in reactor conditions, an uncontrolled reaction is possible due to the unexpectedly high quantity of unreacted material available.

11. Understand the rate of all chemical reactions.

It is not necessary to develop complete kinetic models with rate constants and other details, but you should understand how fast reactants are consumed and generally how the rate of reaction increases with temperature. Thermal hazard calorimetry testing can provide useful kinetic data.

12. Consider possible vapor phase reactions.

These might include combustion reactions, other vapor phase reactions such as the reaction of organic vapors with a chlorine atmosphere, and vapor phase decomposition of materials such as ethylene oxide or organic peroxide.

13. Understand the hazards of the products of both intended and unintended reactions.

For example, does the intended reaction, or a possible unintended reaction, form viscous materials, solids, gases, corrosive products, highly toxic products, or materials which will swell or degrade gaskets, pipe linings, or other polymer components of a system? If you find an unexpected material in reaction equipment, determine what it is and what impact it might have on system hazards. For example, in an oxidation reactor, solids were known to be present, but nobody knew what they were. It turned out that the solids were pyrophoric, and they caused a fire in the reactor.

14. Consider doing a Chemical Interaction Matrix and/or a Chemistry Hazard Analysis.

These techniques can be applied at any stage in the process life cycle, from early research to an operating plant. They are intended to provide a systematic method to identify chemical interaction hazards and hazards resulting from deviations from intended operating conditions.

D.2 Reaction Process Design Considerations

1. Rapid reactions are desirable.

In general, you want chemical reactions to occur immediately when the reactants come into contact. The reactants are immediately consumed, and the reaction energy quickly released, allowing you to control the reaction by controlling the contact of the reactants. However, you must be certain that the reactor is capable of removing all of the heat and any gaseous products generated by the reaction.

2. Avoid batch processes in which all of the potential chemical energy is present in the system at the start of the reaction step.

If you operate this type of process, know the heat of reaction and be confident that the maximum adiabatic temperature and pressure are within the design capabilities of the reactor.

3. Use gradual addition or "semi-batch" processes for exothermic reactions.

The inherently safer way to operate exothermic reaction process is to determine a temperature at which the reaction occurs very rapidly. Operate the reaction at this temperature, and feed at least one of the reactants gradually to limit the potential energy contained in the reactor. This type of gradual addition process is often called "semi-batch." A physical limit to the possible rate of addition of the limiting reactant is desirable – a metering pump, flow limited by using a small feed line, or a restriction orifice, for example. Ideally, the limiting reactant should react immediately, or very quickly, when it is charged. The reactant feed can be stopped if necessary if any kind of a failure occurs (for example, loss of cooling, power failure, loss of agitation) and the reactor will contain little or no potential chemical energy from unreacted material. Some way to confirm actual reaction of the limiting reagent is also desirable. A direct measurement is best, but indirect methods such as monitoring of the demand for cooling from an exothermic batch reactor can also be effective.

4. Avoid using control of reaction mixture temperature as the only means for limiting the reaction rate.

If the reaction produces a large amount of heat, this control philosophy is unstable – an increase in temperature will result in faster reaction and even more heat being released, causing a further increase in temperature and more rapid heat release. If a large amount of potential chemical energy from reactive materials is present, a runaway reaction occurs. This type of process is vulnerable to mechanical failure or operating error. A false indication of reactor temperature can lead to a higher than expected reaction temperature and possible runaway because all of the potential chemical energy of reaction is available in the reactor. Many other single failures could lead to a similar consequence – a leaking valve on the heating system, operator error in controlling reactor temperature, failure of software or hardware in a computer control system.

5. Account for the impact of vessel size on heat generation and heat removal capabilities of a reactor.

Remember that the heat generated by a reactive system will increase more rapidly than the capability of the system to remove heat when the process is operated in a larger vessel. Heat generation increases with the volume of the system – by the cube of the linear dimension. Heat removal capability increases with the surface area of the system, because heat is generally only removed through an external surface of the reactor. Heat removal capability increases with the square of the linear dimension. A large reactor is effectively adiabatic (zero heat removal) over the short time scale (a few minutes) in which a runaway reaction can occur. Heat removal in a small laboratory reactor is very efficient, even heat leakage to the surroundings can be significant. If the reaction temperature is easily controlled in the laboratory, this does not mean that the temperature can be controlled in a plant scale reactor. You need to obtain the heat of reaction data discussed previously to confirm that the plant reactor is capable of maintaining the desired temperature.

6. Use multiple temperature sensors, in different locations in the reactor for rapid exothermic reactions.

This is particularly important if the reaction mixture contains solids, is very viscous, or if the reactor has coils or other internal elements which might inhibit good mixing.

7. Avoid feeding a material to a reactor at a higher temperature than the boiling point of the reactor contents.

This can cause rapid boiling of the reactor contents and vapor generation.

Appendix E – Classifying Process Safety Events Using API RP 754 3rd Edition

E.1 Criterion for PSE

This appendix covers how incidents can be classified using the guidance provided in API RP 754, 3rd Edition as is discussed in Section 9.3. The CCPS Process Safety Incident evaluation app can assist in classification of a PSE. This app is available at app stores.

Classification of a PSE as Tier 1 or 2 requires understanding of several criteria. The first criterion is that there must be a Loss of Primary Containment (LOPC).

> **Loss of Primary Containment (LOPC)** - An unplanned or uncontrolled release of material from primary containment, including non-toxic and non-flammable materials (e.g., steam, hot condensate, nitrogen, compressed CO_2 or compressed air). (CCPS Glossary)
>
> Note: Steam, hot condensate, and compressed or liquefied air are only included in this definition if their release results in one of the consequences other than a threshold quantity release. However, other nontoxic, nonflammable gases with defined UN Dangerous Goods (UNDG) Division 2.2 thresholds (such as nitrogen, argon, compressed CO_2) are included in all consequences including, threshold release. (API RP 754).
>
> **Primary Containment** - A tank, vessel, pipe, transport vessel or equipment intended to serve as the primary container for, or used for the transfer of, a material. Primary containers may be designed with secondary containment systems to contain or control a release from the primary containment. Secondary containment systems include, but are not limited to, tank dikes, curbing around process equipment, drainage collection systems into segregated oily drain systems, the outer wall of double-walled tanks, etc. (CCPS Glossary)
>
> **Process Safety Event** – An event that is potentially catastrophic, i.e., an event involving the release/loss of containment of hazardous materials that can result in large-scale health and environmental consequences. (CCPS 2019)

E.2 Criterion for Classification

Tier 1 and 2 PSEs are LOPCs that occur within any 60-minute window with certain consequences, the severity of which determine the classification tier. These consequences, as shown in Table E.1, include injury; direct cost from resulting fire and explosion damage; community impact; unsafe release from engineered pressure relief or upset emission from a permitted regulated source; and or acute release above a defined threshold quantity.

Table E.1. Tier 1 Level and Tier 2 Level Consequences

(CCPS 2018)

Consequence	Tier 1	Tier 2
Injury	OSHA "days away" Hospitalization Fatality	OSHA Recordable
Fire/Explosion damage	> $100,000 USD	> $2,500 and < $100,000 USD
Community impact	Community evacuation or shelter-in-place	
An engineered pressure relief or upset emission from a permitted or regulated source	Discharge amount, time and rainout Unsafe location Onsite shelter-in-place Public protective measures	Discharge amount, time and rainout Unsafe location Onsite shelter-in-place Public protective measures
Acute release	Above material threshold quantity	

E.2.1 Release Criterion

Release from process. Tier 1 and 2 incidents involve an LOPC from process. Process refers to equipment, storage tanks, active warehouses, ancillary support areas, on-site remediation facilities, and distribution piping under control of the company used in the manufacture of petrochemical and petroleum refining products.

Unplanned and Uncontrolled. Intent of the release is a criterion in Tier determination. PSEs must be unplanned or uncontrolled. It is possible to plan a safe release from primary containment. An example is a safe de-inventory of process in preparation for equipment maintenance. As a planned activity, this type of LOPC is not considered a Tier 1 or 2 indicator.

E.2.2 Outcome Criterion

If the discharge contains one of the four consequences described here, the tier of the release is determined by the release quantify described in Section E.2.3.

OSHA Recordable Injury or Illness - Any work-related injury or illness requiring medical treatment beyond first aid. (adapted from OSHA)

Days Away from Work - An OSHA recordable injury or illness resulting in one or more days away from work, restricted work, or transfer to another job. (adapted from OSHA)

Injury. An LOPC of any material in any amount that results in an injury requiring treatment beyond first aid is considered for Tier 1 or 2 classification. If the LOPC results in an injury classification of OSHA recordable, then the PSE classification is Tier 2. Tier 1 classification results from more serious injury such as an OSHA "days away" case, or hospital admission and/or fatality.

Community Impact. LOPCs that result in an officially declared community evacuation, including precautionary evacuation, or community shelter in place are Tier 1 events.

Direct damage from fire/explosion. Direct damage cost includes the costs to repair equipment or replace it and the costs of clean up and environmental reparations. The fire or explosion must be a result of an LOPC. Direct cost greater than or equal to $100,000 USD are classified as Tier 1 and cost greater than or equal to $2,500 up to $100,000 USD are classified as Tier 2.

Releases from engineered pressure relief and upset emissions from permitted sources. Engineered pressure relief devices include pressure relief device (PRD), rupture disks, Safety Instrumented System (SIS) devices, or manually initiated emergency de-pressure devices. LOPC from these devices are excluded from Tier 1 or Tier 2 classification if the release is as design and proven safe. Safe release can be defined as the maximum concentration of release below ½ the LEL for flammables, or ERPG-3 for toxics at potentially occupied locations around the release. However, if the amount of discharge is greater than or equal to the threshold quantity (described in Section E.2.3) in any one-hour period and results in any one or more of the following consequences: rainout; discharge to a potentially unsafe location, an on-site shelter-in-place or on-site evacuation, excluding precautionary on-site shelter-in-place or on-site evacuation; and/or public protective measures (e.g. road closure) including precautionary public protective measures. These criteria are also applicable for a permitted or regulated source.

E.2.3 Release Quantity Criterion

If the discharge contains one of the four consequences described in Section E.2.2, the tier of the release is determined by the release quantify described as follows.

60 Minute Release. To have a PSE Tier 1 and 2 classification, the acute TQ threshold (described next) must be exceeded any 60-minute window of the release. For a steady state release, the release amount is normalized to the amount released over the 60-minute period. Source models must be developed for non-steady state releases.

Acute release above the threshold quantity (TQ). API classifies materials in eight Tier 1 and Tier 2 threshold release categories. The release categories recognize the relative potential hazardous consequences of release; the higher the potential consequence, the lower the TQ. The category determining the TQ is listed as toxic (toxic inhalation hazard, or TIH Zone), flammability (boiling point and flash point) or corrosivity. When a material can't be classified by these characteristics (in this order), the United Nations Dangerous Goods (UNDG) Packing Group is used. UN Packing Groups can usually be found in section 14, Transportation, of a materials SDS sheet. (Refer to Section 7.3.2)

Indoor v. outdoor release. The Tier 1 and 2 release categories are further subdivided for outdoor and indoor release. The lower indoor quantity accounts for a potentially greater hazardous consequence associated with indoor release. Table E.2 shows the relationship between Tier 1 and 2 outdoor and indoor threshold quantities (TQ).

Table E.2. Threshold quantity relationship

	Outdoor TQ	Indoor TQ
Tier 1	T1	0.10 T1
Tier 2	0.10 T1	0.50 T1

E.3 PSE Tier 1 and Tier 2 Threshold Quantities

Table E.3 is used to determine a material Threshold Quantity (TQ) when determining if the incident is a Tier 1 or Tier 2 PSE. First determine the category in order from Category 1 through 8 by choosing the material classification from the Material Hazard Classification Option 1 or Option 2. Material Hazard Classification Option 2 refers to global harmonization standard which is discussed in API RP 754. Examples of materials in each material classification category are provided in Table E.4. Next, determine if the release is indoor or outdoor and look up the TQ. Option 1 includes the Toxic Inhalation Hazard (TIH) category, flammable characteristic, and United Nations Dangerous Goods (UNDG) packing group which can generally be found in the material's SDS or the chemical list from the CCPS PSIE app.

Table E.3. Material Release Threshold Quantities
(reformatted from API RP 754)

Threshold Release Category	Material Hazard Classification Option 1	Material Hazard Classification Option 2	Threshold Quantity (outdoor)	Threshold Quantity (indoor[b])
TRC-1	TIH Zone A Materials	H330 Fatal if inhaled, Acute toxicity, inhalation (cat 1)	Tier 1: ≥ 5 kg (11 lb) Tier 2: ≥ 0.5 kg (1.1 lb)	Tier 1: ≥ 0.5 kg (1.1 lb) Tier 2: ≥ 0.25 kg (0.55 lb)
TRC-2	TIH Zone B Materials	H330 Fatal if inhaled, Acute toxicity, inhalation (cat 2)	Tier 1: ≥ 25 kg (55 lb) Tier 2: ≥ 2.5 kg (5.5 lb)	Tier 1: ≥ 2.5 kg (5.5 lb) Tier 2: ≥ 1.25 kg (2.75 lb)
TRC-3	TIH Zone C Materials	H331 Toxic if inhaled, Acute toxicity, inhalation (cat 3)	Tier 1: ≥ 100 kg (220 lb) Tier 2: ≥ 10 kg (22 lb)	Tier 1: ≥ 10 kg (22 lb) Tier 2: ≥ 5 kg (11 lb)
TRC-4	TIH Zone D Materials	H332 Harmful if inhaled, Acute toxicity, inhalation (cat 4)	Tier 1: ≥ 200 kg (440 lb) Tier 2: ≥ 20 kg (44 lb)	Tier 1: ≥ 20 kg (44 lb) Tier 2: ≥ 10 kg (22 lb)

Table E.3 continued

Threshold Release Category	Material Hazard Classification Option 1	Material Hazard Classification Option 2	Threshold Quantity (outdoor)	Threshold Quantity (indoor[b])
TRC-5	Flammable Gases	H220 Extremely flammable gas, Flammable gases (cat 1A) H221 Flammable gas, Flammable gases (cat 1B,2)	Tier 1: ≥ 500 kg (1100 lb) Tier 2: ≥ 50 kg (110 lb)	Tier 1: ≥ 50 kg (110 lb) Tier 2: ≥ 25 kg (55 lb)
	Liquids with Normal Boiling Point ≤ 35 °C (95 °F) and Flash Point < 23 °C (73 °F)	H224 Extremely flammable liquid and vapor, Flammable liquids (cat 1)		
	Other Packing Group I Materials (excluding acids/bases)	H228 Flammable solid, Flammable solids (cat 1,2) H230 May react explosively even in the absence of air, Flammable gases (cat A) H231 May react explosively even in the absence of air at elevated pressure and/or temperature, Flammable gases (cat B) H232 May ignite spontaneously if exposed to air, Flammable gases (cat 1A pyrophoric gas) H250 Catches fire spontaneously if exposed to air, Pyrophoric liquids and Pyrophoric solids (cat 1) H310 Fatal in contact with skin, Acute toxicity, dermal (cat 1,2)		

Table E.3 continued

Threshold Release Category	Material Hazard Classification Option 1	Material Hazard Classification Option 2	Threshold Quantity (outdoor)	Threshold Quantity (indoor[b])
TRC-6	Liquids with Normal Boiling Point > 35 °C (95 °F) and Flash Point < 23 °C (73°F)	H225 Highly flammable liquid and vapor, Flammable liquids (cat 2)	Tier 1: ≥ 1000 kg (2200 lb) or ≥ 7 oil bbl Tier 2: ≥ 100 kg (220 lb) or ≥ 0.7 oil bbl	Tier 1: ≥ 100 kg (220 lb) or ≥ 0.7 oil bbl Tier 2: ≥ 50 kg (110 lb) or ≥ 0.35oil bbl
	Crude Oil ≥15 API Gravity (unless actual flashpoint available)	Crude Oil ≥15 API Gravity (unless actual flashpoint available)		
	Other Packing Group II Materials (excluding acids/bases)	H240 Heating may cause an explosion, Self-reactive substances and mixtures and Organic peroxides (type A) H241 Heating may cause a fire or explosion, Self-reactive substances and mixtures and Organic peroxides (type B) H242 Heating may cause a fire, Self-reactive substances and mixtures and Organic peroxides (type C-F) H271 May cause fire or explosion; strong oxidizer, Oxidizing liquids and Oxidizing solids (cat 1) H311 Toxic in contact with skin, Acute toxicity, dermal (cat 3)		

Table E.3 continued

Threshold Release Category	Material Hazard Classification Option 1	Material Hazard Classification Option 2	Threshold Quantity (outdoor)	Threshold Quantity (indoor[b])
TRC-7	Liquids with Flash Point ≥ 23 °C (73 °F) and ≤ 60 °C (140 °F)	H226 Flammable liquid and vapor, Flammable liquids (cat 3)	Tier 1: ≥ 2000 kg (4400 lb) or ≥ 14 oil bbl Tier 2: ≥ 200 kg (440 lb) or ≥ 1.4 oil bbl	Tier 1: ≥ 200 kg (440 lb) or ≥ 1.4 oil bbl Tier 2: ≥ 100 kg (220 lb) or ≥ 0.7 oil bbl
	Liquids with Flash Point > 60 °C (140 °F) released at a temperature at or above Flash Point	H227 Combustible liquid, Flammable liquids (cat 4) [**Released at or above flashpoint**] Liquids with Flash Point > 93 °C (200 °F) released at a temperature at or above Flash Point		
	Crude Oil <15 API Gravity (unless actual flashpoint available)	Crude Oil <15 API Gravity (unless actual flashpoint available)		
	UNDG Class 2, Division 2.2 (non-flammable, non-toxic gases) excluding air	H270 May cause or intensify fire; oxidizer Oxidizing gases (cat1) UNDG Class 2, Division 2.2 (non-flammable, non-toxic gases) excluding air		
	Other Packing Group III Materials (excluding acids/bases)	H272 May intensify fire; oxidizer, Oxidizing liquids and Oxidizing solids (cat 2,3) H312 Harmful in contact with skin, Acute toxicity, dermal (cat 4)		
TRC-8	Liquids with Flash Point > 60 °C (140 °F) and ≤ 93 °C (200 °F) released at a temperature below Flash Point	H227 Combustible liquid, Flammable liquids (cat 4) [**Released below flashpoint**]	Tier 1: N/A Tier 2: ≥ 1000 kg (2200 lb) or ≥ 7 oil bbl	Tier 1: N/A Tier 2: ≥ 500 kg (1100 lb) or ≥ 3.5 oil bbl
	Strong acids/bases (see definition 3.1)	H314 Causes severe skin burns, Skin corrosion/irritation (cat 1A)		
		H370 Causes damage to organs, Specific target organ toxicity, single exposure (cat 1)		

Table E.3 continued

Notes:
1. It is recognized that threshold quantities given in kg and lb or in lb and bbl are not exactly equivalent. Companies should select one of the pair and use it consistently for all recordkeeping activities.
2. Refer to guidance on selecting the correct Threshold Release Category and the use of Material Hazard Classification Option 1 and Option 2.

Table E.4. Examples for material categories

Category	Example Material
1	Br, HCN, Phosgene
2	BF_3, Chlorine, H_2S
3	HCl, HF, SO_2
4	Ammonia, CO, EO
5	Acetylene, Ethylene, Vinyl Acetate Monomer, Aluminum Alkyls
6	Vinyl Acetate Monomer, Benzene, Cyclohexane
7	Diesel, Mineral Oil, Muriatic Acid Nonflammable/nonpoisonous gases

E.4 Classifying PSE Tier 1 and Tier 2 Events

The flowchart in Figure E.1 can be used to classify an LOPC as a PSE Tier 1 or Tier 2.

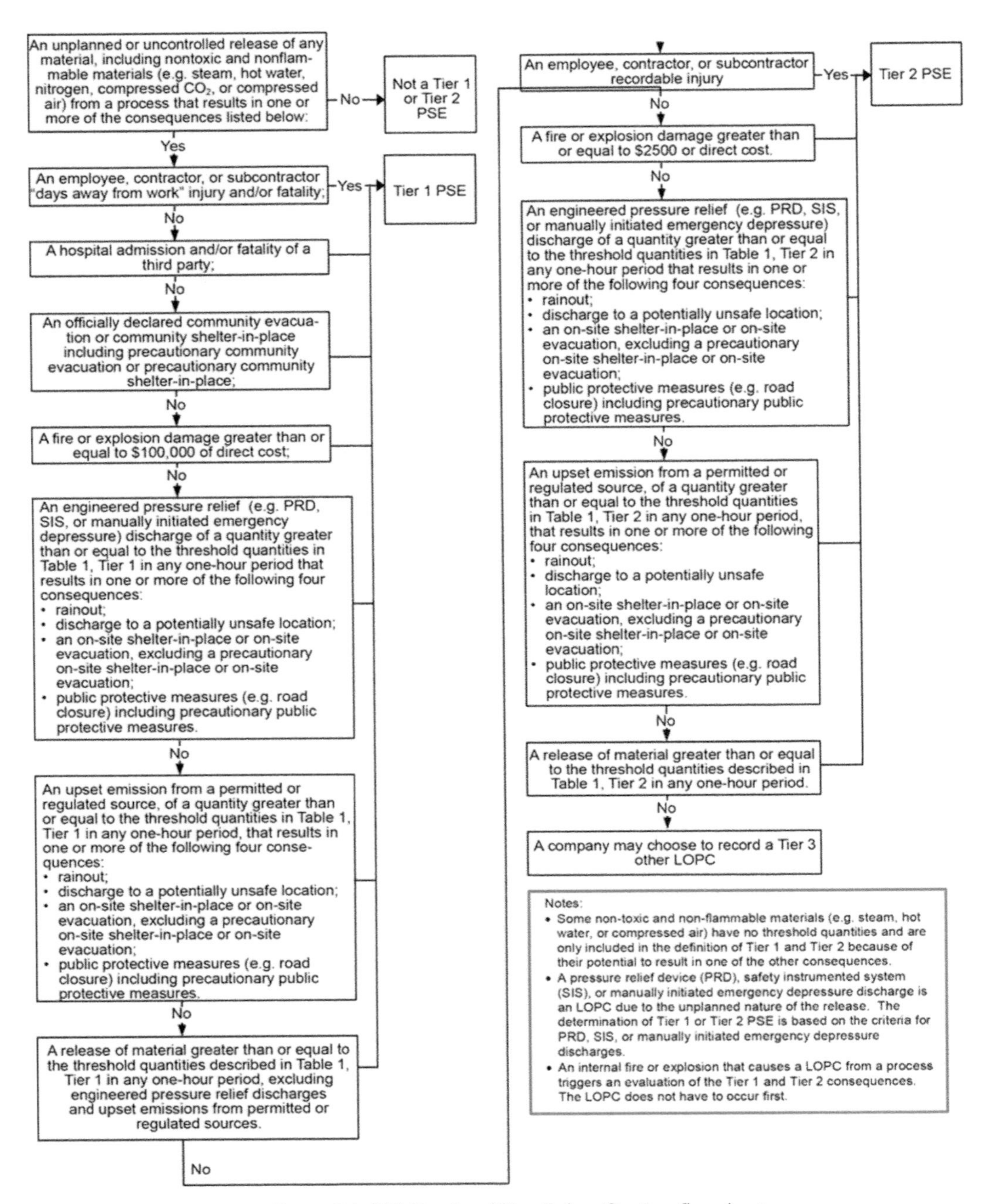

Figure E.1. PSE Tier 1and Tier 2 classification flowchart (API RP 754)

Examples

1. A benzene release occurs from a flange leak. 250 kg is released in 5 minutes. What PSE Tier is this release?
   Answer: Tier 1
2. A flange leak releases 950kg of benzene. What PSE Tier is this release?
   Answer: Tier 2
3. Same as 1, except the benzene is released in 180 minutes. What PSE Tier is this release?
   Answer: 1250kg/180min = 6.9 kg/min 8.3kg/min *60 min = 416 kg
   Tier 2
4. Same as 3, except the release occurs indoors. What PSE Tier is this release?
   Answer: Indoor TQ = 0.1(1000) = 100kg – Tier 1
5. Same as 3, except the release occurs over 1800min. What PSE Tier is this release?
   Answer: 1250/1800*60 = 42 kg
   Not a process safety incident
6. Cyclohexane spills unintentionally from an open valve. 10 kg is released in 20 minutes. What PSE Tier is this release?
   Answer: Not a Tier 1 or 2
7. Same as 6, except the release causes an injury resulting in a 2-day hospitalization. What PSE Tier is this release?
   Answer: Tier 1
8. Same as 1, except the release causes an injury that requires stitches before the worker returns to the job the same day. What PSE Tier is this release?
   Answer: Tier 2
9. Same as 1, except release is safe from a pressure relief valve venting to atmosphere. What PSE Tier is this release?
   Answer: Not a Tier 1 or 2
10. Same as 4, except a community shelter in place is called. What PSE Tier is this release?
    Answer: Tier 1
11. Same as 1, except a jet fire occurs causing $500,000 direct damage. What PSE Tier is this release?
    Answer: Tier 1

E.4 References

API RP 754, API Recommended Practice 754, "Process Safety Performance Indicators for the Refining and Petrochemical Industries", 3rd Edition, American Petroleum Institute, Washington, D.C., April 2021.

CCPS Glossary, "CCPS Process Safety Glossary", Center for Chemical Process Safety, https://www.aiche.org/ccps/resources/glossary.

CCPS 2011, "Process Safety Leading and Lagging Metrics...You Don't Improve What You Don't Measure", Center for Chemical Process Safety, New York, N.Y.

CCPS 2021, "Process Safety Metrics Guide for Selecting Leading and Lagging Indicators, Version 4.0", https://www.aiche.org/ccps/resources/tools/process-safety-metrics

CCPS, https://www.aiche.org/ccps/process-safety-metrics.

OSHA, www.osha.gov/recordkeeping.

Appendix F – Example Process Operations Readings and Evaluations

This appendix includes suggestions on how to create effective process readings and evaluation sheets along with possible examples of what to evaluate.

Developing a Process Evaluation Sheet. Evaluation sheets should be designed with engineering and technician input. They should also be reviewed for effectiveness on a periodic basis. When developing evaluation sheets consider the following.

- Map out and document the route taken by the operator.
- Each mark that an operator makes on a sheet should be part of the evaluation – either no action is taken, or if outside of a soft or hard evaluation criteria, action is taken.
- On return to work, operators review all evaluation sheets for the job they are working that they have not previously reviewed and give positive verification of review.
- Unit supervision and engineers review the operator evaluation sheet daily and ensure action is taken for parameters out of range, equipment that needs maintenance, or other out of range evaluations noted by the operator.
- Each parameter should have the operating range defined and documented, the technical basis, consequence of deviation and pre-defined steps to correct. The operating range and/or expected value should be documented on the evaluation sheet.

Example Process Equipment Evaluations.

Safety Equipment

- Safety showers and eyewashes – flush rust out of system and wash dust off the eyewash. Check for adequate flow and proper temperature. Replace dust caps.
- Have maintenance date tags attached so that everyone knows when the previous tests were run.
- Fire protection equipment – fire extinguishers are in place and not out of date, fire monitors are in service with no leaks. Deluge systems are lined-up and not leaking. Evaluate condition of foam systems.
- Breathing air systems – connections are in good condition.
- Radios – spare batteries available and charged.
- LEL meters - calibrated and charged.

Rotating equipment (pumps, motors, fans, blowers, compressors)

- Listen for unusual sounds.
- Look for unusual vibrations.
- Check for cavitation.
- Ensure coupling guards are in place.
- Evaluate lubrication levels in sight glasses.
- Ensure seal flush flows are adequate.

- Look for packing/ seal leaks.
- Note unusual smells (some leaks may not be visible).
- Look for oil or grease leaks from bearings.
- Check belt and chain condition and that guards are in place.
- Look for smoke from rotating parts.

Electrical and Instrument Boxes/ Motor Control Rooms

- Equipment should be clean and dry.
- All panel covers are in place, closed and sealed.
- Evaluate smoke, smells, and unusual sounds.
- Check for adequate lighting.
- Areas should be free of trash.

Fired Equipment

- Look for gas leaks and/or flames outside of burner box.
- Evaluate normality of flame front.

Flare Operation

- Check seal drum and knock out pot levels.
- Evaluate flow to flare.
- Evaluate flame/ smoke from flare.
- Ensure any required purges are in range.

Hazardous Waste storage evaluation

- Inspect for leaks.
- Observe dike liner conditions.
- Check that dike valves operated closed.
- Ensure that labels are in place and accurate.
- Consider other specific unit considerations.

Cooling Towers

- Observe tower for excessive drift.
- Evaluate tower for broken louvers, and/or packing.
- Look for uniform flow distribution
- Evaluate basin level
- Ensure screens are clear
- Evaluate system for algae & silt buildup
- Observe chemical addition systems for leaks
- Check fans for vibration

Tanks and Vessels

- Observe tanks for leaks when performing walkarounds.
- Review accumulation of liquids in dikes.
- Check that water is drained from the roof of any floating roof tanks.
- Confirm hatches and strapping ports are closed.
- Observe PRDs, especially check for any block valves incorrectly closed isolating the PRD, staining or other evidence that the PRD has lifted, or if the system has an intervening bursting disk check for no pressure reading on the pressure gauge between it and the PRD. Birds are notorious for building nests in PRD outlets.
- Review pad and vent pressures and line-ups

Heat Exchangers

- Check for head, piping or fitting leaks.
- Note unusual or excessive noise or vibration.
- Look for changes in normal differential temperatures and pressures.
- Back flush exchangers when scheduled or needed.

Process filters and Strainers

- Check for leaks and excessive pressure drop.
- Confirm that bypasses are closed and that the offline filter/strainer is ready for operations.

Separators and Knock out Pots

- Monitor levels.
- Check for leaks.
- Ensure level glasses are clean and readable.

Expansion Joints

- Make a general visual inspection ensuring the retaining cable is in place.
- Inspect stay bolts.
- Evaluate no abnormal growth or bulges, and no leaks.

Other things to observe on any plant walk through

- Process leaks
- Missing plugs, caps, or blinds
- Visible Vapor/ odor
- Leaks / Drips
- Oil sheen/ pools
- Steam leaks
- Noise (e.g. nitrogen or air leaks, cavitating pumps, vibrations, knocking, changes in noise levels)

- Oil level, pressure, temperature, flow
- Unit pressure and temperature gauges
- Unit Flows
- Chemical sewers
- Cooling water
- Control valves
- PRD's
- Suction and discharge blocked
- Telltale gauges
- Frost, sweating
- CAR seals (lock out seals for valves) in place
- Guards in place
- Sprinklers
- Gas detectors
- Emergency shutoff devices
- Unit barricades
- Unit signs (e.g., ingress/ egress)
- Area ingress and egress free
- Safety equipment accessible
- Vehicle access, access to equipment
- Field tags – location and date
- Physical evaluation
- Discoloration of paint and equipment
- Housekeeping
- Sample points/ automatic samples
- Filter inspections, e.g., differential pressure, dP
- Dike valve positions
- LOTO and equipment de-energized or out of service
- Head tanks
- Scheduled pump rotations, back flushes, etc.
- Hoses (condition)
- Area lighting

Appendix G – List of CSB Videos

The U.S. Chemical Safety Board (CSB) creates videos of many incidents. As of the publication of this book, these videos are online at www.csb.gov/videos/. Reports can also be searched online at www.csb.gov/investigations. Each video can be accessed from its investigation report website.

The videos listed in Table G.1 are categorized by chapters in this book.

Table G.1. CSB videos (as of January 2021)

Chapter	Video Title	Incident Investigation
1.	Introduction and Regulatory	
2.	Risk Based Process Safety	
3.	Process Safety Regulations, Codes, and Standards	
4.	Fire and Explosion Hazards	
	Inferno: Dust Explosion at Imperial Sugar	Imperial Sugar Company Dust Explosion and Fire
	Combustible Dust: Solutions Delayed Combustible Dust: An Insidious Hazard	AL Solutions Fatal Dust Explosion
	Iron in the Fire	Hoeganaes Corporation Fatal Flash Fires
	Dust Testing	Hoeganaes Corporation Fatal Flash Fires
	Experimenting with Danger	Texas Tech University Chemistry Lab Explosion
	After the Rainbow	Key Lessons for Preventing Incidents from Flammable Chemicals in Educational Demonstrations
	Fire at Formosa Plastics (Texas)	Formosa Plastics Propylene Explosion
	Dangers of Propylene Cylinders	Praxair Flammable Gas Cylinder Fire
	Dangers of Flammable Gas Accumulation	Acetylene Service Company Gas Explosion
	Ethylene Oxide Explosion at Sterigenics	Sterigenics Ethylene Oxide Explosion
	Anatomy of a Disaster	BP America Refinery Explosion
	Updated BP Texas City Animation on the 15th Anniversary of the Explosion	BP America Refinery Explosion
	BP Texas City Animation – English, Spanish, French	BP America Refinery Explosion
	BP Texas City 10 Year Anniversary Safety Message	BP America Refinery Explosion

Table G.1 continued

Chapter	Video Title	Incident Investigation
4. Fire and Explosion Hazards continued		
	Emergency in Apex	EQ Hazardous Waste Plant Explosions and Fire
	Fire from Ice	Valero Refinery Propane Fire
	Static Sparks Explosion in Kansas	Barton Solvents Explosions and Fire
	Little General Store Propane Explosion	Half an Hour to Tragedy
	CSB Interim Animation on Husky Refinery Explosion and Fire	Husky Energy Refinery Explosion and Fire
	Animation of the April 26, 2018 Explosion and Fire at Husky Energy	Husky Energy Refinery Explosion and Fire
	Blowout in Oklahoma	Pryor Trust Fatal Gas Well Blowout and Fire
	Interim Animation of the Pryor Trust Gas Well Blowout and Fire	Pryor Trust Fatal Gas Well Blowout and Fire
	Preliminary Animation of the Philadelphia Energy Solutions Refinery Fire	Philadelphia Energy Solutions (PES) Refinery Fire and Explosions
	Deepwater Horizon Blowout Animation	Macondo Well Blowout
	Fire in Baton Rouge	ExxonMobil Refinery Chemical Release and Fire
	Animation of Fire at ExxonMobil's Baton Rouge Refinery	ExxonMobil Refinery Chemical Release and Fire
	Back to School Safety Message	Key Lessons for Preventing Incidents from Flammable Chemicals in Educational Demonstrations
	Animation of 2015 Explosion of ExxonMobil Refinery in Torrance, CA	ExxonMobil Refinery Explosion
	AirGas Surveillance Footage	AirGas Facility Fatal Explosion
	Combustible Dust: Solutions Delayed	AL Solutions Fatal Dust Explosion
	Silver Eagle Refinery Explosion Surveillance Footage	Silver Eagle Refinery Flash Fire and Explosion and Catastrophic Pipe Explosion
	Animation of Explosion at Tesoro's Anacortes Refinery	Tesoro Refinery Fatal Explosion and Fire
	Tesoro Anniversary Safety Message	Tesoro Refinery Fatal Explosion and Fire

Table G.1 continued

Chapter	Video Title	Incident Investigation
4. Fire and Explosion Hazards continued		
	Falling Through the Cracks	NDK Crystal Inc. Explosion with Offsite Fatality
	Surveillance Video from the August 6 Accident at the Chevron Refinery in Richmond, CA	Chevron Refinery Fire
	Chevron Richmond Refinery Fire Animation	Chevron Refinery Fire
	No Place to Hang Out: The Danger of Oil Sites	Oil Site Safety
5. Reactive Chemical Hazards		
	Preventing Harm from NaHS	
	Reactive Hazards	BP Amoco Thermal Decomposition Incident
		Synthron Chemical Explosion
	Explosion at Formosa Plastics (Illinois)	Formosa Plastics Vinyl Chloride Explosion
	Runaway: Explosion at T2 Laboratories	T2 Laboratories Inc. Reactive Chemical Explosion
	Fire in the Valley	Bayer CropScience Pesticide Waste Tank Explosion
	Animation of the Bayer CropScience Pesticide Waste Tank Explosion	Bayer CropScience Pesticide Waste Tank Explosion
	CSB Video Documenting the Blast Damage in West, Texas	West Fertilizer Explosion and Fire
	Dangerously Close: Explosion in West, Texas	West Fertilizer Explosion and Fire
	Deadly Contract	Donaldson Enterprises, Inc. Fatal Fireworks Disassembly Explosion and Fire
6. Toxic Hazards		
	CITGO Refinery Hydrofluoric Acid Release and Fire	Surveillance video from July 19, 2009, fire and explosion at the CITGO Corpus Christi Refinery
	Mixed Connection, Toxic Result	MGPI Processing Inc. Toxic Chemical Release
	MGPI Processing, Inc. Toxic Chemical Release	MGPI Processing Inc. Toxic Chemical Release

Table G.1 continued

Chapter	Video Title	Incident Investigation
6. Toxic Hazards continued		
	Animation of Chemical Release at DuPont's La Porte Facility	DuPont La Porte Facility Toxic Chemical Release
	Freedom Industries Tank Dismantling	Freedom Industries Chemical Release
	Fatal Exposure: Tragedy at DuPont	DuPont Corporation Toxic Chemical Release
7. Chemical Hazards Data Sources		
8. Other Hazards		
9. Process Safety Incident		
10. Project Design Basics (ISD)		
	Inherently Safer: The Future of Risk Reduction	Bayer CropScience Pesticide Waste Tank Explosion
11. Equipment Failure		
	Falling Through the Cracks	NDK Crystal Inc. Explosion with Offsite Fatality
	Silver Eagle Refinery Explosion Surveillance Footage	Silver Eagle Refinery Flash Fire and Explosion and Catastrophic Pipe Explosion
	Animation of Explosion at Tesoro's Anacortes	Tesoro Refinery Fatal Explosion and Fire
	Behind the Curve	Tesoro Refinery Fatal Explosion and Fire
	The Human Cost of Gasoline	Tesoro Refinery Fatal Explosion and Fire
	Chevron Richmond Refinery Fire Animation	Chevron Refinery Fire
	Surveillance Video from the August 6 Accident at the Chevron Refinery in Richmond, CA	Chevron Refinery Fire
	Freedom Industries Tank Dismantling	Freedom Industries Chemical Release
	Shock to the System	Millard Refrigerated and Ammonia Release
12. Hazard Identification		
13. Consequence Analysis		
14. Risk Assessment		
15. Risk Prevention and Mitigation (Bow tie)		
	Without Safeguards, Pressure Vessels Can be Deadly	Marcus Oil and Chemical Tank Explosion

Table G.1 continued

Chapter	Video Title	Incident Investigation
16. Human Factors		
17. Operational Readiness		
18. Management of Change		
19. Operating Procedures, Safe Work Practices, and Conduct of Operations and Operational Discipline		
	Hazards of Nitrogen Asphyxiation	Valero Refinery Asphyxiation Incident
	Hazards of Nitrogen Asphyxiation	Final Report: Safety Bulletin - Hazards of Nitrogen Asphyxiation
	Death in the Oilfield	Partridge Raleigh Oilfield Explosion and Fire
	Hot Work: Hidden Hazards	E. I. DuPont De Nemours Co. Fatal Hotwork Explosion
	Dangers of Hot Work	Seven Key Lessons to Prevent Worker Deaths During Hot Work In and Around Tanks
	No Escape: Dangers of Confined Spaces	Xcel Energy Company Hydroelectric Tunnel Fire
	Public Worker Safety	Bethune Point Wastewater Plant Explosion
	Deadly Practices	ConAgra Natural Gas Explosion and Ammonia Release
	Deadly Practices	Kleen Energy Natural Gas Explosion
	Blocked In	Williams Olefins Plant Explosion and Fire
	Filling Blind	Caribbean Petroleum Refining Tank Explosion and Fire
	Uncovered Hazards: Explosion at the DeRidder Pulp and Paper Mill	Packaging Corporation of America Hot Work Explosion
	Animation of Explosion at PCA's DeRidder, Louisiana, Pulp and Paper Mill	Packaging Corporation of America Hot Work Explosion
20. Emergency Management		
	Emergency Preparedness: Findings from CSB Accident Investigations	DPC Enterprises Festus Chlorine Release
	Emergency Response Safety Message	Husky Energy Refinery Explosion and Fire
	Winterization Safety Message	Valero Refinery Propane Fire
	Winterization Safety Message	DuPont La Porte Facility Toxic Chemical Release

Table G.1 continued

Chapter	Video Title	Incident Investigation
20. Emergency Management continued		
	Extreme Weather Safety Message	Arkema Inc. Chemical Plant Fire
	Caught in the Storm Extreme Weather Hazards	Arkema Inc. Chemical Plant Fire
	Preliminary 2D Animation of Events Leading to 2017 Fire at Arkema Chemical	Arkema Inc. Chemical Plant Fire
	Prevent Accidents During Subfreezing Weather	
21. People Management Aspects of Process Safety Management		
	Deadly Contract	Donaldson Enterprises, Inc. Fatal Fireworks Disassembly Explosion and Fire
22. Sustaining Process Safety		
23. Process Safety Culture		
	CSB Video Excerpts from Dr. Trevor Kletz	
	Maintain Process Safety During the Recession	

Appendix H – Major Process Safety Incident Vs Root Cause Map

This root cause map is from the IChemE Safety & Loss Prevention Special Interest Group. It is updated routinely and shared via this Group's newsletter. This version is available in full size on the webpage listed in the Online Materials Accompanying this Book section included at the start of this book.

Major Process Safety Incident vs Root Cause Map
(Quick Reference Guide)

IChemE
Safety & Loss Prevention
Special Interest Group

Sector		Date	Event	Country	Type	Fatalities	Injuries	Abnormal Operations	Escalation Potential	Earthquake	Tsunami	Flood	Hurricane	Extreme Cold or Ice	Process Design	Hazard Identification	Equipment/Piping Design	Materials of Construction	Instrumentation	Safety Instrumented Systems	Protective Systems	Plant Layout	Occupied Buildings	Hazard Awareness	Process Control	Process Monitoring	Alarm Management	Creeping Change
								Context		Natural Hazards					Design Factors									Operations Factors				
Oil & Gas	Upstream	06-Jul-88	Piper Alpha	UK	Explosion	167	?										X				X	X						
		27-Jul-05	Mumbai High North	India	Explosion	22	?								X							X		X				
		20-Apr-10	Macondo	USA	Explosion	11	17	X							X	X	X		X	X						X		
		11-Feb-15	Camarupim	Brazil	Explosion	9	26	X							X	X			X			X		X				
	LNG	19-Jan-04	Skikda	Algeria	Explosion	26	74		X						X			?	X			X	X					
	GPP	25-Sep-98	Longford	Australia	Fire	2	8	X								X				X				X		X	X	
	Downstream	04-Jan-66	Feyzin	France	BLEVE	18	84								X		X				X			X				
		23-Jul-84	Romeoville	USA	Explosion	17	22									X		X						X				
		22-Mar-87	Grangemouth	UK	Explosion	1	0	X	X							X	X		X	X	X			X				
		05-May-88	Norco	USA	Explosion	8	18															X	X					
		24-Jul-94	Milford Haven	UK	Explosion	0	26	X	X								X		X							X	X	
		17-Aug-99	Izmit	Turkey	Fire	0	0		X	X					X													
		16-Apr-01	Humber	UK	Explosion	0	5		X							X	X											
		23-Mar-05	Texas City	USA	Explosion	15	180	X							X	X			X				X			X		X
		05-Nov-05	Delaware City	USA	Asphyxiation	2	0																	X				
		16-Feb-07	McKee	USA	Fire	0	4		X					X	X	X				X		X						
		02-Apr-10	Anacortes	USA	Explosion	7	0	X								X	X	X	X							X		X
		06-Aug-12	Richmond	USA	Fire	0	26											X										
		18-Feb-15	Torrance	USA	Explosion	0	4	X	X							X			X									
	Ship	24-Mar-89	Valdez	USA	Pollution	0	0		X					X			X		X							X		
	Rail	06-Jul-13	Lac Mégantic	Canada	Fire	47	?									X	X				X							X
	Terminal	19-Nov-84	San Juan Ixhuatepec	Mexico	BLEVE	542	4,248										X		X	X	X	X						X
		11-Dec-05	Buncefield	UK	Explosion	0	43		X						X	X			X	X		X			X	X		X
		23-Oct-09	Bayamón	Puerto Rico	Explosion	0	?								X	X				X								
Petrochemical	Manufacture	01-Jun-74	Flixborough	UK	Explosion	28	104	X	X													X	X	X				
		23-Oct-89	Pasadena	USA	Explosion	23	314	X									X		X		X	X						
		21-Sep-92	Castleford	UK	Fire	5	201	X														X	X	X				
		01-Feb-94	Ellesmere Port	UK	Fire	0	18		X							X	X							X				
		27-Mar-98	Taft	USA	Asphyxiation	1	1									X								X				
		13-Jun-13	Geismar	USA	BLEVE	2	167	X								X								X				
		03-Jun-14	Moerdijk	Netherlands	Explosion	0	2	X								X			X	X				X	X			
		31-Aug-17	Crosby	USA	Fire	0	21					X	X			X												
Agrochemical	Manuf.	10-Jul-76	Seveso	Italy	Toxic Release	0	462	X							X	X			X	X				X				
		03-Dec-84	Bhopal	India	Toxic Release	2153	>200,000	X							X	X		X						X				
		21-Sep-01	Toulouse	France	Explosion	31	2442		X							X		X				X		X				
		15-Nov-14	La Porte	USA	Toxic Release	4	0	X								X	X				X		X	X			X	
	Store	17-Apr-13	West	USA	Explosion	15	> 100		X							X		X				X		X				
Pharma	Manuf.	04-Jan-92	Grimsby	UK	Runaway	0	0	X	X						X	X			X						X			X
		29-Jan-03	Kinston	USA	Explosion	6	38									X	X				X		X	X				
		28-Apr-08	Cork	Ireland	Runaway	1	1								X	X	X											
Power Gen	Nuclear	28-Mar-79	Three Mile Island	USA	Near Miss	0	0		X						X	X			X			X					X	
		26-Apr-86	Chernobyl	Russia	Explosion	> 30	~ 7,000	X	X						X	X	X			X				X				
		11-Mar-11	Fukushima Daiichi	Japan	Explosion	0	13		X	X	X					X			X			X						
	Coal	08-Apr-99	Gannon	USA	Explosion	3	48																					
		10-Nov-07	Dallman	USA	Explosion	0	0	X	X											X			X		?	?		
Water	Dist	23-May-84	Abbeystead	UK	Explosion	16	28	X							X	X			X					X				
	Treat	06-Jul-88	Camelford	UK	Pollution	0	?									X						X						
		05-Apr-93	Milwaukee	USA	Disease	69	>401,000								X	X									X	X		
Food		03-Sep-93	Hamlet	USA	Fire	25	54									X			X	X		X						
		07-Feb-08	Port Wentworth	USA	Explosion	14	36								X	X								X				

Date	Event	Risk Assessment	Work Planning	Control of Work	Energy Isolation	Housekeeping	Material Degradation	Inspection	Preventative Maintenance	Human Factors	Role Clarity	Personal Protective Equipment	Communication	Procedures	Training	Management of Change	Emergency Preparedness	Failure to Learn	Production over Safety	Normalisation of Deviance	Process Safety Management	Design Standards	Land Use Planning	Regulatory Compliance Audits	Investigation Reports
		Maintenance Factors								Personal			Competency			Culture						Regulator			Links
06-Jul-88	Piper Alpha	X		X	X					X			X	X	X	X	X				X				Click
27-Jul-05	Mumbai High North	X								X	X		X	X	X									X	
20-Apr-10	Macondo	X							X	X	X		X	X	X		X				X	X		X	Click
11-Feb-15	Camarupim	X		X						X	X		X	X	X	X	X		X		X			X	
19-Jan-04	Skikda								X	X					X		X				X	X	X		
25-Sep-98	Longford						X			X			X	X	X		X			X	X			X	Click
04-Jan-66	Feyzin													X	X		X					X			Click
23-Jul-84	Romeoville						X	X						X								X			
22-Mar-87	Grangemouth			X										X	X	X		X	X		X				Click
05-May-88	Norco						X									X							X		
24-Jul-94	Milford Haven	X		X			X	X	X	X			X		X	X	X				X				Click
17-Aug-99	Izmit	X					X								X		X							X	
16-Apr-01	Humber						X	X	X				X			X					X				Click
23-Mar-05	Texas City	X		X			X		X	X	X		X	X	X	X		X	X	X	X			X	Click
05-Nov-05	Delaware City			X						X				X	X										Click
16-Feb-07	McKee	X						X								X						X			Click
02-Apr-10	Anacortes						X							X						X	X	X		X	Click
06-Aug-12	Richmond	X					X	X									X		X		X	X	X		Click
18-Feb-15	Torrance													X		X		X			X				Click
24-Mar-89	Valdez									X			X	X	X		X				X			X	
06-Jul-13	Lac Mégantic	X					X			X				X	X	X	X		X		X			X	
19-Nov-84	San Juan Ixhuatepec	X							X						X	X	X		X		X	X	X		
11-Dec-05	Buncefield	X							X	X			X	X	X	X	X		X		X		X		
23-Oct-09	Bayamón													X	X		X	X							Click
01-Jun-74	Flixborough	X					X	X		X			X	X	X	X	X		X		X		X		
23-Oct-89	Pasadena			X	X				X				X	X	X	X	X	X		X	X				
21-Sep-92	Castleford	X		X						X					X	X	X	X			X				Click
01-Feb-94	Ellesmere Port						X	X	X				X	X			X				X	X			Click
27-Mar-98	Taft			X										X											Click
13-Jun-13	Geismar												X	X		X		X			X				Click
03-Jun-14	Moerdijk									X			X	X	X	X	X	X			X			X	
31-Aug-17	Crosby																X					X	X		Click
10-Jul-76	Seveso									X			X	X		X	X				X	X			
03-Dec-84	Bhopal	X							X	X				X	X	X	X		X		X		X		
21-Sep-01	Toulouse	X				X								X	X		X	X			X		X		Click
15-Nov-14	La Porte	X		X					X	X		X	X	X	X	X	X	X		X	X	X		X	Click
17-Apr-13	West	X													X		X	X					X	X	Click
04-Jan-92	Grimsby	X												X		X				X					
29-Jan-03	Kinston	X				X							X	X	X						X	X			Click
28-Apr-08	Cork	X								X				X	X		X					X			
28-Mar-79	Three Mile Island									X			X	X	X		X							X	
26-Apr-86	Chernobyl									X			X	X	X	X	X	X	X		X			X	
11-Mar-11	Fukushima Daiichi	X												X	X		X				X			X	
08-Apr-99	Gannon			X										X											
10-Nov-07	Dallman								X																
23-May-84	Abbeystead													X	X	X						X			Click
06-Jul-88	Camelford	X								X			X	X	X		X				X				
05-Apr-93	Milwaukee														X		X		X						
03-Sep-93	Hamlet									X			X	X	X		X				X			X	
07-Feb-08	Port Wentworth	X				X				X			X	X	X	X	X	X		X	X			X	Click

Rev. 5 (15-Feb-21)

P. Marsh - XBP Refining Consultants Ltd.

Reference

IChemE Safety & Loss Prevention Special Interest Group Newsletter Issue 2021/01, February 2021

Index